Physiological Mechanisms and Adaptation Strategies in Plants Under Changing Environment

Parvaiz Ahmad • Mohd Rafiq Wani
Editors

Physiological Mechanisms and Adaptation Strategies in Plants Under Changing Environment

Volume 1

Springer

Editors
Parvaiz Ahmad
Department of Botany
Sri Pratap College
Srinagar, Jammu and Kashmir
India

Mohd Rafiq Wani
Department of Botany
Govt. Degree College (Boys)
Anantnag, Jammu and Kashmir
India

ISBN 978-1-4614-8590-2 ISBN 978-1-4614-8591-9 (eBook)
DOI 10.1007/978-1-4614-8591-9
Springer New York Heidelberg Dordrecht London

Library of Congress Control Number: 2013949858

Printed on acid-free paper

Springer is part of Springer Science+Business Media (www.springer.com)

Preface

Food shortage, burgeoning population, and environmental changes have caused grave troubles to mankind globally. Providing food, for the ever-growing population, has become a cumbersome challenge for the governments across the world. On the other hand, changing environment is transforming our cultivated land to wastelands. This adverse impact of climate changes in the form of dwindling rainfall, rising temperatures, increased severity of drought and flooding is bound to threaten the levels of food security and the economy of any nation. Plants on which mankind is directly or indirectly dependent exhibit various mechanisms for their survival. Adaptability of plants, to changing environment, is a matter of colossal concern and almost all the plant biologists worldwide are trying arduously to accomplish the same goal for addressing the key issues vis-à-vis food security. To get more food from the limited resources is rather herculean; however, steady efforts are needed to grow more and more plants in these wastelands to bring them under productive cultivation. To be successful in this campaign of food security, the plant biologists need to be well acquainted with the knowledge of plant adaptability to changing environment. Moreover, consistent research on plant physiological mechanisms under varying and stressed environments will be highly beneficial in the future for reaping out the desired benefits. Therefore, keeping the above facts in mind and problems which are arising out of such environmental stresses every passing day, an endeavour is carried out by the editors to bring out this volume of "Physiological Mechanisms and Adaptation Strategies in Plants Under Changing Environment" *Volume 1* to address these issues and provide some viable solutions in this direction. We hope that our attempt would be a step ahead in generating interest among researchers and students of this field of science worldwide. However, we feel with firm belief that there is no single book which is completely perfect in its contents and matter in the contemporary era because of ultramodern rapid development in the field of science and technology.

The current volume comprises 12 chapters and each chapter has different research dimensions from another having much significance in their respective fields. Chapter 1 deals with the mechanisms and adaptation of plants to environmental stresses: a case of woody species. This chapter throws light on the multiple responses of plants to stresses and the consequent adaptation mechanisms towards the environmental stress. Chapter 2 addresses the drought tolerance, role of organic osmolytes, growth regulators, and mineral nutrients, wherein the authors have comprehensively put in their efforts in elaborating the role of growth regulators, viz., abscisic acid (ABA), ethylene and salicylic acid (SA), and various organic osmolytes, such as proline, glycine betaine, free sugars and polyols, in increasing the plant tolerance to drought stress. Chapter 3 is concerned with influencing the product quality by applying drought stress during the cultivation of medicinal plants. This chapter highlights the cultivation of medicinal plants under drought stress, besides explaining conspicuously the enhanced synthesis and accumulation of natural products in drought-stressed plants.

Chapter 4 is about the water scarcity and water stress in agriculture. The chapter explicitly discusses the approaches and methods for water balance and potential perspectives for avoiding the water stress in European agriculture. Chapter 5 is regarding the use of biotechnology for drought and salinity tolerance of crops. The chapter covers genetic approaches, RNA interference and its applications, transcriptome analysis, proteomic approaches, etc. Chapter 6 describes the effect of salinity on plants and the role of arbuscular mycorrhizal fungi and plant growth promoting rhizobacteria in alleviation of salt stress. Chapter 7 discusses cash crop halophytes—the ecologically and economically sustainable use of naturally salt-resistant plants in the context of global changes. The chapter encases the most important aspects regarding the sustainable use of halophytes including the general problem of soil salinity, the biology and ecology of halophytes, and the prerequisites and possibilities of halophyte utilization and of saline production systems.

Chapter 8 addresses the effects of heat stress on growth and crop yield of wheat (*Triticum aestivum*). The authors have meticulously explained the effect of high-temperature stresses on various metabolic reactions, photosynthetic processes, and the protective mechanisms in plants. Chapter 9 discusses low-temperature stresses in plants: an overview of roles of cryoprotectants in defense. Here, the authors have discussed in detail the role of cryoprotectants in alleviating the low-temperature stress in plants. Chapter 10 deals with abiotic stress and lignins: an overview. The chapter lays an emphasis on the multiple means of lignin composition, content, accumulation, and rearrangements in response to a number of abiotic stresses in the plant kingdom.

Chapter 11 is about humic substances and plant defense metabolism. The chapter highlights the interactions between plant root systems and humic substances (HS), antioxidative responses to HS and protective effects of HS in plants under stress conditions, besides how HS contribute to improve the plant performance through complex metabolic mechanisms. Chapter 12 is on mitochondrial respiration: involvement of the alternative respiratory pathway and residual respiration in abiotic stress responses. The chapter highlights the structural organizations of eukaryotic mitochondria, mitochondrial genome, and role of AOX in integrating the cell metabolism under stress conditions.

This volume is compiled with wealth of knowledge in the field of physiological responses, adaptability of plants towards drought, salt, temperature, and other environmental stresses. Chapters contributed in this book have been published keeping intact author's justifications; however, suitable editorial changes were made wherever considered necessary. In spite of our best efforts, there is a possibility of some errors still creeping in the book for which we seek reader's indulgence and feedback. We wish to express our appreciation to the well-versed contributors who readily accepted our invitation to write the chapters. Moreover, we would like to thank Springer Science+Business Media, LLC, New York, particularly Eric Stannard (Editor Botany, Springer), Andy Kwan (Assistant Editor, Springer), Flora Kim (Developmental Editor, Springer), and all the other staff members of Springer, who were directly or indirectly associated with us in this project for their constant help, valuable suggestions, and efforts in bringing out the timely publication of this volume.

Srinagar, Jammu and Kashmir, India Parvaiz Ahmad
Anantnag, Jammu and Kashmir, India Mohd Rafiq Wani

About the Editors

Dr. Parvaiz Ahmad (Editor) Dr. Parvaiz is Assistant Professor in Botany at Sri Pratap College, Srinagar, Jammu and Kashmir, India. He has completed his post-graduation in Botany in 2000 from Jamia Hamdard, New Delhi India. After receiving Doctorate degree from Indian Institute of Technology (IIT) Delhi, India, he joined International Centre for Genetic Engineering and Biotechnology, New Delhi in 2007. His main research area is Stress Physiology and Molecular Biology. He has published more than 30 research papers in peer reviewed journals and 19 book chapters. He is also an editor of 6 volumes (5 with Springer NY USA and 1 with Studium Press Pvt. India Ltd., New Delhi, India). He is recipient of Junior Research Fellowship and Senior Research Fellowship by CSIR, New Delhi, India. Dr. Parvaiz has been awarded Young Scientist Award under Fast Track scheme in 2007 by Department of Science and Technology (DST), Govt. of India. Dr. Parvaiz is actively engaged in studying the molecular and physio-biochemical responses of different agricultural and horticultural plants under environmental stress.

Dr. Mohd Rafiq Wani (Co-editor) Dr. Mohd Rafiq Wani is currently Assistant Professor in Botany at Government Degree College (Boys), Anantnag, University of Kashmir, India. Dr. Wani did his Masters in Botany in 2003 with specialization in "Genetics and Plant Breeding" from Aligarh Muslim University (AMU), Aligarh, UP, India. After receiving the Degree of Doctorate in 2008 for his research work on "Chemical Mutagenesis in Mungbean" from the same University, joined Department of Higher Education, Jammu and

Kashmir Govt. in 2009. He teaches a range of bioscience related subjects at undergraduate/post graduate levels. At present, his research interests are mainly focused on the improvement of pulses through induced mutations and exploring the physiological and biochemical responses of crop plants to a range of biotic and abiotic stresses. As a part of his research endeavour, Dr. Wani has extensively researched and written on the issues of induced chemo-mutagenesis among the food crops, with special reference to pulses. He has around twenty eight (28) research publications to his credit, published in various international and national journals of repute. Moreover, he has also submitted several book chapters to various research oriented volumes. Dr. Wani, while constantly working for his academic and research interests, is currently in the process of editing many volumes of books on the subjects of plant stress physiology and induced plant mutagenesis with reputed international publishers. In addition, he is an editorial member and reviewer of few online journals pertaining to plant sciences, besides being the fellow of various scientific societies like Indian Society of Pulses Research and Development and Indian Society of Genetics and Plant Breeding.

Contents

Contributors

Mohammad Abass Ahanger School of Studies in Botany, Jiwaji University, Gwalior, Madhya Pradesh, India

Parvaiz Ahmad Department of Botany, Sri Pratap College, Srinagar, Jammu and Kashmir, India

Ricardo L.L. Berbara Soils Department, Federal Rural University of Rio de Janeiro (Universidade Federal Rural do Rio de Janeiro, UFRRJ), Seropédica, Rio de Janeiro, Brazil

Kalpna Bhandari Department of Botany, Panjab University, Chandigarh, India

Renu Bhardwaj Department of Botanical and Environmental Sciences, Guru Nanak Dev University, Amritsar, Punjab, India

Faiçal Brini Plant Protection and Improvement Laboratory, Centre of Biotechnology of Sfax, (CBS)/University of Sfax, Sfax, Tunisia

Azza Chelli-Chaabouni National Institute of Agronomic Research of Tunisia, Ariana, Tunisia

Silvana Chocobar-Ponce Laboratorio de Fisiología Vegetal, Facultad de Ciencias Naturales e IML, Universidad Nacional de Tucumán, Tucumán, Argentina

Andrés C. García Soils Department, Federal Rural University of Rio de Janeiro (Universidade Federal Rural do Rio de Janeiro, UFRRJ), Seropédica, Rio de Janeiro, Brazil

Department of Chemistry, Agricultural University of Havana (Universidad Agraria de La Habana, UNAH), Havana, Cuba

Nicole Geissler Justus Liebig University Giessen, Special Botany, Giessen, Germany

Rosana González Laboratorio de Fisiología Vegetal, Facultad de Ciencias Naturales e IML, Universidad Nacional de Tucumán, Tucumán, Argentina

Neha Handa Department of Botanical and Environmental Sciences, Guru Nanak Dev University, Amritsar, Punjab, India

Shuijin Hu Laboratory of Soil Ecology, Department of Plant Pathology, North Carolina State University, Raleigh, NC, USA

Anjana Jajoo School of Life Science, Devi Ahilya University, Indore, Madhya Pradesh, India

Harpreet Kaur Department of Botanical and Environmental Sciences, Guru Nanak Dev University, Amritsar, Punjab, India

Parminder Kaur Department of Botanical and Environmental Sciences, Guru Nanak Dev University, Amritsar, Punjab, India

Kiyoon Kim Department of Agricultural Chemistry, Chungbuk National University, Cheongju, Chungbuk, Republic of Korea

Maik Kleinwächter Institute for Plant Biology, Technische Universität Braunschweig, Braunschweig, Braunschweig, Germany

Sukhmeen Kohli Department of Botanical and Environmental Sciences, Guru Nanak Dev University, Amritsar, Punjab, India

Hans-Werner Koyro Institute of Plant Ecology, Justus Liebig University Giessen, Giessen, Germany

Vinod Kumar Department of Botanical and Environmental Sciences, Guru Nanak Dev University, Amritsar, Punjab, India

Helmut Lieth Institute of Environmental Systems Research (USF), University of Osnabrück, Osnabrück, Germany

Khaled Masmoudi Plant Protection and Improvement Laboratory, Centre of Biotechnology of Sfax, (CBS)/University of Sfax, Sfax, Tunisia

International Center for Biosaline Agriculture (ICBA), Dubai, United Arab Emirates

Sonal Mathur School of Life Science, Devi Ahilya University, Indore, Madhya Pradesh, India

Harsh Nayyar Department of Botany, Panjab University, Chandigarh, India

Gabrijel Ondrasek Faculty of Agriculture, University of Zagreb, Zagreb, Croatia

Carolina Prado Laboratorio de Fisiología Vegetal, Facultad de Ciencias Naturales e IML, Universidad Nacional de Tucumán, Tucumán, Argentina

Fernando E. Prado Laboratorio de Fisiología Vegetal, Facultad de Ciencias Naturales e IML, Universidad Nacional de Tucumán, Tucumán, Argentina

Mariana Rosa Laboratorio de Fisiología Vegetal, Facultad de Ciencias Naturales e IML, Universidad Nacional de Tucumán, Tucumán, Argentina

Tongmin Sa Department of Agricultural Chemistry, Chungbuk National University, Cheongju, Chungbuk, Republic of Korea

Dirk Selmar Institute for Plant Biology, Technische Universität Braunschweig, Braunschweig, Braunschweig, Germany

Gopal Selvakumar Department of Agricultural Chemistry, Chungbuk National University, Cheongju, Chungbuk, Republic of Korea

Resham Sharma Department of Botanical and Environmental Sciences, Guru Nanak Dev University, Amritsar, Punjab, India

Shiv Ram Tyagi School of Studies in Botany, Jiwaji University, Gwalior, Madhya Pradesh, India

Mohd Rafiq Wani Department of Botany, Government Degree College (Boys), Anantnag, Jammu and Kashmir, India

Chapter 1
Mechanisms and Adaptation of Plants to Environmental Stress: A Case of Woody Species

Azza Chelli-Chaabouni

Contents

1 Introduction

Environmental stresses affect negatively plant growth, productivity, reproductive capacity, and survival. Their effects are predicted to become more pronounced in both duration and severity in the near future (Osakabe et al. 2012; Stella et al. 2013) partly, due to global climatic changes (Niinemets 2010).

A. Chelli-Chaabouni (✉)
National Institute of Agronomic Research of Tunisia,
Hédi Karray Street, 2049 Ariana, Tunisia
e-mail: a_chellichaabouni@yahoo.fr

P. Ahmad and M.R. Wani (eds.), *Physiological Mechanisms and Adaptation Strategies in Plants Under Changing Environment: Volume 1*, DOI 10.1007/978-1-4614-8591-9_1,

Among the total plant kingdom, woody plants and primarily forests greatly contribute to the world carbon biomass stock, biodiversity, and area protection (FAO 2011). Despite afforestation and natural expansion, the world's total forest area that corresponds to 31 % (3.8 billion ha) of the total land area exhibited an increasing reduction—according to the Global Forest Resources Assessment (FAO and JRC 2012) report—due to both anthropogenic and natural actions. The net annual forest loss increased significantly from 1990–2000 to 2000–2005 periods by 3.6 million ha (FAO and JRC 2012).

Environmental stress may result from abiotic factors including drought, salinity, extreme temperature, inadequate or excessive light conditions, ozone, pollution, and radioactivity. It can also be caused by biotic factors resulting from plant interaction with other organisms such as insects, fungi, bacteria, viruses, plant competition, and allelopathy. The occurrence of one abiotic stress may affect the plant functioning mechanisms through the induction of several interrelated changes at the morphological (Karakas et al. 2000; Vollenweider and Günthardt-Goerg 2005), anatomical (Bosabalidis and Kofidis 2002; Chartzoulakis et al. 2002a; Lesniewska et al. 2004; Fortescue and Turner 2005; Junghans et al. 2006; Chelli-Chaabouni et al. 2010; Ennajeh et al. 2010), physiological (Meena et al. 2003; Junghans et al. 2006; Rejskovà et al. 2007), and biochemical levels (Kozlowski 1997; Chelli-Chaabouni et al. 2010; Krasensky and Jonak 2012). However, the effect of extra-optimal environmental factors on plant growth and development is not necessarily harmful. Speed at which the stressful factor installs as well as the intensity and duration of stress determines the beneficial or injuring effect of stress. Hence, the gradual physiological adjustments induced by the slow increase of stress may protect plants from inhibition of growth and/or injury resulting from suddenly imposed stress (Kozlowski and Pallardy 2002).

Woody plants are perennial plants (usually trees, shrubs, or lianas) that are predominately characterized by the production of secondary tissues in stems and roots. In these latter organs, wood is formed, year after year throughout ontogeny, in superposed layers from secondary xylem leading to the reinforcement of the tissue structures. As woody plants have long-lasting biological cycle, they would have to support stress for a long period of time. Therefore, they have evolved specific mechanisms to overcome detrimental injuries resulting from environmental stresses. The situation is more complicated when multiple stresses occur. The mechanisms adopted against one stress factor are not necessarily similar to those evolved to counteract multiple stresses. Plant response may differ when two or more stresses occur successively or simultaneously.

The specificity of woody plants such as longevity and size makes their use in experimental research studies much more complicated and hard but very informative and worthwhile. Most experimental studies on woody plants are performed on seedlings and young trees. However, plant behavior against environmental stresses may significantly differ throughout the successive developmental stages namely germination and both juvenile and adult stages (Ceulemans and Mousseau 1994; Stamp 2003; Boege and Marquis 2005).

As the effects of environmental stresses are well described in the previous sections, this chapter will make only a brief description on the way that various environmental factors affect woody species. We will emphasize on the specific adaptation and resistance of woody plants to extra-optimal conditions and the current methods adopted to assess stress tolerance/resistance.

2 Plant Responses to Environmental Stresses

When subjected to environmental stresses including drought, salinity, frost, and herbivory attacks, woody plants need to reallocate energy in a way allowing stress adaptation (Skirycz and Inzé 2010) but also to maintain growth and productivity. These latter functions are closely related to water movements within the plant which are supported by vascular tissues (Osakabe et al. 2012). Plant controls gaseous exchanges and water loss mainly by the regulation of stomatal movements (Fini et al. 2013; Sapeta et al. 2013). To reach these vital objectives, plant responds by the activation of many metabolic processes controlling photosynthesis, ion homeostasis, and plant hormone signaling that may alter gene expression. These reactions are usually expressed at both phenotypic and genotypic levels.

2.1 Abiotic Stresses

Abiotic stresses may be defined as nonliving factors affecting growth and productivity of living organisms. They may be divided into two main categories: (1) physical stressful factors including drought, flooding, extreme temperature, and inadequate light quality or intensity; and (2) chemical stressful factors including salinity, ozone, elevated CO_2 level, and heavy metal pollution. Plants respond to abiotic stresses through morphological and anatomical symptoms concerning leaves, stems, and roots.

Drought, salinity, and cold are among the major abiotic stresses that cause serious problems to woody plants (Krasensky and Jonak 2012) including low water and nutrient availability, toxic concentrations of salt ions such as sodium (Na) and chlorides (Cl), and may lead to Ca deficiency (Marschner 1995). Apart from low water availability, an excess of water in the soil induces flooding stress for woody plants. The lack of oxygen resulting from root submersion induces physiological (Polacik and Maricle 2013) and metabolic changes that can be expressed through plant injury, changes of plant anatomy, inhibition of seed germination, decrease of vegetative and reproductive growths, early senescence, and even mortality (Kozlowski 1997; Glenz et al. 2008). Environmental pollution mainly results from widespread urbanization, industrialization, and agriculture intensification increased exponentially in a way making several pollutants reaching toxic levels for vegetation. Investigations on woody plant responses to phytotoxic levels of environmental pollutants have gained an interest since the early seventeenths. Due to the important extent of forests in space and duration ecosystems, particular interest on the impact of phytotoxic factors on forest trees increased. Short ultraviolet wavelength radiations (UV-B) increased due to the depletion of stratospheric ozone (O_3) caused by gaseous pollutants, leading, to some extent, to an ecological imbalance and global climatic changes. A 3 years field study on *Pinus taeda* seedlings demonstrated the cumulative inhibitory effect of elevated UV-B on plant growth (Sullivan and Teramura 1992).

Heavy metal concentrations increased at promising toxic levels in the last few years (Sainger et al. 2011). Apart from the direct effect of heavy metals on woody plant growth and survival (Fernández et al. 2013), their effects on soil animals and microorganisms may indirectly decrease organic mineralization (nitrogen and phosphorous mineralization) and subsequently limit nutrient availability to plants (Tyler 1984).

Many morphological and anatomical changes occur in leaves, as they are an important site of photosynthesis and biochemical reactions including defensive mechanisms. Several stresses including drought, salinity, and high light intensity may cause leaf injury varying from chlorosis (Rochdi et al. 2005) to total leaf necrosis (Chelli-Chaabouni et al. 2010). Marginal leaf scorch was associated with sodium accumulation in salt-stressed peach (Karakas et al. 2000). Salt stress may affect cell elongation and expansion inducing a reduction of leaf area (Curtis and Läuchli 1987; Abbruzzese et al. 2009). It affects stomatal characteristics such as stomata density and guard cell length which reflects to stomatal conductance and hydraulic status of the plant (Abbruzzese et al. 2009).

Abiotic stresses could cause severe injuries to woody plant stems. Mechanical effects of strong winds may break twigs and provoke flower bud falls. Epidermis of newly formed shoots may be burned by high solar radiations that occur for a relatively long period of time. At an anatomical level, a reduction of xylem differentiation under salt stress leads to a decrease of vessel lamina due to low nutrient supply to the cambium and low potassium ion (K^+) content in the shoots of salt-sensitive poplar species (Escalante-Pérez et al. 2009).

Roots are affected by abiotic stresses in many ways. Drought and salinity induce a reduction of root system biomass through an increase in root length and width. Cotton seedlings growing in hydroponic salt solution produced less and thinner roots with increasing salinity. Root anatomical analysis showed shorter and more nearly iso-diametrical cortical cells than those of control plants (Kurth et al. 1986). Salt stress may lead to root lignifications in pistachio (Walker et al. 1987). Under soil anaerobiosis conditions (waterlogging or flooding), roots suffer from asphyxia before final death (Kozlowski and Pallardy 2002).

Fruits are also affected by many abiotic stresses either directly or indirectly under climatic conditions favoring pathogen proliferation and growth. Strong or sandy winds and hail falling may be harmful to the fruits especially at the maturation stages. In early stages of fruit development, non-optimal temperatures may affect fruit formation and growth. The ovule of banana tree (*Musa* species) affected by sustained tow temperature before anthesis showed many changes (size reduction, low growth, and more rounded shape) in comparison with unaffected ones (Fortescue and Turner 2005).

Plant growth and productivity may be affected by abiotic stresses such as drought (Picchioni et al. 1990; Ramoliya et al. 2004), salinity (Chelli-Chaabouni et al. 2010;

Akça and Samsunlu 2012), and flooding (Kozlowski and Pallardy 2002; Capon et al. 2009; Glenz et al. 2008). Growth is either inhibited or stimulated according to the nature, severity, and duration of stress. Environmental stresses that induce water and nutrient deficiency affect leaf regeneration and growth by the reduction of new formed leaves and the decrease of leaf area. Plant increases photosynthate allocation to the roots in expense of shoots.

In mature woody plants, the negative correlation between vegetative and reproductive growths (Kozlowski and Pallardy 2002) plays a major role under moderate stress conditions. The vegetative stage of development at which drought occurs is determinant in further growth response. A short period of water deficit may induce stimulation of reproductive growth through flower bud formation, break of flower bud dormancy, and flowering according to the time at which stress happens (Kozlowski and Pallardy 2002). Stimulation of reproductive growth may occur simultaneously with vegetative growth inhibition.

The disruption of stomatal conductance by stress affects directly gaseous exchanges that are mainly related to photosynthesis (CO_2) and photorespiration (O_2). Various stresses such as drought (Angelopoulos et al. 1996; Flexas and Medrano 2001; Faraloni et al. 2011; Fini et al. 2013), salinity (Walker et al. 1988; Chartzoulakis et al. 2002b; Tabatabaei 2006; Abbruzzese et al. 2009), flooding (Glenz et al. 2006; Polacik and Maricle 2013), and sub- (Costa e Silva et al. 2008) or supraoptimal temperatures (Kozlowski and Pallardy 2002) may affect photosynthesis. Plants control transpiration through early stomatal closure to diminish water loss and enhance water use efficiency. The leaf water status affects photosynthesis through the efficiency of the photosystem II (PSII) activity (Fini et al. 2013). For instance, the olive cultivars that maintained high relative water content (RWC) under water deficit conditions sustained a high chlorophyll fluorescence ratio F_v/F_m and showed less injury to PSII performance (Faraloni et al. 2011). Sublethal high temperatures may inhibit photosynthesis and cause membrane injury and protein aggregation and denaturation (Kozlowski and Pallardy 2002). In leaves of poplar, drought induced changes in photosynthetic reactions through a decrease of rubisco content and the changes of light-related and membrane-related proteins (Durand et al. 2011). The progressive decline of carbon assimilation under stress may result from both stomatal and metabolic limitations (Angelopoulos et al. 1996; Flexas and Medrano 2001). Under severe water stress the decrease of non-stomatal component of photosynthesis in olive trees was thought to be due to light-dependent inactivation of the primary photochemistry related to PSII (Angelopoulos et al. 1996). Photosynthetic metabolisms including ribulose 1,5-biphosphate (RuBP) regeneration capacity, adenosine triphosphate (ATP) synthesis, and ribulose 1,5-biphosphate carboxylase/oxidase (Rubisco) activity could be disturbed (Flexas and Medrano 2001). At the onset of flooding in greenhouse experimental conditions, photosynthesis of *Tamarix ramosissima* decreased by non-stomatal limitations while oxygen stress increased (Polacik and Maricle 2013).

The biochemical and enzymatic activities of woody plants may change under stressful conditions. They include antioxidant activities (Zhang et al. 2013), lipid peroxidation (Fernández et al. 2013), proline and glycine-bétaine-metabolizing system (Ahmad et al. 2010), and cellulose biosynthesis (Zhong and Läuchli 1988; Delmer and Armor 1995). Low temperatures induce changes in the concentrations of hormones and metabolites such as sugar and protective proteins and cause alterations in gene expression (Zhu et al. 2007).

Stomatal closure due to many abiotic stresses (including drought, salinity, high light, extra-optimal temperatures, and pathogen attacks) leads to a reduction of assimilation rate and the production of reactive oxygen species (ROS) that are responsible of oxidative stress (Osakabe et al. 2012). ROS are highly destructive to lipids, nucleic acids, and proteins (Türkan and Demiral 2009) and have been shown as important second messengers for stress signal transduction pathways. ROS may affect the cellular activity of the plant through molecular, functional, and structural alterations such as protein, DNA, and lipid oxidative damages (Apel and Hirt 2004). Plant exposure to low temperature may cause mild oxidative stress that show many similarities with plant response to water deficits (Costa e Silva et al. 2008).

2.2 *Biotic Stresses*

Many reports describe the close relationship between plant biotic stresses resulting from pests and diseases and abiotic environmental conditions (Shoeneweiss 1981; Luther et al. 1997). The proliferation of insects, bacteria, fungus, and viruses in woody plants depends on climatic and soil conditions in natural habits but also on orchard management systems of cultivated trees and shrubs (Valdés-Gómez et al. 2011) such as a large amount of coarse woody debris in managed forests that may host an important source of pathogens influencing the stability of forest stands (Santini et al. 2008). Apart from their direct effects, the physico-chemical characteristics of the surrounding environment may have an indirect impact on the plant-parasite biological development and proliferation through their interactions with other insects, fungi, or bacteria. The inter- and intraspecific interactions (symbiotic, parasitic, and synergetic) between these organisms determine the level of stress severity.

The severity of pathogen attacks is also associated to host plant physiology and anatomy (Pérez-Contreras et al. 2008; Rieske and Dillaway 2008; Inbar 2011). The host shifting is a specific trait of phytophagous insects such as aphides and lepidoptera. For examples, aphids of *Pistacia* species assess plant chemical, anatomical, physiological, and structural traits before choosing the adequate host plant (Inbar 2011). Herbivore-induced extensive defoliation of oak forests was correlated with depressed C/N ratio and elevated foliar nitrogen (Rieske and Dillaway 2008). Two main hypotheses are suggested to explain herbivore host plant selection strategy. The "plant-stress hypothesis," mainly adopted by generalist herbivores, states that stressed plants are less able to synthesize defensive chemicals to resist insect attacks and are then more vulnerable (White 1969). The "plant-vigour hypothesis" states that herbivore is indifferent to the level of plant defense and prefers feeding on healthy and vigorous plants (Price 1991). This latter hypothesis is likely preferred by specialist herbivores. However, both main hypotheses may explain plant selection by herbivores for depositing of eggs namely oviposition (Pérez-Contreras et al. 2008).

Biotic stresses affect plant growth through a reduction of photosynthesis (Christen et al. 2007; Bilgin et al. 2010). Bilgin et al. (2010) attempted to understand how do various biotic stresses (including arthropods, fungi, bacteria, and viruses) affect plant photosynthesis activity at a genetic level. They compared transcriptome data from microarray experiments after 22 different forms of biotic damage on eight plant species. Results revealed that regardless of the nature of biotic factor, transcript levels related to photosynthetic activity decreased. The photosynthetic gene down-regulation was accompanied by an up-regulation of genes coding for synthesis of jasmonic acid and those involved in the responses to salicylic acid and ethylene. Authors suggested that these reactions may be a part of defense mechanism.

In natural ecosystems, woody plants may compete with other woody or herbaceous species for resources in different ways such as the large and deep root invasion of the rhizosphere (Schenk 2006) for water and nutrient uptake, the increase of plant height to enhance light availability (Sterck and Bongers 2001), and the release of chemicals namely allelopathy that inhibit or stimulate growth and survival of the neighboring plant and microorganism species (Maclaren 1983). In some cases, different strategies may explain the inter- and intraspecific plant interactions (Inderjit and Mallik 2002). The plant's release of allelo-chemical products may have several effects on individual neighboring plants and, consequently, on organism ecosystem distribution. At the individual plant level, these products may affect the seed germination ability (Blanco 2007), plant growth (Lodhi 1976), physiology (e.g., respiration, photosynthesis, and hormonal and enzymatic processes), the cellular functioning system (e.g., membrane permeability, chloroplast activity, and chlorophyll concentration), the water and nutrient uptake, and transport.

3 Multiple Stresses: Occurrence and Interaction

The occurrence of only one stress at the same time in the field or in the natural conditions is seldom rare. Often, two or more stresses are simultaneously or successively associated. For example, drought stress is closely related to high temperature and luminosity in hot climate arid and semiarid areas. Plants respond to multiple stresses by the activation of one specific mechanism (Rizhsky et al. 2004) through the activation of numerous biochemical and molecular reactions (Osakabe et al. 2012; Perdiguero et al. 2013). The multiple stress combination may lead to a modification in the plant stress susceptibility. The occurrence of one environmental stress may indispose or predispose plant to a second stress. At the genetic level, the over-expression of genes to adapt to a given stress may incur tradeoffs for acclimation to other stresses (Lynch and St Clair 2004). For example, the increase of plant transpiration resulting from pollutants such as sulfur dioxide (SO_2) may expose plant to drought stress (Shoeneweiss 1981). Inversely, stomatal closure in response to many abiotic stresses including drought, flooding, and low atmospheric humidity may lead to higher tolerance to air pollutants (Kozlowski and Pallardy 2002). Similarly, elevated atmospheric CO_2 due to global climatic changes may alleviate oxidative stress (drought and salinity) and enhance plant tolerance through the availability of more energy that can be allocated to defensive mechanisms (Kyoro et al. 2012).

In the particular case of biotic and abiotic stresses association, the occurrence of an abiotic stress can enhance or reduce plant resistance to a pest or pathogen and vice versa (Atkinson and Urwin 2012). Abiotic stresses such as water deficit, salinity, freezing, or heavy metal pollution may weaken plants and make them more vulnerable to the attack of some pests and diseases (Shoeneweiss 1981). For forest stands, this vulnerability may be extended to ecosystem level as reported by Luther et al. (1997) in the case of insect defoliation of balsam fir stands. Under abiotic stress, the normal carbon allocation patterns may be deviated in a way making higher carbohydrate allocation to the root and leaves in favor of carbohydrates left for carbon reserve and defense compound biosynthesis (Luther et al. 1997). At a molecular level, the metabolic signaling pathways of such stress combinations can act antagonistically (Anderson et al. 2004). In light of this, studying plant stress tolerance by imposing each stress individually may not reflect the exact plant response in the field (Mittler and Blumwald 2010).

4 Plant Response to Stress Throughout Ontogeny

Several structural (Loney et al. 2006), physiological, and biochemical (Loney et al. 2006; Juvany et al. 2013) changes occur during plant development. As woody plants develop, they show usually an increase in carbon/nutrient balance and carbon storage capacity (Niinemets 2010) as well as greater accessibility to water, nutrients, and sunlight but also a decrease in growth rate, root/shoot ratio, photosynthesis, stomatal conductance, and metabolic activities (Boege 2005). The tree responses to stress vary throughout the ontogeny (Niinemets 2010) in relation with age-related changes of physiological and biochemical processes controlling carbon assimilation and storage, growth rate, and defensive mechanisms. Compared to seedlings and saplings, many reports described greater resistance of large non-senescent trees to abiotic (Kozlowski 1997; Cavender-Bares and Bazzaz 2000; Rozas et al. 2009) and biotic stresses (Basey et al. 1988; Luther et al. 1997). However, regarding plant resistance to herbivore stresses, both positive (Loney et al. 2006) and negative (Schappert and Shore 2000) relationships between ontogeny and resistance were reported for several woody species (Boege and Marquis 2005). During plant development, Boege and Marquis (2005) proposed a pattern of changes in plant defense and tolerance during ontogeny based on the assumption that plant resources may act as a constraining trait of tolerance and resistance of vegetative tissues. Armas and Pugnaire (2009) findings support this statement as the interaction of the two dominant shrub species *Pistacia lentiscus* and *Juniperus phoenicea* did not benefit any species at seedling stage but when plants became progressively mature, *Pistacia* species gained competitive growth and survival advantage. It should be emphasized that ontogeny has an effect not only on the ability of plant to respond to a given stress but also on the growth recovery after the disappearance of stress (Boege 2005).

At a seedling stage, plants may produce defensive compounds to resist herbivory attacks (Schappert and Shore 2000), but as the plants gain in maturity, the biochemical protective strategy decreases progressively while many protective structural changes (greater leaf thickness, higher lignin and fiber content) occur (Loney et al. 2006). The decrease of plant defensive biochemical synthesis with age was suggested as the result of natural selection based on the reallocation of energy according to the cost/benefit ratio. Plants maximize the production of protective chemicals at developmental stages of great risk of herbivore attack or low tolerance (Schappert and Shore 2000; Stamp 2003).

5 Influence of Biomes on Stress Occurrence and Severity

The severity and timing of stresses vary throughout the growing season according to the nature of a given ecosystem as illustrated by Niinemets (2010). For example, in cool temperate ecosystems, the plant tolerance to frost stress is lower during the early winter period corresponding to the plant dormancy; plants are more susceptible in middle and late winter when temperature becomes progressively warmer (Ögren 1996). In the Mediterranean environments, long-lasting supraoptimal temperatures and light are among the major factors of stress (Angelopoulos et al. 1996). In these areas, plants are exposed to drought and photo-inhibition (Guàrdia et al. 2012) but the severity of stress differs from coastal to high altitudes (Flexas et al. 2001; Yang and Miao 2010). Under progressive drought stress *P. kangdingensis*, originating from higher altitude, displayed superior height growth and leaf development as well as greater increments in soluble proteins, soluble sugars, free proline, and antioxidant enzyme synthesis than *P. cathayana* that grows in lower altitude (Yang and Miao 2010). However, even in Mediterranean areas, woody plants may be subjected to episodic low temperature events (below 0 °C) that limit expansion of species such as *Eucalyptus globulus* (Costa e Silva et al. 2008).

Differing from Mediterranean and cool temperate forest ecosystems, the temperate bogs are subjected to chronic nutrient deficiency and waterlogging (Niinemets 2010).

Climate is an important factor that influences and modulates ecosystem composition and scope. Increasing temperatures may lead some species to move to higher altitudes that are suitable for their growth and development. Increasing plant competition for water and nutrients in arid and semiarid biomes may lead to a dominance of the most adapted species at the expense of species showing lower stress tolerance (Armas and Pugnaire 2009; Eilts and Huxman 2013). Moreover, plant response and susceptibility to various biotic stresses may be affected by the biome (Slippers and Wingfield 2007) as well as by the predicted global climatic changes (Veteli et al. 2002; Allen et al. 2010). Table 1.1 indicates abiotic factors of stress encountered in various temperate climatic conditions.

Table 1.1 Major stress factors encountered in temperate climates

Climate nature	Stress factors	References
Cold and cool temperate climates	Waterlogging and flooding	Kozlowski and Pallardy (2002) and Niinemets (2010)
	Nutrient deficiency	Niinemets (2010)
	Elevated CO_2	Tjoelker et al. (1998)
Arid and semiarid regions and Mediterranean climates	Drought	Chartzoulakis et al. (2002a) and Guàrdia et al. (2012)
	Salinity	Chartzoulakis et al. (2002b) and Lynch and St Clair (2004)
	Episodic low temperatures (<0 °C)	Costa e Silva et al. (2008)
Tropical and subtropical regions	Soil acidity	Lynch and St Clair (2004)
	Mineral toxicity and deficiency	Lynch and St Clair (2004)
	Light deficiency	Sterck and Bongers (2001)

6 Adaptation and Mechanisms of Stress Tolerance/Resistance

The previous three sections have provided insights into specific responses of woody trees to one or multiple stresses according to many intrinsic or extrinsic factors (multi-stress interactions, response throughout ontogeny, biome influence). Plants may activate different mechanisms at various structural and functional levels to overcome possible injury that may be induced by one stressful condition. These mechanisms act simultaneously or successively in relation with the nature of stress, its duration, and severity. As stated by Glenz et al. (2006) for flooding tolerance, specific biotic factors (developed adaptation and capacity of acclimation) related to the woody species influence the final response of plant to a given stress. As one mechanism may be involved by species under various kinds of environmental stresses (for instance, osmotic adjustment is implicated in both drought and salt stress), we will present the main adaptive strategies adopted by plants at structural (morphology and anatomy) as well as functional (growth, physiology, ionic relations, biochemical and enzymatic activities, and genetics) levels.

6.1 *Structural Adaptation*

The major feature that determines the plant stress tolerance is concerned with leaf as it is the principal site of gaseous exchange, photosynthesis, and metabolic activities. In many species exposed to various biotic and abiotic forms of stress such as drought, salinity, wounding, and pathogen attacks, an increase in leaf cuticle thickness (Bacelar et al. 2004; Bosabalidis and Kofidis 2002) and epidermis (Kulkarni et al. 2010), a reduction of size and density of epidermal cells (Chartzoulakis et al. 2002a) and xylem (Bosabalidis and Kofidis 2002; Kulkarni et al. 2010), and a greater cell wall lignification (Niinemets et al. 1999; Osakabe et al. 2011) have been considered as stress-tolerant traits. Drought-stressed olive plants reduced the size of stomata and epidermal and mesophyll cells, increased cell and stomatal density, and reduced the number of nonglandular hairs (Bosabalidis and Kofidis 2002).

These anatomical and morphological modifications reflect a better control of water loss through cuticular transpiration in water deficit conditions. For instance, several olive species tend to diminish leaf area, develop thicker epidermal leaves, and increase trichome density when subjected to drought stress (Ennajeh et al. 2010). The increased leaf fracture toughness seems to be useful in the protection from herbivore damage (Choong et al. 1992) as it improves leaf mechanical properties through the thickening of smaller vein walls forming a venous network (Lucas et al. 1991).

The plant-protective roles of cuticular waxes are widely reported from many species (Shepherd and Griffiths 2006). Cuticular-wax layers are predominately composed from long-chain hydrocarbon compounds, including alkanes, primary alcohols, aldehydes, secondary alcohols, ketones, esters, and other derived compounds (Shepherd and Griffiths 2006). Waxy species are reported to modify reflectance in a way that changes light absorption and consequently photosynthetic activity (Cameron 1970; Holmes and Keiller 2002). The plant reflectance ability plays a major protective role against high radiations in drought conditions and UV-B harmful radiations due to the stratospheric ozone layer damage. Reflectivity is highly influenced by the surface topography of leaves, primary hairs, and cuticular waxes (Shepherd and Griffiths 2006).

The resistance of xylem to cavitation events is an important parameter that determines stress resistance (Tyree and Ewers 1991). The drought-stressed bald cypress plants used their biomass in a way that strengthens the xylem and reduces its vulnerability to cavitation (Stiller 2009). Similarly, woody plants respond to salinity by the reduction of vessel lumina and the increase of wall strength to counteract the reduction of xylem hydraulic conductivity resulting from salt stress. For example, the decrease of the vessel lumina of the salt-resistant poplar species *Populus euphratica* under salt stress conditions was lower than that of the salt-sensitive *Populus × canescens* species (Junghans et al. 2006).

In anoxic conditions such as submersion of the root system, plants develop hypertrophied lenticels that present a pathway for the diffusion of oxygen (O_2) through living bark cells and the release of toxic compounds related to anaerobiosis (Glenz et al. 2006). The presence of lenticels as a morphological adaption to flooding is reported for many Central European trees and shrubs including *Acer campestre*, *Alnus glutinosa*, *Fraxinus excelsior*, *Populus nigra*, and *Salix alba* (Siebel et al. 1998; Hook 1984). In addition of lenticels, aerenchyma tissues are thought to be a morphological adaptation of woody plants to anaerobiosis as they constitute an extensive intercellular air space allowing the diffusion of oxygen from the aerial part of the plant to the roots (Glenz et al. 2006; Wang and Cao 2012).

The variation of cell wall elasticity under stressful conditions was supported as a trait of stress tolerance of some woody species. However, both positive (Chartzoulakis et al. 2002a) and negative (Patakas and Noitsakis 1997) roles of tissue elasticity were suggested as a mechanism of adaptation to stress. As examples of the two

Table 1.2 Morphological and anatomical traits of stress tolerance that have been cited in woody plants

Trait	Stress	Species	References
Increase of leaf cuticle thickness	Drought	*Olea europaea*	Bacelar et al. (2004) and Bosabalidis and Kofidis (2002)
Increase of leaf epidermis		*Ziziphus mauritiana* (Lamk.)	Kulkarni et al. (2010)
Increase of thick palissade mesophyll layers	Salinity	*Olea europaea*	Bacelar et al. (2004)
Reduction of size and density of epidermal cells	Wounding	*Olea europaea*	Chartzoulakis et al. (2002a)
Reduction of size and density of xylem		*Olea europaea*	Bosabalidis and Kofidis (2002) and Kulkarni et al. (2010)
Cell wall lignification	Pathogen attacks	*Ziziphus mauritiana* (Lamk.)	Niinemets et al. (1999)
		Several species	Osakabe et al. (2011)
Great leaf anatomical plasticity	Drought	*Quercus* species (*Q. velutina*, *Q. coccinea*, and *Q. rubra*)	Ashton and Berlyn (1994)
Increased leaf fracture toughness	Herbivore damage	42 tropical tree species	Choong et al. (1992)
Presence of hair and cuticular waxes	High light intensity	*Eucalyptus* species	Cameron (1970)
		45 species including *Citrus*, *Eucalyptus*, *Quercus*, *Prunus*	Holmes and Keiller (2002)
		Woody and non-woody species	Shepherd and Griffiths (2006)
Chloroplast movements	Light limiting conditions	*Tradescantia albiflora*	Park et al. (1996)
		Cissus, *Eustrephus*, *Alocasia*, *Eucalyptus*	Williams et al. (2003) and Way and Pearcy (2012)
Lenticels, aerenchyma tissues, and adventitious roots	Flooding	Central European trees and shrub species	Glenz et al. (2006)
		Taxodium distichum and *Sapium sebiferum*	Wang and Cao (2012)
		Woody species (general review)	Kozlowski and Pallardy (2002)
Resistance to xylem cavitation		Trees and woody plants	Tyree and Ewers (1991)

cases, the reduction of cell wall elasticity was suggested as an efficient mechanism adopted by grapevine to overcome water deficit (Patakas and Noitsakis 1997) and high UV radiation (Lesniewska et al. 2004). Inversely, the increase in tissue elasticity and a decrease of osmotic potential were thought to be likely the two predominant mechanisms of adaptation in avocado (Chartzoulakis et al. 2002a). Table 1.2 shows number of morphological and anatomical traits that have been associated to stress tolerance in woody plants.

6.2 Physical Movements

Plants respond to various abiotic (e.g., drought, salinity, UV radiation, high temperature, and heavy metals) and biotic (e.g., herbivores, bacteria, virus, and fungi) stresses by leaf rolling (Bosabalidis and Kofidis 2002; Kadioglu et al. 2012). This physical movement combined with stomata closure plays a key role in the reduction of water loss and the maintenance of cell turgor in plants subjected to osmotic stress conditions. However, stomata closure leads to a reduction of gaseous exchanges through the leaves and a diminution of plant photosynthetic activity due to a reduction of CO_2 entry. In these conditions, plants optimize carbon uptake by different ways such as the modification of crown architecture and leaf angle positioning to adjust within crown irradiance (Egea et al. 2012) and heterogenic stomatal aperture (Guàrdia et al. 2012). At a cellular level, chloroplast movements operate to adjust photon flux density (Park et al. 1996; Way and Pearcy 2012). At low light conditions, chloroplasts move to increase light absorption. The rapidity of chloroplast rearrangements to periclinally or anticlinally position against leaf surface is a viable mechanism for leaves to reduce excess photon flux density interception (Williams et al. 2003; Way and Pearcy 2012). Within chloroplasts, the size and stacking of thylakoid grana may change within 10 min of high light exposure (Rozak et al. 2002).

6.3 Growth and Physiology

One of the adaptive responses to stress is the plant growth regulation by modulation of both cell division and expansion (Skirycz and Inzé 2010). Woody plants subjected to nutrient deficiency allocate higher amounts of photosynthate to roots (Kozlowski and Pallardy 2002). Tolerant plants use different ways to reactivate root system functioning and compensate from root alteration and mortalities. Under water deficit conditions, *Fagus* species stimulated fine root growth to compensate from root biomass losses during dry mid-summer (Leuschner et al. 2001). Similarly, plants adapt to flooding conditions by the regeneration and growth of adventitious roots to compensate from the alteration of asphyxiated initial roots. Root initiation takes place on the originated root system and/or submerged portion of the stems (Kozlowski and Pallardy 2002). Flood tolerance in woody plants is positively correlated with the ability of adventitious root regeneration, the conversion of toxins produced in the soil to less toxic compounds by rhizosphere oxidation, and the increase of root-synthesized gibberellins and cytokinins (Kozlowski and Pallardy 2002).

Many reports describe a general close association between morphological and anatomical traits and the efficiency of physiological activity in stressed plants (Ashton and Berlyn 1994; Bacelar et al. 2004; Kulkarni et al. 2010). Ashton and Berlyn (1994) found close correlations between anatomical measures of plasticity and physiological measures of plasticity in *Quercus* species. Stomatal density plasticity was correlated with both net photosynthesis and stomatal conductance plasticity.

The most drought-tolerant and higher light demanding *Q. velutina* species exhibited greater anatomical plasticity and higher net photosynthesis than *Q. coccinea* and *Q. rubra* species. *Quercus ilex* L. adapted to severe drought conditions by patchy stomatal closure leading to heterogenic photosynthesis (Guàrdia et al. 2012). Drought-tolerant *Jatropha curcas* accessions maintained high leaf RWC through water-saving strategy involving strict stomatal regulation and growth reduction (Sapeta et al. 2013).

6.4 Cell Homeostasis and Osmotic Adjustment

Drought and salt stress lead to the increase of osmotic pressure in the cytosol due to water deficiency (drought) or the uptake of large amounts of salts (salt stress). Salt-tolerant species develop different strategies either to exclude salt from the cells or to tolerate it within the cells (Kozlowski 1997; Parida and Das 2005). To maintain cell homeostasis and cell turgor and protect metabolic activities, osmotic adjustment mechanism is adopted by tolerant species through two ways: (1) sequestration of excess of salt ions in the vacuole and (2) biosynthesis of organic osmoprotectant compounds (Munns 2002). Osmoprotectants are named compatible solutes as they do not interfere with plant metabolism even if they accumulate to high concentrations. Main biochemical compounds implicated in osmoregulation are proline (Ahmad et al. 2010), glycine-bétaine, and soluble sugars (Kozlowski 1997; Clifford et al. 2002). The nature of biosynthesized osmoprotectants seems to be species-related (Pallardy 2007). The presence of mucilage and polysaccharides in leaves of drought-tolerant *Ziziphus* species is reported to act in osmotic adjustment by the remobilization of solutes allowing better efficiency in water uptake and translocation to the roots and stems before plant defoliation (Clifford et al. 2002). However, in the special case of salt stress, there is uncertainty with regard to the role of proline in salt tolerance. In fact, both positive (Hokmabadi et al. 2005) and negative (Ashraf 1989) associations of proline accumulation with tolerance to salt stress were reported. Similarly, instances of enhanced (Gucci and Tattini 1997) as well as unchanged and reduced soluble sugar contents were reported (Rejskovà et al. 2007).

6.5 Ionic Interactions

Stresses such as drought and salt stress may lead to an ionic imbalance. The disturbance of potassium (K^+) nutrition is a common feature under sodium chloride (NaCl) stress conditions and is often associated with potassium deficiency resulting from potassium–sodium interaction (Cramer et al. 1987; Slama 2004; Parida and Das 2005). Thus, the maintenance of normal cellular functions strongly depends on the K^+ nutrient status and K^+–Na^+ interaction through a selective uptake and transport of K^+ in depend of Na^+ into the shoots (Cramer et al. 1987). A high K^+/Na^+ ratio in the leaves is considered as a salt-tolerant trait in many species (Chartzoulakis et al. 2002b; Meena et al. 2003) as it influences photosynthetic activity and nitrogen assimilation (Tabatabaei 2006). The accumulation of sodium ions in the roots is an adaptive strategy used by several woody species to avoid their toxicity in the shoots (Walker et al. 1987; Picchioni et al. 1990; Gucci and Tattini 1997). On the other hand, salt stress may induce an increase (Ramoliya et al. 2004; De Pascale et al. 2007) or a decrease (Gratten and Grieve 1999; Unno et al. 2002) of calcium ion (Ca^{2+}) content depending on the specific plant physiology as well as the nature and duration of stress. Calcium plays a key role in maintaining the plasma membrane integrity, thus limiting the toxic effect of Na^+ (Rengel 1992; Gucci and Tattini 1997), and acts as a secondary messenger in the regulation of the signal transduction pathways for the response to abiotic stresses (Rengel 1992; Maathuis and Amtmann 1999). It is also reported to control the Na^+ influx via a nonselective ion channel and to influence K^+/Na^+ selectivity. It was suggested that calcium plays a role in salt toxicity alleviation (Cramer et al. 1987; Rengel 1992; Gucci and Tattini 1997; Melgar et al. 2006; Sotiropoulos 2007). Accordingly, the ability of plants to control the root to shoot transport and a relative high Ca^{2+} tissue content can serve as a criterion for stress tolerance. The measurements of K^+/Na^+ and $Ca^{2+/}Na^+$ ratios in the assessment of salt tolerance were suggested in non-woody (Dasgan et al. 2002; Maathuis and Amtmann 1999) and woody plant species (Heimler et al. 1995).

Many conifer species resist very low temperatures (down to −60 °C) through a freeze drying mechanism that allows the progressive expulsion of water from the tissues during cooling (Sakai 1979).

6.6 *Hormonal and Enzymatic Activities*

Environmental stresses can induce an imbalance in hormone physiology of woody species. Plant hormones play a key role in the regulation of the vascular tissue growth and secondary development in woody plants (Osakabe et al. 2012). Auxins are involved in cell division and expansion, apical dominance, root development, and vascular tissue development (Osakabe et al. 2012). They have a role in the regulation of secondary development (Nilsson et al. 2008). In response to developmental and environmental stimuli, trees have developed mechanisms to modulate auxin transport in the vascular meristem (Osakabe et al. 2012). Furthermore, the salt-resistant species may use auxin-conjugates in the stem as a source of auxin to overcome the auxin physiology imbalance due to salt stress. It is the case of the salt-resistant poplar species (*Populus euphratica*) who displayed an increase in IAA-amido-conjugates in the xylem in response to salt stress while the salt-sensitive *Populus × canescens* did not show any variation in this compound (Junghans et al. 2006).

It should be emphasized that the plant tolerance to stress may be accomplished through variable mechanisms within a same species. Working on the drought tolerance of five field-grown olive cultivars, Bacelar et al. (2004) reported different morphological and structural adaptations of three drought-tolerant cultivars. While Manzanilla and Negrinha cultivars tended to increase lamina thickness through building parenchyma tissues and increasing cuticle and epidermis thickness, *Cobrançosa* cultivar enhances protection against water loss through the increase of mesophyll tissues density and thick cuticle and trichome layers. Similarly, the wild almond species *Amygdalus webbii* Spach responded to water deficit by a great reduction of leaf area and low decrease of stomatal frequency whereas the common *Amygdalus communis* L. adopted the inverse behavior (Camposeo et al. 2011).

In woody plants, the assimilated carbon is stored in the secondary cell wall as cellulose and lignin; the disruption of cellulose synthase genes was suggested to affect osmotic stress response (Osakabe et al. 2012). In forest and fruit trees, carbon pool status is involved in environmental stress tolerance.

6.7 Biochemical Release

Plants may produce defensive biochemical compounds to resist herbivore attacks. When tissues are altered, many plants react through the release of a respiratory poison. This phenomenon is called cyanogenesis; a plant biosynthesis of cyanogenic glycosides and cyanolipids which on hydrolysis liberate hydrogen cyanide (Poulton 1990; Miller et al. 2006).

6.8 Adaptation at the Genetic Level

Plants respond to variable stresses by the modification of gene expression. Salt-tolerant woody species respond to high salinity by the regulation of the genes involved in ionic and osmotic homeostasis (Osakabe et al. 2012). Under different levels of drought stress, various genes of loblolly pine (*Pinus taeda*) are differently expressed. These genes include those encoding heat shock proteins (HSPs), late embryogenesis abundant proteins (LEAs), and enzymes in the aromatic acid and flavonoid biosynthetic pathways (Watkinson et al. 2003).

7 Assessment of Stress Tolerance/Resistance in Woody Plants

The long life biological cycle of woody plants and the large tree sizes are the two predominant traits that make environmental stress assessment on woody plant species more complicated and difficult than that of herbaceous species. Most studies in this topic have been performed in controlled growth chamber conditions, in greenhouses, or in vitro. In vitro techniques have been reported to be valuable methods for screening of stress-tolerant lines (Fuller et al. 2006), studying defense mechanisms (Santos et al. 2001), and obtaining tolerant plants through genetic engineering tools. Different plant materials were used varying from whole plant (Shibli and Al-Juboory 2002; Mills et al. 2001; Zhang et al. 2004; Chelli-Chaabouni et al. 2010) to cell suspension cultures. Positive concordances of in vitro plant stress tolerance assessments with those experimented on whole plant in the field were described for several non-woody (Fuller et al. 2006) and woody plants (Vijayan et al. 2003; Faraloni et al. 2011) but relative discordances were also reported (Santos et al. 2001). However, data from in vitro woody plant assessment of stress tolerance should be taken with great caution due to the changes of stress responses as plant develops. When in vitro assessment was made, an additional evaluation in the field is highly recommended before final appreciations.

When woody plant assessments were realized in the field, seedlings and young trees or shrubs were often preferred to adult plant material to reduce the cost of experimentations related to the required spaces and the nature of measurement tools. Consequently, data obtained in these conditions may not be strongly enough correlated with what may happen for woody plants growing in the field or natural conditions due to great differences in spatio-temporal experimental conditions (Cornelissen et al. 2003). In fact, the occurrence of simultaneous or successive stresses as well as the plant interactions with other organisms in the surrounding environment may induce modifications in plant responses at the morphological, anatomical, physiological, biochemical, and molecular levels. Moreover, explanations about the whole plant behavior at both juvenile and reproductive stages are rarely to be realistic due to the variability of plant responses to stress throughout plant ontogeny. In the special case of trees, the within-canopy heterogeneity of organ responses according to sunlight penetration (Way and Pearcy 2012) may influence the whole plant response to stress (Küppers et al. 1996).

Given the above-mentioned specificities of woody plants, the plant response to stress has been generally assessed through multiparameter monitoring (such as growth, survival rate, water and mineral status, photosynthesis, chlorophyll and pigment concentrations, secondary metabolic compounds biosynthesis) using several experimental tools (e.g., growth measurement instruments, mineral analysis methods, sap flow sensors, molecular biology techniques).

8 Conclusion and Future Prospects

In this chapter, we presented an overview of the specific responses of woody plants to various abiotic and biotic stresses. These responses are tightly linked to the plant ontogeny, the nature, the duration and the severity of stress, and the characteristics of the surrounding environment. Woody plant adaptation to stress requires the reallocation of energy in a way allowing the activation of the mechanisms of adaptation and maintaining growth and productivity. They respond to extra-optimal conditions through structural, physiological, biochemical, and genetic modifications that are traduced to one specific mechanism. Stressful conditions may be the results of multiple stress factor effects occurring successively or simultaneously. The occurrence of one stress may lead to more or less adaptation to another stress. Moreover, adaptation to stress may differ according to the plant interactions with the neighboring organisms. It may also be different from the trunk base to the top of canopy in the special case of forest stands and trees.

The above specificities of woody plant responses to environmental stress need to be considered in the assessment of plant stress tolerance/resistance. Furthermore, the impact of the predicted climatic changes on plant susceptibility to stress should also be taken into account to prevent wide damages that may occur to plantation and native forest stands. Well understanding the possible mechanisms evolved to cope with stressful conditions is a key issue for the appropriate management of genetic resources. The use of efficient tools for plant stress investigation and assessment such as new biotechnological methods including proteomics and genetic engineering appears to be promising.

References

Abbruzzese G, Beritognolo I, Muleo R, Piazzai M, Sabatti M, Mugnozza GS, Kuzminsky E (2009) Leaf morphological plasticity and stomatal conductance in three *Populus alba* L. genotypes subjected to salt stress. Environ Exp Bot 66:381–388

Ahmad P, Jaleel CA, Sharma S (2010) Antioxidant defense system, lipid peroxidation, proline-metabolizing enzymes, and biochemical activities in two *Morus alba* genotypes subjected to NaCl stress. Russ J Plant Physiol 57(4):509–517

Akça Y, Samsunlu E (2012) The effect of salt stress on growth, chlorophyll content, proline and nutrient accumulation, and K/Na ratio in walnut. Pak J Bot 44(5):1513–1520

Allen GD, Macalady AK, Chenchouni H, Bachelet D, McDowell N, Vennetier M, Kitzberger T, Rigling A, Breshears DD, Hogg EHT, Gonzalez P, Fensham R, Zhangm Z, Castro J, Demidova N, Lim JH, Allard G, Running SW, Semerci A, Cobb N (2010) A global overview of drought and heat-induced tree mortality reveals emerging climate change risks for forests. Forest Ecol Manag 259(4):660–684

Anderson JP, Badruzsaufari E, Schenk PM, Manners JM, Desmond OJ, Ehlert C, Maclean DJ, Ebert PR, Kazan K (2004) Antagonistic interaction between abscisic acid and jasmonate-ethylene signaling pathways modulates defense gene expression and disease resistance in Arabidopsis. Plant Cell 16:3460–3479

Angelopoulos K, Dichio B, Xiloyannis C (1996) Inhibition of photosynthesis in olive trees (*Olea europea* L.) during water stress and rewatering. J Exp Bot 47(301):1093–1100

Apel K, Hirt H (2004) Reactive oxygen species: metabolism oxidative stress and signal transduction. Annu Rev Plant Physiol Plant Mol Biol 55:373–399

Armas C, Pugnaire FI (2009) Ontogenetic shifts in interactions of two dominant shrub species in a semi-arid coastal sand dune system. J Veg Sci 20:535–546

Ashraf M (1989) The effect of NaCl on water relations, chlorophyll, and protein and proline contents of two cultivars of blackgram (*Vigna mungo* L.). Plant Soil 119:205–210

Ashton PMS, Berlyn GP (1994) A comparison of leaf physiology and anatomy of *Quercus* (Section Erythrobalanus-Fagaceae) species in different light environments. Am J Bot 81(5): 589–597

Atkinson NJ, Urwin PE (2012) The interaction of plant biotic and abiotic stresses: from genes to the field. J Exp Bot 63(10):3523–3543

Bacelar EA, Correira CM, Moutinho-Pereira JM, Gonçalves BC, Lpoes JI, Torres-Pereira JMG (2004) Sclerophylly and leaf anatomical traits of five field-grown olive cultivars growing under drought conditions. Tree Physiol 24:233–239

Basey JM, Jenkins SH, Busher PE (1988) Optimal central-place foraging by beavers: tree-size selection in relation to defensive chemicals of quaking aspen. Oecologia 76:278–282

Bilgin DD, Jorge A, Zavala A, Zhu J, Clough SJ, Ort DR, DeLucia EH (2010) Biotic stress globally downregulates photosynthesis genes. Plant Cell Environ 33:1597–1613

Blanco JA (2007) The representation of allelopathy in ecosystem-level forest models. Ecol Model 209:65–77

Boege K (2005) Influence of plant ontogeny on compensation to leaf damage. Am J Bot 92(10):1632–1640

Boege K, Marquis RJ (2005) Facing herbivory as you grow up: the ontogeny of resistance in plants. Trends Ecol Evol 20(8):441–448

Bosabalidis AM, Kofidis G (2002) Comparative effects of drought stress on leaf anatomy of two olive cultivars. Plant Sci 163:375–379

Cameron RJ (1970) Light intensity and the growth of *Eucalyptus* seedlings II. The effect of cuticular waxes on light absorption in leaves of *Eucalyptus* species. Aust J Bot 18:275–284

Camposeo S, Palasciano M, Vivaldi GA, Godini A (2011) Effect of increasing climatic water deficit on some leaf and stomatal parameters of wild and cultivated almonds under Mediterranean conditions. Sci Hortic 127:234–241

Capon SJ, James CS, Williams L, Quinn GP (2009) Responses to flooding and drying in seedlings of a common Australian desert floodplain shrub: *Muehlenbeckia florulenta* Meisn. (tangled lignum). Environ Exp Bot 66:178–185

Cavender-Bares J, Bazzaz FA (2000) Changes in drought response strategies with ontogeny in Quercus rubra: implications for scaling from seedlings to mature trees. Oecologia 124:8–18

Ceulemans R, Mousseau M (1994) Effects of elevated CO_2 in woody plants. New Phytol 127:425–446

Chartzoulakis K, Patakas A, Kofidis G, Bosabalidis A, Nastou A (2002a) Water stress affects leaf anatomy, gas exchange, water relations and growth of two avocado cultivars. Sci Hortic 95:39–50

Chartzoulakis K, Loupassaki M, Bertaki M, Androulakis I (2002b) Effects of NaCl salinity on growth, ion content and CO_2 assimilation rate of six olive cultivars. Sci Hortic 96:235–247

Chelli-Chaabouni A, Ben Mosbah A, Maalej M, Gargouri K, Gargouri-Bouzid R, Drira N (2010) In vitro salinity tolerance of two pistachio rootstocks: *Pistacia vera* L. and *P. atlantica* Desf. Environ Exp Bot 69:302–312

Choong MF, Lucas PW, Ong JSY, Pereira B, Tan HTW, Turner IM (1992) Leaf fracture toughness and sclerophylly: their correlations and ecological implications. New Phytol 121:597–610

Christen D, Schönmann S, Jermini M, Strasser RJ, Défago G (2007) Characterization and early detection of grapevine (*Vitis vinifera*) stress responses to esca disease by *In situ* chlorophyll fluorescence and comparison with drought stress. Environ Exp Bot 60:504–514

Clifford SC, Arndt SK, Popp M, Jones HG (2002) Mucilages and polysaccharides in *Ziziphus* species (Rhamnaceae): localization, composition and physiological roles during drought-stress. J Exp Bot 53(366):131–138

Cornelissen JHC, Cerabolini B, Castro-Diez P, Villar-Salvador P, Montserrat-Marti G, Puyravod JP, Maestro M, Werger MJA, Aerts R (2003) Functional traits of woody plants: correspondence of species rankings between field adult and laboratory-grown seedlings? J Veg Sci 14:311–322

Costa e Silva F, Shvaleva A, Broetto F, Ortuno MF, Rodrigues ML, Almeida MH, Chaves MM, Pereira JS (2008) Acclimation to short-term low temperatures in two *Eucalyptus globulus* clones with contrasting drought resistance. Tree Physiol 29:77–86

Cramer GR, Lynch J, Laüchli A, Epstein E (1987) Influx of Na^+, K^+, and Ca^{2+} into roots of salt-stressed cotton seedlings: effects of supplemental Ca^{2+}. Plant Physiol 83:510–516

Curtis PS, Läuchli A (1987) The effect of moderate salt stress on leaf anatomy in *Hibiscus cannabinus* (Kenaf) and its relation to leaf area. Am J Bot 74(4):538–542

Dasgan HY, Aktas H, Abak K, Cakmak I (2002) Determination of screening techniques to salinity tolerance in tomatoes and investigation of genotype responses. Plant Sci 163:695–703

De Pascale S, Martino A, Raimondi G, Maggio A (2007) Comparative analysis of water and salt stress-induced modifications of quality parameters in cherry tomatoes. J Hortic Sci Biotechnol 82(2):283–289

Delmer DP, Armor Y (1995) Cellulose biosynthesis. Plant Cell 7:987–1000

Durand TC, Sergeant K, Renault J, Planchon S, Hoffmann L, Carpin S, Label P, Morabito D, Hausman JF (2011) Poplar under drought: comparison of leaf and cambial proteomic responses. J Proteomics 74(8):1396–1410

Egea G, Gonzalez-Real MM, Baille A, Nortes PA, Conesa MR, Ruiz-Salleres I (2012) Effects of water stress on irradiance acclimation of leaf traits in almond trees. Tree Physiol 32(4):450–463

Eilts JA, Huxman TE (2013) Invasion by an exotic, perennial grass alters responses of a native woody species in an arid system. J Arid Environ 88:206–212

Ennajeh M, Vadel AM, Cochard H, Khemira H (2010) Comparative impacts of water stress on the leaf anatomy of a drought-resistant and a drought-sensitive olive cultivar. J Hortic Sci Biotechnol 85(4):289–294

Escalante-Pérez M, Lautner S, Nehls U, Selle A, Teuber M, Schnitzler JP, Teichmann T, Fayyaz P, Hartung W, Polle A, Fromm F, Hedrich R, Ache P (2009) Salt stress affects xylem differentiation of grey poplar (*Populus* x *canescens*). Planta 229:299–309

FAO (2011) States of the world's forests 2011. http://www.fao.org/docrep/013/i2000e/i20000e.pdf

FAO, JRC (2012) Global forest land-use change 1990–2005. In: Lindquist EJ, D'Annunzio R, Gerrand A, MacDicken K, Achard F, Beuchle R, Brink A, Eva HD, Mayaux P, San-Miguel-Ayanz J, Stibig H-J (eds) FAO Forestry Paper No. 169. Food and Agriculture Organization of the United Nations and European Commission Joint Research Centre. FAO, Rome

Faraloni C, Cutino I, Petruccelli R, Leva AR, Lazzeri S, Torzillo G (2011) Chlorophyll fluorescence technique as a rapid tool for in vitro screening of olive cultivars (*Olea europaea* L.) tolerant to drought stress. Environ Exp Bot 73:49–56

Fernández R, Bertrand A, Reis R, Mourato MP, Martins LL, González A (2013) Growth and physiological responses to cadmium stress of two populations of *Dittrichia viscosa* (L.) Greuter. J Hazard Mater 244–245:555–562

Fini A, Bellasio C, Pollastri S, Tattini M, Ferrini F (2013) Water relations, growth, and leaf gas exchange as affected by water stress in *Jatropha curcas*. J Arid Environ 89:21–29

Flexas J, Medrano H (2001) Drought-inhibition of photosynthesis in C3 plants: stomatal and non-stomatal limitations revisited. Ann Bot 89:183–189

Flexas J, Gulias J, Jonasson S, Medrano H, Mus M (2001) Seasonal patterns and control of gas exchange in local populations of the Mediterranean evergreen shrub *Pistacia lentiscus* L. Acta Oecol 22:33–43

Fortescue JA, Turner DW (2005) The association between low temperatures and anatomical changes in preanthetic ovules of *Musa* (Musaceae). Sci Hortic 104:433–444

Fuller MP, Metwali EMR, Eed MH, Jelligs AJ (2006) Evaluation of abiotic stress resistance in mutated populations of cauliflower (*Brassica oleracea* var. botrytis). Plant Cell Tissue Organ Cult 86:239–248

Gaspar T, Franck T, Bibis T, Kevers C, Jouve L, Hausman JF, Dommes J (2002) Concepts in plant stress physiology: application to tissue cultures. Plant Growth Regul 37:263–285

Glenz C, Schlaepfer R, Iorgulescu I, Kienast F (2006) Flooding tolerance of Central European tree and shrub species. Forest Ecol Manag 235:1–13

Glenz C, Iorgulescu I, Kienast F, Schlaepfer R (2008) Modelling the impact of flooding stress on the growth performance of woody species using fuzzy logic. Ecol Model 218:18–28

Gratten SR, Grieve CM (1999) Salinity mineral nutrient relation in horticultural crops. Sci Hortic 78:127–157

Guàrdia M, Fernàndez J, Elena G, Fleck I (2012) Stomatal patchiness in the Mediterranean holm oak (*Quercus ilex* L.) under water stress in the nursery and in the forest. Tree Physiol 32(7):829–838. doi:10.1093/treephys/tps035

Gucci R, Tattini M (1997) Salinity tolerance in olive. In: Janik J (ed) Horticultural reviews. Wiley, New York, pp 177–214

Heimler D, Tatini M, Tici S, Coradeshi MA, Traversi ML (1995) Growth, ion accumulation and lipid composition of two olive genotypes under salinity. J Plant Nutr 18:1723–1734

Hokmabadi H, Arzani K, Grierson PF (2005) Growth, chemical composition, and carbon isotope discrimination of pistachio (*Pistacia vera* L.) rootstock seedlings in response to salinity. Aust J Agric Res 56:135–144

Holmes MG, Keiller DR (2002) Effects of pubescence and waxes on the reflectance of leaves in the ultraviolet and photosynthetic wavebands: a comparison of a range of species. Plant Cell Environ 25:85–93

Hook DD (1984) Adaptations to flooding with fresh water. In: Kozlowski TT (ed) Flooding and plant growth. Academic, Orlando, pp 265–294

Inbar M (2011) Manipulation of *Pistacia* spp. by gall-forming aphids: evolution of complex interactions. Acta Hortic 912:709–715

Inderjit, Mallik AU (2002) Can *Kalmia angustifolia* interference to black spruce (*Picea mariana*) be explained by allelopathy? Forest Ecol Manag 160:75–84

Junghans U, Polle A, Düchting P, Weiller E, Kuhlman B, Gruber F, Teichmann T (2006) Adaptation to high salinity in poplar involves changes in xylem anatomy and auxin physiology. Plant Cell Environ 29(8):1519–1531

Juvany M, Müller M, Munné-Bosch S (2013) Plant age-related changes in cytokinins, leaf growth and pigment accumulation in juvenile mastic trees. Environ Exp Bot 87:10–18

Kadioglu A, Terzi A, Saruhan N, Saglam A (2012) Current advances in the investigation of leaf rolling caused by biotic and abiotic stress factors. Plant Sci 182:42–48

Karakas B, Bianco RL, Rieger M (2000) Association of marginal leaf scorch with sodium accumulation in salt-stressed peach. Hortic Sci 35(1):83–84

Kozlowski TT (1997) Responses of woody plants to flooding and salinity. Tree physiology monograph no. 1. Heron Publishing, Victoria, pp 1–29

Kozlowski TT, Pallardy SG (2002) Acclimation and adaptive responses of woody plants to environmental stresses. Bot Rev 68(2):270–334

Krasensky J, Jonak C (2012) Drought, salt, and temperature stress-induced metabolic rearrangements and regulatory networks. J Exp Bot 63(4):1593–1608

Kulkarni M, Schneider B, Raveh E, Tel-Zur N (2010) Leaf anatomical characteristics and physiological responses to short-term drought in Ziziphus mauritiana (Lamk.). Sci Hortic 124:316–322

Küppers M, Timm H, Orth F, Stegemann J, Stober R, Scneider H, Paliwal K, Karunaichamy KSTK, Ortiz R (1996) Effects of light environment and successional status on lightfleck use by understory trees of temperate and tropical forests. Tree Physiol 16:69–80

Kurth E, Cramer GR, Lauchli A, Epstein E (1986) Effects of NaCl and $CaCl_2$ on cell enlargement and cell production in cotton roots. Plant Physiol 82:1102–1106

Kyoro HW, Ahmad P, Geissler N (2012) Abiotic stress responses in plants. In: Ahmad P, Prasad MNV (eds) Environmental adaptations and stress tolerance of plants in the era of climate change. Springer, New York. doi:10.1007/978-1-4614-0815-4_1

Lesniewska E, Adrian M, Klinguer A, Pugin A (2004) Cell wall modification in grapevine cells in response to UV stress investigated by atomic force microscopy. Ultramicroscopy 100:171–178

Leuschner C, Backes K, Hertel D, Schipka F, Schmitt U, Terborg E, Runge M (2001) Drought responses at leaf, stem and fine root levels of competitive *Fagus sylvatica* L. and *Quercus petraea* (Matt.) Liebl. trees in dry and wet years. Forest Ecol Manag 149:33–46

Lodhi MAK (1976) Role of allelopathy as expressed by dominating trees in a lowland forest in controlling the productivity and pattern of herbaceous growth. Am J Bot 63(1):1–8

Loney PE, McArthur C, Potts BM, Jordan GJ (2006) How does ontogeny in a Eucalyptus species affect patterns of herbivory by Brushtail Possums? Funct Ecol 20:982–988

Lucas PW, Choong MF, Tan HTW, Turner IM, Berrick AJ (1991) The fracture toughness of the leaf of the dicotyledon, *Calophyllum inophyllum* L. (Guttiferae). Philos Trans R Soc Lond B 334(1269):95–106

Luther JE, Franklin SE, Hudak J, Meades JP (1997) Forecasting the susceptibility and vulnerability of balsam fir stands to insect defoliation with Landsat Thematic Mapper data. Remote Sens Environ 59:77–91

Lynch JP, St Clair SB (2004) Mineral stress: the missing link in understanding how global climate change will affect plants in real world soils. Field Crop Res 90:101–115

Maathuis FJM, Amtmann A (1999) K^+ nutrition and Na^+ toxicity: the basis of cellular K^+/Na^+ ratios. Ann Bot 84:123–133

Maclaren P (1983) Chemical welfare in the forest. A review of allelopathy with regard to New Zealand. N Z J For 28(1):75–92

Marschner H (1995) Mineral nutrition of higher plants, 2nd edn. Academic, San Diego

Meena SK, Gupta NK, Gupta S, Khandelwal SK, Sastry EVD (2003) Effect of sodium chloride on the growth and gas exchange of young Ziziphus seedling rootstocks. J Hortic Sci Biotechnol 78(4):454–457

Melgar JC, Benlloch M, Fernandez-Escobar R (2006) Calcium increases sodium exclusion in olive plants. Sci Hortic 109:303–305

Miller RE, Jensen R, Woodrow IE (2006) Frequency of cyanogenesis in tropical rainforests of far north Queensland, Australia. Ann Bot 97(6):1017–1044

Mills D, Zhang G, Benzioni A (2001) Effect of different salts and of ABA on growth and mineral uptake in jojoba shoots grown in vitro. J Plant Physiol 158:1031–1039

Mittler R, Blumwald E (2010) Genetic engineering for modern agriculture: challenges and perspectives. Annu Rev Plant Biol 61:443–462

Munns R (2002) Comparative physiology of water and salt stress. Plant Cell Environ 25:239–250

Niinemets Ü (2010) Responses of forest trees to single and multiple environmental stresses from seedlings to mature plants: past stress history, stress interactions, tolerance and acclimation. Forest Ecol Manag 260:1623–1639

Niinemets Ü, Kull O, Tenhunen JD (1999) Variability in leaf morphology and chemical composition as a function of canopy light environment in coexisting deciduous trees. Int J Plant Sci 160(5):837–848

Nilsson J, Karlberg A, Antti H, Lopez-Vernaza M, Mellerowicz E, Perrot-Rechenmann C, Sandberg G, Bhalerao RP (2008) Dissecting the molecular basis of the regulation of wood formation by auxin in hybrid aspen. Plant Cell 20:843–855

Ögren E (1996) Premature dehardening in *Vaccinium myrtillus* during a mild winter: a case for winter dieback? Funct Ecol 10:724–732

Osakabe Y, Kajita S, Osakabe K (2011) Genetic engineering of woody plants: current and future targets in a stressful environment. Physiol Plant 142:105–117

Osakabe Y, Kawaoka A, Nishikubo N, Osakabe K (2012) Responses to environmental stresses in woody plants: key to survive and longevity. J Plant Res 125(1):1–10. doi:10.1007/s10265-011-0446-6

Pallardy SG (2007) Physiology of woody plants, 3rd edn. Academic, San Diego, 464p

Parida AK, Das AB (2005) Salt tolerance and salinity effects on plant: a review. Ecotoxicol Environ Safe 60:324–349

Park YI, Chow WS, Anderson JM (1996) Chloroplast movement in the shade plant *Tradescantia albiflora* helps protect photosystem II against light stress. Plant Physiol 111:867–875

Patakas A, Noitsakis B (1997) Cell wall elasticity as a mechanism to maintain favorable water relations during leaf ontogeny in grapevines. Am J Enol Vitic 48(3):352–356

Perdiguero P, Barbero MC, Cervera MT, Collada C, Soto A (2013) Molecular response to water stress in two contrasting Mediterranean pines (*Pinus pinaster* and *Pinus pinea*). Plant Physiol Biochem 67:199–208

Pérez-Contreras T, Soler JJ, Soler M (2008) Needle asymmetry, pine vigour and pine selection by the processionnary moth *Thaumetopoae pityocampa*. Acta Oecol 33:213–221

Picchioni GA, Miyamoto S, Storey JB (1990) Salt effects on growth and ion uptake of pistachio rootstock seedlings. J Am Soc Hortic Sci 115:647–653

Polacik KA, Maricle BR (2013) Effects of flooding on photosynthesis and root respiration in saltcedar (*Tamarix ramosissima*), an invasive riparian shrub. Environ Exp Bot 89:19–27

Poulton JE (1990) Cyanogenesis in plants. Plant Physiol 94(2):401–405

Price PW (1991) The plant vigor hypothesis and herbivore attack. Oikos 62:244–251

Ramoliya PJ, Patel HM, Pandey AN (2004) Effect of salinization of soil on growth and macro- and micro-nutrient accumulation in seedlings of *Salvadora persica* (Salvadoraceae). Forest Ecol Manag 202:181–193

Rejskovà A, Patkovà L, Stodulkovà E, Lipavskà H (2007) The effect of abiotic stresses on carbohydrate status of olive shoots (*Olea europaea* L.) under in vitro conditions. J Plant Physiol 164:174–184
Rengel Z (1992) The role of calcium in salt toxicity. Plant Cell Environ 15:625–632
Rieske LK, Dillaway DN (2008) Response of two oak species to extensive defoliation: tree growth and vigor, phytochemistry, and herbivore suitability. Forest Ecol Manag 256:121–128
Rizhsky L, Liang HJ, Shuman J, Shulaev V, Davletova S, Mittler R (2004) The response of Arabidopsis to a combination of drought and heat stress. Plant Physiol 134:1683–1696
Rochdi A, Lemsellek J, Bousarhal A, Rachidai A (2005) Evaluation sous serre de la tolérance à la salinité de quelques porte-greffes d'agrumes: *Citrus aurantium* et deux hybrides de *Poncirus trifoliata* (*Poncirus* x *Citrus sinensis* et *Poncirus* x Mandarinier *sunki*). Biotechnol Agron Soc Environ 9(1):65–73
Rozak PR, Seiser RM, Wacholtz WF, Wise RR (2002) Rapid, reversible alterations in spinach thylakoid appression upon changes in light intensity. Plant Cell Environ 25:421–429
Rozas V, DeSoto L, Olano JM (2009) Sex-specific, age-dependent sensitivity of tree ring-growth to climate in the dioecious tree *Juniperus thurifera*. New Phytol 182:687–697
Sainger PA, Dhankhar R, Sainger M, Kaushik A, Singh RP (2011) Assessment of heavy metal tolerance in native plant species from soils contaminated with electroplating effluent. Ecotoxicol Environ Saf 74:2284–2291
Sakai A (1979) Freezing avoidance mechanism of primordial shoots of conifer buds. Plant Cell Physiol 20:1381–1390
Santini A, Pepori A, Ghelardini L, Capretti P (2008) Persistence of some pine pathogens in coarse woody debris and cones in a *Pinus pinea* forest. Forest Ecol Manag 256:502–506
Santos CL, Campos A, Azevedo H, Caldeira G (2001) In situ and in vitro senescence induced by KCl stress nutritional imbalance, lipid peroxidation and antioxidant metabolism. J Exp Bot 52(355):351–360
Sapeta H, Costa JM, Lourenço T, Maroco J, Van Der Linde P, Oliveira MM (2013) Drought stress response in *Jatropha curcas*: growth and physiology. Environ Exp Bot 85:76–84
Schappert PJ, Shore JS (2000) Cyanogenesis in *Turnera ulmifolia* L. (Turneraceae): II. Developmental expression, heritability and cost of cyanogenesis. Evol Ecol Res 2:337–352
Schenk HJ (2006) Root competition: beyond resource depletion. J Ecol 94(4):725–739
Shepherd T, Griffiths DW (2006) The effects of stress on plant cuticular waxes. New Phytol 171:469–499
Shibli RA, Al-Juboory K (2002) Comparative responses of "Nabali" olive microshoots, callus and suspension cell cultures to salinity and water deficit. J Plant Nutr 25(1):61–74
Shoeneweiss DF (1981) The role of environmental stress on diseases of woody plants. Plant Dis 65(4):308–314
Siebel HN, Wijk MV, Blom CWPM (1998) Can tree seedlings survive increased flood levels of rivers? Acta Bot Neerl 47(2):219–230
Skirycz A, Inzé D (2010) More from less: plant growth under limited water. Curr Opin Biotechnol 21:197–203
Slama F (2004) La salinité et la production végétale. Centre de Publication Universitaire, Tunis, p 163
Slippers B, Wingfield MJ (2007) *Botryosphaeriaceae* as endophytes and latent pathogens of woody plants: diversity, ecology and impact. Fungal Biol Rev 21:90–106
Sotiropoulos TE (2007) Effect of NaCl and $CaCl_2$ on growth and contents of minerals, chlorophyll, proline and sugars in apple rootstock M 4 cultured in vitro. Biol Plant 51(1):177–180
Stamp N (2003) Out of the quagmire of plant defense hypotheses. Q Rev Biol 78(1):23–55
Stella JC, Riddel J, Piégay H, Gagnage M, Trémélo ML (2013) Climate and local geomorphic interactions drive patterns of riparian forest decline along a Mediterranean Basin river. Geomorphology (2013). http://dx.doi.org/10.1016/j.geomorph.2013.01.013
Sterck FJ, Bongers F (2001) Crown development in tropical rain forest trees: patterns with tree height and light availability. J Ecol 89:1–13
Stiller V (2009) Soil salinity and drought alter wood density and vulnerability to xylem cavitation of baldcypress (*Taxodium distichum* (L.) Rich.) seedlings. Environ Exp Bot 67:164–171

Sullivan JH, Teramura AH (1992) The effects of ultraviolet-B radiation on Loblolly pine. 2. Growth of field-grown seedlings. Trees 6:115–120

Tabatabaei SJ (2006) Effects of salinity and N on growth, photosynthesis and N status of olive (*Olea europaea* L.) trees. Sci Hortic 108:432–438

Tjoelker MG, Oleksyn J, Reich PB (1998) Temperature and ontogeny mediate growth response to elevated CO_2 in seedlings in five boreal tree species. New Phytol 140:197–210

Türkan I, Demiral T (2009) Recent developments in understanding salinity tolerance. Environ Exp Bot 67:2–9

Tyler G (1984) The impact of heavy metal pollution on forests: a case study of Gusum, Sweden. Ambio 13(1):18–24

Tyree MT, Ewers FW (1991) The hydraulic architecture of trees and other woody plants. New Phytol 119(3):345–360

Unno H, Maeda Y, Yamamoto S, Okamoto M, Takenaga H (2002) Relationship between salt tolerance and Ca^{2+} retention among plant species. Jpn J Soil Sci Plant Nutr 73:715–718

Valdés-Gómez H, Gary C, Cartolaro P, Lolas-Caneo M, Calonnec A (2011) Powdery mildew development is positively influenced by grapevine vegetative growth induced by different soil management strategies. Crop Prot 30:1168–1177

Veteli TO, Kuokkanen K, Julkunen-Tiitto R, Roininen H, Tahvanainen J (2002) Effects of elevated CO_2 and temperature on plant growth and herbivore defensive chemistry. Glob Chang Biol 8:1240–1252

Vijayan K, Chakraborti SP, Ghosh PD (2003) In vitro screening of mulberry (*Morus* spp.) for salinity tolerance. Plant Cell Rep 22(5):350–357

Vollenweider P, Günthardt-Goerg MS (2005) Diagnosis of abiotic and biotic stress factors using the visible symptoms in foliage. Environ Pollut 137:455–465

Walker RR, Törökfalvy E, Behboodian MH (1987) Uptake and distribution of chloride. Sodium and potassium ions and growth of salt-treated pistachio plants. Aust J Agric Res 38:383–394

Walker RR, Törökfalvy E, Behboodian MH (1988) Photosynthetic rates and solute partitioning in relation to growth of salt-treated pistachio plants (*P. vera* cv. Kerman). Aust J Plant Physiol 15:787–798

Wang GB, Cao FL (2012) Formation and function of aerenchyma in baldcypress (*Taxodium distichum* (L.) Rich.) and Chinese tallow tree (*Sapium sebiferum* (L.) Roxb.) under flooding. S Afr J Bot 81:71–78

Watkinson JI, Sioson AA, Vasquez-Robinet C, Shukla M, Kumar D, Ellis M, Heath LS, Ramakrishnan N, Chevone B, Watson LT, van Zyl L, Egertsdotter U, Sederoff RR, Grene R (2003) Photosynthetic acclimation is reflected in specific patterns of gene expression in drought-stressed loblolly pine. Plant Physiol 133:1702–1716

Way DA, Pearcy RW (2012) Sunflecks in trees and forests: from photosynthetic physiology to global change biology. Tree Physiol 32:1066–1081

White TCR (1969) An index to measure weather-induced stress of trees associated with outbreaks of psyllids in Australia. Ecology 50:905–909

Williams WE, Gorton HL, Witiak SM (2003) Chloroplast movements in the field. Plant Cell Environ 26:2005–2014

Yang F, Miao LF (2010) Adaptive responses to progressive drought stress in two poplar species originating from different altitudes. Silva Fenn 44(1):23–37

Zhang F, Yang YL, He WI, Zhao X, Zhang LX (2004) Effects of salinity on growth and compatible solutes of callus induced from Populus euphratica. In Vitro Cell Dev Biol Plant 40:491–494

Zhang M, Fang Y, Ji Y, Jiang Z, Wang L (2013) Effects of salt stress on ion content, antioxidant enzymes and protein profile in different tissues of *Broussonetia papyrifera*. S Afr J Bot 85:1–9

Zhong H, Läuchli A (1988) Incorporation of ^{14}C glucose into cell wall polysaccharides of cotton roots: effect of NaCI and CaCl2. Plant Physiol 88:511–514

Zhu J, Dong CH, Zhu JK (2007) Interplay between cold-responsive gene regulation, metabolism and RNA processing during plant cold acclimation. Curr Opin Plant Biol 10:290–295

Chapter 2
Drought Tolerance: Role of Organic Osmolytes, Growth Regulators, and Mineral Nutrients

Mohammad Abass Ahanger, Shiv Ram Tyagi, Mohd Rafiq Wani, and Parvaiz Ahmad

Contents

M.A. Ahanger (✉) • S.R. Tyagi
School of Studies in Botany, Jiwaji University, Gwalior 474 011, Madhya Pradesh, India
e-mail: ahangerma@gmail.com

M.R. Wani
Department of Botany, Government Degree College (Boys), Anantnag 192 102, Jammu and Kashmir, India
e-mail: botanyrafiq@gmail.com

P. Ahmad
Department of Botany, Sri Pratap College, Srinagar, 190 001, Jammu and Kashmir, India
e-mail: parvaizbot@yahoo.com

P. Ahmad and M.R. Wani (eds.), *Physiological Mechanisms and Adaptation Strategies in Plants Under Changing Environment: Volume 1*, DOI 10.1007/978-1-4614-8591-9_2,

1 Introduction

World population is increasing at an alarming rate and is expected to reach 8.3 billion by 2030 (FAO 2010). In many densely populated developing countries of the world, expansion of arable land has become more difficult as a result of rapid urbanization, industrialization, and water scarcity (Rengasamy 2010). In future, food grain production has to be increased by 57 % so as to ensure sufficient food for the growing population (Wild 2003). In past few years, no doubt, an increase in productivity of certain major crops has been reported but repeating the same success in future for increased food production seems to be difficult.

Among various abiotic stresses, drought is one of the major environmental constraints limiting crop productivity worldwide (Masoumi et al. 2010; Khamssi et al. 2011; Batlang et al. 2013). About 25 % of the world's agricultural land is affected by drought stress (Jajarmi 2009). Changes in global climate have made this situation even more serious (Anand et al. 2003). Water shortage and soil water losses due to changes in environment and excessive land use are challenges to crop production (Xia et al. 2005). Maintaining higher plant productivity under environmental stresses is the main challenge which modern agriculture is facing (Gill and Tuteja 2010). Drought stress affects both source and sink, thereby causing reduction in yield in a time-dependent manner with respect to the severity of stress and plant developmental stage (Blum 1996). Drought stress imposes osmotic stress leading to loss of turgor and oxidative stress through production of reactive oxygen species (ROS) that results in loss of membrane integrity, protein denaturation, and oxidative damage to other biomolecules. As a consequence of such changes, inhibition of photosynthesis, metabolic dysfunction, and damage to cellular structures occurs causing growth perturbances, reducing fertility, and premature senescence (Munns and Tester 2008). Plants respond differently to water deficiency in different periods of their growth. The generative phase and the beginning of flowering are most frequently the period of the greatest sensitivity to water deficit.

2 Adaptations to Drought Stress

Distribution of plant species depend upon the prevailing environmental conditions. Tolerant plants can survive the extreme harsh environmental conditions at which the growth of sensitive ones is negatively influenced (Munns and Tester 2008). Higher tolerance to adverse environmental conditions is because of different stress response mechanisms. Plants adopt different strategies to cope with drought stress. The strategies adopted include escape, avoidance, and tolerance strategy (Levitt 1980; Chaves et al. 2002; Blum 2005; Ahmad and Sharma 2008; Rasool et al. 2013). Ephemeral plants have rapid phenological development, completing their life cycle during a period of adequate moisture and forming dormant seeds before the onset of dry seasons. Ephemerals never really experience the drought stress. In avoidance strategy, plants somehow reduce the impact of stress factor, even though the stress is present in the environment. Avoidance strategy, generally results in maintaining the favorable internal water content either by conserving water which is brought about by closing the stomata, leaf rolling, and heavy pubescence or by increasing the water uptake through development of deep root system and water spenders (Ruiz-Sanchez et al. 2007). On the other hand, in tolerance strategy, plant endures drought without undergoing injury, retaining the capacity of normal growth and development when rehydrated. In tolerance, plants mitigate the stress by maintaining high water potential through accumulation of compatible osmotic solutes. The accumulation of compatible solutes is well regarded as a basic strategy for the protection and survival of plants under abiotic stress conditions (Chen et al. 2007).

Responses of plant species to drought stress depend on several factors including duration and severity of the drought period as well as its inherent tolerance mechanisms. Severe and prolonged periods of drought stress result in oxidative damage due to the overproduction of ROS (Smirnoff 1993). Among various physiological and developmental mechanisms that a plant species adopts to tolerate periods of water deficit, accumulation of osmotically active solutes is the most commonly reported mechanism. By the accumulation of solutes, turgor and turgor-dependent processes are maintained, thereby allowing cell enlargement and plant growth during water stress and stomata to remain partially open and CO_2 assimilation to continue at low water potentials that are otherwise inhibitory (Pugnaire et al. 1994).

Drought is a multidimensional stress, affecting plants at various levels of their organization (Yordanov et al. 2000). Stress-imposed effects are often manifested at phenological, morpho-physiological, biochemical, and molecular levels (Bahrani et al. 2010). Accumulation of compatible organic solutes (Da Costa and Huang 2009), changes in endogenous levels of certain phytohormones (Seki et al. 2007; Dobra et al. 2010), and overexpression of stress-responsive genes (Xiong and Yang 2003; Jaleel et al. 2006) do occur in response to stress. Most of these responses are directly triggered by the changes in water status of the cell (Chaves et al. 2003). In connection to this, plant hormones such as abscisic acid (ABA), jasmonic acid (JA), and salicylic acid (SA) are involved in a complex signal transduction network, thereby coordinating growth and development with plant responses to the changing environment (Jiang and Zhang 2002; Fujita et al. 2006; Szalai et al. 2010). In order to improve plant tolerance to stress, understanding of complete physio-biochemical responses of plant is pivotal (Jaleel et al. 2006; Ahmad and Sharma 2008; Ahmad et al. 2008a, b, 2010a).

3 Osmoregulation and Osmolytes

Osmoregulation/osmotic adjustment is the general response of plants to water stress so that solute content of the cell is increased. In order to maintain turgor and water uptake for normal growth, plants under stress need to maintain internal water potential well below that of soil which is usually acquired by increasing concentration of cell osmotica, either through uptake of solutes from soil solution or by increased synthesis of compatible solutes (Tester and Davenport 2003; Ahmad and Sharma 2008). Cytoplasm accumulates low molecular mass compounds in order to accommodate the ionic balance in the vacuoles (Zhifang and Loescher 2003). These compatible osmotic solutes do not interfere with normal metabolic reactions but rather, they replace water in these reactions (Ahmad and Sharma 2008; Koyro et al. 2012). Accumulation of these osmolytes is proportional to change in external osmolarity (within species-specific limits), thereby protecting cellular structures and maintaining osmotic balance to support continued water influx (Hasegawa et al. 2000). Majority of these compatible osmolytes are organic solutes, while some are essential ions such as K^+ (Yokoi et al. 2002; Ahmad and Sharma 2008). However, the accumulation varies within the genus as well as plant species. Majority of the organic solutes accumulated are sugars (fructose, glucose, trehalose, and raffinose), sugar alcohols (glycerol and methylated inositols) (Bohnert and Jensen 1996), quaternary amino compounds (proline, glycine betaine, proline betaine, tertiary amines), and sulfonium compounds (choline *O*-sulfate, dimethyl sulfonium propironate) (Yokoi et al. 2002). Osmolyte accumulation is mandatory in plants for osmotic adjustment under water limiting conditions, but osmolyte accumulation mainly depends upon water status, crop growth stage, and cultivar (Shao et al. 2006). Due to accumulation of osmolytes, water status of cell and subcellular structures is maintained and membranes as well as proteins are protected from denaturing effects of osmotic stress (Ashraf and Foolad 2007).

3.1 Proline

Proline is an amino acid that plays multifunctional role in stress defense. It is actively involved in osmoregulation, scavenging of free radicals, and as a molecular chaperone for stabilizing protein structure, thus protects plant cells from damaging effects of various environmental stresses (Verbruggen and Hermans 2008; Ahmad and Sharma 2008; Szabados and Savoure 2010; Koyro et al. 2012; de Carvalho et al. 2013). Accumulation of proline in response to various environmental stresses is well documented (Ahmad 2010; Ahmad et al. 2010b, 2011, 2013; Azooz et al. 2011; Katare et al. 2012; Kim and Nam 2013). Water stress-induced increase in proline has been reported in rice (Pandey and Agarwal 1998), *Medicago sativa* (Slama et al. 2011), wheat (Jatav et al. 2012), and *Arabidopsis* (Ju et al. 2013). Besides its role in stress tolerance, accumulation of proline is possibly a useful drought injury sensor in plants (Zlatev and Stotanov 2005). Tolerant plant genotypes show large accumulation of proline which is often correlated with increased stress tolerance (Katare et al. 2012; Ahmad et al. 2012a, b). As a consequence of drought stress, the concentration of proline in plant leaves increases tenfold in leaves of *Lotus japonicus* (Signorelli et al. 2013).

Biosynthetic and catabolic pathways of proline determine its level (Szabados and Savoure 2010). Proline is biosynthesized from glutamate by sequential action of γ-glutamyl kinase (γ-GK), pyrroline-5-carboxylate synthetase (P5CS), pyrroline-5-carboxylate (P5C), and P5C reductase (P5CR) (Hong et al. 2000; Yamada et al. 2005). In most plant species, P5CS is encoded by two genes and P5CR is encoded by one (Armengaud 2004). Proline can also be synthesized by alternative pathway from ornithine, employing ornithine-delta-aminotransferase (d-OAT)

(Miller et al. 2009). Proline synthesized in cytoplasm or chloroplasts is transported to mitochondria where it is catabolized to P5C through the sequential action of proline dehydrogenase (PDH) or proline oxidase (PROX) and P5C is then converted to glutamate by enzyme P5C dehydrogenase (P5CDH).

Proline level in plants is controlled by two important enzymes, PROX and γ-GK (Girija et al. 2002; Ahmad and Sharma 2008; Koyro et al. 2012). Increased proline accumulation during stress may be due to the activation of proline synthesis through glutamate pathway involving γ-GK, glutamyl phosphate reductases, and P5CR enzymes. During stress, increase in activity of γ-GK and decrease in activity of PROX has been reported by Jaleel et al. (2007) in *Catharanthus roseus* and Ahmad et al. (2010b) in *Morus alba*, thereby helping plants to maintain sufficient levels of proline to combat/ameliorate detrimental effects of stress.

Under stress conditions, proline synthesis is enhanced in plants and on recovery from stress, its catabolism is enhanced. It has been reported that overexpression of P5CS in *Nicotiana* and *Petunia* resulted in increased proline accumulation and enhanced salt and drought tolerance (Hong et al. 2000; Yamada et al. 2005). Rice and tobacco plants overexpressing *Arabidopsis* d-OAT has increased proline levels and greater tolerance to stress (Roosens et al. 2002; Qu et al. 2005).

Proline accumulation is a highly regulated process involving a set of protein kinases that is ubiquitous for stress tolerance including drought. These proteins include SNF-related protein kinases 2 (SnRK2s, i.e., SnRK 2.2, SnRK 2.3, and SnRK 2.6) which are activated on exposure to stress (Boudsocq et al. 2004). It has been reported in *Arabidopsis* mutants that ABA-responsive SNF-related protein kinases 2 (SnRK2)-induced ABA-dependent proline accumulation, therefore imparts more tolerance to osmotic stress (Fujii et al. 2011). Another family of SNF-related protein kinases, that enhance proline levels, is SnRK3s. These are calcineurin B-like (CBL) calcium binding proteins also known as CBL-interacting kinases (CIPKs) and overexpression of OsCIPK03 and OsCIPK12 has been reported to increase tolerance of rice to cold and drought by causing significant increase in proline (Xiang et al. 2007). In addition to the abovementioned protein kinases, *Arabidopsis* calcium-dependent protein kinase 6 (CDPK6) (Xu et al. 2010) and soybean calmodulin GmCAM4 (Yoo et al. 2005) have been reported to contribute positively so as to enhance proline content and stress tolerance in *Arabidopsis*, thereby indicating a key role for intracellular calcium signals in proline metabolism. Variety of abiotic stress responses in plants are regulated through MAPK (mitogen-activated protein kinase) cascades and it has been reported that several MAPKs are activated on exposure to various environmental stresses resulting in increased proline accumulation and tolerance as well (Kong et al. 2011; Zhang et al. 2011). Moreover, genetic manipulation of MAPK signaling pathway results in altered plant stress tolerance (Xiong and Yang 2003; Shou et al. 2004a, b). Proline accumulation can be induced by ABA as well as by other stress-related protein kinases. In addition to the abovementioned positive regulation of proline accumulation by several protein kinases, it shall be pointed out that it may also be regulated negatively, e.g., maize protein phosphatases type 2C (PP2C) regulate various processes of development and responses to environmental stress but have been reported to regulate proline accumulation negatively (Liu et al. 2009; Umezawa et al. 2010).

3.2 Glycine Betaine

Glycine betaine (GB) plays an important role in plant tolerance to stress, enzyme activity, membrane integrity, ROS detoxification, and osmotic adjustment. Glycine betaine (*N*,*N*,*N*-trimethylglycine), one among the quaternary ammonium compounds (glycinebetaine,β-alaninebetaine,choline-*O*-sulfate,and3-dimethylsulfoniopropionate and proline betaine), is dipolar and exists as neutral molecule at physiological pH (Le Rudulier et al. 1984) which is known to play protective role in plants under stress (Yang et al. 2003) and its accumulation has positive associations with stress tolerance (Ashraf and Foolad 2007; Kathuria et al. 2009). GB is mainly found in chloroplasts and plays a pivotal role in protection of thylakoid membranes and other key components of photosynthetic machinery such as ribulose 1,5-bisphosphate carboxylase/oxygenase (Rubisco) and oxygen evolving complex from stress-induced inactivation and dissociation, thereby maintaining the photosynthetic efficiency (Yokoi et al. 2002). Moreover, it stabilizes the association of the extrinsic PS II complex proteins and maintains the highly ordered state of membranes at nonphysiological temperatures and salt concentrations (Papageorgiou and Murata 1995).

Plants synthesize GB in chloroplast from either glycine or choline via two distinct pathways: (1) dehydrogenation of choline or (2) N-methylation of glycine and enzymes involved in choline monooxygenase (CMO) and betaine aldehyde dehydrogenase (BADH) (Nye et al. 1997). Increase in glycine betaine content under stress conditions has been reported in many plants but the increase may be more pronounced in leaves than in roots, e.g., *Haloxylon recurvum* (Wang and Nil 2000). In most of the crop plants, concentration of naturally accumulated GB may not be sufficient enough to mitigate the deleterious effects of various environmental stresses (Subbarao et al. 2001). Accumulation of GB in response to stress has been reported in many crops, e.g., bean (Gadallah 1999), peanut (Girija et al. 2002), sorghum (Yang et al. 2003), and mustard (Ahmad 2010). But the concentrations accumulated vary with plant species, for example, sorghum accumulates manifold more GB than maize (Murata et al. 1992). Stress-tolerant species accumulate GB in high concentrations than sensitive ones (Agastian et al. 2000). Plants that accumulate low concentration of GB, exogenous application can be a useful tool to reduce the adverse effects of environmental stresses (Makela et al. 1998; Yang and Lu 2005). It has been reported that exogenous application of glycine betaine increased tolerance of tomato (Makela et al. 1998) and rice plants to salt stress (Lutts 2000). In maize, exogenous supply of glycine betaine caused considerable increase in yield, RWC, proline, and antioxidant enzyme activity but reduced lipid peroxidation under normal as well as drought conditions (Lv et al. 2007; Anjum et al. 2012). In addition, water-limited conditions increased the yield and yield components have been reported in several other crops such as rice (Rahman et al. 2002), sunflower (Iqbal et al. 2005), maize (Ali and Ashraf 2011), bean (Abou El-Yazied 2011), and *Triticum aestivum* (Aldesuquy et al. 2012).

However, many important crop plants like maize, potato, tomato, and eggplant lack the capability to synthesize GB in adequate amounts (Zwart et al. 2003). In such cases, both the exogenous application of GB and the introduction, via transgenes, of the GB biosynthetic pathway have become imperative to increase their tolerance to different abiotic stresses. This increased tolerance to abiotic stresses will be useful for understanding the mechanisms through which GB protects plants against abiotic stresses. So far, genetically modified plants containing transgenes for production of GB have faced the limitation of being unable to produce sufficient amounts of glycine betaine required to mitigate the stress, but applying glycine betaine exogenously to plants under stress conditions has gained more attention (Ashraf and Foolad 2007). Introduction and overexpression of choline oxidase (Cod A) gene from *Arthrobacter globiformis* and BADH genes from Spinach/*Atriplex* have been widely used for GB production in transgenic plants. Introduction of Cod A gene in *Arabidopsis thaliana* (Hayashi et al. 1997), *Oryza sativa* (Alia and Murata 1998), and *Lycopersicum esculentum* (Kathuria et al. 2009) and BADH gene in *Triticum aestivum* (Wang et al. 2010) have been reported to increase the tolerance to drought, salinity, and cold stresses by increasing the membrane integrity, enzyme activity, photosynthesis regulating ROS detoxification and also yield.

3.3 Polyamines

Polyamines are group of naturally occurring nitrogenous compounds with aliphatic structure that are implicated in several processes such as growth, development as well as responses to various environmental stresses (Ahmad et al. 2012c). Moreover, polyamines due to their hydrophilic properties are involved in the maintenance of pH and in scavenging of active oxygen compounds, therefore are considered as mediators in protective reactions against different stresses (Kovacs et al. 2010). Polyamines protect membrane from disintegration and alleviate oxidative stress (Groppa and Benavides 2008; Alcazar et al. 2011; Hussain et al. 2011; Ahmad et al. 2012c). Putrescine (PUT), spermidine (SPD), and spermine (SPM) are commonly occurring polyamines in higher plants and may exist free or covalently bound to small molecules such as phenolic compounds as well as to macromolecules such as nucleic acids and proteins in soluble-conjugated or insoluble bound forms (Kusano et al. 2007; Duan et al. 2008). In addition to these, uncommon polyamines like homospermidine, cadaverine, and canavalmine have also been reported in several biological systems including plants. At the physiological pH, polyamines usually exist as cations. This polycationic nature of polyamines is one of their important properties affecting their biological activities (Valero et al. 2002). Polyamine levels vary depending on plant species and the stress duration (Liu et al. 2008). It has been suggested that stress-tolerant plants have increased polyamine levels as compared to sensitive ones and polyamines with higher number of amino groups (SPM and SPD) are more effective in scavenging of ROS than the ones with less number of amino groups (PUT) (Kubis 2008).

Polyamines serve as messengers of stress signals (Liu et al. 2007). As a result of acid neutralizing and antioxidant capability, polyamines show antisenescence, anti-stress effects, and membrane and cell wall stabilizing abilities (Zhao and Yang 2008). Role of polyamines in modulating the defensive responses of plants to various environmental stresses is well documented (Alcazar et al. 2011).

Exogenous application of polyamines has been suggested as an effective approach for enhancing stress tolerance of crops and crop productivity as well. Exogenous application of PUT have been successfully utilized in enhancing plant tolerance to high temperature (Murkowski 2001), cold (Nayyar and Chander 2004), osmotic stress (Liu et al. 2004), salinity (Verma and Mishra 2005), drought (Zeid and Shedeed 2006), heavy metals (Wang et al. 2007), water logging (Arbona et al. 2008), and flooding (Yiu et al. 2009). Furthermore, it has been reported that genetic transformation of plants with genes that code for the enzymes involved in polyamine biosynthesis resulted in increased stress tolerance in various plant species (Liu et al. 2007). Transgenic plants overexpressing these genes show increased tolerance to multiple environmental stresses including salinity, drought, and low and high temperatures. This tolerance to multiple abiotic stresses is of practical importance as plants are often encountered by several concurrent forms of environmental stresses during their life cycle (Wi et al. 2006; Prabhavathi and Rajam 2007; Wen et al. 2008).

Plants deficient in arginine decarboxylase (ADC) and spermidine synthase (SPDS) are unable to synthesize sufficient PUT and SPM, respectively, therefore are sensitive to stress (Yamaguchi et al. 2007; Cuevas et al. 2008), whereas overexpression of ADC leads to greater synthesis of PUT and enhanced tolerance to drought (Alcazar et al. 2010; Alet et al. 2011). Tobacco plants overexpressing ornithine decarboxylase (ODC) showed increased tolerance to salt stress (Kumriaa and Rajam 2002). Moreover, it has been reported that *Arabidopsis* plants overexpressing SPDS showed greater tolerance to drought, salinity, and cold stress (Kasukabe et al. 2004). Scaramagli et al. (2000) reported that increase in insoluble-conjugated PUT levels was closely associated with polyethylene glycol-induced stress acclimation in potato. Liu et al. (2004) reported an increase in the polyamine levels in leaves of drought-tolerant wheat seedlings under osmotic stress, indicating the role of polyamines in facilitating osmotic stress tolerance of wheat seedlings.

4 Growth Regulators

4.1 Abscisic Acid

Growth hormones help plants to adapt to changing environments by mediating growth, development, nutrient allocation, and source/sink transitions (Peleg and Blumwald 2011). ABA, a 15-carbon sesquiterpenoid compound resembling terminal portion of some carotenoid molecules, is synthesized in chloroplast and other plastids by mevalonic acid pathway from 40-carbon precursor, zeaxanthin. Zeaxanthin epoxidase (ZEP), 9-cis-epoxycarotenoid deoxygenase (NCED), alcohol dehydrogenase (ABA2), and abscisic aldehyde oxidase (AAO3) are the main enzymes mediating its biosynthesis.

ABA is involved in many cellular processes like germination, gravitropism, and guard cell-mediated stomatal opening (Levitt 1980). ABA plays an important role in the adaptation of plants to environmental stress. Regulation of water balance and osmotic stress tolerance is a well-established function of ABA (Takahashi et al. 2000; Zhu 2002). Under stress conditions, in addition to its well-established role in closing of stomata, ABA increases the ion influx across root cell membrane and also mediates the greater synthesis and accumulation of active osmotic solutes (e.g., proline, trehalose), thereby helping in bringing osmotic adjustment (Nayyar et al. 2005). ABA accumulates under drought stress and degrades gradually upon removal of stress. Since ABA mediates so many stress responses, starting from the perception of stress signal upto changes in gene expression, which ultimately leads to increased ABA in plants (Zhang et al. 2006). Pospisilova et al. (2005) reported that ABA pretreatment further increased the endogenous ABA level in maize seedling. Presoaking seeds with ABA was reported to significantly enhance the activities of antioxidant enzymes in maize seedlings subjected to water stress (Jiang and Zhang 2002). Similarly, Boominathan et al. (2004) found that relative water content of ABA-treated plants was higher under drought stress. Moreover, exogenous application of ABA under water stress increased the grain weight in susceptible wheat cultivars (Nayyar and Walia 2004).

The role of ABA in plants exposed to drought stress has been well studied. ABA has a potential role in regulating the plant water status and growth. Increased expression of genes encoding enzymes and proteins involved in enhancing drought tolerance has also been attributed to increased ABA (Luan 2002; Zhu 2002). During stress, ABA produced in root is transported to shoot for regulating stomatal movements and leaf growth (Zhang et al. 1987; Zhang and Davies 1990a, b). The pH and ionic conditions in the xylem play an important role in this transport (Wilkinson et al. 1998; Bacon et al. 1998; Hartung et al. 2002). Source of ABA appearing in the xylem during drought has been debated. Some are of the opinion that it comes only from the root (Zhang et al. 1987), while some have reported that ABA comes from both root and leaves. Root-sourced ABA is usually involved in the initial sensing of drought to regulate the stomatal conductance so that the excess water loss may be reduced, but under severe and prolonged stress, leaf water deficit becomes unavoidable and older leaves may wilt because of weak hydraulic link or less control over stomatal conductance, increasing ABA concentration in the xylem (Zhang and Davies 1989a, b). Concentration of ABA in xylem has direct influence on leaf conductance and it has been reported that leaf conductance also responds to the flux of ABA into leaves per unit time, indicating its role in regulation of stomatal movements and due to stress-induced changes in transpiration rate, this role may be marginalized (Jarvis and Davies 1997). However, there are certain reports indicating that stomata responds to xylem ABA concentration rather than its flux, e.g., when leaf conductance has decreased considerably as a result of water stress (Jackson et al. 1995). By following the amount of ABA entering the leaves during the process of ABA-induced stomatal closure, it has been reported that changes in leaf conductance are due to xylem ABA and the rapid metabolism of this xylem-derived ABA in the leaves is very essential in order to prevent its accumulation and stomatal movements to be sensitively regulated (Jia and Zhang 1999).

ABA is important for growth and development of plants under water stress (Zhang and Davies 1990b). But increment in the concentration of xylem ABA, beyond certain limits, can restrict shoot growth (Gowing et al. 1990). However, it should be noted that shoot and root respond differently to ABA levels. Sharp et al. (2000) and Spollen et al. (2000) have reported that better growth of roots under water stress is attributed to higher amounts of ABA accumulated in the roots. Maintenance of better root growth under water deficit has positive association with drought tolerance. ABA has dual role in regulating physiology of plant, i.e., inhibitory as well as stimulatory (Finkelstein et al. 2002). Under stress conditions, when it is accumulated beyond certain limits to help plant survival, it may inhibit processes such as stomatal opening and plant size expansion for quite large time, but under normal conditions, when accumulated concentration is normal, it promotes vegetative growth (Sharp et al. 2000; Spollen et al. 2000) and post-germination development (Cheng et al. 2002). He and Cramer (1996) have reported that accumulation of ABA in lower concentrations increased salt tolerance of *Brassica napus*. Excess accumulation of ABA as a result of salinity has often been reported to induce inhibition of leaf expansion in different species (He and Cramer 1996; Montero et al. 1998).

ABA has an important role in signaling plant responses to drought and salt stresses, thereby triggering the expression of drought-responsive genes. As revealed from sequencing studies, among the various stress-responsive genes that are regulated by ABA, only few have been identified for having any probable physiological functions. In ABA-dependent pathway, synthesis of new proteins may or may not be required (Bray 2002). In such pathways, when the synthesis of new proteins is not required, the presence of ABA-responsive element (ABRE) at the promoter domain of ABA-responsive gene is ubiquitous which upon binding to transcription factor (TF) leads to ABA-induced gene expression, e.g., in *Arabidopsis*, ABA-induced expression of dehydration-responsive gene (rd29B) has two ABREs essential and two transcription factors (bZIP) (Uno et al. 2000). However, when synthesis of proteins is required for the ABA-dependent gene expression, de novo synthesis of new proteins is the prerequisite. These genes do not have any ABREs (Leung and Giraudat 1998; Bray 2002). Expression of some genes may be dependent as well as independent of ABA (Shinozaki and Yamaguchi-Shinozaki 1997), e.g., gene rd29A, which is important for water stress, has two types of regulatory *cis*-elements at its promoter, one ABA-dependent and the other is ABA-independent (Ingram and Bartels 1996; Leung and Giraudat 1998).

Under osmotic stress conditions, transcription levels of ABA biosynthetic genes are upregulated. Increased expression of the ZEP gene, under drought stress, has been reported in roots of *Nicotiana plumbaginifolia* (Audran et al. 2001) and leaves of *Arabidopsis* (Xiong et al. 2002). Moreover, stress-induced overexpression of NCED gene is well documented (Thompson et al. 2000; Tan et al. 2003). Under water stress conditions, accumulation of ABA is accompanied by transient increase in NCED transcript and proteins (Qin and Zeevaart 1999). Overexpression of AtNCED3 is highly induced by dehydration, although other NCED genes also contribute positively but their role is minor (Tan et al. 2003). In addition, transgenic

Arabidopsis plants overexpressing NCED show greater ABA levels and increased desiccation tolerance. Similar results have been reported in transgenic tomato and *Nicotiana plumbaginifolia* (Thompson et al. 2000). Exogenous application of ABA has been reported to induce the expression of NCED gene in ABA-deficient mutants (Xiong et al. 2002).

4.2 Salicylic Acid

Salicylic acid (SA) is an endogenous growth regulator of phenolic nature, actively involved in plant growth, development, and several other physiological processes including germination, fruit ripening, flowering, photosynthesis, stomatal conductance, ion uptake and transport (Shakirova 2007), biogenesis of chloroplast, interaction with other organisms, and protection of plants against multiple environmental stresses such as ozone and ultraviolet radiation (Sharma et al. 1996), salinity (Borsani et al. 2001), freezing (Janda et al. 1999), herbicides (Ananieva et al. 2004), heavy metals (Ahmad et al. 2011), osmotic stress (Shi and Zhu 2008), and drought (Sadeghipour and Aghaei 2012). Salicylic acid (SA) acts as a signal involved in the expression of specific responses in plants to biotic and abiotic stresses.

SA induces systemic acquired resistance (SAR) in plants to different pathogens (Metraux 2001). SA has been reported to induce accumulation of lectins in wheat (Shakirova and Bezrukova 1997), synthesis of heat shock proteins, and activation of protein kinase in tobacco exposed to osmotic stress (Burkhanova et al. 1999; Mikolajczyk et al. 2000), suggesting the role of SA in anti-stress mechanisms. Salicylic acid (SA) has long been considered as signal molecule and is known to reduce the oxidative damage caused by salinity stress (Azooz et al. 2011; Sajid and Aftab 2012) and this ability of SA to produce a protective effect in plants under different abiotic stresses has increased the interest of researchers.

Exogenously applied salicylic acid in plants has been reported to enhance the efficiency of several developmental, physiological, and biochemical processes. It has been reported that exogenous application of SA enhances transpiration rate (Rai et al. 1986), seed germination and yield (Raskin 1992), membrane permeability (Barkosky and Einhellig 1993), growth, and photosynthesis (El-Tayeb 2005). Moreover, exogenously applied SA is involved in the defense against pathogen attack and more recently its role has been widely investigated in both biotic and abiotic stresses (Shi et al. 2006). The role of SA in inducing stress tolerance in plants is well documented, e.g., it enhances the resistance of plants against drought and salt stress (Tari et al. 2002) besides metal stress (Ahmad et al. 2011). SA has been found to induce heat stress tolerance in mustard (Dat et al. 1998), chilling tolerance in maize (Janda et al. 1999), drought tolerance in wheat (Singh and Usha 2003), heavy metal stress tolerance in barley (Metwally et al. 2003), and salinity tolerance in barley (El-Tayeb 2005). Singh and Usha (2003) have reported that under drought, application of salicylic acid to wheat increased the moisture content, total chlorophyll content, nitrate reductase activity, carboxylase activity of Rubisco, superoxide dismutase activity, and dry matter accumulation. Moreover, exogenous application of salicylic acid has been reported to maintain the stability of membranes, enhance photosynthetic rate and K^+/Na^+ ratio (Kaydan et al. 2007), and increases proline content and activities of antioxidant enzymes, thereby mitigating the deleterious effects of stress (Shakirova et al. 2003). Agarwal et al. (2005) have reported that under water stress conditions, application of SA to wheat enhanced the chlorophyll and relative water content while caused considerable reduction in hydrogen peroxide and lipid peroxidation.

There are certain reports indicating that exogenous application of SA does not help in mitigation of drought-induced negative effects (Waseem et al. 2006). Nevertheless, it should be noted that before applying SA, one should have a thorough knowledge about the effective means and methods of application so as to increase the efficiency of exogenously applied SA which is believed to depend on several factors including the species, developmental stage, the manner of application, and the concentration of SA as well (Borsani et al. 2001; Horvath et al. 2007; Joseph et al. 2010). Few other methods which have been reported to protect different plant species against abiotic and biotic stresses include presoaking of the seeds, addition of SA to the hydroponic solutions, tissue culture media and spraying with SA solution (Horvath et al. 2007; Sakhanokho and Kelley 2009). In recent years, tissue culture technique has been extensively utilized for screening and developing stress-tolerant plants. Under in vitro conditions, impact of varying SA concentrations on growth and induction of salt tolerance in *Hibiscus* plants have also been reported (Sakhanokho and Kelley 2009). Moreover, salicylic acid has an affinity to bind with the enzymes like catalase, ascorbate peroxidase, and carbonic anhydrase that are involved in metabolism of free radicals and redox homeostasis (Slaymaker et al. 2002). Any kind of imbalance in this homeostasis triggers the induction of defense responses in plants (Torres et al. 2002; Durrant and Dong 2004). Application of salicylic acid has been reported to increase the activities of the antioxidant enzymes in wheat (Agarwal et al. 2005), *Brassica juncea* (Yusuf et al. 2008), and broad bean (Azooz et al. 2011). Under water stress, salicylic acid-induced activity of antioxidant enzymes has also been reported in *Ctenanthe setosa* (Kadioglu et al. 2010). Moreover, exogenously applied salicylic acid to wheat (Shakirova et al. 2003) and *Brassica juncea* (Yusuf et al. 2008) under salinity and water stress, respectively, alleviated the synthesis and accumulation of proline—a good indication of increased stress tolerance.

4.3 Ethylene

In addition to its usual role in plants, ethylene is also involved in defense against a wide variety of environmental stresses (Bleecker and Kende 2000). Increased ethylene biosynthesis is triggered in plants under various environmental stresses including water stress, thereby suggesting its pivotal role in plant acclimation to stress (Gomez-Cadena et al. 1996). The effects of ethylene, whether transitory or long term, vary considerably among species (Hall and Smith 1995). Although examples of abscission of leaves in response to water deficit stress are compelling, exogenous application of ethylene or the ethylene precursor (1-aminocyclopropane-1-carboxylate) enhances leaf abscission, whereas inhibitors of ethylene synthesis (e.g., aminoethoxyvinylglycine and Ag^+) reduce leaf senescence (Taiz and Zeiger 1998). Ethylene is believed to be involved in stomatal closure but seems rather contradictory. In *Arabidopsis*, ethylene has been reported to inhibit ABA-induced stomatal closure (Tanaka et al. 2005) because H_2O_2-induced stomatal closure results in loss of function in *Arabidopsis* mutants, therefore suggesting an important role of ethylene in guard cell ROS signaling and stomatal closure (Desikan et al. 2005).

Ethylene is biosynthesized from methionine. Two main enzymes involved in the biosynthesis of ethylene are 1-aminocyclopropane-1-carboxylic acid (ACC) synthase and ACC oxidase (ACO) that catalyze the conversion of *S*-adenosyl-L-METHIONINE to ACC and the oxidative cleavage of ACC to ethylene, respectively (Zarembinski and Theologis 1994). Normally the activity of ACC synthase (ACS) is very low in tissues that produce less amounts of ethylene, but upon stimulation, its activity is induced rapidly so the reaction catalyzed by this enzyme is considered as rate limiting and regulatory step in induction of ethylene biosynthesis (Kende 1993). Moreover, the activity of this enzyme is also regulated by phosphorylation and dephosphorylation of proteins at the posttranslational level, thereby altering its turnover rate (Chae et al. 2003). In addition to this, induction of genes coding ACS and ACO under changing environmental conditions or endogenous cues, also enhance the ethylene biosynthesis in plants (Wang et al. 2002).

Under stress, plants produce ethylene in greater concentrations and the stress signal perceived triggers cellular responses further downstream. Protein kinases involved in regulation of ethylene synthesis under stress conditions covert these signals into cellular responses, thereby acting as important mediators of signal transduction cascade in cells (Chang and Karin 2001). Studies carried on transgenic tobacco using protein kinase and phosphatase inhibitors reveal that overexpression of NtMEK2 DD causes activation of SIPK, a tobacco MAPK, thereby resulting in increased ethylene production which coincides with the increase in ACS activity, followed by the activation of a subgroup of ACS and ACO genes, suggesting the role of MAPK in activation of ACS and posttranscriptional regulation (Kim et al. 2003).

Stress signaling cascade largely depends on transcription factors and their expression levels have direct bearing with plant adaptation to adverse environment (Schenk et al. 2000). Among several transcription factors that have been identified, ethylene response factors (ERFs) are known to be important (Zhang et al. 2009). ERFs are implicated in biotic and abiotic stress-induced transcription (Riechmann and Meyerowitz 1998; Hu et al. 2008) but among the various ERF genes, only a few are known to mediate responses to abiotic stress (El-Sharkawy et al. 2009). ERFs are DNA binding proteins that are specific to plants (Hao et al. 1998). It has been reported that ERF family genes are implicated in several stress responses like high salinity (Dubouzet et al. 2003), freezing (Yang et al. 2005), and drought (Yamaguchi-Shinozaki and Shinozaki 2006). Overexpression of the tobacco transcription factor NtERF1 leads to increased salt tolerance (Huang et al. 2004). *Arabidopsis thaliana* overexpressing *Helianthus annus* transcription factor Hahb-4 exhibits a characteristic phenotype essential for tolerance to water stress. Moreover this transcription factor is involved in ethylene-mediated signaling pathways and its expression is regulated by water availability (Manavella et al. 2006).

Particular type of ethylene-responsive transcription factor induces/enhances tolerance to stress in particular plant species. For example, expression of tomato ERF5 has been reported to impart regulation of stress responses in *Arabidopsis thaliana* at transcriptional level (Chuang et al. 2010). In transgenic tomato plants, overexpression of SlERF5 resulted in high tolerance to drought and salt stress which was accompanied by increased relative water content (Pan et al. 2012). Zhang et al. (2010) showed that transgenic rice overexpressing JERF3 exhibited better drought and osmotic stress tolerance which is reflected in increase in the contents of soluble sugars and proline. In addition, overexpression of JERF3 led to the upregulated expression of two OsP5CS genes in response to drought stress and also activated the expression of stress-responsive genes, including WCOR413, OsEnol, and OsSPDS2 under normal conditions.

5 Mineral Nutrients

5.1 Calcium

Calcium is one of the macronutrients required for normal growth and development of plants. It is implicated in regulating a number of fundamental cellular processes involving cytoplasmic streaming, thigmotropism, gravitropism, cell division, cell elongation, cell differentiation, cell polarity, photomorphogenesis, plant defense, and stress responses. As a divalent cation (Ca^{2+}), it acts as an intracellular messenger in the cytosol (Marschner 1995; Nobuhiro and Mittler 2006), has structural role in the cell wall and cell membranes and as a counter cation for anions in the vacuoles (White and Broadley 2003). Calcium provides strong structural rigidity to cells by forming cross-links within the pectin polysaccharides (Easterwood 2002). In plants, Ca^{2+} is usually stored as calcium oxalate crystals in plastids. Better plant growth, the structural integrity of stems, and the quality of fruit produced are strongly coupled to Ca^{2+} availability. Calcium is also known to act as an activator of many enzymes like ATPase, phospholipases, amylase, and succinate dehydrogenase. Studies carried on *Phaseolus vulgaris* L. suggested that Ca^{2+} is associated with stomatal closure, decrease of hydraulic conductivity, sap flow, leaf dry weight, leaf K^+ and Mg^{2+} concentrations, and inhibition of CO_2 assimilation (Cabot et al. 2009).

In the absence of a stimulus, the plant cells maintain low cytosolic Ca^{2+} concentration (100 nM), but in response to an external stimuli including light, touch, wind, hormones, and biotic and abiotic stresses, the cytosolic concentration of calcium is rapidly elevated via an increased Ca^{2+} influx due to the release of Ca^{2+} by Ca^{2+} channels from endoplasmic reticulum, plasma membrane, and other cell organelles and then quickly returns to the normal level by Ca^{2+} efflux through Ca^{2+}/H^+ antiporter and Ca^{2+} pumps (Bush 1995). One of the most common signaling pathways causing increase in concentration of calcium in cytosol in response to external stimulus is the phospholipase C enzyme-mediated pathway. Enzyme phospholipase C is activated after the signal perception by cell surface receptors including G protein-coupled receptors and receptor tyrosine kinases. Activated phospholipase C hydrolyzes the membrane phospholipid PIP2 to form 1,4,5-trisphosphate (IP3) and diacylglycerol (DAG) that act as secondary messengers. DAG is also involved in the activation of protein kinase C enzyme, while IP3 after diffusing into the endoplasmic reticulum binds to receptor (IP3 receptor), a Ca^{2+} channel, thus releasing Ca^{2+} from the endoplasmic reticulum. This stimulus-specific increase in cytoplasmic calcium is called as calcium signature (Evans et al. 2001). Current evidences indicate that apart from IP3, cyclic ADP ribose (cADPR) also influences the activity of Ca^{2+} channels and plays an important role in elevating calcium levels in cytosol. The transduction of Ca^{2+} signals into various biochemical and morphological responses is a very complex and specific phenomena that is controlled by several factors. Specific signal induces a specific Ca^{2+} signature in different cell types (Kiegle et al. 2000). Furthermore, in plant cells, Ca^{2+} acts as a secondary messenger, thereby coupling a range of extracellular stimuli with intracellular responses (Sarwat et al. 2013). The specificity of eliciting appropriate physiological response is due to the temporal and spatial changes of Ca^{2+} and the extent of its amplitude as well (McAnish and Hetherington 1998; Allen et al. 1999), while the nature and intensity of stimulus is specified by amplitude and duration of Ca^{2+} transients. Ca^{2+} binding proteins (sensors) help in decoding and transducing the calcium signatures by activating specific targets and pathways (Shao et al. 2008b; Ahmad et al. 2012d).

Increase in the concentration of cytosolic calcium leads to the activation of various Ca^{2+} sensor proteins that convert these signals into a wide variety of biochemical changes. Different Ca^{2+} sensors that exist in higher plants include calmodulin (CaM), calcium-dependent protein kinases (CDPK), and CBL protein which play a crucial role in abiotic stress signaling in plants (Das and Pandey 2010; Ahmad et al. 2012d). Binding of Ca^{2+} to the sensor molecule induces conformational change in the sensor molecule and exposes the hydrophobic pockets, thereby facilitating interactions of the sensor protein with a variety of target proteins. These kinases are reported to play an important role in inhibition of autophosphorylation and enhancing substrate phosphorylation (Patil et al. 1995). Calcium and Ca^{2+} sensor calmodulin (CaM) regulate the expression of structural and regulatory genes by acting on transcription factors (TFs), thereby modulating their activity or Ca^{2+}-loaded CaM may directly bind to promoter sequences to regulate gene expression, thus, indicating the role of CaM as a transcription factor. Ca^{2+}/CaM complex may bind directly with transcription factors so as to regulate their DNA binding affinities or indirectly by associating with complex transcriptional machinery that consists of Ca^{2+}/CaM complex, transcription factor binding protein (TFBP), and transcription factors. TFBP serves as a bridge between Ca^{2+}/CaM and TFs, while Ca^{2+}/CaM complex regulates gene expression by modulating the phosphorylation status of TFs through the activity of CaM binding protein kinase and protein phosphatase (Kim et al. 2009). Moreover, plant species overexpressing these protein kinases are more tolerant to drought, salinity, and cold stresses. In transgenic rice, overexpression of OsCDPK7 enhanced the induction of stress-responsive genes, resulting in increased tolerance to stress (Saijo et al. 2000). In *Arabidopsis*, it has been reported that under drought and salt stress conditions, expression of AtCPK10 and AtCPK11 is induced indicating their possible role in osmotic stress signaling (Urao et al. 1994). *Arabidopsis* plants overexpressing the CBL5 protein showed enhanced tolerance to high salt or drought stress (Cheong et al. 2010).

Ca^{2+} supplementation under drought stress have been reported to enhance water conservation and improve the hydrophobicity of cellular membranes while lowering its permeability through its interaction with the phosphates and proteins in cellular membranes, thus strengthening their stability (Shao et al. 2008a). Ca^{2+} protects membranes from hydration, improves the cohesion of cell walls, maintains protoplasm viscosity, and enhances cellular dehydration resistance. Thus, Ca^{2+} can stabilize plant cells and enhance drought tolerance through its direct effects on structural basis of the plant (Shao et al. 2008b; Ma et al. 2009). Schaberg et al. (2011) reported that addition of calcium increased the concentrations of amino acids, alanine and γ-aminobutyric acid (GABA) and the polyamines, putrescine (PUT) and spermidine (SPD) as well as chlorophyll content in red spruce (*Picea rubens*) under low temperature stress. Abdel-Basset (1998) reported that under drought stress, calcium supplementation caused significant increase in fresh weight, dry weight, chlorophyll, and relative water content, while reduced the membrane leakage in *Vicia faba*.

5.2 *Potassium*

Potassium (K) is an important macronutrient required for growth and development of plants both under normal and stress conditions (Agarwal et al. 2009). Potassium is actively involved in many basic biochemical and physiological functions such as osmoregulation, enzyme activation, and stomatal movements reducing excess uptake of ions such as Na and Fe in saline soils (Epstein and Bloom 2005; Amtmann et al. 2008; Ahmad et al. 2012c; Wang et al. 2013). Potassium is implicated in transport of inorganic anions and metabolites. Moreover, it maintains the transmembrane voltage gradients for cytoplasmic pH homeostasis (White and Karley 2010).

Potassium deficiency causes considerable reduction in leaf area, photosynthesis, and nitrogen metabolism, which ultimately result in reduced plant growth (Ebelhar and Varsa 2000). As a result of reduction in production of photoassimilates due to potassium deficiency, the sink tissues also get restricted supply of photoassimilates. Due to this reduction in partitioning of photoassimilates, the quantity as well as quality of the yield gets affected (Pettigrew and Meredith 1997; Meille and Pellerin 2004). The very first response of plants to potassium deficiency is the reduction in growth rate, and later beginning of chlorosis and necrosis in older leaves (Mengel and Kirkby 2001). Potassium deficiency induces disturbances in turgor, stomatal opening, water relations, and photosynthesis (Marschner 1995; Mengel and Kirkby 2001). Stomatal regulation largely depends upon the distribution of potassium in epidermal cells, guard cells, and leaf apoplast (Shabala et al. 2002). Supplying sufficient potassium levels has been reported to help plants to maintain higher leaf water potential, turgor potential, relative water content, and lower osmotic potential in several crop plants like wheat (Sen Gupta et al. 1989), maize (Premachandra et al. 1991), and *Vigna radiata* (Nandwal et al. 1998) grown under water stress.

Plants suffering from environmental stresses like drought have a larger internal requirement for K and deficiency of potassium results in overproduction of ROS (Cakmak 2005). Potassium has been reported to reduce the detrimental effects of ROS by enhancing photosynthetic electron transport while inhibiting the activity of membrane bound NAD(P)H oxidases.

Potassium promotes root growth under water stress conditions (Sangakkara et al. 1996) because potassium and magnesium enhances transport of sucrose to developing root for their normal growth and development. Moreover, it also enhances the ion uptake, as potassium itself is one among the main constituents of the phloem sap, thus maintaining the osmotica and, thereby mitigating the adverse effects of moisture stress in plants by increasing the translocation and maintaining the water balance within the plants (Jeschke et al. 1997; Walker et al. 1998). In addition to this, activity of several enzymes which are involved in drought resistance is enhanced by the supplementation of appropriate potassium and its adequate concentration in cytoplasm as well (Kant and Kafkafi 2002). Plants with appropriate K levels have enhanced membrane fluidity because potassium maintains the higher ratio of unsaturated to saturated fatty acids in membranes (Wilkinson et al. 2001). Moreover, potassium supplementation has been reported to increase the synthesis of many organic solutes like proline, free sugars, and free amino acids, under normal and water stress conditions, which contribute to osmotic adjustment. Under normal and water stress conditions, potassium-induced increase in proline has been reported in rice (Pandey et al. 2004) and wheat (Jatav et al. 2012). Potassium-induced increase in free sugars under water stress has been reported in rice by Pandey et al. (2004). In sorghum, potassium has been reported to overcome the ill effects of water stress and maintains the higher tissue water content (Umar et al. 1993). Plants accumulate osmolytes under stress conditions, thereby reducing osmotic potential and maintaining RWC (Gupta et al. 2000). Drought-induced proline accumulation and hydrolysis of macromolecules into simpler mono, disaccharides, and amino acids may lead to accumulation of osmotica.

Evidences are emerging from the molecular studies that potassium might be involved in regulating the plant stress responses (Ashley et al. 2006; Wang and Wu 2010). Low potassium status triggers several signaling cascades such as up regulation of K transporters, synthesis of ROS, and phytohormones including jasmonic acid (JA) and ethylene. In addition to up regulation of transport proteins, potassium deficiency also triggers several other responses in roots. All these strategies enable plant species to adapt with the changing environmental conditions. It has been reported that increase in ROS and phytohormone levels is accompanied by transient increase in transcripts coding potassium transporters and channels, suggesting possible regulatory role of potassium in plant stress responses (Cheong et al. 2007; Jung et al. 2009). Cheong et al. (2007) suggested that in K-deficient plants and drought-induced ABA may trigger Ca flow which acts as secondary messenger and initiates the uptake of K by roots and the regulation of stomatal guard cells. Ca signaling, which regulates leaf transpiration and root K uptake, involves membrane localized Ca sensor-interacting proteins. Jung et al. (2009) reported the increased ethylene and ROS production in K-deficient plants. This phytohormone signal is important for changes in root morphology and plant tolerance to low K supply.

6 Conclusion and Future Perspectives

Abiotic stresses cause considerable reduction in yield especially in arid and semi-arid regions of the globe. Around 40–45 % of the agricultural land affected by drought stress and global climate change has made the problem even graver by causing reduction in arable land. Drought stress causes alterations in normal plant metabolism. Exposure to environmental stresses triggers the generation of ROS in different cellular organelles, especially chloroplasts and mitochondria. ROS cause disturbances in normal functioning of the cell by affecting several cellular macro-molecules including DNA, lipids, proteins, carbohydrates, and their overproduction which ultimately can lead to cell death. In order to mitigate the stress, plants have evolved many adaptive mechanisms. Greater synthesis of various organic osmolytes such as proline, glycine betaine, free sugars, and polyols is considered as good indication of greater tolerance to stress. To ameliorate the damaging effects of ROS, plants are well equipped with enzymatic as well as nonenzymatic antioxidants. Positive effects of organic osmoprotectants and antioxidative defense system in combating stress-induced damage are well established. Osmolytes are known to play a role in osmotic adjustment, thereby maintaining the internal water content of cell, besides protecting subcellular structures. Moreover, synthesis of certain growth regulators is upregulated in plants under stress. Growth regulators act as signal molecules and play pivotal role in sensing and combating with the incoming stress. Proper mineral nutrition can enhance the performance of plants, both under normal and stressed conditions by preventing the oxidation of polyunsaturated fatty acids (PUFA), thereby preventing membrane leakage and excessive formation of free radicals.

The biggest challenge to plant scientists is to develop stress-tolerant plant varieties. In order to enhance the stress tolerance, researchers have to look for defined sets of markers. In connection to this, genetic manipulations in important crops are gaining pace, but it should be kept in mind that the genes incorporated should not only enhance the tolerance at certain plant growth stages but at the whole plant level as well. Genomics, proteomics, and ionomics have been contributory towards the understanding of various plant responses to abiotic stresses. These techniques will help to identify unknown links, cross talk across different stress signaling pathways that could be exploited to enhance the plant tolerance to particular abiotic stress. Keeping global climate change in mind, model plants should be developed for increasing the understanding of tolerance mechanisms and interactions with increasing concentration of CO_2 so as to assess them as suitable future crop plants. At the same time, lack of thorough understanding of drought tolerance mechanisms at genetic and physiological levels and their contributions towards the stress tolerance have been a major limitation to develop drought-tolerant plants.

Our present knowledge about causes and consequences of water stress has still many dark areas and we should enhance our efforts towards these issues. Plant biotechnologists have so far been successful to some extent in developing stress-tolerant varieties but there is much more which still remains to be done.

Acknowledgment We are highly thankful to Prof R. M. Agarwal for his valuable suggestions and Prof. Rekha Bhadauria, Head, School of Studies in Botany, Jiwaji University, Gwalior for providing necessary facilities.

References

Abdel-Basset R (1998) Calcium channels and membrane disorders induced by drought stress in *Vicia faba* plants supplemented with calcium. Acta Physiol Plant 2(2):149–153

Abou El-Yazied AA (2011) Foliar application of glycine betaine and chelated calcium improves seed production and quality of common bean (*Phaseolus Vulgaris* L.) under water stress conditions. Res J Agric Biol Sci 7:357–370

Agarwal S, Sairam RK, Srivastava GC, Meena RC (2005) Changes in antioxidant enzymes activity and oxidative stress by abscisic acid and salicylic acid in wheat genotypes. Biol Plant 49:541–550

Agarwal RM, Tomar NS, Singh K, Sharma GL (2009) Potassium induced changes in flowering plants. In: Flower retrospect and prospect (Professor Vishwambhar Puri Birth Centenary Volume). SR Scientific Publication, Delhi, pp 158–186

Agastian P, Kingsley SJ, Vivekanandan M (2000) Effect of salinity on photosynthesis and biochemical characteristics in mulberry genotypes. Photosynthetica 38:287–290

Ahmad P (2010) Growth and antioxidant responses in mustard (*Brassica juncea* L.) plants subjected to combined effect of gibberellic acid and salinity. Arch Agron Soil Sci 56(5):575–588

Ahmad P, Sharma S (2008) Salt stress and phyto-biochemical responses of plants. Plant Soil Environ 54(3):89–99

Ahmad P, Jhon R, Sarwat M, Umar S (2008a) Responses of proline, lipid peroxidation and antioxidative enzymes in two varieties of *Pisum sativum* L. under salt stress. Int J Plant Prod 2(4):353–366

Ahmad P, Sarwat M, Sharma S (2008b) Reactive oxygen species, antioxidants and signaling in plants. J Plant Biol 51(3):167 173

Ahmad P, Jaleel CA, Salem MA, Nabi G, Sharma S (2010a) Roles of enzymatic and non-enzymatic antioxidants in plants during abiotic stress. Crit Rev Biotechnol 30(3):161–175

Ahmad P, Jaleel CA, Sharma S (2010b) Antioxidant defense system, lipid peroxidation, proline metabolizing enzymes and biochemical activities in two *Morus alba* genotypes subjected to NaCl stress. Russ J Plant Physiol 57(4):509–517

Ahmad P, Nabi G, Ashraf M (2011) Cadmium induced oxidative damage in mustard (*Brassica juncea* L. Czern. and Coss.) plants can be alleviated by salicylic acid. S Afr J Bot 77:36–44

Ahmad P, Ozturk M, Gucel S (2012a) Oxidative damage and antioxidants induced by heavy metal stress in two cultivars of mustard (L) plants. Fresen Environ Bull 21(10):2953–2961

Ahmad P, Hakeem KR, Kumar A, Ashraf M, Akram NA (2012b) Salt-induced changes in photosynthetic activity and oxidative defense system of three cultivars of mustard (*Brassica juncea* L.). Afr J Biotechnol 11(11):2694–2703

Ahmad P, Kumar A, Gupta A, Hu X, Hakeem KR, Azooz MM, Sharma S (2012c) Polyamines: role in plants under abiotic stress. In: Ashraf M, Ozturk M, Ahmad MSA, Aksoy A (eds) Crop production for agricultural improvement. Springer, Dordrecht, pp 491–512

Ahmad P, Bhardwaj R, Tuteja T (2012d) Plant signaling under abiotic stress environment. In: Ahmad P, Prasad MNV (eds) Environmental adaptations and stress tolerance of plants in the era of climate change. Springer, New York, pp 297–323

Ahmad P, Ashraf M, Azooz MM, Rasool S, Akram NA (2013) Potassium starvation-induced oxidative stress and antioxidant defense responses in *Brassica juncea*. J Plant Interact (in press). doi:10.1080/17429145.2012.747629

Alcazar R, Altabella T, Marco F, Bortolotti C, Reymond M, Koncz C (2010) Polyamines: molecules with regulatory functions in plant abiotic stress tolerance. Planta 231:1237–1249

Alcazar R, Cuevas JC, Planas J, Zarza X, Bortolotti C, Carrasco P, Salinas J, Tiburcio AF, Altabella T (2011) Integration of polyamines in the cold acclimation response. Plant Sci 180:31–38

Aldesuquy HS, Abbas MA, Abo-Hamed SA, Elhakem AH, Alsokari SS (2012) Glycine betaine and salicylic acid induced modification in productivity of two different cultivars of wheat grown under water stress. J Stress Physiol Biochem 8(2):72–89

Alet AI, Sanchez DH, Cuevas JC (2011) Putrescine accumulation in *Arabidopsis thaliana* transgenic lines enhances tolerance to dehydration and freezing stress. Plant Signal Behav 6:278–286

Ali Q, Ashraf M (2011) Exogenously applied glycine betaine enhances seed and seed oil quality of maize (*Zea mays* L.) under water deficit conditions. Environ Exp Bot 71:249–259

Alia SA, Murata N (1998) Metabolic engineering of rice leading to biosynthesis of glycine betaine and tolerance to salt and cold. Plant Mol Biol 38:1011–1019

Allen GA, Kwak JM, Chu SP, Llopis J, Tsien RY, Harper JF, Schroeder JI (1999) Cameleon calcium indicator reports cytoplasmic calcium dynamics in *Arabidopsis* guard cells. Plant J 19:735–747

Amtmann A, Troufflard S, Armengaud P (2008) The effect of potassium nutrition on pest and disease resistance in plants. Physiol Plant 133:682–691

Anand A, Trick HN, Gill BS (2003) Stable transgene expression and random gene silencing in wheat. Plant Biotechnol J 1(4):241–251

Ananieva EA, Christov KN, Popova LP (2004) Exogenous treatment with salicylic acid leads to increased antioxidant capacity in leaves of barley plants exposed to paraquat. J Plant Physiol 161:319–328

Anjum SA, Saleem MF, Wang L, Bilal MF, Saeed A (2012) Protective role of glycine betaine in maize against drought induced lipid peroxidation by enhancing capacity of antioxidative system. Aust J Crop Sci 6(4):576–583

Arbona V, Hossain Z, Lopez-Climent MF, Perez Clemente RM, Gomez Cadenas A (2008) Antioxidant enzymatic activity is linked to water logging stress tolerance in citrus. Physiol Plant 132:452–466

Armengaud P (2004) Transcriptional regulation of proline biosynthesis in *Medicago truncatula* reveals developmental and environmental specific features. Physiol Plant 120:442–450

Ashley MK, Grant M, Grabov A (2006) Plant responses to potassium deficiencies: a role for potassium transport proteins. J Exp Bot 57:425–436

Ashraf M, Foolad MR (2007) Roles of glycine betaine and proline in improving plant abiotic stress resistance. Environ Exp Bot 59:206–216

Audran C, Liotenberg S, Gonneau M, North H, Frey A (2001) Localization and expression of zeaxanthin epoxidase mRNA in *Arabidopsis* in response to drought stress and during seed development. Aust J Plant Physiol 28:1161–1173

Azooz MM, Youssef AM, Ahmad P (2011) Evaluation of salicylic acid (SA) application on growth, osmotic solutes and antioxidant enzyme activities on broad bean seedlings grown under diluted seawater. Int J Plant Physiol Biochem 3(14):253–264

Bacon MA, Wilkinson S, Davies WJ (1998) pH-regulated leaf cell expansion in droughted plants is abscisic acid dependent. Plant Physiol 118:1507–1515

Bahrani MJ, Bahrami H, Haghighi AAK (2010) Effect of water stress on ten forage grasses native or introduced to Iran. Grassl Sci 56:1–5

Barkosky RR, Einhellig FA (1993) Effects of salicylic acid on plant water relationship. J Chem Ecol 19:237–347

Batlang U, Baisakh N, Ambavaram MM, Pereira A (2013) Phenotypic and physiological evaluation for drought and salinity stress responses in rice. Methods Mol Biol 956:209–225

Bleecker AB, Kende H (2000) Ethylene: a gaseous signal molecule in plants. Annu Rev Cell Dev Biol 16:1–18

Blum A (1996) Crop response to drought and the interpretation of adaptation. J Plant Growth Regul 20:135–148

Blum A (2005) Drought resistance, water-use deficiency, and yield potential—are they compatible, dissonant, or mutually exclusive. Aust J Agric Res 56:1159–1168

Bohnert HJ, Jensen RG (1996) Strategies for engineering water-stress tolerance in plants. Trends Biotechnol 14:89–97

Boominathan P, Shukla R, Kumar A, Manna D, Negi D, Praveen VK, Chattopathy D (2004) Long term transcript accumulation during development of dehydration adaptation in *Cicer arietinum*. Plant Physiol 135:1608–1620

Borsani O, Valpuesta V, Botella MA (2001) Evidence for a role of salicylic acid in the oxidative damage generated by NaCl and osmotic stress in *Arabidopsis* seedlings. J Plant Physiol 126:1024–1030

Boudsocq M, Barbier-Brygoo H, Lauriere C (2004) Identification of nine sucrose nonfermenting 1-related protein kinases 2 activated by hyperosmotic and saline stresses in *Arabidopsis thaliana*. J Biol Chem 279:41758–41766

Bray EA (2002) Abscisic acid regulation of gene expression during water deficit stress in the era of the *Arabidopsis* genome. Plant Cell Environ 25:153–161

Burkhanova EA, Fedina AB, Kulaeva ON (1999) Effect of salicylic acid and (2-5)-oligo-adenylates on protein synthesis in tobacco leaves under heat shock conditions: a comparative study. Russ J Plant Physiol 46:16–22

Bush DS (1995) Calcium regulation in plant cells and its role in signaling. Annu Rev Plant Physiol Plant Mol Biol 46:95–122

Cabot C, Sibole Barcelo JV, Poschenrieder C (2009) Sodium-calcium interactions with growth, water, and photosynthetic parameters in salt treated beans. J Plant Nutr Soil Sci 172:637–643

Cakmak I (2005) The role of potassium in alleviating detrimental effects of abiotic stresses in plants. J Plant Nutr Soil Sci 168:521–530

Chae HS, Faure F, Kieber JJ (2003) The eto1, eto2 and eto3 mutations and cytokinin treatment elevate ethylene biosynthesis in *Arabidopsis* by increasing the stability of the ACS5 protein. Plant Cell 15:545–559

Chang L, Karin M (2001) Mammalian MAP kinase signaling cascades. Nature 410:37–40

Chaves HM, Pereira JS, Maroco J, Rodrigues ML, Ricardo CPP, Osorio ML (2002) How plants cope with water stress in the field. Photosynthesis and growth. Ann Bot 89:907–916

Chaves MM, Maroco JP, Pereira JS (2003) Understanding plant responses to drought from genes to the whole plants. Funct Plant Biol 30:239–264

Chen Z, Cuin TA, Zhou M, Twomey M, Naidu BP, Shabala S (2007) Compatible solute accumulation and stress-mitigating effects in barley genotypes contrasting in their salt tolerance. J Exp Bot 58:4245–4255

Cheng WH, Endo A, Zhou L, Penney J, Chen HC, Arroyo A, Leon P, Nambara E, Asami T, Seo M, Koshiba T, Sheen J (2002) A unique short-chain dehydrogenase/reductase in *Arabidopsis* glucose signaling and abscisic acid biosynthesis and functions. Plant Cell 14:2723–2743

Cheong YH, Pandey GK, Grant JJ, Batistic O, Li L, Kim BK, Lee SC, Kudia J, Luan S (2007) Two calcineurin B-like calcium sensors, interacting with protein kinase CIPK 23, regulate leaf transpiration and root potassium uptake in *Arabidopsis*. Plant J 52:223–239

Cheong YH, Sung SJ, Kim BG, Pandey GK, Cho JS, Kim KN, Luan S (2010) Constitutive overexpression of the calcium sensor CBL5 confers osmotic or drought stress tolerance in *Arabidopsis*. Mol Cells 29:159–165

Chuang HW, Harnrak A, Chen YC, Hsu CM (2010) A harpin-induced ethylene-responsive factor regulates plant growth and responses to biotic and abiotic stresses. Biochem Biophys Res Commun 402:414–420

Cuevas JC, Lopez-Cobollo R, Alcazar R, Zarza X, Koncz C, Altabella T, Salinas J, Tiburcio AF, Ferrando A (2008) Putrescine is involved in *Arabidopsis* freezing tolerance and cold acclimation by regulating abscisic acid levels in response to low temperature. Plant Physiol 148:1094–1105

Da Costa M, Huang B (2009) Physiological adaptations of perennial grasses to drought stress. In: Barrera ED, Smith WK (eds) Perspectives in biophysical plant ecophysiology. Universidad National Autonoma de Mexico, Mexico, pp 169–190. ISBN 987-0-578-00421-1

Das R, Pandey GK (2010) Expressional analysis and role of calcium regulated kinases in abiotic stress signaling. Curr Genomics 11:2–13

Dat JF, Lopez-Delgado H, Foyer CH, Scott IM (1998) Parallel changes in H_2O_2 and catalase during thermo tolerance induced by salicylic acid and heat acclimation of mustard seedlings. Plant Physiol 116:1351–1357

de Carvalho K, de Campos MK, Domingues DS, Pereira LF, Vieira LG (2013) The accumulation of endogenous proline induces changes in gene expression of several antioxidant enzymes in leaves of transgenic *Swingle citrumelo*. Mol Biol Rep 40(4):3269–3279

Desikan R, Hanckok JT, Bright J, Harrison J, Weir I, Hooley R, Neill SJ (2005) A role for ETR1 in hydrogen peroxide signaling in stomatal guard cells. Plant Physiol 137:831–834

Dobra J, Motyka V, Dobrev P, Malbeck J, Prasil IT, Haisel D, Gaudinova A, Havlova M, Gubis J, Vankova R (2010) Comparison of hormonal responses to heat, drought and combined stress in tobacco plants with elevated proline content. J Plant Physiol 167:1360–1370

Duan JJ, Li J, Guo SR, Kang YY (2008) Exogenous spermidine affects polyamine metabolism in salinity-stressed *Cucumis sativus* roots and enhances short term salinity tolerance. J Plant Physiol 165:1620–1635

Dubouzet JG, Sakuma Y, Ito Y, Kasuga M, Dubouzet EG, Miura S, Seki M, Shinozaki K, Yamaguchi-Shinozaki K (2003) OsDREB genes in rice *Oryza sativa* L., encode transcription activators that function in drought, high salt and cold responsive gene expression. Plant J 33:751–763

Durrant WE, Dong X (2004) Systemic acquired resistance. Annu Rev Phytopathol 42:185–209

Easterwood GW (2002) Calcium's role in plant nutrition. Fluid J 1:1–3

Ebelhar SA, Varsa EC (2000) Tillage and potassium placement effects on potassium utilization by corn and soybean. Commun Soil Sci Plant Anal 31:11–14

El-Sharkawy I, Sherif S, Mila I, Bouzayen M, Jayasankar S (2009) Molecular characterization of seven genes encoding ethylene responsive transcriptional factors during plum fruit development and ripening. J Exp Bot 60:907–922

El-Tayeb MA (2005) Response of barley grains to the interactive effect of salinity and salicylic acid. Plant Growth Regul 45:215–224

Epstein E, Bloom AJ (2005) Mineral nutrition of plants: principles and perspectives, 2nd edn. Sinauer Associates, Sunderland, p 400

Evans NH, McAinsh MR, Hetherington AM (2001) Calcium oscillations in higher plants. Curr Opin Plant Biol 4:415–420

FAO (2010) FAO land and plant nutrition management service. http://www.fao.org

Finkelstein RR, Gampala SSL, Rock CD (2002) Abscisic acid signaling in seeds and seedlings. Plant Cell 14:15–45

Fujii H, Verslues PE, Zhu JK (2011) *Arabidopsis* decuple mutant reveals the importance of SnRK2 kinases in osmotic stress responses in vivo. Proc Natl Acad Sci U S A 108:1717–1722

Fujita M, Fujita Y, Noutoshi Y, Takahashi F, Narusaka Y, Yamaguchi-Shinozaki K, Shinozaki K (2006) Crosstalk between abiotic and biotic stress responses: a current view from the points of convergence in the stress signaling networks. Curr Opin Plant Biol 9:436–442

Gadallah MAA (1999) Effects of proline and glycine betaine on *Vicia faba* responses to salt stress. Biol Plant 42:249–257

Gill SS, Tuteja N (2010) Polyamines and abiotic stress tolerance in plants. Plant Signal Behav 5:26–33

Girija C, Smith BN, Swamy PM (2002) Interactive effects of sodium chloride and calcium chloride on the accumulation of proline and glycine betaine in peanut (*Arachis hypogaea* L). Environ Exp Bot 47:1–10

Gomez-Cadena A, Tadeo FR, Primmo-Millo ME (1996) Leaf abscission induced by ethylene in water-stressed intact seedlings of *Cleopatra mandarin* requires previous abscisic acid accumulation in roots. Plant Physiol 112:401–408
Gowing DJ, Davies WJ, Jones HG (1990) A positive root-sourced signal as an indicator of soil drying in apple (*Malus domestica* Bork). Exp Bot 41:1535–1540
Groppa MD, Benavides MP (2008) Polyamines and abiotic stress: recent advances. Amino Acids 34:35–45
Gupta SC, Rathore AK, Sharma SN, Saini RS (2000) Response of chickpea cultivar to water stress. Indian J Plant Physiol 3:274–276
Hall MA, Smith AR (1995) Ethylene and the responses of plants to stress. Bulg J Plant 165:397–406
Hao D, Ohme-Takagi M, Sarai A (1998) Unique mode of GCC box recognition by the DNA-binding domain of ethylene-responsive element-binding factor (ERF domain) in plant. J Biol Chem 273:26857–26861
Hartung W, Sauter A, Hose E (2002) Abscisic acid in the xylem: where does it come from, where does it go to? J Exp Bot 53:27–32
Hasegawa PM, Bressan RA, Zhu JK, Bohnert HJ (2000) Plant cellular and molecular responses to high salinity. Annu Rev Plant Physiol Plant Mol Biol 51:463–499
Hayashi H, Alia Mustardy L, Deshnium P, Ida M, Murata N (1997) Transformation of *Arabidopsis thaliana* with the codA gene for choline oxidase; accumulation of glycine betaine and enhanced tolerance to salt and cold stress. Plant J 12:133–142
He T, Cramer GR (1996) Abscisic acid concentrations are correlated with leaf area reductions in two salt stressed rapid cycling *Brassica* species. Plant Soil 179:25–33
Hong Z, Lakkineni K, Zhang Z, Verma DP (2000) Removal of feedback inhibition of delta pyrroline-5-carboxylate synthetase results in increased proline accumulation and protection of plants from osmotic stress. Plant Physiol 122:1129–1136
Horvath E, Szalai G, Janda T (2007) Induction of abiotic stress tolerance by salicylic acid signaling. J Plant Growth Regul 26:290–300
Hu Y, Zhao L, Chong K, Wang T (2008) Overexpression of OsERF1, a novel rice ERF gene, up-regulates ethylene-responsive genes expression, besides affects growth and development in *Arabidopsis*. J Plant Physiol 165:1717–1725
Huang ZJ, Zhang ZJ, Zhang XL, Zhang HB, Huang DF (2004) Tomato TERF1 modulates ethylene response and enhances osmotic stress tolerance by activating expression of downstream genes. FEBS Lett 573:110–116
Hussain SS, Ali M, Ahmad M, Siddique KH (2011) Polyamines: natural and engineered abiotic and biotic stress tolerance in plants. Biotechnol Adv 29:300–311
Ingram J, Bartels D (1996) The molecular basis of dehydration tolerance in plants. Annu Rev Plant Physiol Plant Mol Biol 47:337–403
Iqbal N, Ashraf MY, Ashraf M (2005) Influence of water stress and exogenous glycine betaine on sunflower achene weight and oil percentage. Int J Environ Sci Technol 2:155–160
Jackson GE, Irvine J, Grace J, Khalil AAM (1995) Abscisic acid concentrations and fluxes in droughted conifer saplings. Plant Cell Environ 18:13–22
Jajarmi V (2009) Effect of water stress on germination indices in seven wheat cultivar. World Acad Sci Eng Technol 49:105–106
Jaleel CA, Gopi R, Manivannan P, Kishorekumar A, Sankar B, Panneerselvam R (2006) Paclobutrazol influences vegetative growth and floral characteristics of *Catharanthus roseus* (L). G. Don. Indian J Appl Pure Biol 21:369–372
Jaleel CA, Gopi R, Sankar B, Manivannan P, Kishorekumar A, Sridharan R, Panneerselvam R (2007) Studies on germination, seedling vigour, lipid peroxidation and proline metabolism in *Catharanthus roseus* seedlings under salt stress. S Afr J Bot 77:190–195
Janda T, Szalai G, Tari T, Paldi E (1999) Hydroponic treatment with salicylic acid decreases the effects of chilling injury in maize (*Zea mays* L) plants. Planta 208:175–180
Jarvis AJ, Davies WJ (1997) Whole plant ABA flux and the regulation of water loss in *Cedrella odorata*. Plant Cell Environ 20:521–527

Jatav KS, Agarwal RM, Singh RP, Shrivastava M (2012) Growth and yield responses of wheat (*Triticum aestivum* L.) to suboptimal water supply and different potassium doses. J Funct Environ Bot 2(1):39–51

Jeschke WD, Kirkby EA, Peuke AD, Pate JS, Hartung W (1997) Effects of P deficiency on assimilation and transport of nitrate and phosphate in intact plants of castor oil bean (*Ricinus communis* L). J Exp Bot 48:75–91

Jia W, Zhang J (1999) Stomatal closure is induced rather by prevailing xylem abscisic acid than by accumulated amount of xylem derived abscisic acid. Physiol Plant 106:268–275

Jiang M, Zhang J (2002) Water stress induced abscisic acid accumulation triggers the increased generation of reactive species and up-regulates the activities of antioxidant enzymes in maize leaves. J Exp Bot 53:2401–2410

Joseph B, Jini D, Sujatha S (2010) Insight into role of exogenous salicylic acid on plants growth under salt environment. Asian J Crop Sci 2(4):226–235

Ju HW, Min JH, Chung MS, Kim CS (2013) The atrzf1 mutation of the novel RING-type E3 ubiquitin ligase increases proline contents and enhances drought tolerance in *Arabidopsis*. Plant Sci 203–204:1–7

Jung JY, Shin R, Schachtman DP (2009) Ethylene mediates response and tolerance to potassium deprivation in *Arabidopsis*. Plant Cell 91:607–621

Kadioglu A, Saruhan N, Saglam A, Terzi R, Tuba A (2010) Exogenous salicylic acid alleviates effects of long term drought stress and delays leaf rolling by inducing antioxidant system. Plant Growth Regul 64:27–37

Kant S, Kafkafi U (2002) Potassium and abiotic stresses in plants. In: Pasricha NS, Bansal SK (eds) Potassium for sustainable crop production. Potash Institute of India, Gurgaon, pp 233–251

Kasukabe Y, He L, Nada K, Misawa S, Ihara I, Tachibana S (2004) Overexpression of spermidine synthase enhances tolerance to multiple environmental stresses and up-regulates the expression of various stress-regulated genes in transgenic *Arabidopsis thaliana*. Plant Cell Physiol 45:712–722

Katare DP, Nabi G, Azooz MM, Aeri V, Ahmad P (2012) Biochemical modifications and enhancement of psoralen content in salt-stressed seedlings of *Psoralea corylifolia* Linn. J Funct Environ Bot 2(1):65–74

Kathuria H, Giri J, Nataraja KN, Murata N, Udayakumar M, Tyagi AK (2009) Glycine betaine induced water stress tolerance in codA-expressing transgenic rice is associated with up regulation of several stress responsive genes. Plant Biotechnol J 7:512–526

Kaydan D, Yagmur M, Okut N (2007) Effects of salicylic acid on the growth and some physiological characters in salt stressed wheat (*Triticum aestivum* L). Tarim Bilimleri Dergisi 13(2):114–119

Kende H (1993) Ethylene biosynthesis. Annu Rev Plant Physiol Plant Mol Biol 44:283–307

Khamssi NN, Golezani KG, Najaphy A, Zehtab S (2011) Evaluation of grain filling rate, effective grain filling period and resistance indices under acclimation to gradual water deficit stress in chickpea cultivars. Aust J Crop Sci 5:1044–1049

Kiegle E, Moore CA, Haseloff J, Tester MA, Knight MR (2000) Cell type specific calcium responses to drought, salt and cold in the *Arabidopsis* root. Plant J 23:267–278

Kim GB, Nam YW (2013) Δ^1-pyrroline-5-carboxylate synthetase gene of *Medicago truncatula* plays a predominant role in stress-induced proline accumulation during symbiotic nitrogen fixation. J Plant Physiol 170(3):291–302

Kim CY, Liu Y, Thorne ET, Yang H, Fukushige H, Gassmann W, Hildebrand D, Sharp RE, Zhang S (2003) Activation of a stress responsive mitogen-activated protein kinase cascade induces the biosynthesis of ethylene in plants. Plant Cell 15:2707–2718

Kim MC, Chung WS, Yun DJ, Cho MJ (2009) Calcium and calmodulin-mediated regulation of gene expression in plants. Mol Plant 2:13–21

Kong X, Pan J, Zhang M, Xing X, Zhou Y, Liu Y, Li D (2011) ZmMKK4, a novel group C mitogen-activated protein kinase in maize (*Zea mays*), confers salt and cold tolerance in transgenic *Arabidopsis*. Plant Cell Environ 34:1291–1303

Kovacs Z, Simon-Sarkadi L, Szucs A, Kocsy G (2010) Different effects of cold, osmotic stress and abscisic acid on polyamine accumulation in wheat. Amino Acids 38:623–631

Koyro HW, Ahmad P, Geissler N (2012) Abiotic stress responses in plants: an overview. In: Ahmad P, Prasad MNV (eds) Environmental adaptations and stress tolerance of plants in the era of climate change. Springer, New York, pp 1–28
Kubis J (2008) Exogenous spermidine differently alters activities of some scavenging system enzymes, H_2O_2 and superoxide radical levels in water stressed cucumber leaves. J Plant Physiol 165:397–406
Kumriaa R, Rajam MV (2002) Ornithine decarboxylase transgene in tobacco affects polyamines, in vitro morphogenesis and response to salt stress. J Plant Physiol 159(9):983–990
Kusano T, Yamaguchi K, Berberich T, Takahashi Y (2007) Advances in polyamine research. J Plant Res 120:345–350
Le Rudulier D, Strom AR, Danekar AM, Smith LT, Valentaine RC (1984) Molecular biology of osmoregulation. Science 224:1064–1068
Leung J, Giraudat J (1998) Abscisic acid signal transduction. Annu Rev Plant Physiol Plant Mol Biol 49:199–222
Levitt J (1980) Responses of plants to environmental stresses. Academic, New York
Liu HH, Dong BH, Zhang YY, Liu ZP, Liu YL (2004) Relationship between osmotic stress and the levels of free, conjugated and bound polyamines in leaves of wheat seedlings. Plant Sci 166:1261–1267
Liu JH, Kitashiba H, Wang J, Ban Y, Moriguchi T (2007) Polyamines and their ability to provide environmental stress tolerance to plants. Plant Biotechnol 24:117–126
Liu JH, Inoue H, Moriguchi T (2008) Salt stress-mediated changes in free polyamine titers and expression of genes responsible for polyamine biosynthesis of apple *in vitro* shoots. Environ Exp Bot 62:28–35
Liu L, Hu X, Song J, Zong X, Li D (2009) Over-expression of a *Zea mays* L. protein phosphatase 2C gene (ZmPP2C) in *Arabidopsis thaliana* decreases tolerance to salt and drought. J Plant Physiol 166:531–542
Luan S (2002) Signaling drought in guard cells. Plant Cell Environ 25:229–237
Lutts S (2000) Exogenous glycine betaine reduces sodium accumulation in salt stressed rice plants. Int Rice Res Notes 25:39–40
Lv S, Yang A, Zhang K, Wang L, Zhang J (2007) Increase of glycine betaine synthesis improves drought tolerance in cotton. Mol Breed 20:233–248
Ma YY, Song WY, Liu ZH, Zhang HM, Guo XL, Shao HB, Ni FT (2009) The dynamic changing of Ca^{2+} cellular localization in maize leaflets under drought stress. Cell Biol 332:35–362
Makela P, Jokinen K, Kontturi M, Peltonen Sainio P, Pehu E, Somersalo S (1998) Foliar application of glycine betaine—a novel product from sugar beet, as an approach to increase tomato yield. Ind Crops Prod 7:139–148
Manavella PA, Arce AL, Dezar CA, Bitton F, Renou JP, Crespi M, Chan RL (2006) Cross talk between ethylene and drought signaling pathways is mediated by the sunflower Hahb-4 transcription factor. Plant J 48:125–137
Marschner H (1995) Mineral nutrition of higher plants, 2nd edn. Academic, London, p 889
Masoumi A, Kafi M, Hamidreza Khazaei H, Davari K (2010) Effect of drought stress on water status, electrolyte leakage and enzymatic antioxidant of kochia (*Kochia scoparia*) under saline condition. Pak J Bot 42(5):3517–3524
McAnish MR, Hetherington AM (1998) Encoding specificity in Ca^{2+} signaling systems. Trends Plant Sci 3:32–36
Meille JL, Pellerin S (2004) Leaf area establishment of a maize (*Zea mays* L.) field crop under potassium deficiency. Plant Soil 265:75–92
Mengel K, Kirkby EA (2001) Principles of plant nutrition, 5th edn. Kluwer Academic, Dordrecht, p 849
Metraux JP (2001) Systemic acquired resistance and salicylic acid: current state of knowledge. Eur J Plant Pathol 107:13–18
Metwally A, Finkmemeier I, Georgi M, Dietz KJ (2003) Salicylic acid alleviates the cadmium toxicity in barley seedlings. Plant Physiol 132:272–281

Mikolajczyk M, Awotunde OS, Muszynska G (2000) Osmotic stress induces rapid activation of a salicylic acid induced protein kinase and a homolog of protein kinase ASK1 in tobacco cell. Plant Cell 12:165–178

Miller G, Honig A, Stein H, Suzuki N, Mittler R, Zilberstein A (2009) Unraveling delta1-pyrroline-5-carboxylate-proline cycle in plants by uncoupled expression of proline oxidation enzymes. J Biol Chem 284:26482–26492

Montero E, Cabot C, Poschenrieder C, Barcelo J (1998) Relative importance of osmotic-stress and ion-specific effects on ABA-mediated inhibition of leaf expansion growth in *Phaseolus vulgaris*. Plant Cell Environ 21:54–62

Munns R, Tester M (2008) Mechanisms of salinity tolerance. Annu Rev Plant Biol 59:651–681

Murata N, Mohanty PS, Hayashi H, Papageorgiou GC (1992) Glycine betaine stabilizes the association of extrinsic proteins with the photosynthetic oxygen evolving complex. FEBS Lett 296:187–189

Murkowski A (2001) Heat stress and spermidine: effect on chlorophyll fluorescence in tomato plants. Biol Plant 44:53–57

Nandwal AS, Hooda A, Datta D (1998) Effect of substrate moisture and potassium on water relations and C, N and K distribution in *Vigna radiata*. Biol Plant 41:149–153

Nayyar H, Chander S (2004) Protective effects of polyamines against oxidative stress induced by water and cold stress in chickpea. J Agron Crop Sci 190:355–365

Nayyar H, Walia DP (2004) Genotypic variation in wheat in response to water stress and abscisic acid-induced accumulation of osmolytes in developing grains. J Agron Crop Sci 190:39–45

Nayyar H, Bains T, Kumar S (2005) Low temperature induced floral abortion in chickpea: relationship to abscisic acid and cryoprotectants in reproductive organs. Environ Exp Bot 1:39–47

Nobuhiro S, Mittler R (2006) Reactive oxygen species and temperature stresses: a delicate balance between signaling and destruction. Physiol Plant 126:45–51

Nye GJ, Scott P, Golbeck H, Hanson AD (1997) Choline monooxygenase, an unusual iron-sulfur enzyme catalyzing the first step of glycine betaine synthesis in plants: prosthetic group characterization and cDNA cloning. Proc Natl Acad Sci U S A 94:3454–3458

Pan Y, Seymour GB, Lu C, Hu Z, Chen X, Chen G (2012) An ethylene response factor (ERF5) promoting adaptation to drought and salt tolerance in tomato. Plant Cell Rep 31:349–360

Pandey R, Agarwal RM (1998) Water stress-induced changes in proline contents and nitrate reductase activity in rice under light and dark conditions. Physiol Mol Biol Plants 4:53–57

Pandey R, Agarwal RM, Jeevaratnam K, Sharma GL (2004) Osmotic stress induced alterations in rice (*Oryza sativa* L.) and recovery. Plant Growth Regul 42:79–87

Papageorgiou GC, Murata N (1995) The unusually strong stabilizing effects of glycine betaine on the structure and function of the oxygen-evolving photosystem II complex. Photosynth Res 44:243–252

Patil S, Takezawa D, Poovaiah BW (1995) Chimeric plant calcium: calmodulin-dependent protein kinase gene with a neural visinin-like calcium-binding domain. Proc Natl Acad Sci U S A 92:4897–4901

Peleg Z, Blumwald E (2011) Hormone balance and abiotic stress tolerance in crop plants. Curr Opin Plant Biol 14:290–295

Pettigrew WT, Meredith WR (1997) Dry matter production, nutrient uptake, and growth of cotton as affected by potassium fertilization. J Plant Nutr 20:531–548

Pospisilova J, Vagner M, Malbeck J, Travnickova A, Batkova P (2005) Interactions between abscisic acid and cytokinins during water stress and subsequent rehydration. Biol Plant 49:533–540

Prabhavathi VR, Rajam MV (2007) Polyamine accumulation in transgenic eggplant enhances tolerance to multiple stresses and fungal resistance. Plant Biotechnol 24:273–282

Premachandra GS, Saneoka H, Ogata S (1991) Cell membrane stability and leaf water relations as affected by potassium nutrition of water stressed maize. J Exp Bot 42:45–73

Pugnaire FI, Endolz LS, Pardos J (1994) Constraints by water stress on plant growth. In: Pessarakli M (ed) Handbook of plant and crop stress. Marcel Dekker, New York, pp 247–259

Qin XQ, Zeevaart JAD (1999) The 9-cis-epoxycarotenoid cleavage reaction is the key regulatory step of abscisic acid biosynthesis in water-stressed bean. Proc Natl Acad Sci U S A 96:15354–15361

Qu LJ, Wu LQ, Fan ZM, Guo L, Li YQ, Chen ZL (2005) Overexpression of the bacterial nhaA gene in rice enhances salt and drought tolerance. Plant Sci 168:297–302

Rahman MS, Miyake H, Takeoka Y (2002) Effects of exogenous glycine betaine on growth and ultra structure of salt-stressed rice seedlings (*Oryza sativa* L.). Plant Prod Sci 5:33–44

Rai VK, Sharma SS, Sharma S (1986) Reversal of ABA induced stomatal closure by phenolic compounds. J Exp Bot 37:129–134

Raskin I (1992) Role of salicylic acid in plants. Annu Rev Plant Physiol Plant Mol Biol 43:439–463

Rasool S, Ahmad A, Siddiqi TO, Ahmad P (2013) Changes in growth, lipid peroxidation and some key antioxidant enzymes in chickpea genotypes under salt stress. Acta Physiol Plant 35(4):1039–1050

Rengasamy P (2010) Soil processes affecting crop production in salt affected soils. Funct Plant Biol 37:613–620

Riechmann JL, Meyerowitz EM (1998) The AP2/EREBP family of plant transcription factors. Biol Chem 379:633–646

Roosens NH, Al Bitar F, Loenders K, Angenon G, Jacobs M (2002) Overexpression of ornithine-delta-aminotransferase increases proline biosynthesis and confers osmotolerance in transgenic plants. Mol Breed 9:73–80

Ruiz-Sanchez MC, Domingo R, Perez-Pastor A (2007) Daily variations in water relations of apricot trees under different irrigation regimes. Biol Plant 51:735–740

Sadeghipour O, Aghaei P (2012) Impact of exogenous salicylic acid application on some traits of common bean (*Phaseolus vulgaris* L.) under water stress conditions. Int J Agric Crop Sci 4(11):685–690

Saijo Y, Hata S, Kyozuka J, Shimamoto K, Izui K (2000) Overexpression of single calcium dependent protein kinase confers both cold and salt/drought tolerance on rice plants. Plant J 28(3): 319–327

Sajid SA, Aftab F (2012) Role of salicylic acid in amelioration of salt tolerance in potato (*Solanum tuberosum* L.) under *in vitro* conditions. Pak J Bot 44:37–42

Sakhanokho HF, Kelley RY (2009) Influence of salicylic acid on in vitro propagation and salt tolerance in *Hibiscus acetosella* and *Hibiscus moscheutos* (cv Luna Red). Afr J Biotechnol 8(8):1474–1481

Sangakkara UR, Hartwig UA, Nosberger J (1996) Soil moisture and potassium affect the performance of symbiotic nitrogen fixation in faba bean and common bean. Plant Soil 184:123–130

Sarwat M, Ahmad P, Nabi G, Hu X (2013) Ca^{2+} signals: the versatile decoders of environmental cues. Crit Rev Biotechnol 33(1):97–109

Scaramagli S, Biondi S, Leone A, Grub S, Torrigiani P (2000) Acclimation to low-water potential in potato cell suspension cultures leads to changes in putrescine metabolism. Plant Physiol Biochem 38:345–351

Schaberg PG, Minocha R, Long S, Halman JM, Hawley GJ, Eagar C (2011) Calcium addition at the Hubbard Brook Experimental Forest increases the capacity for stress tolerance and carbon capture in red spruce (*Picea rubens*) trees during the cold season. Trees 25:1053–1061

Schenk PM, Kazan K, Wilson I, Anderson JP, Richmond T, Somerville SC, Manners JM (2000) Coordinated plant defense responses in *Arabidopsis* revealed by microarray analysis. Proc Natl Acad Sci U S A 97:11655–11660

Seki M, Umezawa T, Urano K, Shinozaki K (2007) Regulatory metabolic networks in drought stress responses. Curr Opin Plant Biol 10:296–302

Sen Gupta A, Berkowitz GA, Pier PA (1989) Maintenance of photosynthesis at low leaf water potential in wheat, role of potassium status and irrigation history. Plant Physiol 89:1358–1365

Shabala S, Schimanski LJ, Koutoulis A (2002) Heterogeneity in bean leaf mesophyll tissue and ion flux profiles: leaf electrophysiological characteristics correlate with the anatomical structure. Ann Bot 89:221–226

Shakirova FM (2007) Role of hormonal system in the manifestation of growth promoting and anti-stress action of salicylic acid. In: Hayat S, Ahmad A (eds) Salicylic acid, a plant hormone. Springer, Dordrecht, p 69

Shakirova FM, Bezrukova MV (1997) Induction of wheat resistance against environmental salinization by salicylic acid. Biol Bull 24:109–112

Shakirova FM, Sakhabutdinova AR, Bezrukova MV, Fatkhutdinova RA, Fatkhutdinova DR (2003) Changes in the hormonal status of wheat seedlings induced by salicylic acid and salinity. Plant Sci 164:317–322

Shao HB, Chen XY, Chu LY, Zhao XN, Wu GY, Yong B (2006) Investigation on the relationship of proline with wheat anti drought under soil water deficits. Colloids Surf B Biointerfaces 53:113–119

Shao HB, Chu LY, Shao MA (2008a) Calcium as a versatile plant signal transducer under soil water stress. Bioessays 30:634–641

Shao HB, Song WY, Chu LY (2008b) Advances of calcium signals involved in plant anti-drought. C R Biol 331:587–596

Sharma YK, Leon J, Raskin I, Davis KR (1996) Ozone-induced responses in *Arabidopsis thaliana*, the role of salicylic acid in the accumulation of defense related transcripts and induced resistance. Proc Natl Acad Sci U S A 93:5099–5104

Sharp RE, Le Noble ME, Else MA, Thorne ET, Gherardi F (2000) Endogenous ABA maintains shoot growth in tomato independently of effects on plant water balance: evidence for an interaction with ethylene. J Exp Bot 51:1575–1584

Shi Q, Zhu Z (2008) Effects of exogenous salicylic acid on manganese toxicity, element contents and antioxidative system in cucumber. Environ Exp Bot 63:317–326

Shi Q, Bao Z, Zhu Z, Ying Q, Qian Q (2006) Effects of different treatments of salicylic acid on heat tolerance, chlorophyll fluorescence and antioxidant enzyme activity in seedlings of *Cucumis sativa* L. Plant Growth Regul 48:127–135

Shinozaki K, Yamaguchi-Shinozaki K (1997) Gene expression and signal transduction in water-stress response. Plant Physiol 115:327–334

Shou H, Bordallo P, Fan JB, Yeakley JM, Bibikova M, Sheen J, Wang K (2004a) Expression of an active tobacco mitogen-activated protein kinase enhances freezing tolerance in transgenic maize. Proc Natl Acad Sci U S A 101:3298–3303

Shou H, Bordallo P, Wang K (2004b) Expression of the *Nicotiana* protein kinase (NPK1) enhanced drought tolerance in transgenic maize. J Exp Bot 55:1013–1019

Signorelli S, Corpas FJ, Borsani O, Barroso JB, Monza J (2013) Water stress induces a differential and spatially distributed nitro-oxidative stress response in roots and leaves of *Lotus japonicus*. Plant Sci 201–202:137–146

Singh B, Usha K (2003) Salicylic acid induced physiological and biochemical changes in wheat seedlings under water stress. Plant Growth Regul 39(2):137–141

Slama I, Tayachi S, Jdey A, Rouached A, Abdelly C (2011) Differential response to water deficit stress in alfalfa (*Medicago sativa*) cultivars: growth, water relations, osmolyte accumulation and lipid peroxidation. Afr J Biotechnol 10(72):16250–16259

Slaymaker DH, Navarre DA, Clark D, Del-Pozo O, Martin GB, Klessig DF (2002) The tobacco salicylic acid binding protein 3 (SABP3) is the chloroplast carbonic anhydrase, which exhibits antioxidant activity and plays a role in the hypersensitive defense response. Proc Natl Acad Sci U S A 99:11640–11645

Smirnoff N (1993) The role of active oxygen in the response of plants to water deficit and desiccation. New Phytol 125:27–58

Spollen WG, Le Noble ME, Sammuels TD, Bernstein N, Sharp RE (2000) Abscisic acid accumulation maintains maize primary root elongation at low water potentials by restricting ethylene production. Plant Physiol 122:967–976

Subbarao GV, Wheeler RM, Levine LH, Stutte GW (2001) Glycine betaine accumulation, ionic and water relations of red-beet at contrasting levels of sodium supply. J Plant Physiol 158:767–776

Szabados L, Savoure A (2010) Proline: a multifunctional amino acid. Trends Plant Sci 15:89–97

Szalai G, Horgosi S, Soos V, Majlath I, Balazs E, Janda T (2010) Salicylic acid treatment of pea seeds induces its de novo synthesis. J Plant Physiol 168:213–219
Taiz L, Zeiger E (1998) Plant physiology, 3rd edn. Panima Publishing Corporation, New Delhi, pp 528–529
Takahashi S, Katagiri T, Yamaguchi-Shinozaki K, Shinozaki K (2000) An *Arabidopsis* gene encoding a Ca^{2+}-binding protein is induced by abscisic acid during dehydration. Plant Cell Physiol 41:898–903
Tan BC, Joseph LM, Deng WT, Liu LJ, Li QB (2003) Molecular characterization of the *Arabidopsis* 9-cis epoxycarotenoid dioxygenase gene family. Plant J 35:44–56
Tanaka Y, Sano T, Tamaoki M, Nakajima N, Kondo N, Hasezawa S (2005) Ethylene inhibits the abscisic acid-induced stomatal closure in *Arabidopsis*. Plant Physiol 138:2337–2343
Tari I, Csiszar J, Szalai G, Horvath F, Pecsvaradi A, Kiss G, Szepesi A, Szabo M, Erdei L (2002) Acclimation of tomato plants to salinity stress after a salicylic acid pre-treatment. Acta Biol Szeged 46:55–56
Tester M, Davenport R (2003) Na^+ tolerance and Na^+ transport in higher plants. Ann Bot 91:503–527
Thompson AJ, Jackson AC, Symonds RC, Mulholland BJ, Dadswell AR (2000) Ectopic expression of a tomato 9-cis-epoxycarotenoid dioxygenase gene causes over-production of abscisic acid. Plant J 23:363–374
Torres MA, Dangl JL, Jones JD (2002) *Arabidopsis* gp91phox homologues AtrbohD and trbohF are required for accumulation of reactive oxygen intermediates in the plant defense response. Proc Natl Acad Sci U S A 99:517–522
Umar S, Rao NR, Sekhon GS (1993) Differential effects of moisture stress and potassium levels on growth and K uptake in sorghum. Indian J Plant Physiol 36(2):94–97
Umezawa T, Nakashima K, Miyakawa T, Kuromori T, Tanokura M, Shinozaki K, Yamaguchi-Shinozaki K (2010) Molecular basis of the core regulatory network in ABA responses: sensing, signaling and transport. Plant Cell Physiol 51(11):1821–1839
Uno Y, Furihata T, Abe H, Yoshida R, Shinozaki K, Yamaguchi Shinozaki K (2000) *Arabidopsis* basic leucine zipper transcription factors involved in an abscisic acid-dependent signal transduction pathway under drought and high-salinity conditions. Proc Natl Acad Sci U S A 97:11632–11637
Urao T, Katagiri T, Mizoguchi T, Yamaguchi-Shinozaki K, Hayashida N, Shinozaki K (1994) Two genes that encode Ca^{2+} dependent protein kinases are induced by drought and high-salt stresses in *Arabidopsis thaliana*. Mol Gen Genet 244:331–340
Valero D, Perez-Vicente A, Martinez-Romero D, Castillo S, Guillen F, Plum SM (2002) Storability improved after calcium and heat postharvest treatments: role of polyamines. J Food Sci 67:2571–2575
Verbruggen N, Hermans C (2008) Proline accumulation in plants: a review. Amino Acids 35:753–759
Verma S, Mishra SN (2005) Putrescine alleviation of growth in salt stressed *Brassica juncea* by inducing antioxidative defense system. J Plant Physiol 162:669–677
Walker CR, Black JA, Miller J (1998) The role of cytosolic potassium and pH in the growth of barley roots. Plant Physiol 118:957–964
Wang Y, Nil N (2000) Changes in chlorophyll, ribulose biphosphate carboxylase-oxygenase, glycine betaine content, photosynthesis and transpiration in *Amaranthus tricolor* leaves during salt stress. J Hortic Sci Biotechnol 75:623–627
Wang Y, Wu WH (2010) Plant sensing and signaling in response to K^+ deficiency. Mol Plant 3:280–287
Wang KLC, Li H, Ecker JR (2002) Ethylene biosynthesis and signaling networks. Plant Cell 14:131–151
Wang X, Shi G, Xu Q, Hu J (2007) Exogenous polyamines enhance copper tolerance of *Nymphoides peltatum*. J Plant Physiol 164:1062–1070
Wang GP, Li F, Zhang J, Zhao MR, Hui Z, Wang W (2010) Over accumulation of glycine betaine enhances tolerance of the photosynthetic apparatus to drought and heat stress in wheat. Photosynthetica 48:30–41

Wang M, Zheng Q, Shen Q, Guo S (2013) The critical role of potassium in plant stress response. Int J Mol Sci 14(4):7370–7390
Waseem M, Athar H, Ashraf M (2006) Effect of salicylic acid applied through rooting medium on drought tolerance of wheat. Pak J Bot 38(4):1127–1136
Wen XP, Pang XM, Matsuda N, Kita M, Inoue H, Hao YJ (2008) Overexpression of the apple spermidine synthase gene in pear confers multiple abiotic stress tolerance by altering polyamine titers. Transgenic Res 17:251–263
White PJ, Broadley MR (2003) Calcium in plants. Ann Bot 92:487–511
White PJ, Karley AJ (2010) Potassium. In: Hell R, Mendel RR (eds) Cell biology of metals and nutrients, plant cell monographs, vol 17. Springer, Berlin, pp 199–224
Wi SJ, Kim WT, Park KY (2006) Overexpression of carnation S-adenosylmethionine decarboxylase gene generates a broad-spectrum tolerance to abiotic stresses in transgenic tobacco plants. Plant Cell Rep 25:1111–1121
Wild A (2003) Soils, land and food: managing the land during the twenty-first century. Cambridge University Press, Cambridge, UK
Wilkinson S, Corlett JE, Oger L, Davies WJ (1998) Effects of xylem pH on transpiration from wild-type and flacca tomato leaves. A vital role for abscisic acid in preventing excessive water loss even from well watered plants. Plant Physiol 117:703–710
Wilkinson S, Clephan AL, Davies WJ (2001) Rapid low temperature-induced stomatal closure occurs in cold-tolerant *Commelina communis* leaves but not in cold-sensitive tobacco leaves, via a mechanism that involves apoplastic calcium but not abscisic acid. Plant Physiol 126:1566–1578
Xia J, Liu MY, Jia SF (2005) Water security problem in North China: research and perspective. Pedosphere J 15:563–575
Xiang Y, Huang Y, Xiong L (2007) Characterization of stress responsive CIPK genes in rice for stress tolerance improvement. Plant Physiol 144:1416–1428
Xiong L, Yang Y (2003) Disease resistance and abiotic stress tolerance in rice are inversely modulated by an abscisic acid inducible mitogen activated protein kinase. Plant Cell 15:745–759
Xiong L, Lee H, Ishitani M, Zhu JK (2002) Regulation of osmotic stress-responsive gene expression by the LOS6/ABA1 locus in *Arabidopsis*. J Biol Chem 277:8588–8596
Xu J, Tian YS, Peng RH, Xiong AS, Zhu B, Jin XF, Gao F, Fu XY, Hou XL, Yao QH (2010) AtCPK6, a functionally redundant and positive regulator involved in salt/drought stress tolerance in *Arabidopsis*. Planta 231:1251–1260
Yamada M, Morishita H, Urano K, Shiozaki N, Yamaguchi-Shinozaki K, Shinozaki K, Yoshiba Y (2005) Effects of free proline accumulation in petunias under drought stress. J Exp Bot 56: 1975–1981
Yamaguchi K, Takahashi Y, Berberich T, Imai A, Takahashi T, Michael AJ, Kusano TA (2007) Protective role for the polyamine spermine against drought stress in *Arabidopsis*. Biochem Biophys Res Commun 352:486–490
Yamaguchi-Shinozaki K, Shinozaki K (2006) Transcriptional regulatory networks in cellular responses and tolerance to dehydration and cold stresses. Annu Rev Plant Biol 57:781–803
Yang X, Lu C (2005) Photosynthesis is improved by exogenous glycine betaine in salt stressed maize. Physiol Plant 124:343–352
Yang WJ, Rich PJ, Axtell JD, Wood KV, Bonham CC, Ejeta G, Mickelbart MV, Rhodes D (2003) Genotypic variation for glycine betaine in sorghum. Crop Sci 43:162–169
Yang Z, Tian L, Latoszek-Green M, Brown D, Wu K (2005) *Arabidopsis* ERF4 is a transcriptional repressor capable of modulating ethylene and abscisic acid responses. Plant Mol Biol 58:585–596
Yiu JC, Juang LD, Fang DYT, Liu CW, Wu SJ (2009) Exogenous putrescine reduces flooding induced oxidative damage by increasing the antioxidant properties of Welsh onion. Sci Hortic 120:306–314
Yokoi S, Quintero FJ, Cubero B, Ruiz MT, Bressan RA, Hasegawa PM, Pardo JM (2002) Differential expression and function of *Arabidopsis thaliana* NHX $Na^+/H+$ antiporters in the salt stress response. Plant J 30:529–539

Yoo JH, Park CY, Kim JC (2005) Direct interaction of a divergent CaM isoform and the transcription factor, MYB2, enhances salt tolerance in *Arabidopsis*. J Biol Chem 280:3697–3706
Yordanov I, Velikova V, Tsone V (2000) Plant response to drought, acclimatation and stress tolerance. Photosynthetica 30:171–186
Yusuf M, Hasan SA, Ali B, Hayat S, Fariduddin Q, Ahmad A (2008) Effect of salicylic acid on salinity-induced changes in *Brassica juncea*. J Integr Plant Biol 50(9):1096–1102
Zarembinski TI, Theologis A (1994) Ethylene biosynthesis and action: a case of conservation. Plant Mol Biol 26:1579–1597
Zeid IM, Shedeed ZA (2006) Response of alfalfa to putrescine treatment under drought stress. Biol Plant 50:635–640
Zhang J, Davies WJ (1989a) Abscisic acid produced in dehydrating roots may enable the plant to measure the water status of the soil. Plant Cell Environ 12:73–81
Zhang J, Davies WJ (1989b) Sequential responses of whole plant water relations towards prolonged soil drying and the mediation by xylem sap ABA concentrations in the regulation of stomatal behaviour of sunflower plants. New Phytol 113:167–174
Zhang J, Davies WJ (1990a) Changes in the concentration of ABA in xylem sap as a function of changing soil water status will account for changes in leaf conductance. Plant Cell Environ 13:277–285
Zhang J, Davies WJ (1990b) Does ABA in the xylem control the rate of leaf growth in soil-dried maize and sunflower plants. J Exp Bot 41:1125–1132
Zhang J, Schurr U, Davies WJ (1987) Control of stomatal behaviour by abscisic acid which apparently originates in roots. J Exp Bot 38:1174–1181
Zhang J, Jia W, Yang J, Ismail AM (2006) Role of ABA in integrating plant responses to drought and salt stresses. Field Crop Res 97:111–119
Zhang G, Chen M, Li L, Xu Z, Chen X, Guo J, Ma Y (2009) Overexpression of the soybean GmERF3 gene, an AP2/ERF type transcription factor for increased tolerances to salt, drought, and diseases in transgenic tobacco. J Exp Bot 60:3781–3796
Zhang H, Liu W, Wan L, Li F, Dai L, Li D, Zhang Z, Huang R (2010) Functional analyses of ethylene response factor JERF3 with the aim of improving tolerance to drought and osmotic stress in transgenic rice. Transgenic Res 19:809–818
Zhang L, Xi D, Li S, Gao Z, Zhao S, Shi J, Wu C, Guo X (2011) A cotton group C MAP kinase gene, GhMPK2, positively regulates salt and drought tolerance in tobacco. Plant Mol Biol 77:17–31
Zhao H, Yang H (2008) Exogenous polyamines alleviate the lipid peroxidation induced by cadmium chloride stress in *Malus hupehensis* Rehd. Sci Hortic 116:442–447
Zhifang G, Loescher WH (2003) Expression of a celery mannose 6-phosphate reductase in *Arabidopsis thaliana* enhances salt tolerance and induces biosynthesis of both mannitol and a glucosyl-mannitol dimmer. Plant Cell Environ 26:275–283
Zhu JK (2002) Salt and drought stress signal transduction in plants. Annu Rev Plant Physiol Plant Mol Biol 53:247–273
Zlatev Z, Stotanov Z (2005) Effect of water stress on leaf water relations of young bean plants. J Cent Eur Agric 6(1):5–14
Zwart FJ, Slow S, Payne RJ, Lever M, George PM, Gerrard JA, Chambers ST (2003) Glycine betaine and glycine betaine analogues in common foods. Food Chem 83:197–204

Chapter 3
Influencing the Product Quality by Applying Drought Stress During the Cultivation of Medicinal Plants

Maik Kleinwächter and Dirk Selmar

Contents

1 Introduction

It is a matter of common knowledge that spice plants grown under Mediterranean or semi-arid conditions, generally are much more aromatic than identical plants of the same species, which however are cultivated in moderate climates. Obviously, under semi-arid conditions the concentrations of aroma relevant natural products are enhanced. Analogous quality differences are observed with regard to medicinal plants, i.e. the content of the corresponding relevant secondary plant products in general is less in plants grown in a moderate Atlantic climate than in those cultivated in semi-arid regions. Commonly, this phenomenon is explained by the unsophisticated statement that plants grown in Southern Europe "are exposed to much more sunlight, resulting in enhanced rates of biosynthesis". Albeit—on the first sight—such assertion appears to be consistent, it has to be considered that sunlight

M. Kleinwächter • D. Selmar (✉)
Institute for Plant Biology, Technische Universität Braunschweig Braunschweig, Mendelssohnstraße 4, 38106 Braunschweig, Germany
e-mail: d.selmar@tu-bs.de

P. Ahmad and M.R. Wani (eds.), *Physiological Mechanisms and Adaptation Strategies in Plants Under Changing Environment: Volume 1*, DOI 10.1007/978-1-4614-8591-9_3,

is not at all a limiting factor for plant growth. Accordingly, even in Central Europe most plants, which grow in open areas without any shading, absorb much more light energy than the plants require and utilize for photosynthetic CO_2-fixation (Wilhelm and Selmar 2011). Notwithstanding, we have to factor that—at least in the subtropics—high irradiation often is co-occurring with water deficiencies. As a consequence, under semi-arid conditions, plants frequently suffer drought stress. Since stress-related metabolism extensively impacts all other metabolic events, the synthesis and accumulation of secondary metabolites also should be affected. Unfortunately, in the past, these coherences have not been considered adequately (Selmar 2008). Just recently, Selmar and Kleinwächter (2013) compiled the relevant literature in order to get a clearer picture of this issue. These authors, for the first time, outlined the metabolic background for the stress-related enhancement of natural product synthesis (Fig. 3.1): Due to water shortage, in combination with high light intensities, stomata are closed. As a result, the uptake of CO_2 is markedly decreased. In consequence, the consumption of reduction equivalents ($NADPH+H^+$) for CO_2-fixation via Calvin cycle declines considerably, generating a massive oversupply of $NADPH+H^+$. Accordingly, all metabolic processes are pushed towards the synthesis of highly reduced compounds like isoprenoids, phenols or alkaloids.

Based on these coherences, impulses for new practical approaches for enhancing the product quality by intentionally applying drought stress during the cultivation of medicinal plants are given. However, as drought stress concurrently leads to massive reductions in biomass production, special emphasis is put on the interference of these both stress-related effects.

2 Enhanced Synthesis and Accumulation of Natural Products in Drought Stressed Plants

The synthesis and accumulation of secondary plant products strongly depend on the growing conditions, e.g. the temperature, the light regime and the nutrient supply (Gershenzon 1984; Falk et al. 2007). Accordingly, much more severe environmental influences (various stress situations), which drastically influence the general metabolism (Bohnert et al. 1995), also will impact on the metabolic pathways responsible for the accumulation of secondary plant products. Indeed, there is a tremendous lot of information dealing with the impact of biological stress (e.g. pathogen or herbivore attack) on the synthesis of secondary plant products (Harborne 1988; Hartmann 2007; Wink 2010). In contrast, corresponding information, how abiotic stress changes the secondary metabolism is rare. Especially, the knowledge on the related biological background is limited (Ramakrishna and Ravishankar 2011; Selmar and Kleinwächter 2013). As the host–pathogen and host–herbivore interactions in general are quite assessable, the situation related to abiotic stress frequently seems to be more complex, since various interferences between numerous factors might co-occur, e.g. the increase in light intensity mostly correlates with elevated temperatures, or, lower water availability, inducing drought stress, often entails higher salt

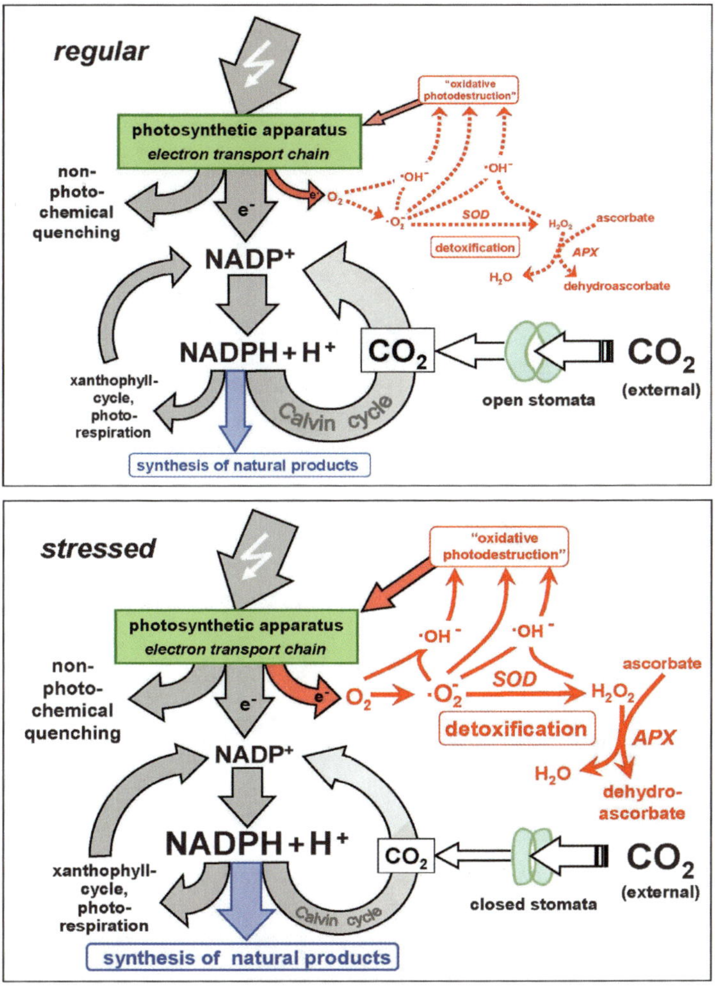

Fig. 3.1 Model scheme for the drought stress-related increase of natural product biosynthesis according to Selmar and Kleinwächter (2013)

concentrations in the soil. Consequently, the results of numerous studies on the impact of a certain abiotic stress on the secondary metabolism lack the distinction to other putative stress factors. Nevertheless, thorough reviewing of literature frequently allows the decisive deductions about the impact of a single factor on the accumulation of natural products. This chapter focuses on drought stress.

Numerous studies revealed that plants exposed to drought stress indeed accumulate higher concentrations of secondary metabolites than those cultivated under well-watered conditions (Table 3.1). Obviously, the drought stress-related concentration increase is a common feature for all different classes of natural products. Corresponding enhancements are reported to occur in the case of simple as well as

Table 3.1 Drought stress-related concentration increase of natural products

Simple phenols			
Helianthus annuus	Chlorogenic acid	Massive increase (tenfold)	del Moral (1972)
Prunus persica	Total phenols	Higher contents	Kubota et al. (1988)
Thymus capitatus	Phenolics	Higher contents	Delitala et al. (1986)
Echinacea purpurea	Total phenols	Strong increase (67 %)	Gray et al. (2003)
Crataegus spp.	Chlorogenic acid	Massive increase (2–6-fold)	Kirakosyan et al. (2004)
Hypericum brasiliense	Total phenols	Strong increase (over 80 %)	de Abreu and Mazzafera (2005)
Trachyspermum ammi	Total phenols	Strong increase (100 %)	Azhar et al. (2011)
Labisia pumila	Total phenols	Significant increase (50 %)	Jaafar et al. (2012)
Complex phenols			
Pisum sativum	Flavonoids	Strong increase (45 %)	Nogués et al. (1998)
Pisum sativum	Anthocyanins	Strong increase (over 80 %)	Nogués et al. (1998)
Crataegus spp.	Catechins/epicatechins	Massive increase (2–12-fold)	Kirakosyan et al. (2004)
Hypericum brasiliense	Rutine/quercetin	Massive increase (fourfold)	de Abreu and Mazzafera (2005)
Hypericum brasiliense	Xanthones	Strong increase (over 80 %)	de Abreu and Mazzafera (2005)
Camellia sinensis	Epicathecins	Higher contents	Hernández et al. (2006)
Salvia miltiorrhiza	Furoquinones	Significant increase	Liu et al. (2011)
Prunella vulgaris	Rosmarinic acid	Slight increase	Chen et al. (2011)
Labisia pumila	Anthocyane/flavonoids	Significant increase	Jaafar et al. (2012)
Monoterpenes/essential oils			
Mentha × piperita ssp.	Essential oils	Significant increase	Charles et al. (1990)
Cymbopogon pendulus	Geraniol and citral	Strong increase	Singh-Sangwan et al. (1994)
Pinus halepensis	α-Pinen, carene	Strong increase	Llusià and Peñuelas (1998)
Cistus monospeliensis	Caryophyllene	Enormous increase	Llusià and Peñuelas (1998)
Satureja hortensis	Essential oils	Increase	Baher et al. (2002)
Picea abies	Monoterpenes	Strong increase	Turtola et al. (2003)
Pinus silvestris	Monoterpenes	Strong increase	Turtola et al. (2003)
Petroselinum crispum	Essential oils	Strong increase (double)	Petropoulos et al. (2008)

Salvia officinalis	Essential oils	Massive increase (2–4-fold)	Bettaieb et al. (2009)
Salvia officinalis	Monoterpenes	Strong increase	Nowak et al. (2010)
Scrophularia ningpoen	Iridoid glycosides	Increase	Wang et al. (2010)
Nepeta cataria	Essential oils	Significant increase	Manukyan (2011)
Ocimum basilicum	Essential oils	Significant increase	Forouzandeh et al. (2012)
Di- and triterpenes			
Solanum tuberosum	Steroid alkaloids	Strong increase	Bejarano et al. (2000)
Hypericum brasiliense	Betulinic acid	Strong increase	de Abreu and Mazzafera (2005)
Bupleurum chinense	Saikosaponin	Significant increase	Zhu et al. (2009)
Prunella vulgaris	Triterpenes	Slight increase	Chen et al. (2011)
Alkaloids			
Senecio longilobus	Pyrrolizidine alkaloids	Strong increase	Briske and Camp (1982)
Lupinus angustifolius	Quinolizidin alkaloids	Strong increase	Christiansen et al. (1997)
Solanum tuberosum	Steroid alkaloids	Strong increase	Bejarano et al. (2000)
Glycine max	Trigonelline	Higher contents	Cho et al. (2003)
Papaver somniferum	Morphine alkaloids	Strong increase	Szabó et al. (2003)
Catharanthus roseus	Indole alkaloids	Strong increase (with Ca^{2+})	Jaleel et al. (2007)
Phellodend amurense	Benzylisoquinolines	Strong increase	Xia et al. (2007)
Senecio jacobaea	Pyrrolizidine alkaloids	Massive increase	Kirk et al. (2010)
Nicotiana tabacum	Nicotiana-alkaloids	Strong increase	Çakir and Çebi (2010)
Various classes			
Manihot esculenta	Cyanogenic glucosides	Strong increase	de Bruijn (1973)
Triglochin maritima	Cyanogenic glucosides	Strong increase	Majak et al. (1980)
Brassica napus	Glucosinolates	Massive increase	Jensen et al. (1996)
Coffea arabica	γ-Aminobutyric acid	Massive increase (tenfold)	Bytof et al. (2005)
Brassica oleracea	Glucosinolates	Significant increase	Radovich et al. (2005)
Brassica carinata	Glucosinolates	Significant increase	Schreiner et al. (2009)
Phaseolus lunatus	Cyanogenic glucosides	Higher content in stressed plants	Ballhorn et al. (2011)

complex phenols and also for the various classes of terpenes (Table 3.1). In the same manner, also nitrogen containing substances, such as alkaloids, cyanogenic glucosides, or glucosinolates, are influenced by drought stress (Table 3.1). Thus, there is no doubt that drought stress frequently enhances the concentration of secondary plant products. However, it has to be taken into consideration that drought stressed plants generally are reduced in their growth. Accordingly, due to the reduction in biomass—even without any increase in the overall amount of natural products—their concentration on dry or fresh weight basis simply is enhanced. Corresponding explanations frequently are reported in the literature. Unfortunately, in most of the studies published, data on the overall biomass of the plants analysed are lacking. One reason for this lack of information is due to the fact that mostly certain plant parts or organs, i.e. roots, leaves or seeds, were only in the centre of focus and the overall content of natural products on whole plant basis was of not much interest. Yet, in some papers, the total contents of secondary plant products are given or could be calculated from the data presented.

In *Hypericum brasiliense*, not only the concentration but also the total content of the phenolic compounds is drastically enhanced under drought stress in comparison to the control plants (de Abreu and Mazzafera 2005). Although the stressed *H. brasiliense* plants had been quite smaller, the product of biomass and concentration of the related phenolics yields in a 10 % higher total amount of these compounds. In the same manner, in stressed peas (*Pisum sativum*), the massive increase in the concentration of phenolic compounds reported by Nogués et al. (1998) resulted in a 25 % higher overall amount of anthocyanins (product of biomass and anthocyanin concentration), despite the fact that the total biomass of the pea plants grown under drought stress is just about one-third of those cultivated under standard conditions. Also Jaafar et al. (2012) reported that not only the concentration but also the overall production of total phenolics and flavonoids per plant is enhanced in plants suffering from drought stress, though the explicit data on biomass per plant are not displayed by the authors. In contrast, the overall yield of flavonoids was nearly the same, when the plants were either grown under drought stress or under well-watered, non-stress conditions. The impact of the drought-related reduction of biomass on the overall content of secondary plant products is even more distinct in red sage (*Salvia miltiorrhiza*): in stressed plants, the overall content of furoquinones is slightly lower than in well-watered controls, although drought stress caused a significant increase of their concentration (Liu et al. 2011).

With regard to terpenoids, there are also some reports documenting an increase in the overall content of these natural products, i.e. the total amount per plant. In sage (*Salvia officinalis*), drought stress induces a massive increase of monoterpenes, which overcompensates the reduction in biomass (Nowak et al. 2010). Accordingly, the entire amount of monoterpenes synthesized in sage is significantly higher in plants suffering moderate drought stress as compared to well-watered controls. In contrast, the slight drought stress-related increase in the concentrations of monoterpenes in catmint and lemon balm plants could not compensate the stress-related detriment of growth. Accordingly, the overall content of terpenoids is lower in the drought stressed plants of *Melissa officinalis*, *Nepeta cataria* and *Salvia officinalis* than in the corresponding controls (Manukyan 2011). In parsley leaves (*Petroselinum crispum*), the drought stress-related concentration enhancement of essential oils is more or less completely compensated by the related loss in biomass, resulting in almost the same overall contents of essential oils in drought stressed and well-watered plants (Petropoulos et al. 2008). In most of the reports dealing with the impact of drought stress on nitrogen containing natural products, the concentration of these compounds in certain organs is in the centre of focus and data on the biomass of the entire plant are not available. Thus, no deductions on the impact of drought on the overall content of natural products could be drawn. Only the stress-related influence on benzylisoquinoline alkaloids in cork tree seedlings (*Phellodendron amurense*) is documented on concentration as well as on total content basis (Xia et al. 2007), whereas the concentration of berberine, jatrorrhizine and palmatine is strongly enhanced by drought stress, their overall content is considerably reduced due to the fact that the biomass of the drought stressed plants accounted only for about one fourth of the control. As cardinal assertion, it could be stated that in nearly all plants analysed, the concentrations of secondary plant products are significantly enhanced under drought stress conditions. Yet, only in few cases, apart from the stress-induced concentration increase, an enhancement of the total content of corresponding natural compounds is reported. This could be either due to the lack of data on the biomass of the corresponding plants, or to the fact that the stress-related decrease in biomass generally overcompensates the increase in the concentration of relevant natural products.

3 Metabolic Background: Higher Reduction Capacity in Drought Stressed Plants

Due to our recurring experience in daily life, we all are aware that energy saving represents one of the most important issues in our subsistence, and corresponding statements have become fundamental. Indeed, on the first sight, it seems reasonable to transfer these considerations also into plant biology. Accordingly, even in reputable textbooks and in scientific publications, corresponding claims and statements can frequently be found. Although we know that light energy, in general, is not the limiting factor of photosynthesis in plants (Wilhelm and Selmar 2011), it is commonly stated: "In order to save energy, plants have evolved a certain mechanism..." or "Due to cost-benefit equations, the energy costs for a certain metabolic process must be minimized". But looked at more closely, it is obvious that in contrast to heterotrophic organisms, quite other cardinal coherences are crucial for autotrophic plants. In general, plants absorb much more energy than that required for photosynthetic CO_2-fixation. This easily can be deduced from the massive enhancement of photosynthesis, when CO_2 concentration is elevated (Fig. 3.2). Under regular environmental conditions, the surplus of energy is dissipated by various mechanisms, i.e. non-photochemical quenching, photorespiration or xanthophyll cycle (Fig. 3.1). However, drought stress-induced stomata closure diminishes the CO_2-influx. As a

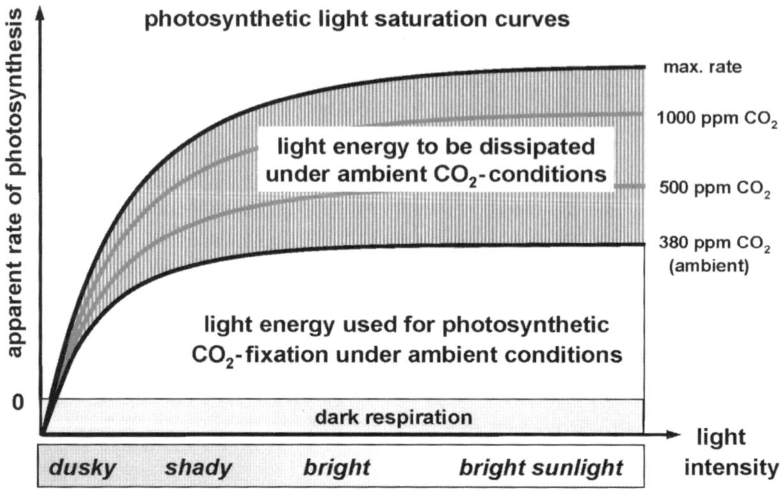

Fig. 3.2 The increase of apparent photosynthesis due to an enhancement of CO_2 concentration illustrates the massive surplus of energy which is absorbed by plants under ambient conditions (Wilhelm and Selmar 2011)

consequence, far less reduction equivalents are consumed (re-oxidized) and although the energy dissipating mechanisms are augmented, the chloroplastic reduction status increases. Apart from the generation of superoxide radicals, which generally are detoxified by superoxide dismutase (SOD) and ascorbate peroxidase (APX), especially the ratio of $NADPH+H^+$ to $NADP^+$ is enhanced. Accordingly, all reactions consuming $NADPH+H^+$, such as the biosynthesis of highly reduced secondary plant products, will be favoured without the need for any change in enzyme activity (Fig. 3.1).

The stress-related increase in the biosynthesis rate of highly reduced natural products consumes large amounts of NADPH+ H^+ and thereby decreases the over-reduced state. With respect to the strong isoprene emission of numerous plants, it was postulated that the energy and reduction equivalents required for the biosynthesis of the isoprene emitted by leaves indeed contribute significantly to the dissipation of the excess of photosynthetic energy (Fall 1999; Sharkey and Yeh 2001). Magel et al. (2006) calculated that under standard conditions the energy consumption for this isoprene biosynthesis accounts for less than 1 %. However, at elevated temperatures, the amount of energy dissipated by the strongly enhanced isoprene emission might rise up to 25 % of the energy supply for net photosynthesis. Such considerations show that the biosynthesis of natural products indeed may represent a relevant system to dissipate a surplus of energy. Accordingly, secondary metabolites, apart from their ecological significance, also could be crucial as a part of the supplemental energy dissipation machinery (Grace and Logan 2000; Wilhelm and Selmar 2011). Such relevance, however, could imply that the corresponding processes should not only be enhanced passively by shifts in the concentration of reduction equivalents, but also be accelerated actively by increasing the corresponding catalytic abilities, i.e. the activities of the enzymes involved in the biosynthesis of the natural products.

4 Interactions with Other Factors

As outlined above, the concentrations of natural products are generally enhanced in plants suffering drought stress. In principle, there are two underlying explanations, either a stress-related change in the benchmark, i.e. the dry or fresh weight used as reference value or a real increase in biosynthesis. In the first case, the drought stress-related decline in biomass production is associated with a more or less unchanged rate of biosynthesis of natural products resulting in an increase of concentration on dry or fresh weight base. In the second case, stress results in an authentic increase of the total content of secondary plant products by an enhanced biosynthesis, which putatively is due to the over-reduction in drought stressed plants, favouring the biosynthesis of highly reduced compounds. However, these simplified causal coherences are much more intransparent, since both issues frequently are overlaid and interfere with numerous factors and side effects. Secondary plant products reveal a high significance for the plants by accomplishing various ecological functions within the complex interactions of plants with their environment, e.g. to repel herbivores, to protect against pathogens or to attract pollinators (Harborne 1988; Hartmann 2007; Wink 2010). Also with respect to abiotic stress, various putative functions for secondary plant products are known, e.g. protection against UV-light or too high light intensities, action as compatible solutes, radical scavenging or reduction of the transpiration (Edreva et al. 2008; Wink 2010). Due to the tremendous progress in molecular biology, we have learned that the synthesis of the corresponding secondary metabolites frequently is induced, modulated and regulated by numerous environmental impacts and abiotic factors, respectively. The situation becomes even more complex, if we consider that phytoalexins are elicited by pathogen attack (Hahlbrock et al. 2003; Saunders and O'Neill 2004) and relevant defense compounds against herbivores are synthesized as result of very complex induction mechanisms (Ferry et al. 2004). As a consequence, the actual synthesis and accumulation of certain natural products are influenced and determined by numerous factors. Moreover, we have to consider that a particular stress situation in general, influences several factors, e.g. a high irradiation is frequently accompanied with elevated temperatures, high irradiation often parallels with UV-radiation and elevated temperatures co-occur with higher evaporation rates. As drought impacts also on the entire ecosystem, it might be associated with a higher herbivore pressure, but a lesser number of pathogens. Consequently, we have to be aware that complex interferences of numerous factors occur, influencing additionally the metabolism of plants exposed to drought stress, and thus, the synthesis and accumulation of natural products. In order to elucidate the corresponding complex metabolic syndrome, suitable and reliable markers are required.

Thorough scientific investigation of drought stress requires the explicit differentiation between the corresponding effects on the osmotic potential, i.e. the water availability within the cell, and the shift in redox potential due to the decrease in CO_2 influx caused by stomata closure. With respect to the decrease in water availability, upon first sight, water potential seems to be an appropriate parameter. However, in response to drought, many plants produce and accumulate osmotic active substances denoted as compatible solutes, which significantly reduce the water potential without changing the actual water content. Thus, the actual water content seems to be a better option to describe the impact of drought stress on water household. Apart from classical gravimetrical methods, just recently a new methodology based on terahertz technology was presented to determine the actual water content (Breitenstein et al. 2011).

Concerning the stress-induced shift in redox potential, the most appropriate markers would either be the enhanced ratio of NADPH + H^+ to $NADP^+$ or the amount of oxygen radicals generated. Unfortunately, the real in situ concentration of both the components could not be quantified due to inappropriate efforts and expenditures. Alternatively, the enzymes responsible for detoxification of toxic reactive oxygen species (ROS) generated, the superoxide dismutases (SODs) and the ascorbate peroxidaes (APXs) frequently are estimated as characteristic stress-related enzymes. Yet, these enzymes occur in various isoforms and are also part of various signal transduction chains. Thus, they do not indicate reliably a specific stress situation. Additionally, glutathione, which also is part of the antioxidative defence against ROS, was used as stress marker. However, the interrelation between the glutathione concentrations and the redox states is not consistently presented in the large number of publications available (Tausz et al. 2004). Moreover, the stress-induced responses of the glutathione system are multilayered and biphasic: an initial response phase is followed by an acclimation phase. Accordingly, the use of the glutathione system as general stress marker for routine analysis is limited (Tausz et al. 2004). As alternative, the occurrence of characteristic stress-metabolites, which are synthesized and accumulated more or less specifically in response to a particular stress situation, could be estimated to determine and quantify its metabolic impact. In this context, proline is in the centre of focus, since this amino acid is accumulated as compatible solute in plants suffering drought stress (Rhodes et al. 1999). Unfortunately, the drought stress-induced proline accumulation does not occur in all plant species. Another option is the quantification of γ-amino butyric acid (GABA), a stress metabolite produced by decarboxylation of glutamic acid (Kinnersley and Turano 2000). Indeed, GABA is produced to a high extent in response to drought stress, but it also is accumulated under various other stress conditions (Satya Narayan and Nair 1990; Bown and Shelp 1997). Consequently, further markers are required. In this context, the most promising complementation is the abundance of dehydrins. These small protective proteins were first discovered in maturing seeds in the course of late embryogenesis. Meanwhile, it is well known that dehydrins also are frequently synthesized in plant cells suffering drought stress (Close 1997; Allagulova et al. 2003; Bouché and Fromm 2004). It is assumed that these small hydrophilic proteins reveal various protective functions in desiccating cells (Hara 2010). Accordingly, the expression of dehydrins is well established to monitor the impact of drought stress on the metabolism. Yet, the complexity of such stress reactions in general was vividly demonstrated by Kramer et al. (2010), who found that the expression of dehydrins and the accumulation of the stress metabolite GABA follow different time patterns in coffee seeds whilst drying. This finding shows that several metabolic responses occur in parallel. In this context, we have to consider (as outlined above) that in leaves exposed to drought stress, apart from the impact of the decrease in water availability, the overreduction due to stomata closure entails numerous metabolic responses. Thus, the elucidation of the entire metabolic syndrome induced in drought stressed medicinal plants requires a combination of several markers. Apart from the accumulation of GABA and the expression of dehydrins, other markers, such as the status of the energy dissipation systems (e.g. the non-photochemical quenching, the xanthophyll cycle or the photorespiration) should be determined.

5 Putative Practical Applications

As mentioned above, the stress-related impacts on secondary metabolism are multifarious. Many mechanisms of elicitation and induction are overlaid, and numerous effects and processes are counteracting. Apart from the influence on secondary metabolism, above all, the consequences and aftereffects of drought stress on the general metabolism are most prominent and cause severe repercussions, such as losses in biomass, retards in development or changes in growth behaviour. As a result, the general metabolic status is altered and frequently the ratio between generative and vegetative characteristics is shifted (Houter and Nederhoff 2007). Accordingly, drought stress could change the *source-sink* properties of the entire plant and thereby—in addition to inducing factors already mentioned, it could impact on the overall performance of the biosynthesis, translocation and accumulation of secondary plant products, too. Thus, apart from the enhanced biosynthesis due to the stress-related over-reduction, stress also impacts on the allocation and accumulation of natural products due to putative changes of the physiological nature of certain organs or tissues. The overall effects of drought stress on secondary metabolism, and thus on the quality of medicinal plants are multilayered and very complex. Indeed, a deliberate application of drought stress during the cultivation of medicinal plants should principally result in an increase of biosynthesis of secondary metabolites. However, due to the other co-occurring impacts and the various effects mentioned above, this enhancement frequently is compensated or even overcompensated. Accordingly, a general recommendation for the deliberate application of drought stress to increase the quality of medicinal plants cannot be made. Nevertheless, in many cases, such approach undoubtedly will be successful. Yet, always the advantages and the drawbacks of a corresponding approach have to be sounded out. In order to facilitate corresponding assessments, Selmar and Kleinwächter (2013) have proposed to answer some simple questions:

1. What kind and which level of stress enhance the accumulation of the desired compounds without causing too high losses in biomass?
2. What is required, a high concentration or a large bulk? (total amount of natural products versus high concentrations in the drug)
3. Are the substances synthesized and accumulated in *source* or in *sink* tissues?
4. Are the substances synthesized and accumulated in generative or vegetative organs?
5. Could the accumulation also be increased by phytohormone treatments (e.g. methyl jasmonate, salicylic acid)?
6. Should the stress be applied only within a certain phase of cultivation or whilst a special developmental phase in order to obtain maximal quality?

The question arises how these objectives could be transferred into appropriate agricultural applications. As most simple approach, the irrigation regime could be altered (Radovich et al. 2005). This, however, in general is restricted to semi-arid regions, where supplemental watering is required. In contrast, in moderate climates, the water supply by rainfall cannot be influenced directly. Nevertheless, the moisture content of the soil could be altered by the choice of the cultivation area or by some simple measures. In this context, it has to be mentioned that the land form directly impacts the drainage properties, e.g. fields with slope will retain markedly less water than flat plains consisting of soils with the same water holding capacity. Even in plains, the drainage effect could be realized artificially by certain cultivation measures. The creation of a furrow and ridge system, which frequently is used for surface irrigation in arid regions, is also appropriate for generating drought stress situations by establishing significant gradients of soil moisture contents. Without any irrigation, the soil water content in the furrows will nearly be the same as in untreated fields, but the moisture content in the ridges strongly decreases due to an enhanced drainage effect. Alternatively, drainage could be achieved by increasing the proportion of sand in the soil. This, however, would irreversibly change the character of the soil and should only be applied in exceptional cases.

A further strategy to influence the product quality corresponds to the application of phytohormones or growth regulators in order to induce the stress-related signal transduction chain for secondary metabolite synthesis. In this context jasmonic acid is of special interest. However, not the active growth regulator is applied, but its volatile methyl ester. After uptake into the cells, it is hydrolysed, and the active jasmonic acid is generated. Such approaches had been successfully employed in numerous tissue and cell culture systems to enhance the concentration of secondary metabolites (Namdeo 2007). Meanwhile, methyl jasmonate was also used for effective elicitation of natural products in intact plants. In this manner, the synthesis of indole alkaloids was increased in seedlings of *Catharanthus roseus* and *Chinona ledgerina* (Aerts et al. 1994). The content of phenols and monoterpenes had been enhanced in *Ocimum basilicum* (Kim et al. 2006) and the concentration of glucosinolates was increased in *Brassica rapa* (Loivamäki et al. 2004). Due to these promising approaches the usage of methyl jasmonate for quality improvement of medicinal and spice plants seems to be an encouraging alternative for a direct drought treatment. In the same manner, salicylic acid, an endogenous key signal substance, responsible for systemic resistance (Durrant and Dong 2004) was also used for impacting on the synthesis and accumulation of secondary metabolites: the content of phenolics could strongly be increased by the application of salicylic acid to *Echinacea purpurea* plants (Kuzel et al. 2009) and the accumulation of glucosinolates in oilseed rape (*Brassica napus*) was markedly enhanced by a salicylic acid treatment (Kiddle et al. 1994). Apart from their effects on secondary metabolism, salicylic acid and methyl jasmonate also impact primary metabolism and developmental processes. In this context, the induction of senescence and retardation of growth are of special interest, when considering the application of growth regulators and phytohormones to increase the product quality. As outlined for the deliberate application of drought stress, also in the case of applying salicylic acid and methyl jasmonate, a thorough balancing of pros and cons is required. This of course necessitates a sound and comprehensive knowledge of the related scientific background.

Apart from the well-known circumstance that the quality of medicinal plants grown under semi-arid conditions is generally higher than that of equivalent plants, cultivated in moderate climates. There are some further examples from daily life that illustrate the relevance of stress induction or omission for the quality of plant-derived commodities. In order to achieve the highest quality of Japanese green teas, *Camellia sinensis* plants are grown under artificial shading. Accordingly, these plants suffer less photo-oxidative stress in comparison to plants, which are exposed to full sunlight. As outlined above, in these plants the synthesis of secondary plant products as result of an over-reduced status should be decreased. Indeed, corresponding analysis revealed that the content of monoterpenes as well as that of coumarin is significantly lower in the non-stressed plants (Shimoda et al. 1995). In the same manner shading, which reduces the stress reactions, impacts the quality of cured tobacco leaves. Tobaccos generally grown in Southern and Middle-America in full sunlight reveal the typical dark colours of the so-called Brazil quality. Shading of the *Nicotiana tabacum* plants, even when grown in Brazil, yields in much brighter hues, comparable to those tobacco leaves grown in Sumatra, where the solar irradiance is much lower due to the foggy climate. Phytochemical analysis revealed that the concentrations of total phenols and alkaloids indeed are lower in tobacco leaves grown in shade (Andersen et al. 1985), and thus, confirming once again the interrelation between oxidative stress and synthesis of secondary plant products.

6 Conclusion and Future Perspective

As conclusion of our deliberations, it can be stated that the generation of drought stress situations, e.g. directly by water shortage, or indirectly via the application of growth regulators, indeed could lead to an enhancement of the contents of secondary plant products and thereby increasing the quality of plant-derived drugs. However, due to numerous interactions, the related increase could be compensated by other metabolic responses, such as growth reduction, decrease in biomass production or an onset of senescence. Accordingly, for each particular case, a thorough balancing of pros and cons is necessary. For this, however, much more research is required in order to elucidate comprehensively the entire issue.

References

Aerts RJ, Gisi D, de Carolis E, de Luca V, Baumann TW (1994) Methyl jasmonate vapor increases the developmentally controlled synthesis of alkaloids in *Catharanthus* and *Cinchona* seedlings. Plant J 5(5):635–643

Allagulova CR, Gimalov FR, Shakirova FM, Vakhitov VA (2003) The plant dehydrins: structure and putative functions. Biochemistry 68:945–951

Andersen RA, Kasperbauer MJ, Burton HR (1985) Shade during growth—effects on chemical composition and leaf color of air-cured burley tobacco. Agron J 77:543–546

Azhar N, Hussain B, Ashraf YM, Abbasi KY (2011) Water stress mediated changes in growth, physiology and secondary metabolites of desi ajwain (*Trachyspermum ammi*). Pak J Bot 43(SI):15–19

Baher ZF, Mirza M, Ghorbanli M, Rezaii MB (2002) The influence of water stress on plant height, herbal and essential oil yield and composition in *Satureja hortensis* L. Flav Frag J 17: 275–277

Ballhorn DJ, Kautz S, Jensen M, Schmitt S, Heil M, Hegeman AD (2011) Genetic and environmental interactions determine plant defences against herbivores. J Ecol 99:313–326

Bejarano L, Mignolet E, Devaux A, Espinola N, Carrasco E, Larondelle Y (2000) Glycoalkaloids in potato tubers: the effect of variety and drought stress on the α-solanine and α-chaconine contents of potatoes. J Sci Food Agric 80:2096–2100

Bettaieb I, Zakhama N, Aidi Wannes W, Kchouk ME, Marzouk B (2009) Water deficit effects on *Salvia officinalis* fatty acids and essential oils composition. Sci Hortic 120(2):271–275

Bohnert HJ, Nelson DE, Jensen RG (1995) Adaptations to environmental stresses. Plant Cell 7: 1099–1111

Bouché N, Fromm H (2004) GABA in plants: just a metabolite. Trends Plant Sci 9:100–115

Bown AW, Shelp BJ (1997) The metabolism and functions of γ-aminobutyric acid. Plant Physiol 115:1–5

Breitenstein B, Scheller M, Shakfa MK, Kinder T, Müller-Wirts T, Koch M, Selmar D (2011) Introducing terahertz technology into plant biology: a novel method to monitor changes in leaf water status. J Appl Bot Food Qual 82(2):158–161

Briske DD, Camp BJ (1982) Water stress increases alkaloid concentrations in threadleaf groundsel (*Senecio longilobus*). Weed Sci 30(1):106–108

Bytof G, Knopp S-E, Schieberle P, Teutsch I, Selmar D (2005) Influence of processing on the generation of γ-aminobutyric acid in green coffee beans. Eur Food Res Technol 220:245–250

Çakir R, Çebi U (2010) The effect of irrigation scheduling and water stress on the maturity and chemical composition of Virginia tobacco leaf. Field Crop Res 119:269–276

Charles DJ, Joly RJ, Simon JE (1990) Effects of osmotic stress on the essential oil content and composition of peppermint. Phytochemistry 29:2837–2840

Chen Y, Guo Q, Liu L, Liao L, Zaibiao Z (2011) Influence of fertilization and drought stress on the growth and production of secondary metabolites in *Prunella vulgaris* L. J Med Plants Res 5(9):1749–1755

Cho Y, Njitiv N, Chen X, Lightfood DA, Wood AJ (2003) Trigonelline concentration in field-grown soybean in response to irrigation. Biol Plant 46(3):405–410

Christiansen JL, Jørnsgard B, Buskov S, Olsen CE (1997) Effect of drought stress on content and composition of seed alkaloids in narrow-leafed lupin, *Lupinus angustifolius* L. Eur J Agron 7:307–314

Close TJ (1997) Dehydrins: a commonality in the response of plants to dehydration and low temperature. Physiol Plant 100:291–296
de Abreu IN, Mazzafera P (2005) Effect of water and temperature stress on the content of active constituents of *Hypericum brasiliense* Choisy. Plant Physiol Biochem 43:241–248
de Bruijn GH (1973) The cyanogenic character of cassava (*Manihot esculenta*). In: Nestel B, MacIntyre R (eds) Chronic cassava toxicity. International Development Research Centre, Ottawa, pp 43–48
del Moral R (1972) On the variability of chlorogenic acid concentration. Oecologia 9:289–300
Delitala I-F, Gessa C, Solinas V (1986) Water stress and flexibility of phenolic metabolism in *Thymus capitatus*. Fitoterapia 57(6):401–408
Durrant WE, Dong X (2004) Systemic acquired resistance. Annu Rev Phytopathol 42:185–209
Edreva A, Velikova V, Tsonev T, Dagnon S, Gürel A, Aktaş L, Gesheva E (2008) Stress-protective role of secondary metabolites: diversity of functions and mechanisms. Gen Appl Plant Physiol 34(1–2):67–78
Falk KL, Tokuhisa JG, Gershenzon J (2007) The effect of sulfur nutrition on plant glucosinolate content: physiology and molecular mechanisms. Plant Biol 9:573–581
Fall R (1999) Biogenic emissions of volatile organic compounds from higher plants. In: Hewitt CN (ed) Reactive hydrocarbons in the atmosphere. Academic, New York, pp 41–95
Ferry N, Edwards MG, Gatehouse JA, Gatehouse AMR (2004) Plant–insect interactions: molecular approaches to insect resistance. Curr Opin Biotechnol 15:155–161
Forouzandeh M, Fanoudi M, Arazmjou E, Tabiei H (2012) Effect of drought stress and types of fertilizers on the quantity and quality of medicinal plant Basil (*Ocimum basilicum* L.). Ind J Innov Dev 1(10):734–737
Gershenzon J (1984) Changes in the levels of plant secondary metabolites under water and nutrient stress. Recent Adv Phytochem 18:273–320
Grace SC, Logan BA (2000) Energy dissipation and radical scavenging by the plant phenylpropanoid pathway. Philos Trans R Soc Lond B Biol Sci 355:1499–1510
Gray DE, Pallardy SG, Garrett HE, Rottinghaus G (2003) Acute drought stress and plant age effects on alkamide and phenolic acid content in purple coneflower roots. Planta Med 69(1):50–55
Hahlbrock K, Bednarek P, Ciolkowski I, Hamberger B, Heise A, Liedgens H, Logemann E, Nürnberger T, Schmelzer E, Somssich IE (2003) Non-self recognition, transcriptional reprogramming, and secondary metabolite accumulation during plant/pathogen interactions. Proc Natl Acad Sci U S A 100(2):14569–14576
Hara M (2010) The multifunctionality of dehydrins—an overview. Plant Signal Behav 5:1–6
Harborne JB (1988) Introduction to ecological biochemistry, 3rd edn. Academic, London
Hartmann T (2007) From waste products to ecochemicals: fifty years research of plant secondary metabolism. Phytochemistry 68:2831–2846
Hernández I, Alegre L, Munné-Bosch S (2006) Enhanced oxidation of flavan-3-ols and proanthocyanidin accumulation in water-stressed tea plants. Phytochemistry 67:1120–1126
Houter B, Nederhoff E (2007) Controlling plant growth—Part 3. Plant balance: 'vegetative' or 'generative'. Grower 62(2):42–44
Jaafar HZ, Ibrahim MH, Mohamad Fakri NF (2012) Impact of soil field water capacity on secondary metabolites, phenylalanine ammonia-lyase (PAL), maliondialdehyde (MDA) and photosynthetic responses of Malaysian kacip fatimah (*Labisia pumila* Benth). Molecules 17(6): 7305–7322
Jaleel CA, Manivannan P, Sankar B, Kishorekumar A, Gopi R, Somasundaram R, Panneerselvam R (2007) Induction of drought stress tolerance by ketoconazole in *Catharanthus roseus* is mediated by enhanced antioxidant potentials and secondary metabolite accumulation. Colloids Surf B 60(2):201–206
Jensen CR, Mogensen VO, Mortensen G, Fieldsend JK, Milford GFJ, Andersen MN, Thage JH (1996) Seed glucosinolate, oil and protein contents of field-grown rape (*Brassica napus* L.) affected by soil drying and evaporative demand. Field Crop Res 47:93–105
Kiddle GA, Doughty KJ, Wallsgrove RM (1994) Salicylic acid-induced accumulation of glucosinolates in oilseed rape (*Brassica napus* L.) leaves. J Exp Bot 45(278):1343–1346

Kim HJ, Chen F, Wang X, Rajapakse NC (2006) Effect of methyl jasmonate on secondary metabolites of sweet basil (*Ocimum basilicum* L.). J Agric Food Chem 54(6):2327–2332
Kinnersley AM, Turano FJ (2000) Gamma aminobutyric acid (GABA) and plant responses to stress. Crit Rev Plant Sci 19:479–509
Kirakosyan A, Kaufman P, Warber S, Zick S, Aaronson K, Bolling S, Chul Chang S (2004) Applied environmental stresses to enhance the levels of polyphenolics in leaves of hawthorn plants. Physiol Plant 121(2):182–186
Kirk H, Vrieling K, van der Meijden E, Klinkhamer PGL (2010) Species by environment interactions affect pyrrolizidine alkaloid expression in *Senecio jacobaea*, *Senecio aquaticus*, and their hybrids. J Chem Ecol 36:378–387
Kramer D, Breitenstein B, Kleinwächter M, Selmar D (2010) Stress metabolism in green coffee beans (*Coffea arabica* L.): expression of dehydrins and accumulation of GABA during drying. Plant Cell Physiol 51(4):546–553
Kubota N, Mimura H, Shimamura K (1988) The effects of drought and flooding on the phenolic compounds in peach fruits. Sci Rep Fac Agric Okayama Univ 171:17–21
Kuzel S, Vydra J, Triska J, Vrchotova N, Hruby M, Cigler P (2009) Elicitation of pharmacologically active substances in an intact medical plant. J Agric Food Chem 57(17):7907–7911
Liu H, Wang X, Wang D, Zou Z, Lianga Z (2011) Effect of drought stress on growth and accumulation of active constituents in *Salvia miltiorrhiza* Bunge. Ind Crop Prod 33:84–88
Llusià J, Peñuelas J (1998) Changes in terpene content and emission in potted Mediterranean woody plants under severe drought. Can J Bot 76:1366–1373
Loivamäki M, Holopainen JK, Nerg AM (2004) Chemical changes induced by methyl jasmonate in oilseed rape grown in the laboratory and in the field. J Agric Food Chem 52(25): 7607–7613
Magel E, Mayrhofer S, Müller A, Zimmer I, Hampp R, Schnitzler JP (2006) Photosynthesis and substrate supply for isoprene biosynthesis in poplar leaves. Atmos Environ 40:138–151
Majak W, McDiarmid RE, Hall JW, van Ryswyk AL (1980) Seasonal variation in the cyanide potential of arrowgrass (*Triglochin maritima*). Can J Plant Sci 60(4):1235–1241
Manukyan A (2011) Effect of growing factors on productivity and quality of lemon catmint, lemon balm and sage under soilless greenhouse production: I. Drought stress. Med Aromat Plant Sci Biotechnol 5(2):119–125
Namdeo AG (2007) Plant cell elicitation for production of secondary metabolites: a review. Pharmacogn Rev 1(1):69–79
Nogués S, Allen DJ, Morison JIL, Baker NR (1998) Ultraviolet-B radiation effects on water relations, leaf development, and photosynthesis in droughted pea plants. Plant Physiol 117:173–181
Nowak M, Manderscheid R, Weigel H-J, Kleinwächter M, Selmar D (2010) Drought stress increases the accumulation of monoterpenes in sage (*Salvia officinalis*), an effect that is compensated by elevated carbon dioxide concentration. J Appl Bot Food Qual 83:133–136
Petropoulos SA, Daferera D, Polissiou MG, Passam HC (2008) The effect of water deficit stress on the growth, yield and composition of essential oils of parsley. Sci Hortic 115:393–397
Radovich TJK, Kleinhenz MD, Streeter JG (2005) Irrigation timing relative to head development influences yield components, sugar levels, and glucosinolate concentrations in cabbage. J Amer Soc Hortic Sci 130(6):943–949
Ramakrishna A, Ravishankar GA (2011) Influence of abiotic stress signals on secondary metabolites in plants. Plant Signal Behav 6(11):1720–1731
Rhodes D, Verslues PE, Sharp RE (1999) Role of amino acids in abiotic stress resistance. In: Singh BK (ed) Plant amino acids: biochemistry and biotechnology. Marcel Dekker, New York, pp 319–356
Satya Narayan V, Nair PM (1990) Metabolism, enzymology and possible roles of 4-aminobutyrate in higher plants. Phytochemistry 29:367–375
Saunders J, O'Neill NR (2004) The characterization of defense responses to fungal infection in alfalfa. Biocontrol 49:715–728
Schreiner M, Beyene B, Krumbein A, Stützel H (2009) Ontogenetic changes of 2-propenyl and 3-indolylmethyl glucosinolates in *Brassica carinata* leaves as affected by water supply. J Agric Food Chem 57(16):7259–7263

Selmar D (2008) Potential of salt and drought stress to increase pharmaceutical significant secondary compounds in plants. Landbauforsch Völk 58:139–144

Selmar D, Kleinwächter M (2013) Influencing the product quality by deliberately applying drought stress during the cultivation of medicinal plants. Ind Crop Prod 42:558–566

Sharkey TD, Yeh S (2001) Isoprene emission from plants. Annu Rev Plant Phys 52:407–436

Shimoda M, Shigematsu H, Hideki S, Yutaka O (1995) Comparison of volatile compounds among different grades of Green Tea and their relations to odor attributes. J Agric Food Chem 43:1621–1625

Singh-Sangwan N, Abad Farooqi AH, Sangwan RS (1994) Effect of drought stress on growth and essential oil metabolism in lemongrasses. New Phytol 128(1):173–179

Szabó B, Tyihák E, Szabó LG, Botz L (2003) Mycotoxin and drought stress induced change of alkaloid content of *Papaver somniferum* plantlets. Acta Bot Hung 45(3/4):409–417

Tausz M, Šircelj H, Grill D (2004) The glutathione system a stress marker in plant ecophysiology: is a stress-response concept valid? J Exp Bot 55(404):1955–1962

Turtola S, Manninen A-M, Rikala R, Kainulainen P (2003) Drought stress alters the concentration of wood terpenoids in Scots pine and Norway spruce seedlings. J Chem Ecol 29:1981–1995

Wang DH, Du F, Liu HY, Liang ZS (2010) Drought stress increases iridoid glycosides biosynthesis in the roots of *Scrophularia ningpoensis* seedlings. J Med Plants Res 4(24):2691–2699

Wilhelm C, Selmar D (2011) Energy dissipation is an essential mechanism to sustain the viability of plants: the physiological limits of improved photosynthesis. J Plant Physiol 168(2):79–87

Wink M (2010) Introduction: biochemistry, physiology and ecological functions of secondary metabolites. In: Wink M (ed) Biochemistry of plant secondary metabolism. Wiley-Blackwell, Oxford, pp 1–19

Xia L, Yang W, Xiufeng Y (2007) Effects of water stress on berberine, jatrorrhizine and palmatine contents in amur corktree seedlings. Acta Ecol Sin 27(1):58–64

Zhu Z, Liang Z, Han R, Wang X (2009) Impact of fertilization an drought response in the medicinal herb *Bupleurum chinense* DC.: growth and saikosaponin production. Ind Crop Prod 29(2–3):629–663

Chapter 4
Water Scarcity and Water Stress in Agriculture

Gabrijel Ondrasek

Contents

1 Introduction

During the last decades, quality and availability of the most important environmental resources for food production, such as arable land and freshwater, have been exposed to extreme pressures and non-sustainable management practices (contamination, overexploitation, urbanisation, deforestation) all over the world. Such pressures disturb cycling of crucial natural components (H_2O, carbon) and in the forthcoming era of global processes (more frequent and intensive drought cycles, growth of humans in water-stressed regions) could further reduce availability of fresh hydro-resources for basic human requirements and increased food demands.

G. Ondrasek (✉)
Faculty of Agriculture, University of Zagreb,
Svetosimunska 25/10000, Zagreb, Croatia
e-mail: gondrasek@agr.hr

P. Ahmad and M.R. Wani (eds.), *Physiological Mechanisms and Adaptation Strategies in Plants Under Changing Environment: Volume 1*, DOI 10.1007/978-1-4614-8591-9_4,

Water, being the crucial component in most ecosystems, can severely impact food productivity in the case of reduced availability (stress).

Agriculture is not only the main source of the world's food supply but also the main consumer of renewable freshwater resources, using annually ~7 trillion m^3 of water, either in rain-fed (~4 trillion m^3) or irrigated (~3 trillion m^3) crop production (e.g., IWMI 2007; UNEP 2008). On ~1.5 billion ha of arable/permanently cropped agricultural areas (ICID 2009), current annual global food production reaches ~4 billion tonnes, out of which ~40 % is produced on less than 0.3 billion ha of irrigated land (IME 2013). Over the last 5 decades, agricultural food production has grown ~3-fold, whereas cultivated areas increased by only 12 % (but declined sharply per person, from ~0.44 to ~0.25 ha) in the same period (FAO 2012), confirming strong intensification of agriculture. Given increased environmental awareness, a further significant expansion into nonarable areas (e.g., forest ecosystems) and/or switching from rain-fed to more productive irrigated agriculture seems unlikely. Accordingly, with ~0.9 billion people (~13 % of population) currently exposed to hunger and taking into account relatively realistic midterm scenarios about population growth and climate variability (FAO 2012), it will be extremely difficult (in some cases impossible) to meet demands for food (water).

Moreover, some of the most recent reports (FAO 2011, IME 2013) warn about crushing facts related to food (water) issues, i.e., that 30–50 % (1.2–2 Gt) of globally produced food (i.e., even more than is currently produced on all irrigated areas) is wasted on the field-to-plate route, mostly due to inappropriate practices in harvesting, storage and transport of food as well as relatively high market and consumer standard criteria. In some Asian countries, where rice cultivation in low-efficiency irrigation systems may use up to 5,000 m^3 water per kilogram of product (ICID 1997), losses of the rice grain can reach up to 80 % of the total production, mostly due to poor efficiency of (post) harvesting practices and a lack of infrastructure for food storage/transport (IME 2013). Also, in some high-income consumer societies, a lot of horticultural produce (up to 30 % in the UK) remains unharvested in paddocks, never reaching markets, because it does not fit existing market (consumer) standards (IME 2013; Fig. 4.1a) and/or because of periodic low demands/prices for specific foodstuff (Fig. 4.1b). For instance, in lowlands of the Neretva River Estuary (one of the most intensive horticultural regions in Croatia), a double-cropping annual pattern is very common, comprising early spring transplantation of watermelon/muskmelon (using plastic mulch and drip fertigation) and thereafter midsummer transplantation of *Brassicaceae* (with micro irrigation). In spite of the fact that water consumption is on average 600 m^3 (cabbage) and >2,000 m^3 (muskmelon) per ha, it is common that because of relatively low marketable fruit quality (fruit size/shape, colour) (Fig. 4.1a), and temporally low market demands and prices (Fig. 4.1b), a large amount of produced foodstuff from these areas serves as organic soil amendments, never reaching the consumers. These examples clearly confirm that in avoiding water scarcity (stress) in agriculture to meet increasing food (water) demands, in addition to numerous relatively expensive strategies for enhancing water and land productivity (Sect. 5), fundamental changes in minimising foodstuff wastes could also be an effective approach.

Fig. 4.1 Foodstuff losses (photos taken 22/03/2012 in the Neretva River Valley). Unharvested produce was ploughed in shortly after the photos were taken. (**a**) Unharvested plantation of muskmelon planted in April 2011. (**b**) Broccoli paddocks planted in August 2011

2 Variability and Availability of Hydro-Resources for a Potential Use in Agriculture

Agricultural food production is highly depended on availability (variability) and quality of natural resources, principally waters and soils. Although (fresh) water as a renewable (vs. nonrenewable soil) resource exists in a continuous state of flux among its phases in the hydrological cycle (Table 4.1) and in general is constant on a global scale, significant exacerbation of different anthropogenic pressures in the last century has substantially depleted the quantity of good-quality water in particular areas, making the water scarcity a factor not just impeding the economic development but human existence areas as well. Freshwater (surface/ground) resources are mainly refreshed by precipitations with a specific temporal and spatial distribution pattern, and thus hydro-climatology of a particular area determines the dynamics of water availability/variability (Table 4.1). Depending on climate, the most dominant components of water cycling (precipitation, ET, run-off and infiltration) have variable contributions to (1) renewing water in aquifers and surface sources and (2) ensuring net primary productivity, i.e., food production.

An annual global hydrological cycle involves ~580,000 km^3 of freshwater (precipitation/vapour), out of which ~512,000 km^3 of water vapour is released from

Table 4.1 Estimated amounts of freshwater in the annual terrestrial hydrocycle and water storage in various pools globally (adapted according to Young et al. 2004; UNEP 2008 and references therein)

Annual hydrological cycle in terrestrial ecosystems (km^3)				Global fresh hydro-resources (km^3)			
Evapotranspiration	Runoff	Precipitation	Continent/Area	Ice & Snow	Surface w.	Groundwater	Total
17,840	4,460	22,300	Africa	0.2	31,776	5,500,000	5,531,776
4,602	2,478	7,080	Australia	180	221	1,200,000	1,200,401
17,710	14,490	32,200	Asia	60,984	30,622	7,800,000	7,891,606
5,389	2,902	8,290	Europe	18,216	2,529	1,600,000	1,620,745
10,065	8,235	18,300	North America	90,000	27,003	4,300,000	4,417,003
16,188	12,212	28,400	South America	900	12,849	3,000,000	3,013,749
			Antarctica/Arctic/Greenland	24,024,000			24,024,000
71,794	44,777	116,570	Total	24,194,280	105,000	23,400,000[a]	47,699,280
			Subtotal (only[b])			12,870,000[b]	37,169,280
			Subtotal (only[c])			10,530,000[c]	34,829,280

[a] Assumes a sum of [b] and [c]

[b] Saline groundwater resources up to 2,000 m depth

[c] Non saline groundwater resources

seas and the land surface hydrosphere to the atmosphere, to be returned via precipitation into the saline hydrosphere [~460,000 km^3 (90 %)] and terrestrial ecosystems (~117,000 km^3) (UNEP 2008; Table 4.1). Water transfer from land and vegetation to the atmosphere, as global evapotranspiration (ET), reaches ~60 % (~72,000 km^3) of annual globally renewable freshwater. From the total freshwater resources of ~47.7, ~24.2 million km^3 represents ice/permanent snow covers, ~23.4 million km^3 is stored water in shallow and deep (up to 2,000 m) aquifers, and a minuscule portion of 105,000 km^3 (0.2 %) is surface freshwater (lakes, wetlands, rivers) (Table 4.1). So, from the prospective use of water in agriculture, most freshwater is (1) inaccessible frozen water and (2) deep and quality-restricted groundwater. In addition to the rate of renewal and storage volume, of crucial importance is water quality, given that >50 % of relatively deep groundwater sources are mineralised and thus restricted (e.g., >20 mM dissolved salts in irrigation water is causing stress to most cultivated species) and/or prohibited (e.g., in the case of excessive concentration of some elements, such as F and As) (e.g., Romic et al. 2012). Irrespective of that, groundwater is globally one of the most exploited raw material (especially in arid areas) whose withdrawal rate is currently estimated at 600–700 billion m^3/year for municipal (65 %) and agricultural use (20 %) (UNEP 2008).

The annual per capita availability of fresh hydro-resources sharply decreased in the past 50 years (from ~13,000 m^3 in 1970 to ~7,000 m^3 in 2000) and is expected to reach only ~5,000 m^3 by 2025 (UNEP 2008). According to the same source, from 1900 to the present, the total global freshwater consumption grew linearly. In all spheres of human activities, water withdrawals per capita increased many fold, relatively less in the agri-sector (from ~550 to ~2,800 km^3) than the municipal (from ~100 to ~800 km^3) and industrial sectors (from ~50 to ~500 km^3). Modern agriculture is a predominant user of freshwater for irrigated food/feed production, consuming ~70 % of total freshwater withdrawn globally per year (currently ~3,000 km^3), with substantial differences among countries. In high-income countries, water use by agri-sector is around 40 % (of total use), i.e., double lower than in developing countries where agriculture accounts for more than 80 % of water consumption (UNDP 2006).

A global spread of irrigation areas was relatively steady over the last decade compared to almost linear expansion that started in the mid of the last century, which has recorded average annual increase of ~3 Mha (periodically up to 5 Mha) for about 50 years (Fig. 4.2). Also, irrigated areas on many continents (Europe, North/Central America) were decreased in comparison to 1995, or in less than a decade (Australia) (Fig. 4.2), mostly due to physical and/or quality degradation of terrestrial and aquatic resources. Recent data of International Commission on Irrigation and Drainage (ICID 2009) show that global irrigated areas are spreading on around 18 % (~280 Mha) of totally cultivated land areas (1,540 Mha); out of that, Asian countries irrigate almost 70 % (200 Mha): India (61 Mha), China (58 Mha), Pakistan (20 Mha), Iran (9 Mha) and Turkey and Thailand (5 Mha).

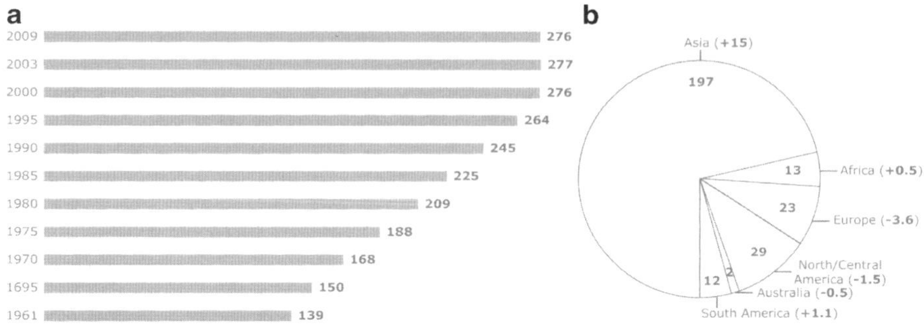

Fig. 4.2 Global irrigated land, all in Mha (adapted from FAOSTAT 2006; ICID 2009; Australian Bureau of Statistics 2011). (**a**) Total irrigated land in the 1961–2009 period. (**b**) Total irrigated land per continent in 2009 (in the brackets: changes in irrigated land in comparison to 1995)

2.1 Availability of Land and Water Resources for Irrigation in Croatia

Comprising ~56.5 km^2 of land with diverse climate and geomorphology (Pannonian Plain with continental climate 55 %, coastal part with Mediterranean climate 32 % and mountain part with humid climate 13 %) and only 4.3 million inhabitants (CBS 2011), Croatia belongs to the richest European countries according to renewable water sources, either per unit area or per capita (SUV 2008). The average annual rainfall varies between 650 and 3,500 mm depending on the area, with an average for the whole country of 1,162 mm (CBS 2009; Fig. 4.3). Annual rainfall is almost 66 km^3 (14,806 m^3 per capita), with the total surface run-off of 26.1 km^3 (Fig. 4.3). Due to a negative population trend over the last 20 years (a reduction by almost 0.5 million people) (CBS 2011), water availability per capita has increased in Croatia by ~12 %. Total renewable water resources are estimated at 165.4 km^3 (37,286 m^3 per capita) out of which 86 % are rivers and the rest is groundwater (Fig. 4.3). Total annual water use in Croatia in the last decade was around 0.5 or 0.3 km^3 less than during 1980s. For instance, in 2006 the total water withdrawal was 0.52 km^3 out of which 0.4 km^3 went mostly to industry (56 %) and the rest (44 %) to the public (domestic) water supply (Fig. 4.3). Water consumption for irrigation, given a small portion of area being irrigated (<1 % of arable land; see below), is practically negligible. Of the total agricultural land of 3.1 Mha, around 50 % represent pastures (1.2 Mha) and meadows (0.4 Mha), and the rest is arable land with domination of crops and gardens (1.5 Mha), than orchards (0.07 Mha) and vineyards (0.06 Mha) (Romic et al. 2005).

Construction of land amelioration systems (channels, dams, pump stations), mostly for drainage and less for irrigation, has started in river estuaries at the beginning of the last century. These works were in most cases carried out in parallel with flood protection activities or immediately afterwards. Until the 1960s, only open-drainage channel networks were constructed, but thereafter there was increased construction of subsoil (pipe) drainage and irrigation systems. The current state of

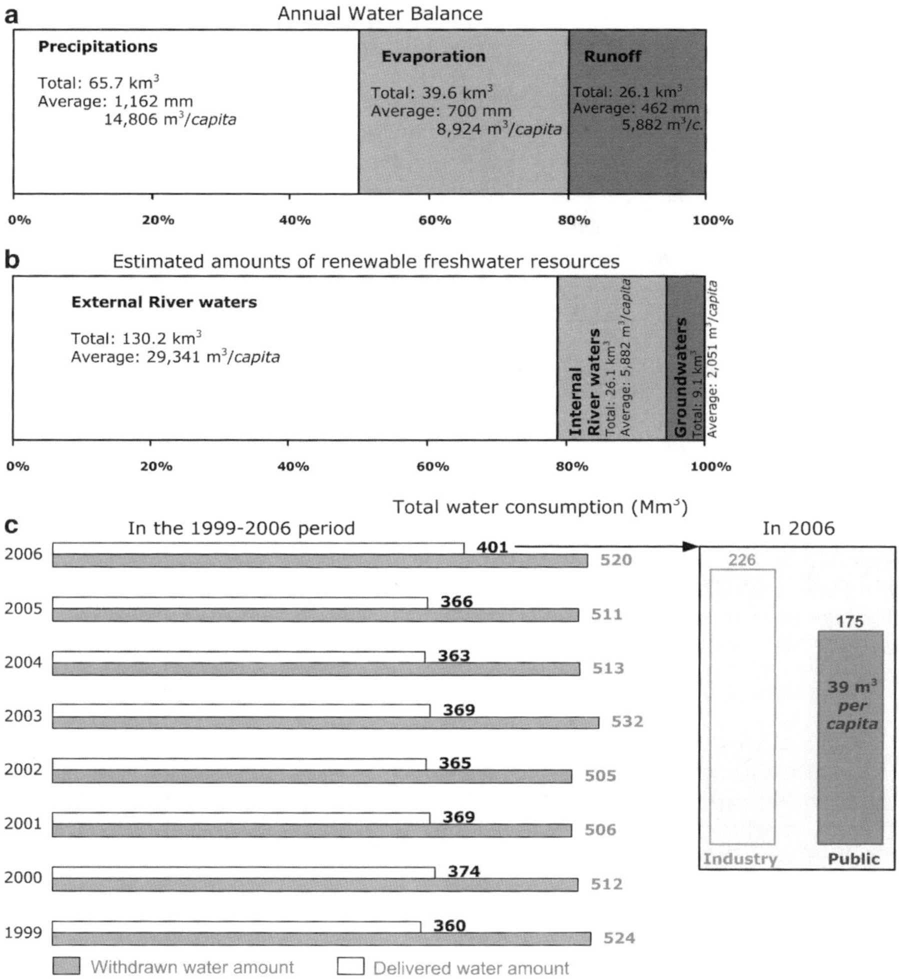

Fig. 4.3 Annual water balance (**a**), freshwater resources (**b**) and their use (**c**) in Croatia (adapted from SUV 2008; CBS 2009)

the amelioration (irrigation/drainage) systems at the national level includes the following: (1) completely constructed surface, open-drainage (channel) systems on ~725,000 ha, (2) partly constructed surface drainage systems on ~325,000 ha and (3) subsurface drainage systems (plastic pipes) on ~150,000 ha (Petosic et al. 2012) and implemented modern (sprinkler and micro) irrigation systems on ~20,000 ha (Ondrasek unpublished). In the recent national project for irrigation and management by agricultural land and water in Croatia (Romic et al. 2005), analysing the most important environmental (soil/water) resources, balance of hydro-resources as well as the socio-economic aspects, it was documented that there is >500,000 ha with a (very) high potential for irrigation nationally and that in midterm period irrigation could be increased manifold, i.e., to 65,000 ha by 2020.

3 Water Scarcity and Water Stress in Agriculture

In agriculture water scarcity is often equalised with drought. However, while drought (agronomical, hydrological or meteorological) implies a temporary decrease in water availability (e.g., due to precipitation deficiency and/or reduced groundwater/soil moisture), water scarcity assumes that water demands exceed the sustainable exploitation of available hydro-resources in the long term. More specifically, water scarcity would be a gap between available supplies of and demands for (principally fresh) surface/underground water resources within a certain domain (region, river basin) under prevailing institutional arrangements and infrastructural conditions (FAO 2012).

Seckler et al. (1998) defined the most important types of water scarcity: (1) physical (if there is not enough water to meet all demands, including environmental flows, with consequences such as severe environmental degradation, declining groundwater levels and water allocations that favour some activities/uses over others) and (2) economic (imposed by a lack of infrastructural investments or institutional factors limiting access to water and can be obvious in scant infrastructure development for instance). Approximately 1.2 billion people (in ~30 countries) already experience physical water scarcity, and another 0.5 billion are approaching this situation, whereas another ~1.6 billion people face economic water shortage (UNDP 2006; UNEP 2008). The same sources estimate that under the existing climate scenario, in the next 20 years ~50 % of the world's population in around 50 countries could be exposed to water stress/scarcity. One of the most used and reliable criteria of water scarcity/stress in a certain domain, as a ratio between annual renewable freshwater and human population (m^3/per capita) (Falkenmark and Widstrand 1992), distinguishes the following categories: (1) absolute (extreme) water scarcity (<500), (2) chronic water shortage (500–1,000), (3) regular water stress (1,000–1,700), (4) occasional water stress, i.e., vulnerable area (1,700–2,500), and sufficient (abundant) renewable water per person (>2,500 m^3). This classification, as many others, does not consider local factors determining access to water, as well as the feasibility of various solutions in different locations with prevailing climatic conditions and their variability, issues of water access, competition among sectors, potential for recycling (reuse) and/or implementation of unconventional water resources (Molle and Mollinga 2003). Of crucial importance is water quality, given that in many areas water quality may impose and/or exacerbate scarcity, even in the cases where fresh hydro-resources are sufficient for all domains/sectors. It is estimated that around 75 % of all industrial and up to 95 % of sewage effluents in the developing countries (which are most affected by water scarcity) are discharged into surface waters without any treatment (UNDP 2006), thus further escalating the water crisis.

In agricultural food/feed production, water scarcity assumes dysfunction (stress) in accessing (readily available) rhizosphere moisture, its uptake and transport through the plant and finally through stomata to the atmosphere (ET). Water stress can be caused by some above-explained causes and many other natural/anthropogenic (a)biotic factors, such as soil salinity, weeds and pests. For instance, increased soil salinity (causing decreased osmotic pressure gradient at the soil-root interface) restricts, or in extreme cases blocks, water uptake by roots, i.e., induces water stress initially (Ondrasek 2013) and then imposes numerous other physiological dysfunctions, i.e., secondary (oxidative) stresses (Ahmad et al. 2012; Chandna et al. 2013). Water stress caused by excessive soil salinity is potentially present on ~1 billion ha worldwide (Ondrasek et al. 2011), and some appropriate long-term strategies could help in remediation or alleviation of this situation (Sect. 5).

4 Approaches and Methods for Water Balance in Agriculture

Water balance (ΔW) of the certain system represents a temporal difference between total water inputs (W_{in}) and total water outputs (W_{out}), i.e., positive/negative changes in incoming and outgoing water (liquid and/or gas) flows and can be expressed as follows:

$\Delta W = W_{in}(P+I+Cr+Sin+SUBin) - W_{out}(E+T+R+Rp+SUBout)$ (Fig. 4.4).

Some components of water balance, like subsurface inflows/outflows and capillary rise (Cr), under certain pedological conditions (lighter sandy and/or coarser/rocky soils developed on uniform relief) may have short retention times in, or may bypass, the rhizosphere without significant moistening and thus can be disregarded. Also, in agroecosystems on plains, run-off is negligible as well as subsurface inflows/outflows and can be extremely difficult to measure/assess and in short time periods

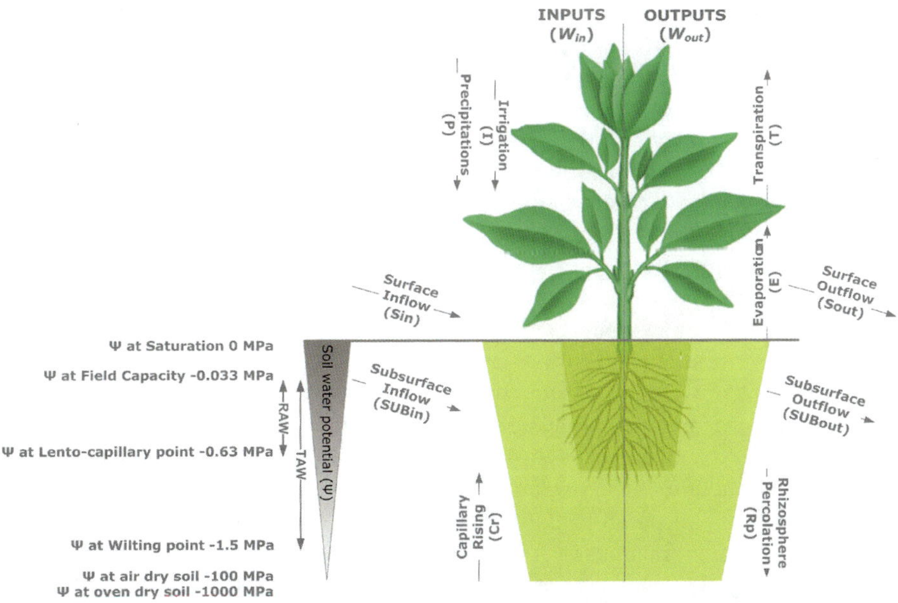

Fig. 4.4 Components of water balance in a soil-plant-atmosphere continuum (adapted from Allen et al. 1998)

(less than a week) usually is not worth considering (e.g., Allen et al. 1998). Accordingly, in a soil-plant-atmosphere continuum, remaining parameters such as precipitation (P) and/or irrigation (I) as well as evaporation and transpiration (evapotranspiration—ET) have the dominant influence on water balance (Fig. 4.4).

Precipitation measured by using different standard/automated gauges installed at the nearest meteorological station (or at the irrigated field), ensures the most reliable data for calculation of water balance. Given that total precipitation may not be fully effective in refilling the soil available water reserves (Fig. 4.4), water balance must consider the so-called effective precipitation (Peff), detected/assessed by appropriate (semi)empirical methods (Bos et al. 2009).

In irrigated agroecosystems, irrigation (I) value (irrigation efficiency) is very important, given significant diversity among irrigation methods/systems (Sect. 5). On the other side, ET is usually the most difficult output element to determine in water balance (Fig. 4.4), but makes the greatest contribution to the crop water demands (see next section). As elaborated above, ET from irrigated agricultural land represents by far the largest consumption of water withdrawn for human use and food production. There are different methods for direct (in situ) measurement or indirect estimation of the ET (and some other parameters in water balance), and some of them are shortly described in the next section.

Eddy covariance method represents a micrometeorologically based approach for measuring very rapid turbulent water vapour fluxes (CO_2, CH_4) at the atmosphere-canopy interface. The method is comprises highly sophisticated, state-of-the art in situ instrumental measurements of different fluxes and their components (e.g., 3D air components, concentrations, temperature, humidity), which are then integrated via algorithms and numerical approaches (e.g., eddy covariance equation) for detection of water fluxes over a particular ecosystem. Accordingly, mass flow of water vapour (ET), expressed by eddy flux (covariance) equation (Fig. 4.5), represents the product of mean air density (ρ_a) and a covariance between instantaneous fluctuations in vertical wind speed (w') and instantaneous deviations in gas concentration, i.e., mixing ratio of H_2O in the air (s').

Although eddy covariance method is routinely used in estimating terrestrial H_2O/C budgets, testing/calibration ecosystem models and predicting their responses to climate changes (e.g., Gu et al. 2012), further development and improvements should ensure its application for flux measurements (water balancing) in topographically/floristically more complex ecosystems.

Lysimetry measurement methods comprised a wide range of types/techniques for direct detection of water losses from the land surface overlain by vegetation (ET) and also for soil water flow/sampling along the soil solum. Lysimeters enable in situ ET detection under nondisturbed/disturbed paedological conditions (Fig. 4.6). Among numerous variations, modern large weighing monolithic lysimeters (stations) equipped with the most advanced monitoring and sampling installations ensure the most accurate direct method for measuring ET and also represent one of the most appropriate ways for studying water/solute transport from topsoil to subsoil horizons and/or groundwater (Fig. 4.6). Lysimeter balances can record water losses precisely (10 g or 0.01 mm) within the short time period (down to several seconds) as needed

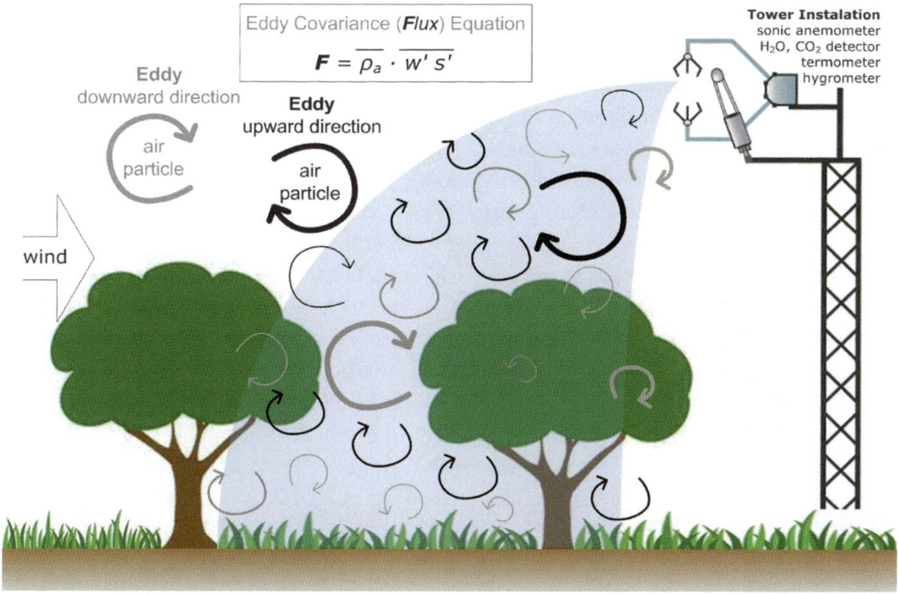

Fig. 4.5 Principles of the eddy covariance method for evapotranspiration calculation (adapted from Baldocchi 2003; Burba and Anderson 2010)

Fig. 4.6 A lysimetry station (**a**, **c**; INRA, Angers area/France, 2005) and installation of the lysimeter in disturbed soil conditions (**b**, **d**; Varazdin area/Croatia, 2011). *Arrows* show direction of the preferential water flows in-between lysimeter walls and a soil column (**c**) and in-between of the soil volume with disturbed and undisturbed soil structure (**d**)

for ET determination (Unold and Fank 2008). Recent weighing lysimeters posses patented systems for installation of soil monoliths, with an aim to minimise preferential water flows along the lysimeter walls (Fig. 4.6), which are often a problem. On the other side, one of the main disadvantages of weighing lysimeters is their non-suitability for water balancing on slopped terrains and in forest ecosystems. However, quality and reliability of data obtained from weighing lysimetry stations in determining reference ET (ETo), potential/actual ET, crop coefficients and water stress coefficients are irreplaceable (as explained in the following sections).

Among numerous indirect methods for ET estimation, initial Penman equation is probably the most modified one. One of its modifications, the Penman-Monteith (P-M) approach, is the most used mathematical approach for ET determination, accepted by environmental scientists in their research as well as in practice of water management and planning. As proposed by the FAO expert consultation team, the combined P-M approach has become a standard method for estimation of the reference ET (ETo). Using that approach and assuming certain constants (hypothetical grassland of 12 cm height with fixed surface resistance of 70 s m^{-1} and reflectance of 0.23), FAO's experts derived the FAO P-M equation, which uses standard meteorological data and is recognised as a globally valid approach for ETo, i.e., crop water requirements estimation (Allen et al. 1998). The (FAO) P-M method can be successfully applied for ET calculations and water management not just in the field conditions, but also in protected environments (e.g., Seginer 2002; Ondrasek et al. 2007).

The ET (actual, potential, reference) rates, either directly measured or indirectly estimated, are of crucial importance for determining crop (irrigation) water requirements (next section). Many computer programs, such as CROPWAT (Smith 1992; FAO 2013), PROREG (Teixeira et al. 1995) and CRIWAR (Bos et al. 2009), have incorporated algorithms for calculating particular water balance components (ET, Peff, TAW) and can be useful tools for simulating water management scenarios (yield reduction, water losses, irrigation efficiency) at different spatial/temporal scales under a wide range of environmental conditions.

4.1 Crop Water Requirements in Irrigated Agroecosystems: Croatian Example

A difference between effective precipitation (Peff) and potential ET of crops (ETcrop) represents the irrigation (crop) water requirements (Allen et al. 1998). Conditions for obtaining the potential ET can be achieved under well-watered rhizosphere (with enough readily available soil water; RAW), whereas in other cases, the actual value of ET will be lower than the (maximal) potential (Bos et al. 2009). ETcrop data may be estimated by one of the (in)direct approaches (explained above) or by the so-called the (single) crop coefficient approach for calculating ETcrop under standard conditions (Allen et al. 1998) as used for Croatian agroecosystems (Table 4.2).

Using this approach, ETcrop is estimated from the relation of the reference ET (ETo) and single crop coefficient (Kc) as follows (see Allen et al. 1998): $ETcrop = ETo \times Kc$. All calculations (ETo according to P-M approach, Peff by

Table 4.2 Crop (irrigation) water requirements and yield reduction for the average and the dry (25 % probability of precipitation) conditions (1981–2000) in continental (based on 9 official meteorological stations) and Mediterranean (based on 6 official meteorological stations) Croatian ecosystems. Yield reductions (in the case without irrigation practice) were calculated for two soil types: texture-lighter (total available soil moisture = 80 mm up to 1 m depth) and texture-heavier (total available soil moisture = 140 mm up to 1 m depth)

				Yield reduction %			
		Irrigation requirements mm		Texture-lighter soil 80 mm/m		Texture-heavier soil 140 mm/m	
Ecosystem	Crop	Average	Dry	Average	Dry	Average	Dry
Continental	Corn	81–191	168–314	9–36	33–67	2–28	24–61
ETo=690–820 mm	Sugar beet	116–260	226–383	10–38	34–64	3–30	26–58
Peff=521–890 mm	Tomato	98–191	174–286	14–39	35–59	8–30	28–54
	Apple	59–196	164–328	3–27	24–55	0–23	19–48
Mediterranean	Corn	219–511	306–634	38–75	59–95	31–70	53–92
ETo=883–1390 mm	Sugar beet	286–606	381–725	39–71	58–87	33–66	53–83
Peff=650–1085 mm	Tomato	229–478	294–578	41–69	56–85	35–65	49–82
	Apple	214–503	315–644	28–54	47–71	24–51	42–67

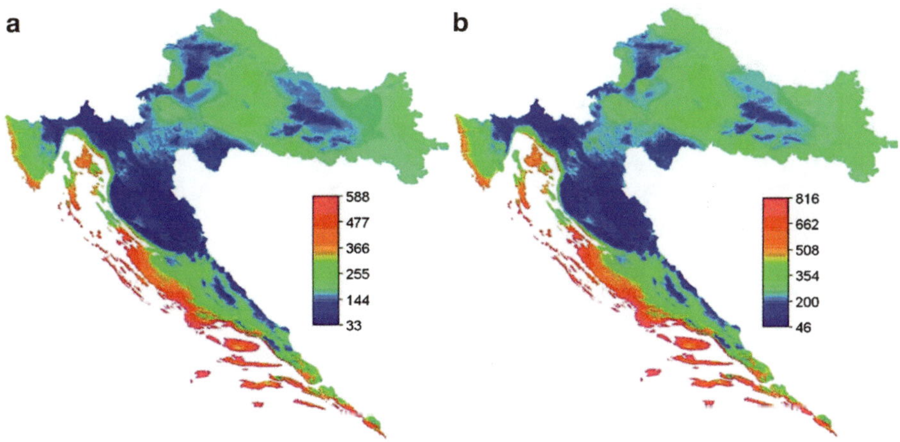

Fig. 4.7 Water deficit maps for tomato (**a**) and apple grown with grassed mulch cultivation (**b**) for average (1981–2000) Croatian climate (source Romic et al. 2005)

USBR method and related crop yield reductions without irrigation practice) were performed in CROPWAT (Smith 1992) for a 20-year period (1981–2000) under the average and the dry (25 % probability of precipitation) conditions in two Croatian agroecosystems: relatively cold continental and warm Mediterranean (Table 4.2, Fig. 4.7). Figure 4.7 shows a modelled map of water deficit for tomatoes and apples (grown with grass mulch in-between apple rows) using the average Croatian climate conditions in the 1981–2000 period. For spatial distribution of tomato/apple, water deficit interpolation was done by ordinary kriging based on three predictors (land slope, digital elevation model and average air temperature, *T*) to be obtained the exponential model for (1) tomato = exp(−5.0404 + 0.273*T)/(1 + exp(−5.0404 + 0.273*T))*2,000 (Fig. 4.7a; R^2 = 0.77) and (2) apple = exp(−5.285 + 0.2578*T)/(1 + exp(−5.285 + 0.2578*T))*2,000 (Fig. 4.7b; R^2 = 0.82) (Romic et al. 2005).

5 Perspectives for Avoiding Water Stress (Scarcity) in Agriculture

Overexploitation of hydro-resources and agricultural intensification (double/triple cropping) pose a threat to the sustainability of (agro)ecosystems and food production. In the recent past, saline aquifers as well as channel waters loaded by potentially harmful substances (metals, nutrients, polynuclear aromatic hydrocarbons) have become an important and common water sources in intensive irrigated horticultural production (Romic et al. 2012; Savic et al. 2013), whereas wastewater and different industrial/municipal effluents are widely used for the same purpose in many countries affected by water crisis. For instance, wastewater is used in irrigating about 20 million ha of land globally (UNDP 2006), i.e., about 7 % of total irrigated area.

Water quality has become one of the most critical environmental issues. Majority of developed (vs. developing/undeveloped) countries implement technological, economical and regulative frameworks to support the most effective methods for using hydro-resources and reusing wastewater as a possible solution for increasing water demands in agriculture and other sectors (Table 4.3). For example, desalination

Table 4.3 Water scarcity (stress) risk factors and possible perspectives for their mitigation

Risk factor	Perspectives
Global climate variability	Implementation of water-efficient irrigation systems
	Implementation of water/land conservation technologies
Increased frequency of weather extremes (droughts, heatwaves)	Implementation of modern irrigation systems
	Cropping pattern adaptation (introducing early-maturing genotypes, choosing sowing dates to avoid periods with the highest ET demands)
	Introduction to drought-tolerant varieties
	Implementation of water/land conservation technologies
Regional/Global political and/or economic crises	Investments by individual farmers/private sector as well as support by public investments
	Support investments in more sustainable systems
	Support research & innovation in sustainable management of land/water resources
Increasing human population	Support investments/technologies for (re)using of *marginal* waters
	Desalination
	Improve capacity for produce handling on the farm-to-plate route
	Reassessment (change) of consumer/market food quality criteria
Environmental degradation processes (salinity/contamination/soil organic matter depletion)	Adaptation (modernisation) of irrigation systems
	Introduction of salt-tolerant crop genotypes
	Implementation of water/land conservation technologies

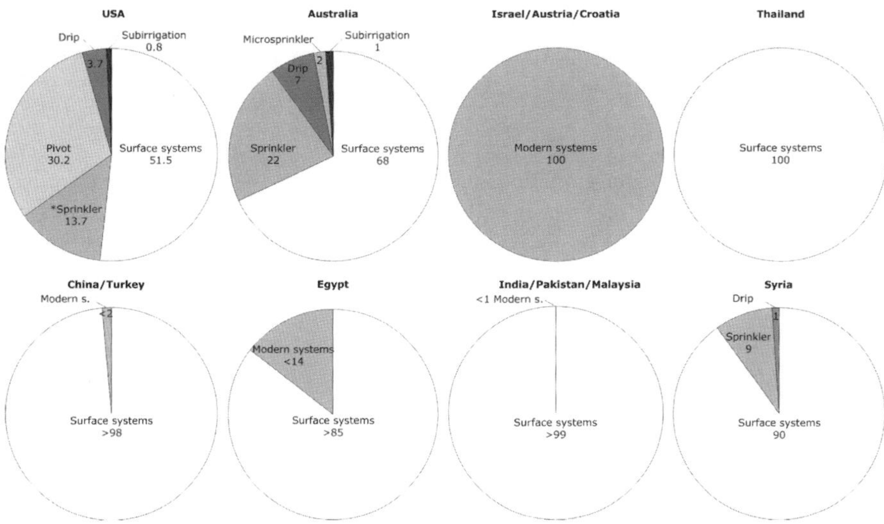

Fig. 4.8 Distribution (%) of irrigation systems in various countries (where modern systems are not specified, assume sprinkler and/or micro/subirrigation. *Asterisk* assumes permanent, hand-/mechanical-move sprinkler systems. Adapted from ICID 1997; Varela-Ortega and Sagardoy 2001; Howell 2001; Australian Bureau of Statistics 2007; Ondrasek unpublished)

currently contributes 0.2 % to the global water withdrawals, and although it is widely used for meeting water demands in many countries, due to high costs it is unlikely to resolve the fundamental mismatch between water supply and demand (UNDP 2006), especially in agriculture.

5.1 Implementation of (Modern) Irrigation Systems

Water can be utilised by crops only if in the rhizosphere soil. Water relations in the soil have been shown to explain >50 % of yield variability in the field (e.g., Romic et al. 2005 and references therein); therefore, temporal/spatial management of soil water may significantly improve water use by crops. Irrigation is one of the prerequisites for stable and high-quality crop yields, principally during drought seasons and for arid areas. There are big differences among countries in implementation of irrigation systems, but generally in low-income countries (and also in Australia, the USA, etc.), technically simple surface irrigation and high-pressure sprinkler systems are likely to be used (Fig. 4.8).

In the recent 20-year period (1979–1998) in the USA, the total irrigated area stayed rather stable (~20 Mha), but a portion under surface gravity systems has declined by 22 % (from 12.6 to 10.2 Mha), whereas during the same period, the areas under (1) sprinkler systems increased by 25 % (from 7.5 to 9.3 Mha), (2) subirrigation by >120 % (from 0.08 to 0.2 ha) and (3) trickle systems by >550 % (from 0.12 to 0.85 Mha) (USDA 1999). In Australia irrigated area grew steadily in

1920–1950 period and increased substantially until the mid-1990s, to be between 1.7 and 2.5 Mha over the last decade (Australian Bureau of Statistics 2011). One of the main reason for prevalence of surface irrigation systems in Australia is a structure of crop production on the irrigated area, with relatively low-income pastures (40 %) and cereals (13 %) occupying more than half, whereas high-income horticulture (more appropriate for implementation of modern systems) occupied around 22 % of total irrigated area during 2009–2010 (Australian Bureau of Statistics 2011). Based on graphical representation (Fig. 4.8) and related references, at the global scale it can be assumed that ~95 % of irrigated area is under surface and only ~5 % under modern (e.g., ~3 % sprinkler and ~2 % micro) irrigation systems.

Among traditional and modern irrigation methods/systems, there are many significant differences in their operational (technical) and environmentally related characteristics, as well in water consumption, i.e., water-use efficiency (WUE). Modern irrigation systems have reduced the water requirement by almost 5,000 m^3/ha, whereas traditional systems have water consumption at least twofold higher than the modern systems (ICID 1997). Globally, there are big differences in irrigation practices (Fig. 4.8); in Asian countries with large proportions of agricultural area under irrigation, 96 % of their irrigated area use traditional methods, contrary to European countries where >80 % of their irrigated area adopt modern methods (ICID 1997). Traditional surface gravity flow (furrow, borders, contours) methods (vs. modern drip and/or low-pressure sprinklers) have substantially lowered (up to twofold) WUE (see down).

Enhancing WUE is contingent on two critical measures: optimising the amount of water applied to crops and minimising the water losses, both aimed at increasing crop productivity (food production). From the water source to the field application, either in traditional or modern system, there are many possibilities to increase the overall WUE, including storage, delivery (distribution) and/or application efficiency. In most surface irrigation networks, <50 % of the extracted water actually benefits crops, whereas the rest are losses due to seepage through unlined/lined canal network, evaporation, percolation below the rhizosphere and/or surface run-off (ICID 1997). Therefore, in such systems WUE can be enhanced by (1) appropriate land treatment and management measures (improved on-farm water conveyance systems, canal lining, partial pressurising, precision field levelling, shortened water runs, alternative furrow irrigation, surge flow and cablegation, tailwater reuse, etc.) (see the next section) and/or (2) replacing a traditional surface system by a modern one. Field application efficiencies in traditional surface and sprinkler systems typically range from 40 to 65 % and 75 to 85 %, respectively, (USDA 1999), whereas modern systems due to sophisticated applicators (emitters) may achieve efficiencies of >95 % (e.g., drip irrigation). Also, there is possibility to modify/upgrade existing (relatively modern) systems for improved (>85 %) WUE. For instance, some high-pressure (>6 bars) sprinkle systems (centre pivot, linear-move, hand-move, travelling big gun) can be upgraded to low-energy/low-pressure (~1 bar) applicators, precisely delivering water in controlled amounts near the ground and reducing water losses from evaporation and wind drift, i.e., increasing application uniformity (Fig. 4.9). Application efficiency (the ratio of irrigation water infiltrated/stored in

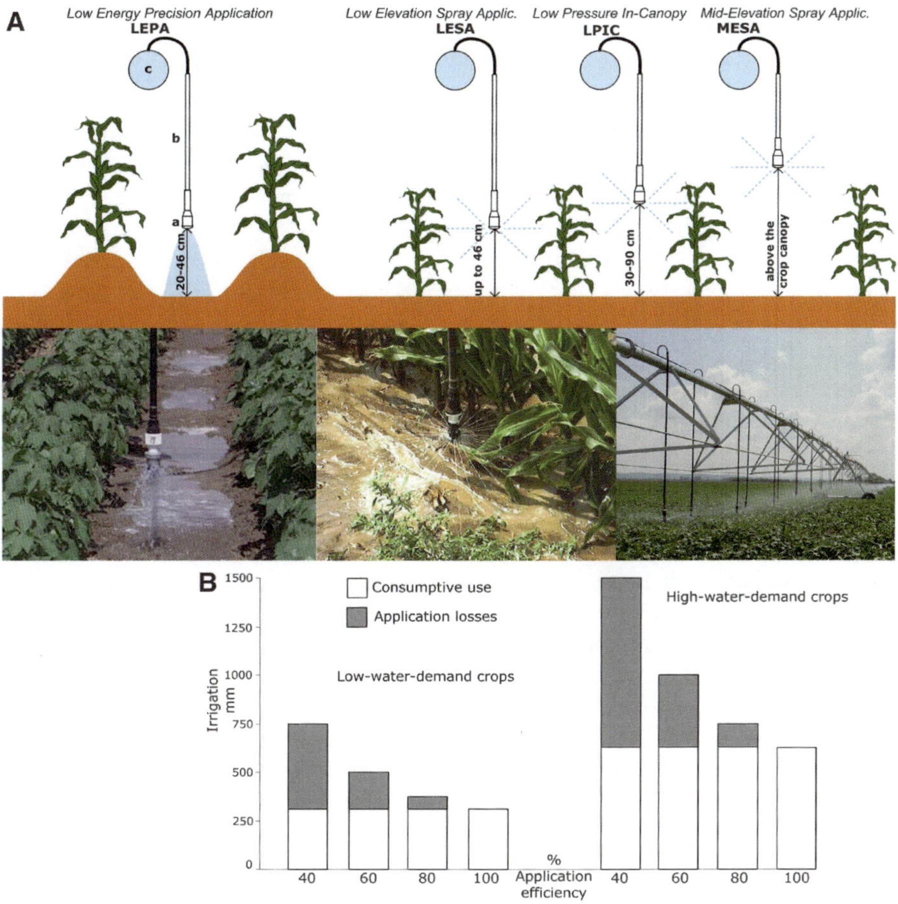

Fig. 4.9 (**A**) Shows modern low-energy/low-pressure irrigation (systems) applicators (**a**) connected over connection set (**b**) to the pivot's mainline (**c**) (Ondrasek unpublished). (**B**) Shows possible irrigation water conservation due to crop selection and enhanced application efficiency (adapted from USDA 2006)

the rhizosphere for crop use to the average amount of applied water) in the most sophisticated systems may be significantly improved from 80 to 90 % in LESA to 90–95 % in LEPA or up to 98 % in drip systems (Fig. 4.9).

Improved irrigation technology can reduce off-site water quantity (quality) issues and may substantially increase WUE on the field scale (USDA 1999). A 50 % increase in application efficiency (from 40 to 60 %) may reduce water application by ~33 % and generally is achievable through relatively inexpensive system modifications and management adjustments (explained above). However, in more efficient (modern) systems, a comparable increase in efficiency (from 60 to 80 %) will result in lower water savings (by ~20 %), and thus appropriate technologies and/or management practices (crop selection, mulching, weed/pest control) are required to achieve the additional water savings (Fig. 4.9).

5.2 *Possible Other (Integrative) Measures for Avoiding Water Stress (Scarcity) in Agriculture*

In either irrigated or rain-fed agriculture, there are many possibilities and strategies to improve water management (e.g., Table 4.3). Improved WUE in the irrigated fields can be obtained by choosing optimised irrigation management strategy, which assumes maintenance of the rhizosphere moisture: (1) close to the range of readily available water (around the field capacity) to meet the full crop ET demands (e.g., plant demand strategy) or (2) at relatively low soil water potential (e.g., <60 % of the field capacity), i.e., below the full crop ET (e.g., deficit/supplemental irrigation strategies). The second group of irrigation strategies can be successfully applied in cash crops horticultural production (e.g., olives, grapes) in (semi)arid climates. For instance, regulated deficit irrigation (DI) and partial root zone drying (PRD) irrigation strategy can improve WUE by 45–50 % compared to irrigation strategy where soil water was maintained >80 % of the field capacity (Leib et al. 2006). These two strategies are based on different concepts: in DI water application is manipulated temporally and in PRD spatially and temporally. In humid to temperate areas, supplemental irrigation strategy is used as a tactical measure to complement reasonably ample rainfall and stabilise production (Hsiao et al. 2007) and generally may be one of the precautions ensuring stable and continuous yield in next vegetation (e.g., irrigation during the drought period at time of initialisation of generative buds will positively impact the yield in fruit crops next year).

Regardless of the irrigation strategy, implementing some rhizosphere moisture measuring (in situ/laboratory) techniques (gravimetric, CFD/TDR probes, tensiometers, gypsum blocks) is likely to optimise irrigation timing and the amount of water applied to crops, which usually take up 90 % of the water needed from the top 75 % of the rhizosphere, so irrigating the soil (beyond field capacity) in the bottom 25 % of the root zone is likely to cause deep percolation (water losses) rather than increasing yields (USDA 2006). Also, irrigation scheduling (timing) based on soil water monitoring rather than some approximate modelling approach may result in significant (by 25–30 %) water conservation, as confirmed by Leib et al. (2006) during the 3-year field trials.

Some improvements in land management or shifting to new cropping systems is another possible and widely used strategy in combating water stress and some other accompanied soil constrains (low water retention capacity, high ET demands) or land degradations (organic matter depletion, desertification, salinisation) on (non) irrigated areas (Ondrasek et al. 2011). For instance, changing crop establishment technique from transplanted to direct seeded systems (dry seeding) is a successful strategy for reducing the non-productive water consumption in many Asian countries (ICID 1997). Implementation of some land conservation practices (from reduced or minimum to zero tillage) are increasingly used worldwide (>70 Mha) (Ondrasek and Rengel 2012) with an aim to restrict loses of topsoil water and/or alleviate some other soil constrains. By leaving at least 30 % of the crop biomass on the soil surface, conservation tillage prevents soil wind/water erosion and positively impacts most soil characteristics (organic matter, soil compaction, soil structure/texture, etc.), which is important for the soil water retention as well. Stable soil structure (crucial for subsurface water flow) and medium soil texture (crucial for good water-holding capacity) are key determinants of water infiltration rates and the amount of readily available water within the root zone. Organic matter generally improves soil structure and water-holding capacity, but also its functions in food safety/security as well protection of hydro-resources from metal contamination might be of great interest (Ondrasek et al. 2012).

One of the widely used techniques in precision agriculture is laser land levelling, which grades fields to contour the land for different irrigation practices, conserving water by reducing run-off and allowing uniform distribution of water applied. Laser levelling can reduce water use by 20–30 % and increase crop yields by 10–20 % (USDA 2006). Also, furrow diking conserves rain/irrigation water in small earthen dams along the furrows, and that water slowly infiltrates into the soil, increasing soil moisture and reducing/eliminating run-off. Furrow dikes are applicable in the rain-fed as well as modern (e.g., LEPA) irrigated agriculture.

One of the most advanced concepts in modern agriculture (chemigation, i.e., simultaneous application of water soluble chemicals/gasses with irrigation) can contribute to efficient nutrient use, improve weed/pest management and in combination with mulch cropping system (e.g., drip irrigation with black plastic mulch) may result in increased yield and quality with a decreased water input (Romic et al. 2008). Application of polyacrylamide (PAM), with irrigation or as soil amendment, is widely practised in the USA (e.g., >0.4 million irrigated hectares by 1999) to stabilise soil and water-borne sediment under irrigation (USDA 2006). According to the same source, PAM can reduce soil erosion in furrow systems (in medium- and fine-textured soils), enhance water infiltration, improve uptake of nutrients/pesticides, reduce furrow-reshaping operations and reduce sediment control requirements downhill from the field.

One of the most recent and promising strategies for WUE improvement is application of O_2-enriched irrigation water (oxygenation). Oxygenation, as a common (obligate) practice in the most sophisticated crop technologies such as hydroponics, ensures rhizosphere aeration, i.e., prevents the hypoxia. As a new innovation in open-field irrigation, oxygenation has been effective in increasing WUE and crop yield and could also be instrumental in ameliorating some other soil constrains such as compaction and salinity (Chen et al. 2011; Bhattarai et al. 2006).

6 Conclusion and Future Perspectives

Over the last several decades, the availability of freshwater in many rain-fed/irrigated agricultural areas of our planet decreased to critical levels. In the next few decades, given the global climate variability (likely to comprise more frequent and more severe droughts and heat waves, higher variability in precipitation distribution), accompanied by growing human population, the most relevant projections predict altered hydrological regimes in the terrestrial ecosystems, i.e., reduced availability of freshwater for agroecosystems. Besides numerous water/land management strategies for alleviating water scarcity (stress) in food production (implementation of the most efficient irrigation systems, land conservation techniques, precise farming, etc.), a great potential exists in (1) enhancing the capacity for food handling (harvest, storage, transport) and (2) serious reconsidering (changing) market-consumer relations, to potentially save a larger amount of water (or food) than is currently used (or produced) in the irrigated agriculture globally.

Acknowledgements The author is grateful to Winthrop Professor Zed Rengel (University of Western Australia) for valuable discussion, comments and text improvement. This work was partly supported by Croatian Ministry of Agriculture—Council for Investigation in Agriculture (VIP), Contract 2012-11-02 and by Zagreb County.

References

Ahmad P, Ashraf M, Younis M, Hu X, Kumar K, Akram NA, Al-Qurainy F (2012) Role of transgenic plants in agriculture and biopharming. Biotechnol Adv 30(3):524–540

Allen RG, Pereira LS, Raes D, Smith M (1998) Crop evapotranspiration: guidelines for computing crop water requirements. Irrigation and drainage paper 56. United Nations FAO, Rome, Italy, 300p

Australian Bureau of Statistics (2007) Water use on Australian farms. ABS, Canberra, p 52

Australian Bureau of Statistics (2011) Water use on Australian farms. ABS, Canberra, p 36

Baldocchi DD (2003) Assessing the eddy covariance technique for evaluating carbon dioxide exchange rates of ecosystems: past, present and future. Glob Change Biol 9:479–492

Bhattarai SP, Pendergast L, Midmore DJ (2006) Root aeration improves yield and water use efficiency of tomato in heavy clay and saline soils. Sci Hortic 108:278–288

Bos MG, Kselik RAL, Allen RG, Molden DJ (2009) Water requirements for irrigation and the environment. Springer, New York, p 174

Burba GG, Anderson DJ (2010) A brief practical guide to eddy covariance flux measurements: principles and workflow examples for scientific and industrial applications. LI-COR Biosciences, Lincoln, p 211

Chandna R, Azooz MM, Ahmad P (2013) Recent advances of metabolomics to reveal plant responses during salt stress. In: Ahmad P, Azooz MM, Prasad MNV (eds) Salt stress in plants: signalling, omics and adaptations. Springer, New York, pp 1–14

Chen X, Dhungel J, Bhattarai SP, Torabi M, Pendergast L, Midmore DJ (2011) Impact of oxygation on soil respiration, yield and water use efficiency of three crop species. J Plant Ecol 4(4): 236–248

Croatian Bureau of Statistics (CBS) (2009) Statistical yearbook. Available at http://www.dzs.hr/Hrv_Eng/ljetopis/2009/PDF/00-sadrzaj.pdf. Accessed 20 Mar 2013

Croatian Bureau of Statistics (CBS) (2011) Population, by age and sex (population pyramides), 1953-2011 censuses. Available at http://www.dzs.hr/default_e.htm. Accessed 12 Jan 2013

Falkenmark M, Widstrand C (1992) Population and water resources: a delicate balance. Population Bulletin, Washington

FAO (2011) Global food losses and food waste. In: Gustavsson J, Cederberg C, Sonesson U, Otterdijk R, Meybeck A (eds) Extent, causes and prevention. FAO, Rome. Available at http://www.fao.org/docrep/014/mb060e/mb060e00.pdf. Accessed 22 Dec 2012

FAO (2012) FAO water reports 38. Coping with water scarcity. An action framework for agriculture and food security. FAO, Rome, p 100

FAO (2013) CropWat 8.0 for Windows. Avialable at http://www.fao.org/nr/water/infores_databases_cropwat.html. Accessed 25 Mar 2013

FAOSTAT (2006) http://faostat.fao.org/faostat/collections?subset=agriculture. Accessed 17 Aug 2006

Gu L, Massman WJ, Leuning R, Pallardy SG, Meyers T, Hanson PJ, Riggs JS, Hosman KP, Yang B (2012) The fundamental equation of eddy covariance and its application in flux measurements. Agr Forest Meteorol 152:135–148

Howell TA (2001) Enhancing water use efficiency in irrigated agriculture. Agron J 93(2):281–289

Hsiao TC, Steduto P, Fereres E (2007) A systematic and quantitative approach to improve water use efficiency in agriculture. Irrig Sci 25:209–231

Institute of Mechanical Engineers (IME) (2013) Global food: waste not, want not. Report 35 p. Available at www.imeche.org. Accessed 10 Jan 2013

International Commission on Irrigation and Drainage (ICID) (1997) Watsave scenario. ICID, New Delhi. Available at http://www.icid.org/ws_pdf.html. Accessed 17 Jan 2009

International Commission on Irrigation and Drainage in Agriculture (ICID) (2009) Important data of ICID member countries of irrigated area. Available from http://www.icid.org/imp_data.pdf. Accessed 10 July 2010

International Water Management Institute (IWMI) (2007) Comprehensive assessment of water management in agriculture. Water for food, water for life: a comprehensive assessment of water management in agriculture. IWMI, London. ISBN 978-1-84407-396-2

Leib BG, Caspari HW, Redulla CA, Andrews PK, Jabro JJ (2006) Partial root zone drying and deficit irrigation of "Fuji" apples in a semi-arid climate. Irrig Sci 24:85–99

Molle F, Mollinga P (2003) Water poverty indicators: conceptual problems and policy issues. Water Policy 5(5–6):529–544

Ondrasek G (2013) The responses of salt-affected plants to cadmium. In: Ahmad P, Azooz MM, Prasad MNV (eds) Salt stress in plants: signalling, omics and adaptations. Springer, New York, pp 439–463

Ondrasek G, Rengel Z (2012) The role of soil organic matter in trace elements bioavailability and toxicity. In: Ahmad P, Prasad MNV (eds) Abiotic stress responses in plants: metabolism productivity and sustainability. Springer, New York, pp 403–423

Ondrasek G, Romic D, Borosic J, Romic M, Stricevic I (2007) Comparison of transpiration models in tomato soilless culture. ACS 72(2):113–116

Ondrasek G, Rengel Z, Veres S (2011) Soil salinisation and salt stress in crop production. In: Shanker AK, Venkateswarlu B (eds) Abiotic stress in plants: mechanisms and adaptations, pp 171–190. Available at http://www.intechopen.com/articles/show/title/soil-salinisation-and-salt-stress-in-crop-production. Accessed 12 Jan 2013

Ondrasek G, Rengel Z, Romic D, Savic R (2012) Salinity decreases dissolved organic carbon in the rhizosphere and increases trace elements phyto-accumulation. Eur J Soil Sci 63(5):685–693

Petosic D, Marusic J, Tomic F, Mustac I, Stricevic I (2012) The role and importance of melioration drainage in flood protection of R. Croatia. In: Biondic D, Holjevic D (eds) Flood protection in Croatia, conference proceedings, pp 207–221. ISBN 978-953-7672-06-5

Romic D, Marusic J, Petosic D, Husnjak S et al (2005) National project for irrigation and management by agricultural land and water in Croatia (In Croatian only). University of Zagreb, Zagreb, p 250

Romic D, Ondrasek G, Romic M, Josip B, Vranjes M, Petosic D (2008) Salinity and irrigation method affect crop yield and soil quality in watermelon (*Citrullus lanatus* L.) growing. Irrig Drain 57(4):463–469

Romic D, Romic M, Rengel Z, Zovko M, Bakic H, Ondrasek G (2012) Trace metals in the coastal soils developed from estuarine floodplain sediments in the Croatian Mediterranean region. Environ Geochem Health 34(4):399–416

Savic R, Maksimovic L, Cimpeanu S, Bucur D, Ondrasek G, Vasin J, Harja M, Milic S (2013) Hazardous and harmful substances in sediments of the Jegricka stream. J Food Agric Environ 11(1):1152–1156

Seckler D, Amerasinghe U, Molden D, De Silva R, Barker R (1998) World water demand and supply, 1990 to 2025: scenarios and issues. IWMI research report 19, Colombo, Sri Lanka.

Seginer I (2002) The Penman-Monteith evapotranspiration equation as an element in greenhouse ventilation design. Biosyst Eng 82(4):423–439

Smith M (1992) Cropwat. A computer program for irrigation planning and management. FAO irrigation and drain paper 46. FAO, Rome, 126p

Teixeira JL, Farrajota MP, Pereira LS (1995) PROREG simulation software to design demand in irrigation projects. In: Pereira LS, van den Broek BJ, Kabat P, Allen RG (eds) Crop-water simulation models in practice. Wageningen, The Netherlands, pp 273–285

U.S. Department of Agriculture (USDA) (1999) Farm and Ranch irrigation survey (1998), vol 3, Special studies Part 1 of the 1997 Census of Agriculture, AC97-SP-1

U.S. Department of Agriculture (USDA) (2006) Agricultural water conservation practices. Available at http://www.nrcs.usda.gov/Internet/FSE_DOCUMENTS/nrcs143_023630.pdf. Accessed 20 Oct 2012

UNDP (2006) Human development report—beyond scarcity: power, poverty and the global water crisis. United Nations Development Programme, New York. ISBN 0-230-50058-7

UNEP (2008) Vital water graphics—an overview of the state of the world's fresh and marine waters, 2nd edn. United Nations Environment Programme, Nairobi. ISBN 92-807-2236-0

Unold G, Fank J (2008) Modular design of field lysimeters for specific application needs. Water Air Soil Pollut 8:233–242

Varela-Ortega C, Sagardoy JA (2001) The utilization of water resources for agriculture: analysis of the current regime and policy. Assistance in Institutional Strengthening and Agricultural Policy—Syria. FAO GCP/SYR/006/ITA

Water management strategy (SUV) (In Croatian only) (2008) Biondic D (ed) ISBN 978-953-7672-00-3, 180p. Available at http://www.voda.hr/Default.aspx?sec=569. Accessed 17 Feb 2010

Young GJ, Dooge JCI, Rodda JC (2004) Global water resource issues. Cambridge University Press, Cambridge, p 216

Chapter 5
Biotechnology for Drought and Salinity Tolerance of Crops

Faiçal Brini and Khaled Masmoudi

Contents

1 Introduction

There is a growing imbalance between supply and demand of the major cereals, viz., wheat, rice and maize, which together provide 70 % of the calorific intake for the world's population. Whilst in recent years, genetic and agronomic developments

F. Brini (✉)
Plant Protection and Improvement Laboratory, Centre of Biotechnology of Sfax (CBS)/University of Sfax, B.P. "1177" 3018, Sfax, Tunisia
e-mail: faical.brini@cbs.rnrt.tn

K. Masmoudi
Plant Protection and Improvement Laboratory, Centre of Biotechnology of Sfax (CBS)/University of Sfax, B.P. "1177" 3018, Sfax, Tunisia

International Center for Biosaline Agriculture (ICBA), P.O. Box 14660, Dubai, United Arab Emirates

P. Ahmad and M.R. Wani (eds.), *Physiological Mechanisms and Adaptation Strategies in Plants Under Changing Environment: Volume 1*, DOI 10.1007/978-1-4614-8591-9_5,

have steadily increased the production of these crops, the rate of increase is still less than that needed to match the requisite demand. This has caused price volatility and fuelled concerns over long-term food security. The demand for cereals is increasing in response to increased population and wealth. However, the loss of land for crop production due to urbanisation, degradation and alternate uses (e.g. for bioenergy crops or leisure) and the projected changes in climate are major obstacles against further increases in production.

The availability of water is a major determinant of plant production, and shortages of water are recognised as major threats to food security (Parry et al. 2005). In areas with low rainfall and high evapotranspiration (semiarid areas), plant growth can also be decreased by soil salinisation, a problem which is exacerbated by irrigation of poorly draining soils with low-quality water (Tardieu 2013). Developing high-yielding crops for water-limited environments is a major challenge. Conventional breeding for yield under any conditions is difficult because of the complex nature of the trait, which is determined by multiple genes, the large size of cereal genomes and the comparatively limited gene pool available for breeders (Malik et al. 2003). There is now genome mapping and sequence data available for some major food crops such as rice (Sakata et al. 2002) and sorghum (Paterson et al. 2009). However, exploitation of genomic data for improved crop performance under drought is limited by the complexity of the underlying traits which are often determined by multiple genes (Parry et al. 2005; Parry and Reynolds 2007) and the seasonal and year-on-year variation of water availability. However, biotechnological tools including plant transformation, random and targeted mutagenesis, transposon/T-DNA tagging and RNA interference (RNAi) permit the linking of genes to their biological function, thereby elucidating their contribution to traits, in ways not previously possible (closing the genotype to phenotype gap) (Pérez-Clémente et al. 2013). With this information, biotechnology has the potential to deliver higher and more stable yields for saline and water-limited environments (Fig. 5.1).

Genetic engineering has been used successfully to improve agronomically important traits in cereals (Vasil 2007). Herbicide-tolerant and insect-resistant genetically modified (GM) maize varieties have been in use since the mid-1990s, and insect-resistant GM rice varieties have been approved for commercialisation in China. The market for 'biotech maize' is now well established worldwide, with significant cultivation even in Europe (Halford 2006). The market for wheat biotechnology has proved more difficult to establish, but experimental GM lines have been produced with improved end-use quality traits (Shewry 2007; Tamás et al. 2009). Here we review the biotechnological methodologies that are available and the prospects for their successful application for improving drought and salt tolerance in cereals.

2 Genetic Approaches

Genetic analysis has played a role in wheat breeding for more than a century, and by the 1970s, the chromosomal locations had been established for major genes controlling dwarfing, spike morphology, grain colour and hardness, the major

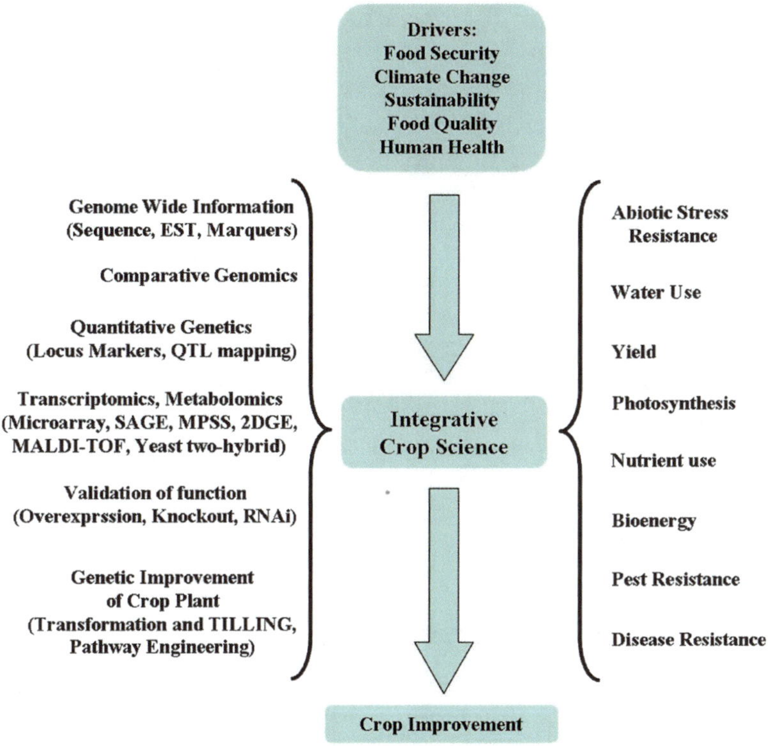

Fig. 5.1 Targets and approaches for improving crop performance under stress conditions. *2DGE* two-dimensional gel electrophoresis; *EST* expression sequence tag; *MALDI-TOF* matrix-assisted laser desorption/ionisation time of flight; *MPSS* massively parallel signature sequencing; *QTL* quantitative trait locus; *SAGE* serial analysis of gene expression

classes of storage proteins, vernalisation and photoperiod response (Snape 1998). Traits that are controlled by multiple genes and loci (quantitative traits), including yield and drought tolerance, have been more intractable, and it was not until the last decade of the twentieth century that progress began to accelerate. This was brought about by the development of genetic maps based on markers; initially these markers were based on restriction fragment length polymorphisms (RFLP), but subsequently a range of markers have been developed, including amplified fragment length polymorphisms (AFLP), random amplification of polymorphic DNA (RAPD), variable number tandem repeats (VNTR), microsatellite polymorphisms based on simple sequence repeats (SSR), single-nucleotide polymorphism (SNP), single feature polymorphism (SFP) and restriction site-associated DNA markers (RAD). Researchers and breeders have been able to construct genetic maps using these markers that enable a trait that segregates in a cross, to be attributed to a specific location in the genome, even if the exact gene responsible is not known. The locus that is identified is known as a quantitative trait locus, or QTL.

2.1 Targeting QTL for Tolerance to Drought and Salinity

Marker-assisted selections of target QTLs are powerful support for improving productivity under drought and/or saline conditions which will assist selection in the breeding process. One of the major difficulties in drought QTLs identification in crops in general and wheat in particular is the identification of the key physiological and morphological determinants of drought tolerance. Most QTLs for drought tolerance in wheat have been identified through yield and yield measurement under water-limited conditions (Maccaferri et al. 2008). Considerable progress has already been made in deconvoluting traits related to water use and in identifying variation in component traits (e.g. root traits—Clark et al. 2008; Courtois et al. 2009; leaf traits—Khowaja and Price 2008; Khowaja et al. 2009). The component traits may have direct impacts on the uptake and use of water or affect these processes indirectly, for example, in some water-limited environments, a shorter life cycle may enable a crop to escape from water limitation. Yields can be increased if such traits are strategically targeted and effectively selected for drought stress tolerance (Richards et al. 2010).

2.2 Mutagenesis and TILLING

Genetic mutation is a powerful tool that establishes a direct link between the biochemical function of a gene product and its role in vivo. Chemical mutagens have been used for forward genetic screens in a variety of organisms. Compounds, such as EMS (ethyl methanesulfonate) and DMS (dimethyl sulphate), are used to generate mutants. This class of mutagens causes a large number of random point mutations in the genome, thus theoretically multiple allele of any gene can be obtained in the population (Greene et al. 2003). Despite the clear advantages of EMS mutagenesis, until recently, it has been useful as a tool for reverse genetics because of the lack of high-throughput techniques for detecting point mutations.

Modern genomics makes reverse genetics possible as large amounts of genomic and expressed sequence information become available. In the last few years, the TILLING method (for Targeting Induced Local Lesions in Genomes; McCallum et al. 2000) have been developed. TILLING has been used successfully as a functional genomics discovery platform in model organisms such as *Arabidopsis* (McCallum et al. 2000; Till et al. 2006) and in plant systems including rice, barley, maize, wheat and soybean (Caldwell et al. 2004; Till et al. 2004; Slade et al. 2005; Anai 2012; Chen et al. 2012). TILLING has several advantages over other techniques used to detect single-bp polymorphism. Alleles generated by TILLING can be readily used in traditional breeding programmes, since the technology is non-transgenic and the mutations are stably inherited. This makes TILLING an attractive strategy not only for functional genomics but also for agricultural applications. It can be predicted that more and more direct or indirect benefits will be revealed through continuous applications of TILLING in the near future.

3 RNA Interference and Its Application in Cereals

RNAi is a potent and highly specific gene-silencing phenomenon that is based on sequence-specific RNA degradation following by the formation of double-stranded (dsRNA) homologous in sequence to the targeted gene (Marx 2000; Baulcombe 2004).

The natural function of RNAi and its related processes seem to be protection of the genome against invasion by mobile genetic elements such as transposons and viruses. Given the gene-specific feature of RNAi, it is conceivable that this method will play an important role in therapeutic application. RNAi has proven to be very efficient in interfering with gene expression in various plant systems such as *Arabidopsis thaliana* and rice (Chuang and Meyerowitz 2000; Miki et al. 2005).

Functional genomics using RNAi is particularly an attractive technique for genomic mapping and annotation in plants. RNAi has been successfully used for functional genomics studies in bread wheat (Travella et al. 2006) as well as plant model systems such as *Arabidopsis* and maize (McGinnis et al. 2005). To develop RNAi technology for functional genomics, there is a need to characterise, in molecular detail, the silencing of homologous genes as well as the inheritance of RNAi-induced phenotype (Travella et al. 2006).

4 Transcriptome Analyses of Plant Drought and Salt Stress Response

The transcriptomics approach deals with comprehensive analysis of gene expression in a cell. Understanding the transcriptome is essential for analysing the genomic function and the molecular constituents of cells and tissues. Different technologies have been developed to study the transcriptome, including northern hybridisation and quantitative real-time PCR (Q-RT-PCR). The above low-throughput techniques are still used for validating the results obtained from global approaches. Advances in genomics technologies allow measurement of transcript levels of thousands of genes at the same time. The DNA microarray, using the principle of nucleic acid hybridisation of mRNA or cDNA fragments, is among these techniques.

4.1 DNA Microarrays

Microarray technology is a powerful tool for analysing the expression profiles of many genes (Richmond and Somerville 2000; Seki et al. 2004). Basically, there are two types of microarray formats: cDNA arrays and oligoarrays. Despite its power and usefulness, microarray technology is both expensive and time intensive. Besides several technical problems such as contamination of DNA in spots on arrays, uneven hybridisation and spurious hybridisation, it requires multiple biological and technical replications for generating reliable data. Microarray technology has been applied to the analysis of expression profiles in response to abiotic stresses, such as drought, high salinity and cold (Kawasaki et al. 2001; Seki et al. 2001, 2002; Chen et al. 2002; Fowler and Thomashow 2002; Kreps et al. 2002; Lee et al. 2005). Stress-responsive genes have been identified in many plant species, such as *Arabidopsis* (Fowler and Thomashow 2002; Lee et al. 2005), *Arabidopsis*-related halophyte, *Thellungiella halophila* (Inan et al. 2004; Taji et al. 2004; Gong et al. 2005; Wong et al. 2006), rice (Kawasaki et al. 2001; Rabbani et al. 2003; Lan et al. 2005), barley (Oztur et al. 2002), wheat (Gulick et al. 2005), maize (Wang et al. 2003; Yu and Setter 2003), pine (Watkinson et al. 2003), hot pepper (Hwang et al. 2005), potato (Rensink et al. 2005), poplar (Gu et al. 2004; Brosche et al. 2005) and sorghum (Buchanan et al. 2005).

4.2 High-Throughput Approaches for the Identification of Drought and Salt-Tolerance Genes in Plants

The development of automated sequencing technologies has led to the production of sequencing machines with dramatically lower costs and higher throughput than the technology of just 2 years ago. The high-throughput sequencing technologies opened new view into the fields, thus allowing scientists to decode the genomes of many organisms (Soon et al. 2013). Various methods have been developed previously to directly determine cDNA sequences, based mostly around traditional (and more expensive) Sanger sequencing, whilst others include methodologies such as serial analysis of gene expression (SAGE) (Velculescu et al. 1995; Harbers and Carninci 2005), cap analysis gene expression (CAGE) (Nakamura and Carninci 2004; Shiraki et al. 2003; Kodzius et al. 2006) and massively parallel signature sequencing (MPSS) (Peiffer et al. 2008; Reinartz et al. 2002; Brenner et al. 2000). Recently, mapping and quantifying of transcriptomes can be easily done with the development of novel high-throughput DNA sequencing methods. This method, termed as RNA-Seq (RNA sequencing), has clear advantages over existing approaches and is expected to revolutionise the manner in which transcriptomes are analysed. It has already been applied to *Saccharomyces cerevisiae*, *Schizosaccharomyces pombe*, *Arabidopsis thaliana* and mouse and human cells (Wilhelm et al. 2008; Nagalakshmi et al. 2008; Lister et al. 2008; Mortazavi et al. 2008; Cloonan et al. 2008; Marioni et al. 2008). The unbiased information on transcript sequence abundance and unparalleled ability of HTS to quantitative yield has afforded some remarkable new insights into transcriptome complexity and regulation. RNA-Seq provides quantitative readout and extremely reproducible transcript abundance (e.g. Li et al. 2008; Marioni et al. 2008; Pan et al. 2008; Wang et al. 2008a). RNA-Seq offers a large dynamic range of expression levels and a high-level reproducibility and less RNA sample than either large-scale Sanger expressed sequenced tag (EST) sequencing or tiling arrays (Soneson and Delorenzi 2013). Transcriptome Sequencing (RNA-Seq) can be done with a variety of platforms to test many ideas and hypotheses such as HiSeq (Illumina, formerly Solexa), 5500xl SOLiD System (Live Technologies) and 454 Genome Analyzer FLX (Roche).

The various technologies differ in the procedures used to array the DNA fragments. There are two key features that determine which sequencing platform is best suited for each experiment: the length of sequenced reads and the total number of sequenced reads output. In general, the 454 Genome Analyzer FLX sequencer generates reads of up to 200–300 bp and is currently best suited for applications involving de novo genome and transcriptome assemblies. In contrast, HiSeq and SOLiD generate approximately 35 bp reads and are best suited for resequencing or applications such as gene profiling where the short length of the microread is not a concern.

4.3 Gene Expression Profiling for Abiotic Stress Tolerance in Crops

Several new stress-related pathways, in addition to the previously well-described stress-related genes, have been related to abiotic stress transcriptome profiling in model species such as *Arabidopsis* and rice (Desikan et al. 2001; Kreps et al. 2002; Chen et al. 2002; Seki et al. 2002; Oh et al. 2005; Wang et al. 2011). ESTs are currently used as an efficient and fast method for profiling genes expressed in various tissues, cell types or stages of development (Andrews et al. 2001). Based on the research results, estimates of gene number in the cereals are very similar to other complex organisms; for example, a total of approximately 13,000 abiotic stress-related ESTs were reported in barley and rice (Zhang et al. 2004) and approximately 21,000 ESTs in wheat (Mochida et al. 2004). The clustering of ESTs sequence generated from abiotic stress-treated cDNA libraries provides information on gene number and gene families involved in stress responses. Gene expression profiling using cDNA macroarrays or microarrays will provide an opportunity for the discovery of higher number of transcripts and pathways related to stress tolerance mechanisms. There are few published reports on the use of barley or wheat chips for studying altered gene expression in response to abiotic stress.

5 Proteomic Approaches for Abiotic Stress Response

The importance of protein profiling has long been acknowledged in plant abiotic stress studies. Proteomics not only involves large-scale identification of proteins but also deals with analysis of all protein isoforms and post-translational modifications, protein-protein interactions, enzymatic assays for the functional determination, localisation studies of gene products and promoter activity and structural information of protein complexes (Wilkins et al. 1996; Brosche et al. 2005). The advancement in MS techniques (O'Farrell 1975) coupled with database searching have played a crucial role in proteomics for proteins identification. Databases have been constructed containing all expressed proteins from plant organs and cell organelles of various species (Table 5.1).

Table 5.1 Websites for plant omics research

Transcriptomics-related websites	
Genevestigator	http://www.genevestigator.com/gv/index.jsp
Gene expression omnibus	http://www.ncbi.nlm.nih.gov/projects/geo
Stanford Microarray Database	http://smd.stanford.edu/index.shtml
ArrayExpress	http://www.ebi.ac.uk/arrayexpress
PLEXdb	http://www.barleybase.org/plexdb/html/index.php
TIGR *Arabidopsis* arrays	http://www.jcvi.org/arabidopsis/qpcr/
Rice transcriptional database	http://microarray.rice.dna.affrc.go.jp
Rice Expression Database (RED)	http://red.dna.affrc.go.jp/RED/
BarleyBase	http://www.barleybase.org
Zeamage	www.maizearray.org
TIGR Solanaceae Genomics Resource	http://www.jcvi.org/potato/
Soybean Genomics and Microarray Database	http://psi081.ba.ars.usda.gov/SGMD/default.htm
Tomato Expression Database	http://ted.bti.cornell.edu
Genomics-related websites	
EMBL nucleotide sequence database	http://www.ebi.ac.uk/embl
National Center for Biotechnology Information	http://www.ncbi.nlm.nih.gov
Gramene	http://www.gramene.org
GrainGenes	http://wheat.pw.usda.gov
Gene Ontology	www.geneontology.org
The Arabidopsis Information Resource (TAIR)	http://arabidopsis.org/index.jsp
Rice Genome Project (RGP)	http://rgp.dna.affrc.go.jp/
RiceGE	http://signal.salk.edu/cgi-bin/RiceGE
Oryzabase	http://www.shigen.nig.ac.jp/rice/oryzabase/top/top.jsp
Maize Sequence	http://maizesequence.org/index.html
Maize genome resources	http://www.maizegenome.org/
Maize Genetics and Genomics Database	http://www.maizegdp.org/genome/
Sorghum Genomics	http://sorghoblast3.tamu.edu
Proteomics-related websites	
Proteome analysis at EBI	http://www.ebi.ac.uk/proteome/
Swiss-Prot	http://us.expasy.org/sprot/
Arabidopsis Membrane Protein Library	http://www.cbs.umn.edu/arabidopsis/
Database for *A. thaliana* annotation	http://luggagefast.Stanford.EDU/group/arabprotein/
ExPASy *A. thaliana* 2D-proteome database	http://expasy.ch/cgi-bin/map2/def?ARABIDOPSIS
PlantsP: Functional Genomics of Plant Phosphorylation	http://PlantsP.sdsc.edu/

A major limitation of the current technology is the reduced coverage and inability to detect low abundance proteins. High-resolution 2D gels can resolve about 1,000 proteins that are highly abundant in a crude mixture. Even under optimal conditions, approximately 25 % of the proteome may be observed (Zivy and deVienne 2000). However, development in direct mass spectrometric analysis is increasing sensitivity, robustness and data handling (Wilkins et al. 1996). A number of proteome-wide platforms have been developed to complement mass spectrometric platforms. Yeast two-hybrid systems (Unlu et al. 1997) can detect weak interactions between low abundance proteins. Analogous to DNA microarrays, protein microarrays (Bayer et al. 2005) allow rapid interrogation of protein activity. The intensity or identity of resulting protein-protein interactions may be determined by fluorescence imaging or mass spectrometry.

New insights have been obtained on plant adaptation to abiotic stresses through application of proteomics approach to organelles and tissues in several plant species (Eldakak et al. 2013). Proteomics provided excellent opportunities to study the response of plants to stresses caused by heat, drought, salinity, ozone, heavy metals, UV light, nutrient deficiencies and elevated CO_2 conditions (Majoul et al. 2000). Proteome of poplar leaves (MacBeath 2002), rice anthers and leaves (Taylor et al. 2005; Renaut et al. 2006) and mitochondria of *Pisum sativum* (Fields and Song 1989) have been analysed to study plant response to cold stress. The effect of salinity stress, especially in crops plants, was investigated by comparative proteome studies in various tissue types in rice (Imin et al. 2004; Cui et al. 2005; Parker et al. 2006; Chitteti and Peng 2006), wheat (Yan et al. 2005; Zhang et al. 2009) and barley (Wang et al. 2008b).

Although proteomics has been exploited in abiotic stress tolerance studies in plants, large-scale proteomics studies are still limited. Application of proteomic approach particularly the comparative proteomics studies provided essential information about stress-induced alterations in protein quantity and quality and specific modifications of proteome (Abbasi and komatsu 2004).

6 Genetic Transformations of Cereals

Cereal improvement by genetic engineering requires the delivery, integration and expression of defined foreign genes into suitable regenerable explants. The available technologies and approaches used for production of transgenic cereal crops are complicated, and their efficiency is low. Moreover, different varieties of the same cereal crop and even different explants of the same variety would often require different methods for transformation. Two main methods are widely used for cereal transformation: (1) DNA transfer via particle bombardment developed by Sanford (1988) based on the use of the helium-driven PDS-1000/He particle gun and (2) *Agrobacterium*-based systems which exploit the ability to transfer a particular T-DNA on the tumour-inducing (Ti) plasmid into the nucleus of infected cells where it is then stably integrated into the host genome. Both of these methods involve delivery of the transgene to callus tissue, followed by selection of transformed cells and regeneration of plantlets carrying the gene of interest. Each method has its advantages and limitations: biolistic transformation facilitates a broad variety of transformation strategies with a wide range of gene expression, has no host limitations or biological constraints, and diverse cell types can be targeted efficiently for foreign DNA delivery (Altpeter et al. 2005). The main limitations of the bombardment approach include the insertion of backbone vector DNA and the insertion of multiple copies and fragmentation of the DNA during bombardment (Hu et al. 2003; Janakiraman et al. 2002). Numerous downstream breeding cycles are needed to select out those transgenic plants with good insertions and then to regenerate the homozygous lines used in breeding programmes for the development of a commercial product. Whereas *Agrobacterium*-mediated method is a simple, low cost alternative to particle bombardment. In addition, the *Agrobacterium*-mediated transformation system facilitates the precise integration of a small number of gene copies into the plant genome and shows a greater degree of stability for the transgene. Unlike microprojectile bombardment, *Agrobacterium* method seems to induce less transgene silencing, since introduced genes remain transcriptional active and has higher transformation efficiency than the microprojectile bombardment method. For all these reasons, *Agrobacterium*-mediated transformation has been adopted as the method of choice for most cereals. Therefore, main focus in this review will be on this transformation method.

The first success in cereal transformation using *Agrobacterium* was reported by Hiei et al. (1994) for stable transformation of rice. *Agrobacterium*-mediated transformation of other agronomically important cereal crop species, such as barley (Tingay et al. 1997), maize (Ishida et al. 1996) and wheat (Cheng et al. 1997), has now succeeded. For wheat, most of the research effort has focused on the model spring genotype 'Bobwhite' (Cheng et al. 1997, 2003; Haliloglu and Baenziger 2003; Hu et al. 2003), but reports of *Agrobacterium* transformation have been made using other spring varieties, Verry5, Cadenza, Fielder (Jones et al. 2005; Weir et al. 2001; Khanna and Daggard 2003; Wu et al. 2003), and the winter-type Florida (Wu et al. 2003; Jones et al. 2005).

Most of the protocols, efficiently used for cereal transformation, generally rely on the use of hypervirulent *Agrobacterium* strains such as AGL-0 and AGL-1 in wheat and barley (Tingay et al. 1997; Wu et al. 2003; Hensel et al. 2008), EHA101 and EHA105 in maize (Hood et al. 1986) as well as hypervirulent derivatives of LBA4404 in barley, maize and wheat (Khanna and Daggard 2003; Kumlehn et al. 2006; Hensel et al. 2008).

The recovery of stable plant cells after *Agrobacterium*-mediated transformation remains however influenced by many factors such as *Agrobacterium* strain, *Agrobacterium* density and surfactants, genotype, explant, binary vector, selectable marker gene and promoter, inoculation and coculture medium, inoculation and coculture conditions, regeneration medium, desiccation, osmotic treatment and tissue culture (Shrawat and Lorz 2006; Cheng et al. 2004). Some of these factors represent a drawback in extending the *Agrobacterium*-mediated transformation system to elite cultivars of economically important cereals. These limitations inspired some investigators to search for new alternative transformation procedures such as in planta transformation which involves no in vitro culture of plant cells or tissue (Supartana et al. 2005; Lin et al. 2009) or the flower dipping method originally developed for *Arabidopsis thaliana* transformation (Zale et al. 2009). Although these alternative methods seem simple and straightforward, yet they are technically challenging, and the results are not always convincing.

7 Conclusions and Future Perspective

Plants are often exposed to multiple abiotic stresses. Considerable advances have been made in understanding the plant's adaptation in stress environments and complex genetics involving multitude of gene and stress tolerance mechanisms. There is a great potential of genetic breeding for drought and salinity tolerance through the contribution of wild relatives to the identification of drought and salinity QTLs and functional markers. Gene expressing profiling has been widely used to understand mechanisms involved in the response of plants to abiotic stresses. Its application will determine a new revolution in crop research as technologies with lower costs. Future research effort should be directed using the omics approaches to elucidate plant's response to abiotic stresses. High-throughput omics technologies coupled with easily accessible integrated databases should now facilitate the elucidation of the complex stress regulatory network and their components to understand the mechanism of stress tolerance. The real benefits of these technologies, however, will only be realised when the knowledge and the tools resulting from the advances in omics field are translated into a product with improved abiotic stress tolerance in field environment.

Acknowledgments The authors would like to thank Dr. Parvaiz Ahmed and Dr. Mohd Rafiq Wani from the Department of Botany, Govt. Degree College (Boys), Anantnag, Jammu and Kashmir, India, for their invitation to contribute to this volume. This chapter was supported by grants from the Ministry of Higher Education and Scientific Research, Tunisia.

References

Abbasi F, Komatsu S (2004) A proteomic approach to analyze salt responsive proteins in rice leaf sheath. Proteomics 4:2072–2081

Altpeter F, Baisakh N, Beachy R, Bock R, Capell T, Christou P, Daniell H, Datta K, Datta S, Dix PJ, Fauquet C, Huang N, Kohli A, Mooibroek H, Nicholson L, Nguyen TT, Raemakers K, Romano A, Somers DA, Stoger E, Taylor N, Visser R (2005) Particle bombardment and the genetic enhancement of crops: myths and realities. Mol Breed 15:305–327

Anai T (2012) Potential of a mutant-based reverse genetic approach for functional genomics and molecular breeding in soybean. Breed Sci 61(5):462–467

Andrews MR, Mitra PP, DeCarvalho R (2001) Tripling the capacity of wireless communication using electromagnetic polarization. Nature 409:316–318

Baulcombe DC (2004) RNA silencing in plants. Nature 431:356–363

Bayer E, Bottrill AR, Walshaw J, Vigouroux M, Naldrett MJ, Thomas CL, Mule AJ (2005) *Arabidopsis* cell wall proteome defined using multidimensional protein identification technology. Proteomics 6:301–311

Brenner S, Johnson M, Bridgham J, Golda G, Lloyd DH, Johnson D, Luo SJ, McCurdy S, Foy M, Ewan M, Roth R, George D, Eletr S, Albrecht G, Vermaas E, Williams SR, Moon K, Burcham T, Pallas M, DuBridge RB, Kirchner J, Fearon K, Mao J, Corcoran K (2000) Gene expression analysis by massively parallel signature sequencing (MPSS) on microbead arrays. Nat Biotechnol 18:630–634

Brosche M, Vinocur B, Alatalo ER, Lamminmaki A, Teichmann T, Ottow EA, Djilianov D, Afif D, Bogeat-Triboulot MB, Altman A, Polle A, Dreyer E, Rudd S, Paulin L, Auvinen P, Kangasjarvi

J (2005) Gene expression and metabolite profiling of *Populus euphratica* growing in the Negev desert. Genome Biol 6:R101
Buchanan C, Lim S, Salzman RA, Kagiampakis I, Klein RR, Pratt LH, Cordonnier-Pratt M-M, Klein PE, Mullet JE (2005) *Sorghum bicolor*'s transcriptome response to dehydration, high salinity and ABA. Plant Mol Biol 58:699–720
Caldwell DG, McCallum N, Shaw P, Muehlbauer GJ, Marshall DF, Waugh R (2004) A structured mutant population for forward and reverse genetics in Barley (*Hordeum vulgare* L.). Plant J 40:143–150
Chen W, Provart NJ, Glazebrook J, Katagiri F, Chang HS, Eulgem T, Mauch F, Luan S, Zou G, Whitham SA, Budworth PR, Tao Y, Xie Z, Chen X, Lam S, Kreps J, Harper J, Si-Ammour A, Mauch-Mani B, Heinlein M, Kobayashi K, Hohn T, Dangl J, Wang X, Zhu T (2002) Expression profile matrix of Arabidopsis transcription factor genes suggests their putative functions in response to environmental stresses. Plant Cell 14:559–574
Chen L, Huang L, Min D, Philips A, Wang S, Madgwick PJ, Parry MAJ, Hu YG (2012) Development and characterization of a new TILLING population of common bread wheat (*Triticum aestivum* L.). PLoS One 7(7):e41750
Cheng M, Fry JE, Pang SZ, Zhou HP, Hironaka CM, Duncan DR, Conner TW, Wan YC (1997) Genetic transformation of wheat mediated by *Agrobacterium tumefaciens*. Plant Physiol 115:971–980
Cheng M, Hu TC, Layton J, Liu CN, Fry JE (2003) Desiccation of plant tissues post Agrobacterium infection enhances T-DNA delivery and increases stable transformation efficiency in wheat. In Vitro Cell Dev Biol Plant 39:595–604
Cheng M, Lowe BA, Spencer TM, Ye XD, Armstrong CL (2004) Factors influencing *Agrobacterium*-mediated transformation of monocotyledonous species. In Vitro Cell Dev Biol Plant 40:31–45
Chitteti BR, Peng ZH (2006) Proteome and phosphoproteome differential expression under salinity stress in rice (*Oryza sativa*) roots. J Proteome Res 6:1718–1727
Chuang CF, Meyerowitz EM (2000) Specific and heritable genetic interference by double-stranded RNA in *Arabidopsis thaliana*. Proc Natl Acad Sci U S A 97:4985–4990
Clark LJ, Price AH, Steele KA, Whalley R (2008) Evidence from near-isogenic lines that root penetration increases with root diameter and bending stiffness in rice. Funct Plant Biol 35:1163–1171
Cloonan N, Forrest AR, Kolle G, Gardiner BB, Faulkner GJ, Brown MK, Taylor DF, Steptoe AL, Wani S, Bethel G, Robertson AJ, Perkins AC, Bruce SJ, Lee CC, Ranade SS, Peckham HE, Manning JM, McKernan KJ, Grimmond SM (2008) Stem cell transcriptome profiling via massive-scale mRNA sequencing. Nat Methods 5:613–619
Courtois B, Ahmadi N, Khowaja F, Price A, Rami JF, Frouin J, Hamelin C, Ruiz M (2009) Rice root genetic architecture: meta-analysis from a QTL database improves resolution to a few candidate genes. Rice 2:115–128
Cui S, Huang F, Wang J, Ma X, Cheng Y, Liu J (2005) A proteomic analysis of cold stress responses in rice seedlings. Proteomics 5:3162–3172
Desikan R, Mackerness SAH, Hancock JT, Neill SJ (2001) Regulation of the Arabidopsis transcriptome by oxidative stress. Plant Physiol 127:159–172
Eldakak M, Milad SIM, Nawar AI, Rohila JS (2013) Proteomics: a biotechnology tool for crop improvement. Front Plant Sci 4:35
Fields S, Song O (1989) A novel genetic system to detect protein-protein interactions. Nature 340:245–246
Fowler S, Thomashow MF (2002) *Arabidopsis* transcriptome profiling indicates that multiple regulatory pathways are activated during cold acclimation in addition to the CBF cold response pathway. Plant Cell 14:1675–1690
Gong Q, Li P, Ma S, Rupassara SI, Bohnert HJ (2005) Salinity stress adaptation competence in the extremophile *Thellungiella halophila* in comparison with its relative *Arabidopsis thaliana*. Plant J 44:826–839
Greene EA et al (2003) Spectrum of chemically induced mutations from a large-scale reverse-genetic screen in *Arabidopsis*. Genetics 164:731–740

Gu R, Fonseca S, Puskas LG, Hackler L Jr, Zvara A, Dudits D, Pais M (2004) Transcript identification and profiling during salt stress and recovery of *Populus euphratica*. Tree Physiol 24:265–276
Gulick PJ, Drouin S, Yu Z, Danyluk J, Poisson G, Monroy AF, Sarhan F (2005) Transcriptome comparison of winter and spring wheat responding to low temperature. Genome 48:913–923
Halford NG (2006) From primitive selection to genetic modification, ten thousand years of plant breeding. In: Halford N (ed) Plant biotechnology: current and future applications of genetically modified crops. Wiley, London, pp 3–27
Haliloglu K, Baenziger PS (2003) *Agrobacterium tumefaciens* mediated wheat transformation. Cereal Res Commun 31:9–16
Harbers M, Carninci P (2005) Tag-based approaches for transcriptome research and genome annotation. Nat Methods 2:495–502
Hensel G, Valkov V, Middlefell-Williams J, Kumlehn J (2008) Efficient generation of transgenic barley: the way forward to modulate plant-microbe interactions. J Plant Physiol 165:71–82
Hiei Y, Ohta S, Komari T, Kumashiro T (1994) Efficient transformation of rice (*Oryza sativa* L) mediated by *Agrobacterium* and sequence-analysis of the boundaries of the T-DNA. Plant J 6:271–282
Hood EE, Helmer GL, Fraley RT, Chilton MD (1986) The hypervirulence of *Agrobacterium tumefaciens* A281 is encoded in a region of pTiBo542 outside of T-DNA. J Bacteriol 186:1291–1301
Hu T, Metz S, Chay C, Zhou HP, Biest N, Chen G, Cheng M, Feng X, Radionenko M, Lu F, Fry J (2003) *Agrobacterium*-mediated large-scale transformation of wheat (*Triticum aestivum* L.) using glyphosate selection. Plant Cell Rep 21:1010–1019
Hwang EW, Kim KA, Park SC, Jeong MJ, Byun MO, Kwon HB (2005) Expression profiles of hot pepper (Capsicum annum) genes under cold stress conditions. J Biosci 30:657–667
Imin N, Kerim T, Rolfe BG, Weinman JJ (2004) Effect of early cold stress on the maturation of rice anthers. Proteomics 4:1873–1882
Inan G, Zhang Q, Li P, Wang Z, Cao Z, Zhang H, Zhang C, Quist TM, Goodwin SM, Zhu J, Shi H, Damsz B, Charbaji T, Gong Q, Ma S, Fredricksen M, Galbraith DW, Jenks MA, Rhodes D, Hasegawa PM, Bohnert HJ, Joly RJ, Bressan RA, Zhu JK (2004) Salt cress. A halophyte and cryophyte Arabidopsis relative model system and its applicability to molecular genetic analyses of growth and development of extremophiles. Plant Physiol 135:1718–1737
Ishida Y, Saito H, Ohta S, Hiei Y, Komari T, Kumashiro T (1996) High efficiency transformation of maize (*Zea mays* L.) mediated by *Agrobacterium tumefaciens*. Nat Biotechnol 14(6):745–750
Janakiraman V, Steinau M, Mccoy SB, Trick HN (2002) Recent advances in wheat transformation. In Vitro Cell Dev Biol Plant 38:404–414
Jones HD, Doherty A, Wu H (2005) Review of methodologies and a protocol for the *Agrobacterium-mediated* transformation of wheat. Plant Methods 1:5
Kawasaki S, Borchert C, Deyholos M, Wang H, Brazille S, Kawai K, Galbraith D, Bohnert H (2001) Gene expression profiles during the initial phase of salt stress in rice. Plant Cell 13:889–905
Khanna HK, Daggard GE (2003) *Agrobacterium tumefaciens* mediated transformation of wheat using a superbinary vector and a polyamine-supplemented regeneration medium. Plant Cell Rep 21:429–436
Khowaja FS, Price AH (2008) QTL mapping rolling, stomatal conductance and dimension traits of excised leaves in the Bala x Azucena recombinant inbred population of rice. Field Crop Res 106:248–257
Khowaja FS, Norton NJ, Courtois B, Price AH (2009) Improved resolution in the position of drought-related QTLs in a single mapping population of rice by meta-analysis. BMC Genomics 10:279. doi:10.1186/1471-2164-10-276
Kodzius R, Kojima M, Nishiyori H, Nakamura M, Fukuda S, Tagami M, Sasaki D, Imamura K, Kai C, Harbers M, Hayashizaki Y, Carninci P (2006) CAGE: cap analysis of gene expression. Nat Methods 3:211–222
Kreps JA, Wu YJ, Chang HS, Zhu T, Wang X, Harper JF (2002) Transcriptome changes for Arabidopsis in response to salt, osmotic, and cold stress. Plant Physiol 130:2129–2141

Kumlehn J, Serazetdinova L, Hensel G, Becker D, Loerz H (2006) Genetic transformation of barley (*Hordeum vulgare* L.) via infection of androgenetic pollen cultures with *Agrobacterium tumefaciens*. Plant Biotechnol J 4:251–261
Lan L, Li M, Lai Y, Xu W, Kong Z, Ying K, Han B, Xue Y (2005) Microarray analysis reveals similarities and variations in genetic programs controlling pollination/fertilization and stress responses in rice (*Oryza sativa* L.). Plant Mol Biol 59:151–164
Lee BH, Henderson DA, Zhu JK (2005) The *Arabidopsis* cold-responsive transcriptome and its regulation by ICE1. Plant Cell 17:3155–3175
Li H, Lovci MT, Kwon YS, Rosenfeld MG, Fu XD, Yeo GW (2008) Determination of tag density required for digital transcriptome analysis: application to an androgen-sensitive prostate cancer model. Proc Natl Acad Sci U S A 105:20179–20184
Lin J, Zhou B, Yang Y, Mei J, Zhao X, Guo X, Huang X, Tang D, Liu X (2009) Piercing and vacuum infiltration of the mature embryo: a simplified method for *Agrobacterium*-mediated transformation of indica rice. Plant Cell Rep 28:1065–1074
Lister R, O'Malley RC, Tonti-Filippini J, Gregory BD, Berry CC, Millar AH, Ecker ER (2008) Highly integrated single base resolution maps of the epigenome of *Arabidopsis*. Cell 133:523–536
MacBeath G (2002) Protein microarrays and proteomics. Nat Genet 32:526–532
Maccaferri M, Sanguineti MC, Corneti S, Ortega JL, Salem MB, Bort J, DeAmbrogio E, del Moral LF, Demontis A, El-Ahmed A, Maalouf F, Machlab H, Martos V, Moragues M, Motawaj J, Nachit M, Nserallah N, Ouabbou H, Royo C, Slama A, Tuberosa R (2008) Quantitative trait loci for grain yield and adaptation of durum wheat (*Triticum durum* Desf.) across a wide range of water availability. Genetics 178:489–511
Majoul T, Chahed K, Zamiti E, Ouelhazi L, Ayadi A, Brulfert J (2000) Analysis by two dimensional-electrophoresis of the effect of salt stress on the polypeptide patterns in roots of a salt-tolerant and salt-sensitive cultivar of wheat. Electrophoresis 21:2562–2565
Malik ST, Hamid R, Tayyaba Y, Minhas NM (2003) Effects of 2, 4-D on callus induction from mature wheat (*Triticum aestivum* L.) seeds. Int J Agric Biol 6:156–159
Marioni J, Mason C, Mane S, Stephens M, Gilad Y (2008) RNA-seq: an assessment of technical reproducibility and comparison with gene expression arrays. Genome Res 18(9):1509–1517. doi:10.1101/gr.079558.108
Marx J (2000) Interfering with gene expression. Sciences 288:1370–1372
McCallum CM, Comai L, Greene EA, Henikoff S (2000) Targeted screening for induced mutations. Nat Biotechnol 18(4):455–457
McGinnis K, Chandler V, Cone K, Kaeppler H, Kaeppler S, Kerschen A, Pikaard C, Richards E, Sidorenko L, Smith T, Springer N, Wulan T (2005) Transgene-induced RNA interference as a tool for plant functional genomics. Methods Enzymol 392:1–24
Miki D, Itoch R, Shimamoto K (2005) RNA silencing of single and multiple members in a gene family of rice. Plant Physiol 138:1903–1913
Mochida K, Yamazaki Y, Ogihara Y (2004) Discrimination of homologous gene expression in hexaploid wheat by SNP analysis of contigs grouped from a large number of expressed sequence tags. Mol Genet Genome 270:371–377
Mortazavi A, Williams BA, McCue K, Schaeffer L, Wold B (2008) Mapping and quantifying mammalian transcriptomes by RNA-Seq. Nat Methods 5:621–628
Nagalakshmi U, Wang Z, Waern K, Shou C, Raha D, Gerstain M, Snyder M (2008) The transcriptional landscape of the yeast genome defined by RNA sequencing. Sciences 320:1344–1349
Nakamura M, Carninci P (2004) [Cap analysis gene expression: CAGE]. Tanpakushitsu Kakusan Koso 49:2688–2693 (in Japanese)
O'Farrell PH (1975) High resolution 2-dimensional electrophoresis of proteins. J Biol Chem 250:4007–4021
Oh SJ, Song SI, Kim YS, Jang HJ, Kim SY, Kim M, Kim YK, Nahm BH, Kim JK (2005) *Arabidopsis* CBF3/DREB1A and ABF3 in transgenic rice increased tolerance to abiotic stress without stunting growth. Plant Physiol 138:341–351

Oztur ZN, Talame V, Deyholos M, Michalowski CB, Galbraith DW, Gozukirmizi N, Tuberosa R, Bohnert HJ (2002) Monitoring large-scale changes in transcript abundance in drought- and salt-stressed barley. Plant Mol Biol 48:551–573

Pan Q, Shai O, Lee LJ, Frey BJ, Blencowe BJ (2008) Deep surveying of alternative splicing complexity in the human transcriptome by high-throughput sequencing. Nat Genet 40:1413–1415

Parker R, Flowers TJ, Moore AL, Harpham NJV (2006) An accurate and reproducible method for proteome profiling of the effects of salt stress in the rice leaf lamina. J Exp Bot 57:1109–1118

Parry MAJ, Reynolds MP (2007) Improving resource use efficiency. Ann Appl Biol 151: 133–135

Parry MAJ, Flexas J, Medrano H (2005) Prospects for crop production under drought: research priorities and future directions. Ann Appl Biol 147:211–226

Paterson AH, Bowers JE, Bruggmann R, Dubchak I, Grimwood J, Gundlach H, Haberer G, Hellsten U, Mitros T, Poliakov A, Schmutz J, Spannagl M, Tang HB, Wang XY, Wicker T, Bharti AK, Chapman J, Feltus FA, Gowik U, Grigoriev IV, Lyons E, Maher CA, Martis M, Narechania A, Otillar RP, Penning BW, Salamov AA, Wang Y, Zhang LF, Carpita NC, Freeling M, Gingle AR, Hash CT, Keller B, Klein P, Kresovich S, McCann MC, Ming R, Peterson DG, Mehboob-ur R, Ware D, Westhoff P, Mayer KFX, Messing J, Rokhsar DS (2009) The Sorghum bicolor genome and the diversification of grasses. Nature 457:551–556

Peiffer JA, Kaushik S, Sakai H, Arteaga-Vazquez A, Sanchez-Leon N, Ghazal H, Vielle-Calzada JP, Meyers BC (2008) A spatial dissection of the *Arabidopsis* floral transcriptome by MPSS. BMC Plant Biol 8:43

Pérez-Clémente RM, Vives V, Zandalinas SI, Lopèz-Climent MF, Munoz V, Cadenas AG (2013) Biotechnological approaches to study plant responses to stress. Biomed Res Int 2013:654120

Rabbani MA, Maruyama K, Abe H, Khan MA, Katsura K, Ito Y, Yoshiwara K, Seki M, Shinozaki K, Yamaguchi-Shinozaki K (2003) Monitoring expression profiles of rice (*Oryza sativa* L.) genes under cold, drought and high-salinity stresses, and ABA application using both cDNA microarray and RNA gel blot analyses. Plant Physiol 133:1755–1767

Reinartz J, Bruyns E, Lin JZ, Burcham T, Brenner S, Bowen B, Kramer M, Woychik R (2002) Massively parallel signature sequencing (MPSS) as a tool for in-depth quantitative gene expression profiling in all organisms. Brief Funct Genomic Proteomic 1:95–104

Renaut J, Hausman JF, Wisniewski ME (2006) Proteomics and low temperature studies: bridging the gap between gene expression and metabolism. Physiol Plant 126:97–109

Rensink WA, Lobst S, Hart A, Stegalkina S, Liu J, Buell CR (2005) Gene expression profiling of potato responses to cold, heat, and salt stress. Funct Integr Genomics 5:201–207

Richards A, Rebetzke GJ, Watt M, Saad RB, Zouari N, Ramdhan WB (2010) Breeding for improved water productivity in temperate cereals: phenotyping, quantitative trait loci, markers and the selection environment. Funct Plant Biol 37:85–97

Richmond T, Somerville S (2000) Chasing the dream: plant EST microarrays. Curr Opin Plant Biol 3:108–116

Sakata K, Nagamura Y, Numa H, Antonio BA, Nagasaki H, Idonuma A, Watanabe W, Shimizu Y, Horiuchi I, Matsumoto T, Sasaki T, Higo K (2002) RiceGAAS: an automated annotation system and database for rice genome sequence. Nucleic Acids Res 30:98–102

Sanford JC (1988) The biolistic process. Trends Biotechnol 6:299–302

Seki M, Narusaka M, Abe H, Kasuga M, Yamaguchi-Shinozaki K, Carninci P, Hayashizaki Y, Shinozaki K (2001) Monitoring the expression pattern of 1300 *Arabidopsis* genes under drought and cold stresses using a full-length cDNA microarray. Plant Cell 13:61–72

Seki M, Narusaka M, Kamiya A, Ishida J, Satou M, Sakurai T, Nakajima M, Enju A, Akiyama K, Oono Y, Muramatsu M, Hayashizaki Y, Kawai J, Carninci P, Itoh M, Ishii Y, Arakawa T, Shibata K, Shinagawa A, Shinozaki K (2002) Functional annotation of a full-length *Arabidopsis* cDNA collection. Science 296:141–145

Seki M, Satou M, Sakurai T, Akiyama K, Iida K, Ishida J, Nakajima M, Enju A, Narusaka M, Fujita M, Oono Y, Kamei A, Yamaguchi-Shinozaki K, Shinozaki K (2004) RIKEN *Arabidopsis* full-length (RAFL) cDNA and its applications for expression profiling under abiotic stress conditions. J Exp Bot 55:213–223

Shewry P (2007) Improving the protein content and composition of cereal grain. J Cereal Sci 46:239–250

Shiraki T, Kondo S, Katayama S, Waki K, Kasukawa T, Kawaji H, Kodzius R, Watahiki A, Nakamura M, Arakawa T, Fukuda S, Sasaki D, Podhajska A, Harbers M, Kawai J, Carninci P, Hayashizaki Y (2003) Cap analysis gene expression for high-throughput analysis of transcriptional starting point and identification of promoter usage. Proc Natl Acad Sci U S A 100:15776–15781

Shrawat AK, Lorz H (2006) *Agrobacterium*-mediated transformation of cereals: a promising approach crossing barriers. Plant Biotechnol J 4:575–603

Slade AJ, Fuerstenberg SI, Loeffler D, Steine MN, Faciotti D (2005) A reverse genetic, nontransgenic approach to wheat crop improvement by TILLING. Nat Biotechnol 23:75–81

Snape JW (1998) Golden calves or white elephants? Biotechnologies for wheat improvement. Euphytica 100:207–217

Soneson C, Delorenzi M (2013) A comparison of methods for differential expression analysis of RNA-seq data. BMC Bioinformatics 14(1):91

Soon WW, Hariharan M, Snyder M (2013) High throughput sequencing for biology and medicine. Mol Syst Biol 9:640

Supartana P, Shimizu T, Shioiri H, Nogawa M, Nozue M, Kojima M (2005) Development of simple and efficient in planta transformation method for rice (*Oryza sativa* L.) using *Agrobacterium tumefaciens*. J Biosci Bioeng 100:391–397

Taji T, Seki M, Satou M, Sakurai T, Kobayashi M, Ishiyama K, Narusaka Y, Narusaka M, Zhu JK, Shinozaki K (2004) Comparative genomics in salt tolerance between *Arabidopsis* and *Arabidopsis*-related halophyte salt cress using Arabidopsis microarray. Plant Physiol 135:1697–1709

Tamás C, Kisgyörgy BN, Rakszegi M, Wilkinson MD, Yang MS, Láng L, Tamás L, Bedő Z (2009) Transgenic approach to improve wheat (*Triticum aestivum* L.) nutritional quality. Plant Cell Rep 28:1085–1094

Tardieu F (2013) Plant response to environmental conditions: assessing potential production, water demand, and negative effects of water deficit. Front Physiol 4:17

Taylor NL, Heazlewood JL, Day DA, Millar AH (2005) Differential impact of environmental stresses on the pea mitochondrial proteome. Mol Cell Proteomics 4:1122–1133

Till B, Reynolds S, Weil C, Springer N, Burtner C, Young K, Bowers E, Codomo C, Enns L, Odden A, Greene E, Cormai L, Henikoff S (2004) Discovery of induced point mutations in maize genes by TILLING. BMC Plant Biol 4:12

Till BJ, Zerr T, Comai L, Henikoff S (2006) A protocol for TILLING and ecotilling in plants and animals. Nat Protoc 1(5):2465–2477

Tingay S, McElroy D, Kalla R, Fieg S, Wang MB, Thornton S, Brettell R (1997) Agrobacterium tumefaciens-mediated barley transformation. Plant J 11:1369–1376

Travella S, Klimm TE, Keller B (2006) RNA interference-based gene silencing as an efficient tool for functional genomics in hexaploid bread wheat. Plant Physiol 142:6–20

Unlu M, Morgan ME, Minden JS (1997) Difference gel electrophoresis: a single gel method for detecting changes in protein extracts. Electrophoresis 18:2071–2077

Vasil IK (2007) Molecular genetic improvement of cereals: transgenic wheat (*Triticum aestivum* L.). Plant Cell Rep 26:1133–1154

Velculescu VE, Zhang L, Vogelstein B, Kinzler KW (1995) Serial analysis of gene expression. Sciences 270:484–487

Wang H, Miyazaki S, Kawai K, Deyholos M, Galbraith DW, Bohnert HJ (2003) Temporal progression of gene expression responses to salt shock in maize roots. Plant Mol Biol 52:873–891

Wang ET, Sandberg R, Luo S, Khrebtukova I, Zhang L, Mayr C, Kingsmore SF, Schroth GP, Burge CB (2008a) Alternative isoform regulation in human tissue transcriptomes. Nature 456:470–476

Wang MC, Peng ZY, Li CL, Li F, Liu C, Xia GM (2008b) Proteomic analysis on a high salt tolerance introgression stain of *Triticum aestivum/Thinopyrum ponticum*. Proteomics 8:1470–1489

Wang D, Pan Y, Zhao X, Zhu L, Fu B, Li Z (2011) Genome-wide temporal-spatial gene expression profiling of drought responsiveness in rice. BMC Genomics 12:149

Watkinson JI, Sioson AA, Vasquez-Robinet C, Shukla M, Kumar D, Ellis M, Heath LS, Ramakrishnan N, Chevone B, Watson LT, Zyl LV, Egertsdotter U, Sederoff RR, Grene R (2003) Photosynthetic acclimation is reflected in specific patterns of gene expression in drought stressed loblolly pine. Plant Physiol 133:1702–1716

Weir B, Gu X, Wang MB, Upadhyaya N, Elliott AR, Brettell RIS (2001) *Agrobacterium tumefaciens-mediated* transformation of wheat using suspension cells as a model system and green fluorescent protein as a visual marker. Aust J Plant Physiol 28:807–818

Wilhelm BT, Marguerate S, Watt S, Schubert F, Wood V, Goodhead I, Penkett CJ, Rogers J, Bahler J (2008) Dynamic repertoire of a eukaryotic transcriptome surveyed at single-nucleotide resolution. Nature 453:1239–1243

Wilkins MR, Pasquali C, Appel RD, Ou K, Golaz O, Sanchez JC, Yan JX, Gooley AA, Hughes G, Humphery-Smith I, Williams KL, Hochstrasser DF (1996) From proteins to proteomes: large scale identification by two-dimensional electrophoresis and amino acid analysis. Biotechnology 14:61–65

Wong CE, Li Y, Labbe A, Guevara D, Nuin P, Whitty B, Diaz C, Golding GB, Gray GR, Weretilnyk EA, Griffith M, Moffatt BA (2006) Transcriptional profiling implicates novel interactions between abiotic stress and hormonal responses in *Thellungiella*, a close relative of *Arabidopsis*. Plant Physiol 140:1437–1450

Wu H, Sparks C, Amoah B, Jones HD (2003) Factors influencing successful *Agrobacterium-mediated* genetic transformation of wheat. Plant Cell Rep 21:659–668

Yan S, Tang Z, Su W, Sun W (2005) Proteomic analysis of salt stress-responsive proteins in rice roots. Proteomics 5:235–244

Yu LX, Setter TL (2003) Comparative transcriptional profiling of placenta and endosperm in developing maize kernels in response to water deficit. Plant Physiol 131:568–582

Zale JM, Agarwal S, Loar S, Steber M (2009) Evidence for stable transformation of wheat by floral dip in Agrobacterium tumefaciens. Plant Cell Rep 28:903–913

Zhang H, Sreenivasulu N, Weschke W, Stein N, Rudd S, Radchuk V, Potokina E, Scholz U, Schweizer P, Zierold U, Langridge P, Varshney RK, Wobus U, Graner A (2004) Large-scale analysis of the barley transcriptome based on expressed sequence tags. Plant J 40:276–290

Zhang L, Tian LH, Zhao JF, Song Y, Zhang CJ, Guo Y (2009) Identification of an apoplastic protein involved in the initial phase of salt stress response in rice root by two-dimensional electrophoresis. Plant Physiol 149:916–928

Zivy M, deVienne D (2000) Proteomics: a link between genomics, genetics and physiology. Plant Mol Biol 44:575–580

Chapter 6
Effect of Salinity on Plants and the Role of Arbuscular Mycorrhizal Fungi and Plant Growth-Promoting Rhizobacteria in Alleviation of Salt Stress

Gopal Selvakumar, Kiyoon Kim, Shuijin Hu, and Tongmin Sa

Contents

G. Selvakumar • K. Kim • T. Sa (✉)
Department of Environmental and Biological Chemistry, Chungbuk National University, Cheongju, Chungbuk 361-763, Republic of Korea
e-mail: tomsa@chungbuk.ac.kr

S. Hu
Laboratory of Soil Ecology, Department of Plant Pathology, North Carolina State University, Raleigh, NC 27695 7616, USA

P. Ahmad and M.R. Wani (eds.), *Physiological Mechanisms and Adaptation Strategies in Plants Under Changing Environment: Volume 1*, DOI 10.1007/978-1-4614-8591-9_6,

1 Introduction

Soil salinity has received great attention from agriculturalists all over the world due to its adverse effects on plant growth and yield. More than 800 million hectares (around 6 %) of the total surface land of earth is affected by salinity (Arzani 2008). Among the 1.5 billion hectares of total cultivable land present in the world, about 5 % which is approximately 77 million hectares are affected by salinity (Evelin et al. 2009). In particular arid and semiarid areas are facing major ecological and agronomical problems (Ruiz-Lozano et al. 2012) due to salinity. Soil salinity that occurs naturally is commonly known as primary salinity, while soil salinity that occurs or induced by human activities is known as secondary salinity. The most common causes of secondary salinity are improper irrigation system and dry land. Approximately 20 % of the irrigated cultivable land is severely affected by soil salinity (Yamaguchi and Blumwald 2005; Wu et al. 2010), and the percentage is increasing due to improper irrigation practices (Jun-li and Yue-hu 2009; Bothe 2012). Soil salinity control methods like land reclamation and improved irrigation system provide only short-term solution and are often expensive (Arzani 2008).

1.1 What Is Salinity?

Proper evaluation of the condition of the soil is necessary before methods to alleviate soil salinity can be done. Depending on the concentration of salt, soils can be categorized as saline and sodic. Saline soils have a high electrical conductivity (EC>4 dS m^{-1}), with sodium absorption ratio of less than 13 and with a pH value less than 8.5. Sodic soil has less electrical conductivity (EC<4 dS m^{-1}) compared to saline soil, with sodium absorption ratio of more than 13 and with a pH value more than 8.5.

1.2 Salinity Effect on Plants

Plants exposed to soil salinity are negatively affected with reduction in establishment, growth, and yield (Zahran 1999; Ruiz-Lozano et al. 2012). Salinity present in the soil accumulates in the root zone of plants which eventually damages the crop and reduction in plant yield (Al-Karaki 2000). Depending on the concentration of salt in soil, the ability of plants to withstand salinity and effects on growth and yield differ. Although salinity differs widely in all agricultural land, there is no certain level of concentration which can inhibit plant growth or make it unproductive. Plants that grow under increasing saline conditions become chlorotic and eventually die. Table 6.1 shows plant responses to different salinity levels with conductivity ranges based on the Food and Agriculture Organization (FAO).

Salinity affects plant in three different aspects (Arzani 2008; Colla et al. 2008; Porcel et al. 2012; Ruiz-Lozano et al. 2012). Firstly, the low osmotic potential of the soil reduces the available water in the soil causing physiological water deficit

Table 6.1 Effects of different level of salinity on plant physiology and yield

Soil salinity class	ECe (dS m^{-1})[a]	Effect on crop plants
Non-saline	0–2	Salinity effects negligible
Slightly saline	2–4	Yields of sensitive crops may be restricted
Moderately saline	4–8	Yields of many crops are restricted
Strongly saline	8–16	Only tolerant crops yield satisfactorily
Very strongly saline	>16	Only a few very tolerant crops yield satisfactorily

[a]ECe—conductivity of the saturation extract

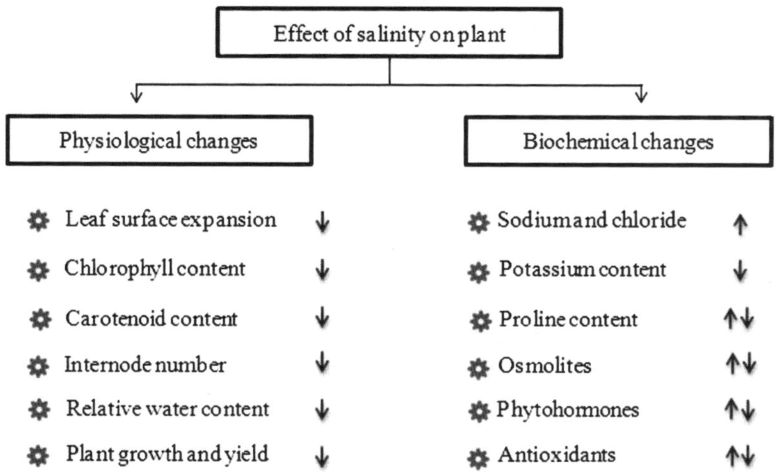

Fig. 6.1 Schematic representation of effects of salinity on plants. *Down arrows* indicate reduction in activity/content. *Up arrows* indicate increase in activity/content

in the plants. To overcome this, plants must refrain from losing water from their roots to the soil by maintaining lower osmotic potential (Feng et al. 2002; Jahromi et al. 2008). Secondly, excessive toxic effects of ions specifically sodium and chloride lead to the disruption of plasma membrane; damage to cell organelles and other macromolecules; reduction in photosynthetic activity, respiration, and protein synthesis; and inhibition of various enzymatic activities (Feng et al. 2002). Finally, salinity decreases nutrient uptake, thus leading to nutrient imbalance in the plant and/or transport to the shoot (Evelin et al. 2009). Physiological and biochemical changes in plants during salt stress are presented in Fig. 6.1.

1.3 Plant Responses to Salinity

The effects of high salt concentration on plants and various responses of plants to salinity have been studied for several decades (Flowers et al. 1977; Greenway and Munns 1980; Hasegawa et al. 2000; Porcel et al. 2012). Munns (1993) suggested that salinity may indirectly reduce plant growth by affecting turgor, photosynthesis, or enzyme activities. Direct effects of salinity include accumulation of salt in old leaves which may hasten leaf death. This prevents the supply of assimilates or hormones to the growing regions which eventually affects the plant growth.

1.3.1 Growth and Development

High concentrations of Na^+ and Cl^- ions in the soil or water reduce water potential which leads to initial growth reduction and low productivity (Parida and Das 2005). Suppression of plant growth occurs in all plants, but their tolerance level to salinity and the rate of growth reduction differ among plant species. Salinity severely affects the early stages of seedling growth. Seed germination is significantly reduced by salinity (Chartzoulakis and Loupassaki 1997). Moreover, early responses of plants to salinity include reduction in leaf surface expansion followed by cessation of expansion (Andriolo et al. 2005; Saied et al. 2005; Colla et al. 2006). Low osmotic potential reduces plant water uptake. Sairam et al. (2002) reported that plants exposed to salinity showed decreased relative water content. A reduction in the number of internodes in grafted tomato plants was under salinity stress over control (Voutsela et al. 2012). Root dry weight and shoot dry weight also significantly reduced under salinity (Tejera et al. 2006). Colla et al. (2006) reported that total fruit yield, mean fruit mass, fruit dry matter, total soluble solids, leaf area, pulp, seeds, and relative amounts of peel were significantly affected by salinity.

1.3.2 Ion Levels

Plants exposed to salt stress especially NaCl uptake more Na^+ and Cl^- ions. The cytoplasm cannot tolerate large amounts of salt, so in order to facilitate their metabolic activity, either they restrict the excess salts in the vacuole or compartmentalize the ions in other plant tissues (Zhu 2003). Under normal conditions, plant cytosol contains high K^+ and low Na^+ concentrations, giving negative electrical potential to the cell (Blumwald et al. 2000). When plants are exposed to salt stress, low and high affinity K^+ carriers transport the Na^+ ions from the external environment to the plant cytosol. Excess amounts of sodium ions in the vacuole causes osmotic stress to the cell. High accumulation of K^+ concentration in the vacuole lowers the osmotic potential of the cell (Bohnert and Sheveleva 1998). Plant can tolerate the salinity up to a certain level by excluding the toxic ions at the roots, sequestering the ions in lower leaves so that the ion toxicity is prohibited in newly developing leaves and reproductive tissues. In halophyte plants, the uptake of abundantly available Na^+ is effectively partitioned and confined to the vacuole by sodium/proton antiporters. In a study by Turner et al. (2013), the mean concentrations of sodium, potassium, and chloride, and particularly chloride ions, in the youngest fully expanded leaf were significantly higher in chickpea plants exposed to long-term salinity. Similarly, Sarhadi et al. (2012) found that Na^+ concentration in anthers exposed to salinity was fivefolds higher than non-saline-treated plants. In different olive cultivars, Na^+ concentration increased gradually due to ion exclusion mechanism when plants were exposed to salinity up to 80 mM, but when salinity increased to 160 mM, the capacity of this mechanism was limited because Na^+ concentration sharply increased (Mousavi et al. 2008).

The effect of salinity on plants may vary depending on the following: (1) the level or the concentration (mild, moderate, severe) of the stress exposed to; (2) the variety of plant species or the plant organ investigated; (3) and duration of the stress, plant developmental stage, and calcium or potassium ions activity in the saline root medium (Wu et al. 1996). Different levels of salinity and their effect on various plant species are listed in Table 6.2.

1.3.3 Photosynthetic Pigments

Photosynthesis is the most dominant physiological process in plants. Biomass production is a measure of net photosynthesis, thus factors affecting plant biomass also affects photosynthetic activity. Under saline conditions, chlorophyll and total carotenoid contents of leaves decreased (Parida and Das 2005). When soil salinity increased from 40 to 160 mM NaCl, chlorophyll a, chlorophyll b, and chlorophyll a+b decreased in olive trees (Mousavi et al. 2008). Under chloride-dominated conductivity level, photosynthesis was reduced more than sulfate-dominated conductivity level, but chlorophyll content decreased under both salinity levels (Datta and Sharma 1990). Chlorophyll contents were significantly affected by salinity with 56–59.8 % reduction in total Chl a+b over the control. Moreover, the Chl a+b/carotenoid ratio decreased with increasing salinity levels (Jampeetong and Brix 2009). When *Sesbania grandiflora* seedlings were exposed to low-level salinity, leaf pigments increased; however when exposed to salinity level up to 50 mM, leaf contents significantly reduced. The total chlorophyll content (chlorophyll a and b) of the seedlings were also significantly reduced under salt stress (Dhanapackiam and Ilyas 2010).

1.3.4 Proline and Other Osmolyte Content

Osmotic stress induces plant cells to lose water under salinity. Plants accumulate many metabolites in the cytoplasm known as compatible solutes to increase their tolerance against water loss. In halophytes, this function is accomplished by simple sugars (glucose, fructose, and sucrose), complex sugars (trehalose, raffinose, and fructans), proline, and betaines (Porcel et al. 2012). Proline and glycine betaine (*N*, *N*, *N*-trimethylglycine betaine) are two major osmoprotectant osmolytes, which are produced by many plant species under stress conditions. Proline is a proteinogenic amino acid with an exceptional conformational rigidity (Szabados and Savoure 2009). Proline plays a vital role in protecting the subcellular structures, macromolecules, scavenging free radicals, and buffering cellular redox potential under

Table 6.2 Response of various plant species to different levels of salinity

Crop	Salinity treatment	Plant response	References
Wheat (*Triticum* sp.)	S_1—NaCl 50 mmol L^{-1} Na_2SO_4 25 mmol L^{-1} $CaCl_2$ 25 mmol L^{-1} S_2—NaCl 50 mmol L^{-1} Na_2SO_4 25 mmol L^{-1} $CaCl_2$ 25 mmol L^{-1}	Relative water content decreased Grain yield reduced up to 83.4 % Glutathione reductase activity significantly increased	Sairam et al. (2002)
Eggplant (*Solanum melongena* L.)	0, 10, 25, 50, 100, and 150 mmol NaCl to half-strength Hoagland solution	Seed germination reduced significantly Radical elongation severely affected Yield decreased	Chartzoulakis and Loupassaki (1997)
Rice (*Oryza sativa*)	EC in nutrient solution 1.0, 3.9, and 6.5 dS m^{-1}	Grain yield and most yield components were significantly reduced	Zeng and Shannon (2000)
Watermelon (*Citrullus lanatus* L.)	EC in nutrient solution 2.0 and 5.2 dS m^{-1}	Total fruit yield, mean fruit mass, fruit dry matter, total soluble solids, leaf area, and seeds were significantly reduced	Colla et al. (2006)
Lettuce (*Lactuca sativa*)	EC content—0.80, 1.93, 2.81, 3.73, and 4.72 dS m^{-1}	Fresh weight and leaf area were reduced	Andriolo et al. (2005)
Canola (*Brassica napus* L.)	EC of irrigated water—1.2, 2.1, 3.9, 6.0, 7.9, and 9.7 dS m^{-1}	Plant height reduced 40–50 % Seed yields reduced significantly	Francois (1994)
Chickpea (*Cicer arietinum* L.)	EC of saline solutions—0, 4, 6, and 8 dS m^{-1}	Plant growth and grain yield reduced. Root weight, shoot weight and flower production were decreased	Singla and Garg (2005)
Chickpea (*Cicer arietinum* L.)	Nutrient solution with NaCl concentrations of 0, 50, 75, and 100 mM	Root and shoot dry weight decreased up to 23 % and 40 %, respectively. Nodule number was reduced about 70 %	Tejera et al. (2006)
Barley (*Hordeum vulgare* cv)	Hoagland's nutrient solution containing 0, 50, 100, and 200 mM NaCl	Germination percentage and photosynthetic pigments significantly reduced with increasing NaCl Total free amino acids and proline contents progressively increased	El-Tayeb (2005)

Madagascar periwinkle (*Catharanthus roseus* L.)	40, 60, 80, 100, and 120 mM NaCl solutions	Significantly decreased plant height, root length, leaf area, fresh weight, and dry weight	Jaleel et al. (2007)
		Nonenzymatic antioxidants like AA and GSH were affected significantly	
Maize (*Zea mays* L.)	40 mM NaCl	Dry yield decreased	Gunes et al. (2007)
		Membrane permeability increased	
Barley (*Hordeum vulgare* L.)	Hoagland's solution with Na^+ and Cl^-—0, 10, 20, 40, 60, 80, 100, 120, 140, and 160 mM	Dry weight reduced at high NaCl	Tavakkoli et al. (2011)
		K and gas exchange parameters were decreased	
Chickpea (*Cicer arietinum* L.)	20 mM NaCl, 40 mM NaCl, 60 mM NaCl	Seed yield significantly decreased. The mean concentrations of Na^+, K, and Cl^-, and particularly chloride ions, in the fully expanded leaf, were significantly higher	Turner et al. (2013)
Rice (*Oryza sativa* L.)	0 and 100 mM NaCl	Pollen viability reduced up to 83 %	Sarhadi et al. (2012)
		Na^+ concentration considerably higher	
Faba bean (*Vicia faba*)	100 mmol kg^{-1} Na^+	Dry weight, plant height reduced significantly	Tavakkoli et al. (2010)
	100 mmol kg^{-1} Cl^-		
	100 mmol kg^{-1} NaCl	Stomatal conductance was more sensitive to salinity treatments	
Strawberry (*Fragaria x ananassa*)	EC—0.3, 2.6, and 5.1 dS m^{-1}	Leaf area reduced	Saied et al. (2005)
		Leaf water potential and cell sap osmotic potential decreased	
		Fruit yield per plant and fruit number were significantly reduced	

osmotic stress (Kavi Kishor et al. 2005; Porcel et al. 2012). Proline accumulation in plants serves as a storage site for carbon and nitrogen as stress leads to slower growth (Bohnert and Jensen 1996).

Increased proline content was positively correlated with increasing salt stress on the first week of exposure of salt-sensitive cultivars of rice. But after 2 weeks of stress, root proline level reached high amounts and decreased suddenly. Both salt-sensitive and salt-resistant cultivars accumulated proline mainly in young leaves (Lutts et al. 1996). Amirjani (2010) reported that when soybeans were exposed to different levels of salinity, proline content increased up to 20-folds. Proline content increased three times higher when *Salvinia natans* was exposed to 150 mM concentration (Jampeetong and Brix 2009). Canola leaf and root proline content was increased significantly when plant are exposed to high salinity (Nazarbeygi et al. 2011).

1.3.5 Abscisic Acid Content

Abscisic acid (ABA) is a phytohormone which increases when plants are subjected to various environmental stresses including salt stress. Like proline, ABA also increases in roots, xylem sap, and shoots (Dodd and Perez-Alfocea 2012) under stress conditions. ABA plays an important role in plant growth and development processes like stomatal opening, seed development, embryo morphogenesis, dormancy, and synthesis of storage proteins and lipids (Porcel et al. 2012; Sreenivasulu et al. 2012). After 2 h of 200 mM NaCl exposure, ABA strongly accumulated in maize roots but gradually decreased over time (Zhu et al. 2005), indicating an early recovery from water deficit. ABA concentration significantly increased in leaves of the salt-resistant plants (Zorb et al. 2013). Albacete et al. (2008) reported that ABA concentration of both tomato roots and leaves increased on the first day of the salinity with up to 8-fold in roots, 4-fold in leaves, and 2.3-fold in xylem sap increment over control plants. Similarly, when tomato plants were treated with 120 mM NaCl for 7 days, ABA concentration in the xylem sap increased fivefold over the control, whereas leaf exhibited threefold increase (Mulholland et al. 2003).

1.3.6 Cytokinin Production

Generally cytokinins are considered to be antagonists of ABA, since they have opposing effects on most developmental stages of plants including stomatal opening, cotyledon expansion, cell division, apical dominance, chloroplast biogenesis, nutrient mobilization, leaf senescence, vascular differentiation, photomorphogenic development, shoot differentiation, and seed germination (Thomas 1992; Javid et al. 2011). It is widely accepted that cytokinins are produced in the root tips and in developing seeds (Zahir et al. 2001). The xylem translocates the cytokinin from roots to shoots where they influence the regulation of development and senescence. Cytokinins involves in plant growth regulation including cell division, apical dominance, nutrient mobilization, chloroplast development, senescence, and flowering

(Hare and Van Staden 1997). Under stress conditions, cytokinin production generally decreases, which means that the reduction in cytokinin supply from roots tends to alter the gene expression in shoots thereby inducing the appropriate response to tolerate external or environmental stresses (Hare et al. 1997).

1.3.7 Ethylene Production

Ethylene is produced naturally by plants. Ethylene exerts a variety of roles in the development of plant, including chlorosis, leaf abscission, epinasty, stimulation of adventitious roots, and fruit ripening. Ethylene also causes increased permeability of cell membranes. Certain amount of ethylene is required for seed germination by many plant species, with the level enhanced during germination and seedling growth (Zahir et al. 2011). Although low levels of ethylene were found to help in root initiation and growth, high level of ethylene produced by fast-growing roots can lead to reduction in root elongation (Ma et al. 1998). 1-Aminocyclopropane-1-carboxylic acid (ACC) is the immediate precursor of ethylene, derived from methionine in plants (Yang and Hoffman 1984). Biotic and abiotic stress conditions also elicit ethylene synthesis (Bleecker and Kende 2000). ACC concentration increased in response to salinity in roots and leaves after 15 days of exposure and reached up to 2.4-fold in roots, 12-fold in leaves, and 30-fold in xylem sap of tomato (Albacete et al. 2008).

1.3.8 Auxins

Auxins naturally occur as indole acidic acid (IAA), indole-3-butyric acid (IBA), 4-chloroindole-3-acetic acid, and phenylacetic acid. Among these, IAA is the most abundant and widely studied auxin. Auxin is synthesized mainly in young aerial tissues and roots, particularly in primary root tips (Ljung et al. 2005). Auxin is transported by both active and passive mechanisms, for instance, the phloem distributes the auxin passively from source tissues to roots, shoot tips, and other sink tissues (Tognetti et al. 2012). Auxins play an important role in plant growth and developmental processes, such as vascular development, apical dominance, tropic responses, and organ pattern formation (Woodward and Bartel 2005; Simon and Petrasek 2011). Under stress conditions, auxin production tends to increase in response to protect the plant from external stress. Exogenous application of IAA increased the leaf auxin content under induced salinity (Veselov et al. 2008). In contrast, Wang et al. (2001) reported that long-term exposure to salinity decreased the IAA content in young leaves of *Iris hexagona*, but high IAA content was observed in seeds than in pods. Moreover, it was characterized that the initial seed germination was enhanced by IAA. When canola was exposed to salinity, fresh and dry weight of leaves and roots reduced but IAA oxidase activity increased. Cytoplasm showed higher IAA oxidase activity than it showed in cell wall (Bybordi et al. 2010). Salinity decreased IAA concentration dramatically up to 80 % on the

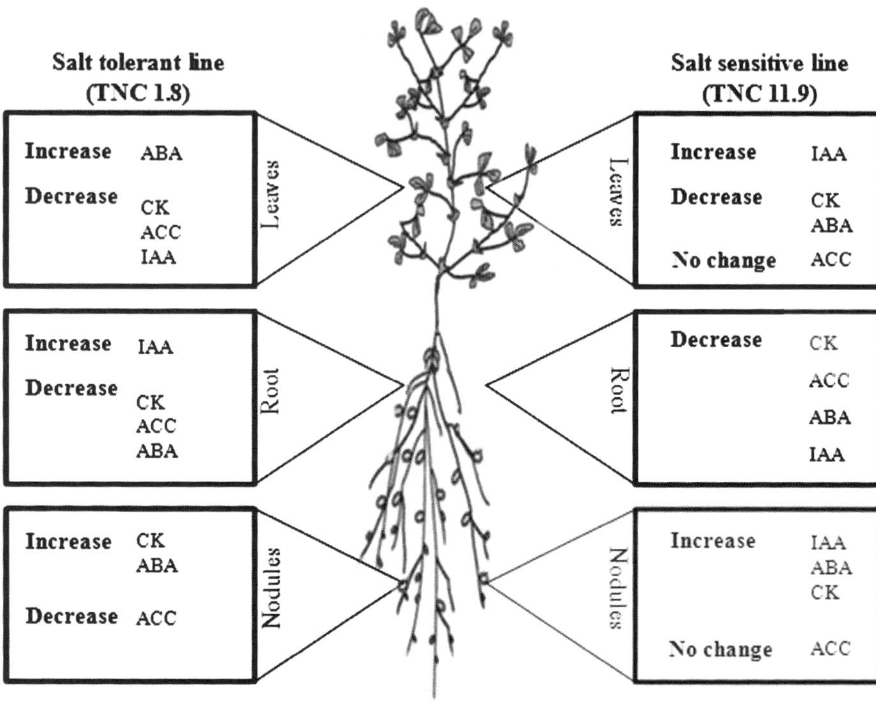

Fig. 6.2 Differential hormonal changes in leaves, roots, and nodules of salt-tolerant (TNC 1.8) and salt-sensitive (TNC 11.9) lines of *Medicago ciliaris* under 100 mM NaCl concentration (modified from Salah et al. 2013)

first day of the treatment which lasted throughout the experiment (Ghanem et al. 2008). Though some reports argue that IAA concentration increases under salinity, most of the data support that IAA concentration decreases corresponding to salinity. Zorb et al. (2013) reported that the free auxin (IAA) was significantly lower in roots of salt-sensitive hybrid Lector, whereas no change in IAA content was found in salt-resistant plants. Figure 6.2 illustrates the changes in hormonal concentrations of salt-tolerant (TNC 1.8) and salt-sensitive (TNC 11.9) lines of *Medicago ciliaris* under 100 mM NaCl concentration. Data presented here are based on the results obtained by Salah et al. (2013).

1.3.9 Antioxidant Systems

Salinity causes water deficit due to osmotic effects on various metabolic activities of plants. This water deficit leads to oxidative stress because of the formation of reactive oxygen species such as superoxides and hydroxy and peroxy radicals. Reactive oxygen species are by-products of osmotic and ionic stresses which causes membrane dysfunction and cell death (Bohnert and Jensen 1996). Plants possess defense mechanisms for this reactive oxygen species by enhancing the activities of certain antioxidative enzymes such as catalase, peroxidase, glutathione reductase, and superoxide dismutase (SOD), which scavenge reactive oxygen species (Parida and Das 2005). For instance, Bohnert and Jensen (1996) described the scavenging mechanism for hydroxyl radical, where the photosynthetic electron transport system is main source for reactive oxygen species. Regulation of the photosynthetic electron flow controls the assembly and disassembly of these radical. SODs enzyme dismutates the superoxide into O_2 and H_2O_2, which can possibly inhibit CO_2 fixation. Superoxide reacts with H_2O_2 to generate the most potent oxidant of hydroxyl radical. This radical damages the cellular macromolecules, degenerates lesions in the DNA, affects protein synthesis and stability which eventually end in metabolic dysfunction and cell death.

The activities of oxidative enzymes such as Cu/ZnSOD and MnSOD increased significantly under salt stress in all tissues of chickpea, while FeSOD decreased up to 81 % after 2 days of stress duration. Ascorbate peroxidase (APX) and glutathione reductase (GR) activities were higher in leaf whereas catalase (CAT) activity was decreased after 4 days of stress duration (Eyidogan and Oz 2007). When halophyte plant *Nitraria tangutorum* Bobr. was subjected to 200 mM NaCl for 3, 6 and 9 days SOD, APX, CAT activities and H_2O_2 level significantly increased but this change was not observed with peroxidase (POD) (Yang et al. 2010), instead POD decreased under salinity. They also found that the application of dimethylthiourea (DMTU), an H_2O_2 scavenger, could alleviate the salinity-induced elevation of APX activity. Wheat plants exposed to salinity level up to 200 mM showed increased antioxidant activities such as SOD, ascorbate peroxidase (APX) glutathione reductase (GR), and contents of hydrogen peroxide, thiobarbituric acid-reactive substances (TBARS) but decreased ascorbic acid (AA) content (Sairam et al. 2005).

2 Microorganisms in Alleviation of Salt Stress

Salinity affects plants physiologically and biochemically and it triggers production of various enzymes and hormones. Salinity also remarkably affects metabolic pathways by damaging protein structure, lesion of DNA and inhibiting enzyme activity, which eventually lead to plant growth suppression or plant death. In order to improve the ability of plants to tolerate external stresses like salinity, physiologists have been trying to find suitable solutions. In the natural environment, plants can be colonized both by external and internal microorganisms. Some beneficial fungi and bacteria can improve plant performance under stress environments (Evelin et al. 2009). It has been widely studied that microorganisms can ameliorate salinity stress in various plant species. Arbuscular mycorrhizal fungi (AMF) and plant growth-promoting rhizobacteria (PGPR) tend to increase plants capability to withstand various environmental stresses. Here we focus mainly on the roles of AMF and PGPR in alleviating salt stress in plants. Figure 6.3 illustrates the role of AMF and PGPR in alleviation of salt stress in plants.

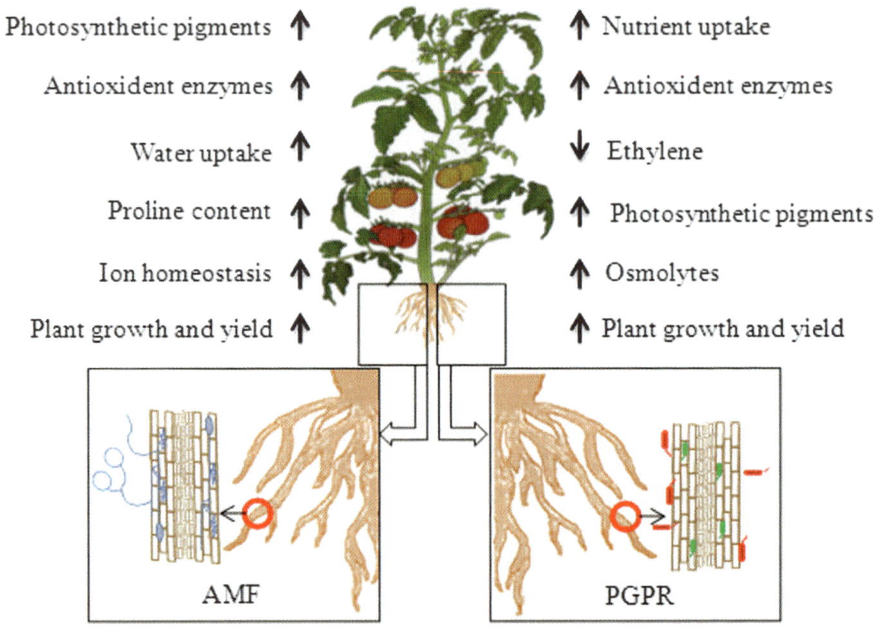

Fig. 6.3 Improvement of plant adaptation mechanisms against salt stress by arbuscular mycorrhizal fungal and plant growth-promoting bacterial inoculation

3 Arbuscular Mycorrhizal Fungi

Mycorrhizal fungi are endophytic and biotrophic fungi, which form mutualistic symbiosis with several plant species. Among the different types mycorrhizal fungi, endo (arbuscular mycorrhizal fungi) and ectomycorrhizal fungi are important in agriculture and forestry. AMF are well known for their efficiency in plant growth promotion especially in arid and semiarid lands. AMF might have appeared on earth before 400 million years ago (Remy et al. 2004), and they are grouped into the phylum Glomeromycota. AMF mainly depend on host plants for their growth and reproduction. Many studies have reported that AMF enhance plant growth and yield by increasing the supply of phosphorus to the host plant.

3.1 Interaction Between AMF and Plants

AMF are abundant in terrestrial ecosystems. AMF form symbiotic or mutualistic associations with around 80 % of the plant species and 92 % of the plant families (Wang and Qiu 2006) by colonizing plant roots (Bothe 2012). AMF spores present in the soil form hyphae, which can be either intraradical hyphae or extraradical hyphae.

Intraradical hyphae infect the plant roots, by breaking through the epidermal cells and then colonizing the cortical cells. The unique property of AMF is the formation of arbuscules in the infected cortical cells. Both host and AMF benefit from this colonization and cause no detrimental effect on each other. Inside the root, AMF form intercellular hyphae and intracellular hyphae. Intercellular hyphae are formed between the cortical cells, and these hyphae can form vesicles which can act as nutrient storage sites. Spores belonging to *Glomus* sp. are well documented to form these vesicles. Intracellular hyphae are formed inside the cortical cell by infecting the cell wall without lysis. These form arbuscules, which are nutrient exchange sites where the fungi supply nutrients like phosphorus and nitrogen to plants and plants supply carbon to fungi (Bonfante and Genre 2010).

3.2 Why AMF Are Important Under Stress Conditions?

The main and foremost application of AMF in agriculture is for enhancement of plant growth and yield. Under stress conditions, generally, plant root growth is inhibited, and they cannot spread widely in the soil which leads to less nutrient uptake. Phosphate ion is very poorly mobile in most soils. After a certain period of time, the soil surrounding the root zone becomes a phosphate-depleted zone (Helgason and Fitter 2009) because of continuous nutrient uptake by root. After successful colonization, the extraradical hyphae of AMF play a vital role in nutrient uptake from soil. Extraradical hyphae penetrate through the soil and spread more widely in nutrient-undepleted zone. Nutrients taken up by the hyphae from nutrient-undepleted zone are transferred to host plant at the membrane surrounding the arbuscule (Bago et al. 2002). Under stress conditions, mycorrhizal plant can absorb or accumulate remarkably more nutrients especially phosphorus (Parniske 2008) from the soil than nonmycorrhizal plants. In addition to that, extraradical hyphae in soil produce hydrolytic enzymes which hydrolyze the macromolecules like lignin, chitin, protein, and nucleic acid into simple monomers. AMF tend to uptake the nutrients as monomers more efficiently.

3.3 AMF and Salinity

Various studies have reported that AMF exist in saline soil (Evelin et al. 2009; Bothe 2012; Porcel et al. 2012). Although the exact mechanism of AMF mitigate growth reduction by salinity is not clear, most of the studies reveal that AMF improve the plant growth and yield by enhancing nutrient uptake. Aliasgharzadeh et al. (2001) observed that under severe saline soil (EC—162 dS m^{-1}), AMF species like *Glomus intraradices*, *Glomus versiforme*, and *Glomus etunicatum* were most predominant. The authors also found that AMF spore did not decrease rather increased relatively; this may be due to the fact that sporulation is stimulated under salt stress (Tressner and Hayes 1971). High-level salinity negatively affects spore germination, colonization capacity, and hyphal growth (Giri et al. 2007; Sheng et al. 2008), indicating that salinity can suppress the formation of arbuscular mycorrhiza. Though under severe salinity spore germination is delayed, it is not prevented or inhibited (Cantrell and Linderman 2001; Juniper and Abbott 2006; Sheng et al. 2008). *Glomus mosseae* isolated from saline soil had lower capacity to alleviate the saline stress compared to one isolated from non-saline soil in cotton plants (Tian et al. 2004). In another work, Porras-Soriano et al. (2009) tested the efficiency of three species of AMF—*G. mosseae*, *G. intraradices*, and *Glomus claroideum*—to alleviate salt stress in olive trees under greenhouse condition. The authors found that *G. mosseae* was most efficient in protecting plants against detrimental effects of salinity. These experiment results suggest that the ability of AMF in protecting plants from salt stress may depend on the behavior of each species. Salt tolerance in plants is determined by multiple biochemical pathways which include retention and/or acquisition of water and ion homeostasis and protect chloroplast functions. The important roles of AMF in alleviation of salt stress in plants are discussed below.

3.3.1 Water Uptake

AMF are known to improve physiological processes like water absorption capacity of plants by increasing hydraulic conductivity and favorably adjusting the osmotic balance (inorganic ions mainly K^+ or Cl^- uncharged organic compounds like proline and betaine) and composition of carbohydrates like sucrose, pinitol, or mannitol (Ruiz-Lozano 2003). Leaf relative water content of salinity-stressed mycorrhizal peanut plants was significantly higher than salinity-stressed nonmycorrhizal plants (Al-Khaliel 2010). Lettuce plants inoculated with *G. intraradices* maintained higher water content than uninoculated plant regardless of salt level (Jahromi et al. 2008). AMF plants are found to exhibit a higher stomatal conductance thereby increasing the demand for transpiration (Jahromi et al. 2008; Sheng et al. 2008). Higher turgor potential and lower water saturation deficit in mycorrhizal plants also improve the water status of the plant.

3.3.2 Balancing of Ion Imbalance

It is well known that salinity causes nutrient imbalance in plants. AMF help plants to uptake more nutrient. Colla et al. (2008) reported that AMF colonization increased leaf concentrations of N, P, and K by 9 %, 221 %, and 17 %, respectively. Moreover, Na accumulation in the leaves was limited compared to nonmycorrhizal zucchini plants. AMF colonization in wheat significantly increased the shoot concentrations of P, K, and Zn whereas decreased Na and Cl concentrations (Daei et al. 2009). Na^+ accumulation dramatically increased in nonmycorrhizal *Lotus glaber*, whereas *G. intraradices* colonized plants counteracted such accumulation (Sannazzaro et al. 2006). Barley inoculated with *G. intraradices* showed relatively increased uptake of macronutrients such as iron (Fe), manganese (Mn), copper (Cu), and zinc (Zn) under slightly and moderately saline soil, whereas such uptake was decreased under highly saline soil (Mohammad et al. 2003). Under irrigated saline water stress, AMF inoculation increased leaf N, P, K, Ca, Mg, and S content (Abd El-Wahab et al. 2011). Talaat and Shawky (2011) reported that AMF alleviated the adverse effect of salt stress and enhanced N, P, and K accumulation under salt-stressed conditions.

3.3.3 Photosynthetic Pigments

Inoculation with AMF influences photosynthetic pigments in both normal and saline condition (Abdel-Fattah and Asrar 2012). Inoculation of *G. mosseae* significantly increased leaf chlorophyll content in peanut plants under salinity stress (Al-Khaliel 2010). In grape plants, inoculation of AMF also significantly increased chlorophyll content at 3,000 ppm of irrigated saline water (Abd El-Wahab et al. 2011). Tomato plants treated with salt exhibited higher amount of chlorophyll a and b, total chlorophyll contents, and carotenoid content with inoculation of AMF (Basak et al. 2011).

3.3.4 Proline and Other Osmolytes

The influences of AMF in proline accumulation may vary depending upon the crop. At anthesis stage, leaves of wheat plants colonized by *G. mosseae* and *Glomus deserticola* had higher proline content; such increase of proline content was related to the rate of mycorrhizal colonization (Abdel-Fattah and Asrar 2012). Increased proline content was also observed in salinity-stressed mycorrhizal peanuts (Al-Khaliel 2010). Under nonmycorrhizal salinity stress, proline accumulation was increased with increasing salinity; however, with AMF association, pigeon pea plant accumulated significantly higher proline content at the saline dosage of 8 dS m^{-1} (Garg and Manchanda 2009). The authors also reported that the mycorrhizae further induced the accumulation of glycine betaine in stressed plants. In contrast, Selvakumar and Thamizhiniyan (2011) reported that proline content was reduced in *Capsicum annuum* plants colonized by *G. intraradices*.

3.3.5 Antioxidant Enzymes

During salt stress, plant cell minimizes its ability to protect from oxidative damage; thus it produces reactive oxygen species. The induction of ROS-scavenging enzymes such as SOD, CAT, POD, and APX is the most common mechanism for detoxifying reactive oxygen species. Wu et al. (2010) observed that the leaf H_2O_2 content was significantly lower in *G. versiforme*- or *G. mosseae*-inoculated trifoliate orange seedlings. They also suggested that lower accumulation of ROS may reduce the membrane damage. Abdel Latef and Chaoxing (2011) reported that in tomato plants, AMF inoculation accompanied for the enhancement of activities of SOD, CAT, POD, and APX by 56 %, 37 %, 28 %, and 33 %, respectively, at 100 mM NaCl. The authors also suggested that lower accumulation of lipid peroxidation indicates lower oxidative damage in the colonized plants. Pepper plants inoculated with *G. intraradices* had increased SOD and APX activities, although remarkable increase was not observed with GR and CAT activity (Cekic et al. 2012). Arbuscular mycorrhizal symbiosis clearly shows that increments in enzymatic antioxidant production may in turn enhance salinity resistance in AMF plants.

3.3.6 Plant Growth and Yield

Under salt stress, root colonization by *G. mosseae* enhanced maize growth regardless of the P level; shoot and root dry weight were increased (Feng et al. 2002). Mycorrhization increased shoot and root growth of salt-tolerant plants, whereas it decreased the inhibition of growth reduction in salt-sensitive plants (Sannazzaro et al. 2006). AMF inoculated wheat plants significantly improved productivity. Mycorrhization by *Glomus* sp. increased grain yield by 83–155 % in wheat varieties under 3.13–9.38 dS m^{-1} salinity levels (Talaat and Shawky 2011). Pigeon pea inoculated with *G. mosseae* improved root and shoot dry weights under saline conditions (Garg and Manchanda 2009). Under irrigated saline water stress, grape plants inoculated with AMF increased vegetative growth parameters like shoot length, shoot diameter, number of leaves per plant, average leaf area, shoot biomass, and root biomass (Abd El-Wahab et al. 2011).

3.3.7 Other Benefits of AMF

In addition to the above-described parameters, AMF also play a vital role in improving soil quality and health. AMF hyphae in soil produce glomalin, a glycoprotein which helps in soil aggregation (Rillig 2004). Although the exact mechanism or the gene responsible for the production of glomalin is still unknown, many studies have reported that glomalin and its related soil proteins produced by AMF could lead to the formation of "sticky" string bag of hyphae that would stabilize aggregates (Rillig et al. 2002; Borie et al. 2008; Singh 2012). Various benefits have been proposed for mycorrhizal inoculation under salinity; however, their efficiency in alleviating salinity stress is dependent on the AMF inocula and on the plant species. Various plant species inoculated with AMF and their responses to salinity are listed in Table 6.3.

4 Bacteria in Alleviation of Salt Stress

Naturally bacteria can colonize the plants as endophytes or just adhere to the roots. To adapt to extreme environments, microorganisms developed different adaptation mechanisms to combat the stress. For instance, in bacteria optimum metabolic processes like enzymatic activity and membrane stability occur at high temperature or

Table 6.3 Effect of AMF inoculation on different plant species under salinity stress

AMF	Crop	Salinity level	Plant responses	References
Glomus mosseae	Cotton (*Gossypium arboreum* L.)	0–3 g kg^{-1} NaCl	Increased shoot dry weight up to 68 % and root dry weight by 27 %	Tian et al. (2004)
			Dependency of cotton plants on mycorrhizal fungi was increased	
Glomus intraradices	Carnation (*Dianthus caryophyllus* L.)	EC of irrigation water—1, 3, and 6 dS m^{-1}	Flower dry weight and the total number of flowers per plant increased	Navarro et al. (2012)
			Number of buds and flowers were increased	
Glomus intraradices	Tomato (*Solanum lycopersicum* L.)	EC—0.63 dS m^{-1} 5 dS m^{-1}	Na uptake in the inoculated plants lower compared to control	Hajiboland et al. (2010)
		10 dS m^{-1}	AMF plants had greater values for K/Na and Ca/Na in both shoots and roots	
Glomus mosseae, *Glomus versiforme*	Orange (*Poncirus trifoliata* L.)	100 mmol NaCl	Accumulation of ROS and membrane damage reduced. SOD activity was largely induced	Wu et al. (2010)
Glomus mosseae	Tomato (*Lycopersicon esculentum*)	0, 50, and 100 mM NaCl (EC –2.2, 7.0, and 12.0 dS m^{-1})	Improved fruit fresh weight and fruit yield	Abdel Latef and Chaoxing (2011)
			The activity of POD and APX was markedly increased	
Glomus mosseae, *Glomus intraradices*, *Glomus claroideum*	Olive (*Olea europaea*)	EC—4.0 dS m^{-1}	AMF colonization was more effective under saline condition	Porras-Soriano et al. (2009)
			Shoot and root dry weight was increased	
			Plant growth increased.	
			K concentration increased in shoot	

(continued)

Table 6.3 (continued)

AMF	Crop	Salinity level	Plant responses	References
Glomus intraradices	Sweet Basil (*Ocimum basilicum* L.)	3 g L^{-1} NaCl (EC=5.6 mS cm^{-1})	Reduced Na concentration in the plants Mycorrhizal plants had a tendency to grow faster	Zuccarini and Okurowska (2008)
Glomus sp.	Blue-green saltbush (*Atriplex nummularia*)	2 g kg^{-1} NaCl (EC—12 dS m^{-1})	Shoot dry weight increased significantly Plant growth and nutrient uptake increased	Asghari et al. (2005)
Glomus clarum	Pepper (*Capsicum annum*)	EC—2.15, 7.15, and 12.15 dS m^{-1}	Significantly improved shoot, root dry matter, and fruit yield Improved chlorophyll concentration Proline concentration was lower	Kaya et al. (2009)
Glomus mosseae, *Paraglomus occultum*	Citrus (*Citrus tangerine*)	100 mM NaCl (EC 9.7 dS m^{-1})	Leaf number, leaf area, shoot, and root dry weights were increased Relative water content was increased Root concentrations of K^+, Ca^{2+}, and Mg^{2+} were higher	Wu and Zou (2009)

salinity in order to survive in such conditions (Madigan and Oren 1999). Under stress conditions, to maintain cytoplasmic osmolarity, rhizobacteria produce osmoprotectants such as glutamate, trehalose, proline, glycine betaine, proline betaine, and ectoine (Talibart et al. 1994). Some bacteria produce exopolysaccharides under stress conditions in order to prevent hydric stress and fluctuations in water potential by enhancing water retention and regulating the diffusion of carbon sources in microbial environment (Sandhya et al. 2009). In addition to aiding plants to survive under stress, bacteria also enhance plant growth and yield. Certain groups of bacteria known as PGPR promote plant growth by improving nutrient uptake like phosphorous and by producing auxins like IAA.

4.1 Nutrient Uptake

Phosphorus is a major nutrient for plant growth, but less than 10 % of soil phosphate is available to plants. The remaining soil phosphates react with the soil and become progressively less available for plant use (Chaiharn and Lumyong 2011). P-solubilizing bacteria produce and release metabolites such as organic acids. The hydroxyl and carboxyl groups of these metabolites chelate cations which are bound to the phosphates. Eventually, the insoluble form of phosphate is converted into soluble forms (Vassilev et al. 2012). Bacteria which were isolated from saline soil increased soluble P content of soil (Patel et al. 2012). Therefore, utilization of halotolerant P-solubilizing bacteria can be a potential approach in improving P nutrition under salinity stress.

4.2 Photosynthetic Pigments

Photosynthesis is a key process in plants. Thus, factors that may affect the photosynthetic apparatus or pathways are important. High salinity levels have adverse effects on photosynthesis; however, several reports of inoculation of bacteria showed positive influence on photosynthetic pigments of various plants under salinity stress. Under 80 mM NaCl concentrations, radish inoculated with PGPR strains *Staphylococcus kloosii* EY37 and *Kocuria erythromyxa* EY43 showed significant increase in chlorophyll contents (Yildirim et al. 2008). Maize plants inoculated with bacterial strains of *Azotobacter* sp. increased chlorophyll content to up to sixfold (Rojas-Tapias et al. 2012). ACC deaminase-producing rhizobacterial strains inoculated with maize increased chlorophyll a and b at different levels of salinity (Nadeem et al. 2006). In another study, Mohamed and Gomaa (2012) found that under salinity, photosynthetic pigments are reduced or affected severely in radish, but inoculation of bacterial strains *Bacillus subtilis* and *Pseudomonas fluorescens* significantly improved the photosynthetic pigments compared to uninoculated plants treated with the same level of salinity. Chlorophyll a, chlorophyll b, and carotenoid contents were significantly increased in *Lens esculenta* inoculated with *Oceanobacillus profundus* and *Staphylococcus saprophyticus* under 200 mM NaCl stress (Qurashi and Sabri 2011).

4.3 Ethylene Reduction

As described earlier, fast-growing roots will produce high amount of ethylene which eventually leads to inhibition of root elongation. Under pathogen attack or abiotic stress, ethylene acts as an important signaling molecule, and it inhibits plant growth (Viterbo et al. 2010). Ethylene production in plant increases due to high salt concentrations. ACC deaminase catalyzes the cleavage of ACC to α-ketobutyrate and ammonia; ACC is the precursor of ethylene (Onofre-Lemus et al. 2009; Li et al. 2011). Several bacteria with ACC deaminase have been identified as PGPR. ACC deaminase-producing bacteria attach to plant cells, act as a sink for plant ACC, and cleave this ACC, thus reducing plant ACC concentration which can be elevated under stress conditions (Glick et al. 1998). Using ACC as a nitrogen source, the following halotolerant bacterial strains showed relatively high ACC deaminase activities: *Brevibacterium epidermidis*, *Brevibacterium iodinum*, *Bacillus licheniformis*, *Bacillus stratosphericus*, *Brevibacterium epidermidis*, and *Brevibacterium epidermidis* (Siddikee et al. 2010). *Achromobacter piechaudii* ARV8 significantly reduced the ethylene production in tomato seedlings grown under 207 mM NaCl concentrations (Mayak et al. 2004). ACC concentrations were reduced in tissue extracts of pepper seedlings inoculated with ACC deaminase-producing halotolerant bacteria *Brevibacterium epidermidis* RS15, *Bacillus aryabhattai* RS341, and *Micrococcus yunnanensis* RS222 with 47 %, 48 %, and 55 % reduction, respectively, compared to positive control (Siddikee et al. 2012). These reductions in ACC concentration consequently reduced ethylene production in tomato plants under salinity stress. This indicates that ACC deaminase-producing bacteria can mitigate the production of salinity stress-induced ethylene in plants, thus reducing the inhibitory effects of stress ethylene on plant growth and development under salinity stress.

4.4 Antioxidant Enzyme Activity

Several antioxidant enzymes in the plants are responsible in protecting plants under unfavorable conditions due to biotic and abiotic factors. SODs are metalloenzymes which are usually considered into three classes depending on the active site of metal cofactors (Mn, Fe, or CuZn). SODs play a key role in the protection against oxidative stress, nitrogen fixation, and delayed senescence (Santos et al. 2000; Moran et al. 2003). Peroxidases are associated with biochemical or physiological processes during plant growth such as cell formation, fruit development, ethylene biosynthesis, and response to various stresses (Matamoros et al. 2003). APX is an important peroxidase involved in H_2O_2 detoxification. Ascorbate has a reducing power which catalyzes the reduction of H_2O_2 to water (Noctor and Foyer 1998). Inoculation with *Rhizobium tropici* increased nodule SOD activity in *Phaseolus vulgaris* BAT477 genotype. Nodule SOD activity increased by 26 % under 25 mM NaCl and about 50 % under 50 mM NaCl concentrations. In terms of peroxidase activity, inoculation also produced higher activity in BAT477, whereas APX activity was found higher in *Phaseolus vulgaris* COCOT genotype under salinity stress (Jebara et al. 2010). Increasing salinity stress significantly increased enzyme activity in lettuce leaf inoculated with PGPR strains *Serratia proteamaculans* ATCC35475 (SP) and *Rhizobium leguminosarum* bv.128C56G (RL) (Han and Lee 2005).

4.5 Plant Growth Promotion

Canola seedlings inoculated with *Bacillus stratospshericus* RS616, *Planococcus rifietoensis* RS18, and *Exiguobacterium acetylicum* RS19 had increased root lengths (29–47 %) and dry weights (35 and 43 %) after 7 days at 150 mM NaCl concentration compared to uninoculated plants (Siddikee et al. 2010). *Achromobacter piechaudii* ARV8 significantly increased the fresh weight and dry weight of the tomato plants after 7 weeks exposure to the different salinity concentrations (Mayak et al. 2004). Dry matter yield of roots and shoots was increased up to 522 and 281 % when wheat plants were inoculated with exopolysaccharide-producing bacterial strains *Aeromonas hydrophila/caviae*, *Bacillus insolitus*, and *Bacillus* sp. under saline conditions (Ashraf et al. 2004). The reductive effect of salt stress on seed germination of wheat was successfully alleviated by IAA producing bacterial strains *Pseudomonas aureantiaca* TSAU22, *Pseudomonas extremorientalis* TSAU6, and *Pseudomonas extremorientalis* TSAU20 (Egamberdieva 2009).

4.6 Other Beneficial Effects of Bacterial Inoculation

As described early, plants accumulate osmolytes as a protective mechanism under stress conditions. Bacterial inoculation helps the plants accumulate osmolytes under various abiotic stresses. Qurashi and Sabri (2011) reported that *Lens esculenta* plants inoculated with bacterial strains *O. profundus* and *S. saprophyticus* increased endogenous proline accumulation by 68 and 51 % at 100 and 200 mM NaCl concentration. Bacteria can also be inoculated as biocontrol agents against some pathogenic fungi. Fischer et al. (2010) reported that *Pseudomonas* sp. SF4c and *Pseudomonas* sp. SF10b showed antagonistic activity against phytopathogenic fungi and suggested that this inhibition could be due to the production of extracellular enzymes, hydrogen cyanide, or siderophores. Further they recommended that these microorganisms might be applied in agriculture to minimize the utilization of chemical pesticides and fertilizers. Although bacterial inoculation exhibited positive mechanisms for amelioration of plant growth under salt stress, their efficiency in alleviating salt stress varies with different plant species (Table 6.4).

Table 6.4 Effect of PGPR inoculation on different plant species under salinity stress

Bacteria	Crop	Salinity level	Plant responses	References
Azotobacter sp.: C5, C7, C8, and C9	Maize (*Zea mays*)	NaCl—0, 2.93, and 5.85 g L^{-1}	Plant growth, root dry weight, and chlorophyll content increased Na^+ content in shoots was decreased	Rojas-Tapias et al. (2012)
Pseudomonas sp.	Wheat (*Triticum aestivum* L.)	100 mM NaCl	Seed germination percentage increased Seedling root growth up to 40 % and shoot growth up to 52 % increased	Egamberdieva (2009)
Raoultella planticola	Cotton (*Gossypium hirsutum* L.)	5 g L^{-1} of NaCl	The dry weight, fresh weight, plant height, and germination rate significantly increased IAA content increased. Ethylene production reduced	Wu et al. (2012)
Pseudomonas sp. and *Serratia* sp.	Wheat (*Triticum aestivum* L.)	EC—1, 5, 10, and 15 dS m^{-1}	Plant height and root dry weight were increased. Straw yield and chlorophyll content increased	Zahir et al. (2009)
Brevibacterium iodinum, *Bacillus licheniformis*, *Zhihengliuela alba*	Red pepper (*Capsicum annuum* L.)	150 mM NaCl	Both length and dry weight of root and shoot were increased Tolerance indices were significantly increased In plant, N, P, K, Ca contents were increased, whereas Na content reduced	Siddikee et al. (2011)
Pseudomonas putida, *Pseudomonas pseudoalcaligenes*	Chickpea (*Cicer arietinum* L.)	300 mM NaCl	Number of leaves increased up to 31 % Number of branches increased up to 8.6 % Increased number of fruits up to 112 %	Patel et al. (2012)
Brachybacterium saurashtrense, *Brevibacterium casei*, *Haererohalobacter*	Peanut (*Arachis hypogaea* L.)	100 mM NaCl	Plant length, shoot length, root length, shoot dry weight, root dry weight, total biomass, and percentage water content were significantly higher	Shukla et al. (2012)
Serratia proteamaculans, *Rhizobium leguminosarum*	Lettuce (*Lactuca sativa* L.)	EC—1.5 and 7.0 dS m^{-1}	Fresh weight increased by 6.8–12.9 % Leaf area and chlorophyll content increased. Photosynthetic rate and stomatal resistance showed higher activity	Han and Lee (2005)

5 Conclusion and Future Perspective

Biotechnological approaches and conventional plant breeding were able to obtain genetically modified plants which are more resistant to salinity stress; however, the transfer of these results to agriculture is still lagging behind. In the field, plants are subjected to multiple stresses, varying in intensity and duration of stress. Salinity effect on plant performance depends on the plant developmental stage. AMF and PGPR can withstand under various environmental stresses including salinity, and they form symbiotic relationship with most plant species. AMF and PGPR help plants thrive under salinity stress by stimulating the plant defense system and improving plant growth and yield. In addition to the amelioration of salt stress in plants, these microorganisms also aid in improved nutrient uptake, increase auxin production and photosynthetic activity, and even protect plants from some pathogen attack. Likewise, the high concentration of Na^+ and Cl^- ions accumulated in the cytosol is reduced or regulated by microorganisms which eventually leads to improved or uninhibited growth and yield of plants.

Although the performance of AMF and PGPR under salt stress is well studied under controlled environment and field level, researches on the molecular mechanisms governing the process of salt amelioration in plants are limited. To improve the resistance of plants to salinity, the mechanisms of symbiosis between plants and microorganisms need to be addressed which can lead to improved selection of efficient microorganisms for specific plant species. In addition to this, studying microbe–microbe interaction in soil may reveal the mechanisms behind their interaction among each other and their co-inoculation effects which may eventually lead to improved selection of effective microorganisms. The utilization of effective AMF and PGPR strains for the amelioration of plant growth under salt stress is a potential approach which can be used in synergy with other known technologies to increase crop productivity in salt-affected areas.

References

Abd El-Wahab MA, El-Helw HA, Tolba HI (2011) Physiological studies on the effect of inoculation with arbuscular mycorrhizae (AM) fungi on superior grape rootings under salt stress conditions. Nat Sci 9(1):85–100

Abdel Latef AAF, Chaoxing H (2011) Effect of arbuscular mycorrhizal fungi on growth, mineral nutrition, antioxidant enzymes activity and fruit yield of tomato grown under salinity stress. Sci Hortic 127:228–233

Abdel-Fattah GM, Asrar AA (2012) Arbuscular mycorrhizal fungal application to improve growth and tolerance of wheat (*Triticum aestivum* L.) plants grown in saline soil. Acta Physiol Plant 34:267–277

Albacete A, Ghanem ME, Martinez-Andujar C, Acosta M, Sanchez-Bravo J, Martinez V, Lutts S, Dodd IC, Perez-Alfocea F (2008) Hormonal changes in relation to biomass partitioning and shoot growth impairment in salinized tomato (*Solanum lycopersicum* L.) plants. J Exp Bot 59(15):4119–4131

Aliasgharzadeh N, Saleh Rastin N, Towfighi H, Alizadeh A (2001) Occurrence of arbuscular mycorrhizal fungi in saline soils of the Tabriz Plain of Iran in relation to some physical and chemical properties of soil. Mycorrhiza 11:119–122

Al-Karaki GN (2000) Growth of mycorrhizal tomato and mineral acquisition under salt stress. Mycorrhiza 10:51–54

Al-Khaliel AS (2010) Effect of salinity stress on mycorrhizal association and growth response of peanut infected by *Glomus mosseae*. Plant Soil Environ 56(7):318–324

Amirjani MR (2010) Effect of salinity stress on growth, mineral composition, proline content, antioxidant enzymes of soybean. Am J Plant Physiol 5(6):350–360

Andriolo JL, Luz GL, Witter MH, Godoi RS, Barros GT, Bortolotto OC (2005) Growth and yield of lettuce plants under salinity. Hortic Bras 23(4):931–934

Arzani A (2008) Improving salinity tolerance in crop plants: a biotechnological view. In Vitro Cell Dev Biol Plant 44:373–383

Asghari HR, Marschner P, Smith SE, Smith FA (2005) Growth response of *Atriplex nummularia* to inoculation with arbuscular mycorrhizal fungi at different salinity levels. Plant Soil 273:245–256

Ashraf M, Hasnain S, Berge O, Mahmood T (2004) Inoculating wheat seedlings with exopolysaccharide-producing bacteria restricts sodium uptake and stimulates plant growth under salt stress. Biol Fertil Soils 40:157–162

Bago B, Pfeffer PE, Zipfel W, Lammers P, Shachar-Hill Y (2002) Tracking metabolism and imaging transport in arbuscular mycorrhizal fungi. Plant Soil 244:189–197

Basak H, Demir K, Kasim R, Okay FY (2011) The effect of endo-mycorrhiza (VAM) treatment on growth of tomato seedling grown under saline conditions. Afr J Agric Res 6(11):2532–2538

Bleecker AB, Kende H (2000) Ethylene: a gaseous signal molecule in plants. Annu Rev Cell Dev Biol 16:1–18

Blumwald E, Aharon GS, Apse MP (2000) Sodium transport in plant cells. Biochim Biophys Acta 1465:140–151

Bohnert HJ, Jensen RG (1996) Strategies for engineering water stress tolerance in plants. Trends Biotechnol 14:89–97

Bohnert HJ, Sheveleva E (1998) Plant stress adaptations-making metabolism move. Curr Opin Plant Biol 1:267–274

Bonfante P, Genre A (2010) Mechanisms underlying beneficial plant-fungus interactions in mycorrhizal symbiosis. Nat Commun 1:1–11

Borie F, Rubio R, Morales A (2008) Arbuscular mycorrhizal fungi and soil aggregation. J Soil Sc Plant Nutr 8(2):9–18

Bothe H (2012) Arbuscular mycorrhiza and salt tolerance of plants. Symbiosis 58(1–3):7–16

Bybordi A, Tabatabaei SJ, Ahmedov A (2010) Effect of salinity on the growth and peroxidase and IAA oxidase activities in canola. J Food Agric Environ 8(1):109–112

Cantrell IC, Linderman RG (2001) Preinoculation of lettuce and onion with VA mycorrhizal fungi reduces deleterious effects of soil salinity. Plant Soil 233:269–281

Cekic FO, Unyayar S, Ortas I (2012) Effects of arbuscular mycorrhizal inoculation on biochemical parameters in *Capsicum annuum* grown under long term salt stress. Turk J Bot 36:63–72

Chaiharn M, Lumyong S (2011) Screening and optimization of indole-3-acetic acid production and phosphate solubilization from rhizobacteria aimed at improving plant growth. Curr Microbiol 62:173–181

Chartzoulakis KS, Loupassaki MH (1997) Effects of NaCl salinity on germination, growth, gas exchange and yield of greenhouse eggplant. Agric Water Manage 32:215–225

Colla G, Roupahel Y, Cardarelli M (2006) Effect of salinity on yield, fruit quality, leaf gas exchange, and mineral composition of grafted watermelon plants. HortScience 41(3):622–627

Colla G, Rouphael Y, Cardarelli M, Tullio M, Rivera CM, Rea E (2008) Alleviation of salt stress by arbuscular mycorrhizal in zucchini plants grown at low and high phosphorus concentration. Biol Fertil Soils 44:501–509

Daei G, Ardekani MR, Rejali F, Teimuri S, Miransari M (2009) Alleviation of salinity stress on wheat yield, yield components, and nutrient uptake using arbuscular mycorrhizal fungi under field conditions. J Plant Physiol 166:617–625

Datta KS, Sharma KD (1990) Effect of chloride and sulphate types of salinity on characteristics of chlorophyll content, photosynthesis and respiration of chickpea (*Cicer arietinum* L.). Biol Plant 32(5):391–395

Dhanapackiam S, Ilyas MHM (2010) Effect of salinity on chlorophyll and carbohydrate contents of *Sesbania grandiflora* seedlings. Indian J Sci Technol 3(1):64–66

Dodd IC, Perez-Alfocea F (2012) Microbial amelioration of crop salinity stress. J Exp Bot 63(9):3415–3428

Egamberdieva D (2009) Alleviation of salt stress by plant growth regulators and IAA producing bacteria in wheat. Acta Physiol Plant 31:861–864

El-Tayeb MA (2005) Response of barley grains to the interactive effect of salinity and salicylic acid. Plant Growth Regul 45:215–224

Evelin H, Kapoor R, Giri B (2009) Arbuscular mycorrhizal fungi in alleviation of salt stress: a review. Ann Bot 104:1263–1280

Eyidogan F, Oz MT (2007) Effect of salinity on antioxidant responses of chickpea seedlings. Acta Physiol Plant 29:485–493

Feng G, Zhang FS, XI L, Tian CY, Tang C, Rengel Z (2002) Improved tolerance of maize plants to salt stress by arbuscular mycorrhiza is related to higher accumulation of soluble sugars in roots. Mycorrhiza 12:185–190

Fischer SE, Jofre EC, Cordero PV, Gutierrez Manero FJ, Mori GB (2010) Survival of native *Pseudomonas* in soil and wheat rhizosphere and antagonist activity against plant pathogenic fungi. Antoni van Leeuwenhoek 97:241–251

Flowers TJ, Troke PF, Yeo AR (1977) The mechanism of salt tolerance in halophytes. Annu Rev Plant Physiol 28:89–121

Francois LE (1994) Growth, seed yield, and oil content of canola grown under saline conditions. Agron J 86:233–237

Garg N, Manchanda G (2009) Role of arbuscular mycorrhizae in the alleviation of ionic, osmotic and oxidative stresses induced by salinity in *Cajanus cajan* (L.) millsp. (pigeonpea). J Agron Crop Sci 195:110–123

Ghanem ME, Albacete A, Martinez-Andujar C, Acosta M, Romero-Aranda R, Dodd IC, Lutts S, Perez-Alfocea F (2008) Hormonal changes during salinity-induced leaf senescence in tomato (*Solanum lycopersicum* L.). J Exp Bot 59(11):3039–3050

Giri B, Kapoor R, Mukerji KG (2007) Improved tolerance of *Acacia nilotica* to salt stress by arbuscular mycorrhiza, *Glomus fasciculatum* may be partly related to elevated K/Na ratios in root and shoot tissues. Microb Ecol 54:753–760

Glick BR, Penrose DM, Li J (1998) A model for the lowering of plant ethylene concentrations by plant growth promoting bacteria. J Theor Biol 190:63–68

Greenway H, Munns R (1980) Mechanisms of salt tolerance in nonhalophytes. Annu Rev Plant Physiol 31:149–190

Gunes A, Inal A, Alpaslan M, Eraslan F, Bagci EG, Cicek N (2007) Salicylic acid induced changes on some physiological parameters symptomatic for oxidative stress and mineral nutrition in maize (*Zea mays* L.) grown under salinity. J Plant Physiol 164:728–736

Hajiboland R, Aliasgharzadeh N, Laiegh SF, Poschenrieder C (2010) Colonization with arbuscular mycorrhizal fungi improves salinity tolerance of tomato (*Solanum lycopersicum* L.) plants. Plant Soil 331:313–327

Han HS, Lee KD (2005) Plant growth promoting rhizobacteria effect on antioxidant status, photosynthesis, mineral uptake and growth of lettuce under soil salinity. Res J Agric Biol Sci 1(3):210–215

Hare PD, Van Staden J (1997) The molecular basis of cytokinin action. Plant Growth Regul 23:41–78

Hare PD, Cress WA, van Staden J (1997) The involvement of cytokinins in plant responses to environmental stress. Plant Growth Regul 23:79–103

Hasegawa PM, Bressan RA, Zhu JK, Bohnert HJ (2000) Plant cellular and molecular responses to high salinity. Annu Rev Plant Physiol Plant Mol Biol 51:463–499

Helgason T, Fitter AH (2009) Natural selection and evolutionary ecology of the arbuscular mycorrhizal fungi (phylum Glomeromycota). J Exp Bot 60(9):2465–2480

Jahromi F, Aroca R, Porcel R, Ruiz-Lozano JM (2008) Influence of salinity on the *in vitro* development of *Glomus intraradices* and on the *in vivo* physiological and molecular responses of mycorrhizal lettuce plants. Microb Ecol 55:45–53

Jaleel CA, Gopi R, Manivannan P, Panneerselvam R (2007) Responses of antioxidant defense system of *Catharanthus roseus* (L.) G. Don. to paclobutrazol treatment under salinity. Acta Physiol Plant 29:205–209

Jampeetong A, Brix H (2009) Effects of NaCl salinity on growth, morphology, photosynthesis and proline accumulation of *Salvinia natans*. Aquat Bot 91:181–186

Javid MG, Sorooshzadeh A, Moradi F, Modarres Sanavy SAM, Allahdadi I (2011) The role of phytohormones in alleviating salt stress in crop plants. Aust J Crop Sci 5(6):726–734

Jebara S, Drevon JJ, Jebara M (2010) Modulation of symbiotic efficiency and nodular antioxidant enzyme activities in two *Phaseolus vulgaris* genotypes under salinity. Acta Physiol Plant 32:925–932

Juniper S, Abbott LK (2006) Soil salinity delays germination and limits growth of hyphae from propagules of arbuscular mycorrhizal fungi. Mycorrhiza 16:371–379

Jun-li T, Yue-hu K (2009) Changes in soil properties under the influences of cropping and drip irrigation during the reclamation of severe salt-affected soils. Agric Sci China 8(10): 1228–1237

Kavi Kishor PB, Sangam S, Amrutha RN, Sri Laxmi P, Naidu KR, Rao KRSS, Rao S, Reddy KJ, Theriappan P, Sreenivasulu N (2005) Regulation of proline biosynthesis, degradation, uptake and transport in higher plants: Its implications in plant growth and abiotic stress tolerance. Curr Sci 88(3):424–438

Kaya C, Ashraf M, Sonmez O, Aydemir S, Tuna AL, Cullu MA (2009) The influence of arbuscular mycorrhizal colonization on key growth parameters and fruit yield of pepper plants grown at high salinity. Sci Hortic 121:1–6

Li Z, Chang S, Lin L, Li Y, An Q (2011) A colorimetric assay of 1-aminocyclopropane-1-carboxylate (ACC) based on ninhydrin reaction for rapid screening of bacteria containing ACC deaminase. Lett Appl Microbiol 53:178–185

Ljung K, Hull AK, Celenza J, Yamada M, Estelle M, Normanly J, Sandberg G (2005) Sites and regulation of auxin biosynthesis in arabidopsis roots. Plant Cell 17:1090–1104

Lutts S, Kinet JM, Bouharmont J (1996) Effects of salt stress on growth, mineral nutrition and proline accumulation in relation to osmotic adjustment in rice (*Oryza sativa* L.) cultivars differing in salinity resistance. Plant Growth Regul 19:207–218

Ma JH, Yao JL, Cohen D, Morris B (1998) Ethylene inhibitors enhance *in vitro* root formation from apple shoot cultures. Plant Cell Rep 17:211–214

Madigan MT, Oren A (1999) Thermophilic and halophilic extremophiles. Curr Opin Microbiol 2:265–269

Matamoros MA, Dalton DA, Ramos J, Clemente MR, Rubio MC, Becana M (2003) Biochemistry and molecular biology of antioxidants in the rhizobia-legume symbiosis. Plant Physiol 133:499–509

Mayak S, Tirosh T, Glick BR (2004) Plant growth-promoting bacteria confer resistance in tomato plants to salt stress. Plant Physiol Biochem 42:565–572

Mohamed HI, Gomaa EZ (2012) Effect of plant growth promoting *Bacillus subtilis* and *Pseudomonas fluorescens* on growth and pigment composition of radish plants (*Raphanus sativus*) under NaCl stress. Photosynthetica 50(2):263–272

Mohammad MJ, Malkawi HI, Shibli R (2003) Effects of arbuscular mycorrhizal fungi and phosphorus fertilization on growth and nutrient uptake of barley grown on soils with different levels of salts. J Plant Nutr 26(1):125–137

Moran JF, James EK, Rubio MC, Sarath G, Klucas RV, Becana M (2003) Functional characterization and expression of a cytosolic iron-superoxide dismutase from cowpea root nodules. Plant Physiol 133:773–782

Mousavi A, Lessani H, Babalar M, Talaei AR, Fallahi E (2008) Influence of salinity on chlorophyll, leaf water potential, total soluble sugars, and mineral nutrients in two young olive cultivars. J Plant Nutr 31:1906–1916

Mulholland BJ, Taylor IB, Jackson AC, Thompson AJ (2003) Can ABA mediate responses of salinity stressed tomato. Environ Exp Bot 50:17–28

Munns R (1993) Physiological processes limiting plant growth in saline soil: some dogmas and hypotheses. Plant Cell Environ 16:15–24

Nadeem SM, Hussain I, Naveed M, Asghar HN, Zahir ZA, Arshad M (2006) Performance of plant growth promoting rhizobacteria containing ACC-deaminase activity for improving growth of maize under salt-stressed conditions. Pak J Agri Sci 43:114–121

Navarro A, Elia A, Conversa G, Campi P, Mastrorilli M (2012) Potted mycorrhizal carnation plants and saline stress: growth, quality and nutritional plant responses. Sci Hortic 140:131–139

Nazarbeygi E, Yazdi HL, Naseri R, Soleimani R (2011) The effects of different levels of salinity on proline and A-, B-chlorophylls in canola. Am Eurasian J Agric Environ Sci 10(1):70–74

Noctor G, Foyer CH (1998) Ascorbate and glutathione keeping active oxygen under control. Annu Rev Plant Physiol Plant Mol Biol 49:249–279

Onofre-Lemus J, Hernandez-Lucas I, Girard L, Caballero-Mellado J (2009) ACC (1 aminocyclopropane-1-carboxylate) deaminase activity, a widespread trait in B*urkholderia* species, and its growth-promoting effect on tomato plants. Appl Environ Microbiol 75(20): 6581–6590

Parida AK, Das AB (2005) Salt tolerance and salinity effects on plants: a review. Ecotox Environ Safe 60:324–349

Parniske M (2008) Arbuscular mycorrhiza: the mother of plant root endosymbiosis. Nat Rev Microbiol 6:763–775

Patel D, Jha CK, Tank N, Saraf M (2012) Growth enhancement of chickpea in saline soils using plant growth-promoting rhizobacteria. J Plant Growth Regul 31:53–62

Porcel R, Aroca R, Ruiz-Lozano JM (2012) Salinity stress alleviation using arbuscular mycorrhizal fungi. A review. Agron Sustain Dev 32:181–200

Porras-Soriano A, Soriano-Martin ML, Porras-Piedra A, Azcon R (2009) Arbuscular mycorrhizal fungi increased growth, nutrient uptake and tolerance to salinity in olive trees under nursery conditions. J Plant Physiol 166:1350–1359

Qurashi AW, Sabri AN (2011) Osmoadaptation and plant growth promotion by salt tolerant bacteria under salt stress. Afr J Microbiol Res 5(21):3546–3554

Remy W, Taylor TN, Hass H, Kerp H (2004) Four hundred-million-year-old vesicular arbuscular mycorrhizae. Proc Natl Acad Sci U S A 91:11841–11843

Rillig MC (2004) Arbuscular mycorrhizae, glomalin, and soil aggregation. Can J Soil Sci 84:355–363

Rillig MC, Wright SF, Eviner VT (2002) The role of arbuscular mycorrhizal fungi and glomalin in soil aggregation: comparing effects of five plant species. Plant Soil 238:325–333

Rojas-Tapias D, Moreno-Galvan A, Pardo-Diaz S, Obando M, Rivera D, Bonilla R (2012) Effect of inoculation with plant growth-promoting bacteria (PGPB) on amelioration of saline stress in maize (*Zea mays*). Appl Soil Ecol 61:264–272

Ruiz-Lozano JM (2003) Arbuscular mycorrhizal symbiosis and alleviation of osmotic stress. New perspectives for molecular studies. Mycorrhiza 13:309–317

Ruiz-Lozano JM, Porcel R, Azcon C, Aroca R (2012) Regulation by arbuscular mycorrhizae of the integrated physiological response to salinity in plants: new challenges in physiological and molecular studies. J Exp Bot 63(11):4033–4044

Saied AS, Keutgen AJ, Noga G (2005) The influence of NaCl salinity on growth, yield and fruit quality of strawberry cvs. 'Elsanta' and 'Korona'. Sci Hortic 103:289–303

Sairam RK, Veerabhadra Rao K, Srivastava GC (2002) Differential response of wheat genotypes to long term salinity stress in relation to oxidative stress, antioxidant activity and osmolyte concentration. Plant Sci 163:1037–1046

Sairam SGC, Agarwal S, Meena RC (2005) Differences in antioxidant activity in response to salinity stress in tolerant and susceptible wheat genotypes. Biol Plant 49(1):85–91

Salah IB, Albacete A, Messedi D, Gandour M, Andujar CM, Zribi K, Martinez V, Abdelly C, Perez-Alfocea F (2013) Hormonal responses of nodulated *Medicago ciliaris* lines differing in salt tolerance. Environ Exp Bot 86:35–43

Sandhya V, Ali SKZ, Grover M, Reddy G, Venkateswarlu B (2009) Alleviation of drought stress effects in sunflower seedlings by the exopolysaccharides producing *Pseudomonas putida* strain GAP-P45. Biol Fertil Soils 46:17–26

Sannazzaro AI, Ruiz OA, Alberto EO, Menendez AB (2006) Alleviation of salt stress in *Lotus glaber* by *Glomus intraradices*. Plant Soil 285:279–287

Santos R, Herouart D, Puppo A, Touati D (2000) Critical protective role of bacterial superoxide dismutase in Rhizobium-legume symbiosis. Mol Microbiol 38:750–759

Sarhadi E, Bazargani MM, Sajise AG, Abdolahi S, Vispo NA, Arceta M, Nejad GM, Singh RK, Salekdeh GH (2012) Proteomic analysis of rice anthers under salt stress. Plant Physiol Biochem 58:280–287

Selvakumar G, Thamizhiniyan P (2011) The effect of the arbuscular mycorrhizal (AM) fungus *Glomus intraradices* on the growth and yield of chilli (*Capsicum annuum* L.) under salinity stress. World Appl Sci J 14(8):1209–1214

Sheng M, Tang M, Chen H, Yang B, Zhang F, Huang Y (2008) Influence of arbuscular mycorrhizae on photosynthesis and water status of maize plants under salt stress. Mycorrhiza 18:287–296

Shukla PS, Agarwal PK, Jha B (2012) Improved salinity tolerance of *Arachis hypogaea* (L.) by the interaction of halotolerant plant-growth-promoting rhizobacteria. J Plant Growth Regul 31:195–206

Siddikee MA, Chauhan PS, Anandham R, Han GH, Sa T (2010) Isolation, characterization, and use for plant growth promotion under salt stress, of ACC deaminase-producing halotolerant bacteria derived from coastal soil. J Microbiol Biotechnol 20(11):1577–1584

Siddikee MA, Glick BR, Chauhan PS, Yim WJ, Sa T (2011) Enhancement of growth and salt tolerance of red pepper seedlings (*Capsicum annuum* L.) by regulating stress ethylene synthesis with halotolerant bacteria containing 1-aminocyclopropane-1-carboxylic acid deaminase activity. Plant Physiol Biochem 49:427–434

Siddikee MA, Chauhan PS, Sa T (2012) Regulation of ethylene biosynthesis under salt stress in red pepper (*Capsicum annuum* L.) by 1-aminocyclopropane-1-carboxylic acid (ACC) deaminase-producing halotolerant bacteria. J Plant Growth Regul 31:265–272

Simon S, Petrasek J (2011) Why plants need more than one type of auxin. Plant Sci 180:454–460

Singh PK (2012) Role of glomalin related soil protein produced by arbuscular mycorrhizal fungi: a review. Agric Sci Res J 2(3):119–125

Singla R, Garg N (2005) Influence of salinity on growth and yield attributes in chickpea cultivars. Turk J Agric For 29:231–235

Sreenivasulu N, Harshavardhan VT, Govind G, Seiler C, Kohli A (2012) Contrapuntal role of ABA: does it mediate stress tolerance or plant growth retardation under long-term drought stress? Gene 506:265–273

Szabados L, Savoure A (2009) Proline: a multifunctional amino acid. Trends Plant Sci 15(2):89–97

Talaat NB, Shawky BT (2011) Influence of arbuscular mycorrhizae on yield, nutrients, organic solutes, and antioxidant enzymes of two wheat cultivars under salt stress. J Plant Nutr Soil Sci 174:283–291

Talibart R, Jebbar M, Gouesbet G, Himdi-Kabbab S, Wróblewski H, Blanco C, Bernard T (1994) Osmoadaptation in rhizobia: ectoine-induced salt tolerance. J Bacteriol 176(17):5210–5217

Tavakkoli E, Rengasamy P, McDonald GK (2010) High concentrations of Na^+ and Cl^- ions in soil solution have simultaneous detrimental effects on growth of faba bean under salinity stress. J Exp Bot 61(15):4449–4459

Tavakkoli E, Fatehi F, Coventry S, Rengasamy P, McDonald GK (2011) Additive effects of Na^+ and Cl^- ions on barley growth under salinity stress. J Exp Bot 62(6):2189–2203

Tejera NA, Soussi M, Lluch C (2006) Physiological and nutritional indicators of tolerance to salinity in chickpea plants growing under symbiotic conditions. Environ Exp Bot 58:17–24

Thomas TH (1992) Some reflections on the relationship between endogenous hormones and light-mediated seed dormancy. Plant Growth Regul 11:239–248

Tian CY, Feng G, Li XL, Zhang FS (2004) Different effects of arbuscular mycorrhizal fungal isolates from saline or non-saline soil on salinity tolerance of plants. Appl Soil Ecol 26:143–148

Tognetti VB, Mühlenbock P, Breusegem FV (2012) Stress homeostasis—the redox and auxin perspective. Plant Cell Environ 35:321–333
Tressner HD, Hayes JA (1971) Sodium chloride tolerance of terrestrial fungi. Appl Microbiol 22(2):210–213
Turner NC, Colmer TD, Quealy J, Pushpavalli R, Krishnamurthy L, Kaur J, Singh G, Siddique KHM, Vadez V (2013) Salinity tolerance and ion accumulation in chickpea (*Cicer arietinum* L.) subjected to salt stress. Plant Soil 365(1–2):347–361
Vassilev N, Eichler-Löbermann B, Vassileva M (2012) Stress-tolerant P-solubilizing microorganisms. Appl Microbiol Biotechnol 95:851–859
Veselov DS, Sabirzhanova IB, Sabirzhanov BE, Chemeris AV (2008) Changes in expansin gene expression, IAA content, and extension growth of leaf cells in maize plants subjected to salinity. Russ J Plant Physiol 55(1):101–106
Viterbo A, Landau U, Kim S, Chernin L, Chet I (2010) Characterization of ACC deaminase from the biocontrol and plant growth-promoting agent *Trichoderma asperellum* T203. FEMS Microbiol Lett 305:42–48
Voutsela S, Yarsi G, Petropoulos SA, Khan EM (2012) The effect of grafting of five different rootstocks on plant growth and yield of tomato plants cultivated outdoors and indoors under salinity stress. Afr J Agric Res 7(41):5553–5557
Wang B, Qiu YL (2006) Phylogenetic distribution and evolution of mycorrhizas in land plants. Mycorrhiza 16:299–363
Wang Y, Mopper S, Hasenstein KH (2001) Effects of salinity on endogenous ABA, IAA, JA, and SA in *Iris hexagona*. J Chem Ecol 27:327–342
Woodward AW, Bartel B (2005) Auxin: regulation, action, and interaction. Ann Bot 95:707–735
Wu QS, Zou YN (2009) Arbuscular mycorrhizal symbiosis improves growth and root nutrient status of citrus subjected to salt stress. ScienceAsia 35:388–391
Wu SJ, Ding L, Zhu JK (1996) SOS1, a genetic locus essential for salt tolerance and potassium acquisition. Plant Cell 8:617–627
Wu QS, Zou YN, Liu W, Ye XF, Zai HF, Zhao LJ (2010) Alleviation of salt stress in citrus seedlings inoculated with mycorrhiza: changes in leaf antioxidant defense systems. Plant Soil Environ 56(10):470–475
Wu Z, Yue H, Lu J, Li C (2012) Characterization of rhizobacterial strain Rs-2 with ACC deaminase activity and its performance in promoting cotton growth under salinity stress. World J Microbiol Biotechnol 28:2383–2393
Yamaguchi T, Blumwald E (2005) Developing salt-tolerant crop plants: challenges and opportunities. Trends Plant Sci 10(12):615–620
Yang SF, Hoffman NE (1984) Ethylene biosynthesis and its regulation in higher plants. Annu Rev Plant Physiol 35:155–189
Yang Y, Shi R, Wei X, Fan Q, An L (2010) Effect of salinity on antioxidant enzymes in calli of the halophyte *Nitraria tangutorum* Bobr. Plant Cell Tissue Organ Cult 102:387–395
Yildirim E, Turan M, Donmez MF (2008) Mitigation of salt stress in radish (*Raphanus sativus* L.) by plant growth promoting rhizobacteria. Roum Biotechnol Lett 13(5):3933–3943
Zahir ZA, Asghar HN, Arshad M (2001) Cytokinin and its precursors for improving growth and yield of rice. Soil Biol Biochem 33:405–408
Zahir ZA, Ghani U, Naveed M, Nadeem SM, Asghar HN (2009) Comparative effectiveness of *Pseudomonas* and *Serratia* sp. containing ACC-deaminase for improving growth and yield of wheat (*Triticum aestivum* L.) under salt-stressed conditions. Arch Microbiol 191:415–424
Zahir ZA, Zafar-ul-Hye M, Sajjad S, Naveed M (2011) Comparative effectiveness of *Pseudomonas* and *Serratia* sp. containing ACC-deaminase for coinoculation with *Rhizobium leguminosarum* to improve growth, nodulation, and yield of lentil. Biol Fertil Soils 47:457–465
Zahran HH (1999) Rhizobium-legume symbiosis and nitrogen fixation under severe conditions and in an arid climate. Microbiol Mol Biol Rev 63(4):968–989
Zeng L, Shannon MC (2000) Effects of salinity on grain yield and yield components of rice at different seeding densities. Agron J 92:418–423
Zhu JK (2003) Regulation of ion homeostasis under salt stress. Curr Opin Plant Biol 6:441–445

Zhu C, Schraut D, Hartung W, Schaffner AR (2005) Differential responses of maize MIP genes to salt stress and ABA. J Exp Bot 56(421):2971–2981

Zorb C, Geilfus CM, Muhling KH, Ludwig-Muller J (2013) The influence of salt stress on ABA and auxin concentrations in two maize cultivars differing in salt resistance. J Plant Physiol 170:220–224

Zuccarini P, Okurowska P (2008) Effects of mycorrhizal colonization and fertilization on growth and photosynthesis of sweet basil under salt stress. J Plant Nutr 31:497–513

Chapter 7
Cash Crop Halophytes: The Ecologically and Economically Sustainable Use of Naturally Salt-Resistant Plants in the Context of Global Changes

Nicole Geissler, Helmut Lieth, and Hans-Werner Koyro

Contents

Abbreviations

FAO	Food and Agriculture Organization of the United Nations
IPPC	Intergovernmental Panel on Climate Change
ppm	Parts per million
UNEP	United Nations Environment Programme

N. Geissler (✉)
Justus Liebig University Giessen, Special Botany,
Heinrich-Buff-Ring 38, 35392 Giessen, Germany
e-mail: Nicole.Geissler@bot2.bio.uni-giessen.de

H. Lieth
Institute of Environmental Systems Research (USF),
University of Osnabrück, 49069 Osnabrück, Germany

H.-W. Koyro
Institute of Plant Ecology, Justus Liebig University Giessen,
Heinrich-Buff-Ring 26, 35392 Giessen, Germany

P. Ahmad and M.R. Wani (eds.), *Physiological Mechanisms and Adaptation Strategies in Plants Under Changing Environment: Volume 1*, DOI 10.1007/978-1-4614-8591-9_7,

1 Introduction: Population Growth and Climate Change as Challenges for the Future

Feeding the rapidly growing human population is one of the biggest challenges we (will) face on our planet today and in future. During the last 12 years, the world population has increased from 6 to 7 billion (Deutsche Stiftung Weltbevölkerung 2012a). By 2050, 9.3 billion people are expected to live on earth, while the African population will probably double within the next 40 years (Deutsche Stiftung Weltbevölkerung 2012b). Especially in arid and semiarid regions, the increasing population density is collaterally catalysed by the consequences of global climate change which is caused by anthropogenic emissions of trace gases (IPCC 2007). One of the most important greenhouse gases is CO_2, the atmospheric concentration of which has increased from 280 ppm to approximately 380 ppm since the beginning of industrialisation and will continue to rise in future (IPCC 2007). Global climate change manifests itself in extreme weather events such as flooding, storms or drought, as well as in rising global temperatures which have increased by about 1 °C over the course of the last century and will likely rise even more rapidly in coming decades. Scientists predict that temperatures could rise by another 3–9 °C by the end of this century with far-reaching effects (IPCC 2007). Increased drought, desertification, salinisation of arable land and freshwater scarcity are expected to have devastating effects on global croplands and food production (Wang et al. 2003). Abiotic stresses such as drought and salinity are already the primary reason for loss of agricultural productions worldwide, reducing average yields for most major crop plants by more than 50 % (Bray et al. 2000; Wang et al. 2003). Yield losses will soon become even more severe as desertification increases, so the current amount of annual loss of arable land may double by the end of the century because of global warming (Evans 2005; Vinocur and Altman 2005).

One of the major consequences of the increasing freshwater scarcity regarding agriculture in arid climate zones is the frequent use of saline irrigation water in an unprofessional manner, which often leads to further degradation of arable land. If a field is irrigated inaccurately and evapotranspiration exceeds total precipitation, water will rise by capillarity in the soil, and the eluviation of salts will be prevented. Thus the soluble salts present in the soil will be relocated to the upper soil layers together with water and will accumulate there. This process leads to secondary salinisation (Schubert 1999). According to the Food and Agriculture Organization of the United Nations (FAO), more than 30 % of the global irrigated land area (equal to six million km^2 farmland) is already affected by salinity today (Hussin et al. 2013).

Apart from desertification and salinisation, food availability is also limited by the decreasing reserves of fossil resources because energy requirements are partly covered by the energetic exploitation of food plants or the cultivation of energy crops on arable land even today. Food scarcity is further reinforced by ageing societies, especially in western countries (more food has to be provided for a person's lifetime), and by a higher meat consumption. According to the FAO, meat consumption has globally increased by one third since 1980 (industrial countries, from 76 to

82 kg/head; developing countries, from 14 to 30 kg/head), which enhances the need for biomass. Presuming the same energy yield, the acreage needed for animal food is ten times larger than the one needed for vegetable food.

However, there are various approaches to solve the problems mentioned above, including:

- A more equal food distribution and a direct or indirect reduction of water consumption during food production. This is part of the United Nations Millennium Development Goals proclaimed in September 2000.
- An intensified application of pesticides to maintain or (as far as possible) increase agricultural productivity.
- An increased application of genetic engineering to enhance plant stress resistance and productivity.
- The screening of biological resources and the development of new crop plants which consume less water and thrive in very dry habitats. An example is the so-called Groasis system in California, which allows the cultivation of vine with very low water consumption. This system is currently in an experimental phase in the Western Sahara (www.groasis.com).
- The environmentally sustainable and lasting use of alternative water resources such as saline and brackish water.

In order to reclaim degraded land and to prevent the wasting of precious freshwater for agriculture in the future, we need to rethink our use of water, especially the use of alternative water resources, which has a high potential for the extension of arable land. So, many scientists are currently striving to develop salt-resistant plants as crops suitable for agriculture (Rozema and Flowers 2008). The sustainable use of so-called cash crop halophytes is very well suited to ameliorate the negative impacts of global change on agriculture and environment while simultaneously counteracting its cause (due to their ability to sequester CO_2). These and other aspects of cash crop halophytes, and their utilisation will be presented in more detail in the following chapters.

2 Halophytes: Promising Crop Plants for the Future

More than 80 % of the earth's water resources are sea water. However, only few plants can tolerate sea water conditions—i.e. a salinity of about 3.5 % (sea water salinity, sws)—without serious damage. Though the evolution of plant life started in saline sea water three billions of years ago, this primary adaptation was lost during the development of terrestrial plants about 450 million years ago (Flowers et al. 2010). Today, 99 % of the terrestrial plant species are salt-sensitive meso- and glycophytes which are not able to grow and/or reproduce near coastlines or in saline inlands, including our contemporary crop plants. Their salt resistance is comparatively low, at most 20 % sea water salinity. Therefore, scientists have tried to adapt crop plants such as rice, maize or wheat to adjust to high soil salinity for approximately 20 years. Yet, attempts to increase their salt resistance significantly via plant breeding or molecular genetics has not been very successful up to now (Koyro et al. 2008). This is probably due to the fact that NaCl simultaneously affects several aspects of plant physiology (Koyro et al. 2013a) and to the polygenic inheritance of traits which confer salt resistance (Flowers 2004). Consequently, conventional crop plants are not well suited for production systems using saline irrigation.

Therefore, more focus should be laid on the utilisation potential of plants which can naturally survive in salty environments (Koyro et al. 2013b). They are called halophytes (from ancient Greek halos = salt and phytón = plant) which comprise roughly 2.600 species (Lieth 1999). Halophytes belong to the extremophytes and include annuals and perennials, mono- and dicotyledoneae, herbs, shrubs and trees. Except the polar Arctic and Antarctic regions, halophytes are distributed on all continents. They are found along European seashores as well as in tropical rain forests, arid salt steppes, in alpine regions near saline lakes and salt springs up to 2.500 m above sea level, or in deserts (e.g. Sahara) on sodium-rich sands. Although halophytes are present in all nonpolar climate zones, the tropical and temperate zones are their main distribution range.

Most halophytes prefer a saline environment but can also survive in freshwater habitats while using water very efficiently (Schimper 1898, 1903; Dansereau 1957). They grow preferentially in brackish water which is about half as saline as sea water. However, wide differences in salt resistance can be observed within the group of halophytes. Depending on the level of salt resistance, they can be classified into three categories of halophilia (Koyro and Lieth 1989; Le Houerou 1993): (a) low-salt resistance (up to 7 g NaCl/L), (b) transition zone (up to 25 g NaCl/L) and (c) high salt resistance (up to 65 g NaCl/L) (as a comparison, sea water contains approximately 30 g NaCl/L). Additionally, obligate, facultative or site-indifferent halophytes can be distinguished. Obligate halophytes such as *Sesuvium portulacastrum* and *Tecticornia indica* are salt-loving plants (Rabhi et al. 2011). They grow only in salty habitats and show a clear optimisation of their development through an increased salt supply in experiments (von Sengbusch 2003). Salinity levels of more than 0.5 % lead to growth stimulation and an increased vitality of these plants, whereas salt deficiency causes stunted growth (Künnemann and Gad 1997). Facultative halophytes are able to settle on salty soils, but their optimum lies in a salt-free or at least low-salt milieu (von Sengbusch 2003). They show better growth on salt-free or low-salt soils (physiological optimum) than in saline environments. Site-indifferent halophytes have a big advantage regarding the competition with other plants because on the one hand they can compete with glycophytes in salt-free locations and on the other hand they can grow on saline soils.

The above considerations show that halophytes include a variety of plants that prefer different abiotic conditions. In general, they are very promising future crops because halophytes are already adapted to extreme habitat conditions associated with global change. Additionally, elevated CO_2 can enhance their salt resistance and/or productivity, so that they are likely to thrive in a future with rising atmospheric CO_2 concentration (see sub-title 5 for more details). They are thus well suited to ameliorate the consequences of global change.

3 Utilisation of Halophytes

Halophytes have been considered as almost valueless up to now. At the Danish, German and Dutch coastlines, salt marshes have been destroyed by diking, while in the tropics, mangrove marshes have been reduced by half in order to use this land for agriculture or tourism. In fact halophytes are very valuable because the negative impacts of global change on agriculture and environment in arid regions could be solved or ameliorated by a targeted cultivation of these plants, owing to their diverse options for use. Thereby, we must pursue the goal to grow halophytes in an ecologically and economically sustainable manner. Cultivation should be done at locations which—due to increasing drought and/or improper use of saline irrigation water—are no longer suitable for the production of conventional crops (phytoremediation) or which actually do not host any plants, such as coasts, deserts or regions with saline ground water resources.

The ecologically sustainable use is an important prerequisite for preventing the expansion of salinised areas. It depends on the selection of suitable plant species which should be chosen according to the following principles:

- The species used should tolerate at least 15 g NaCl/L, i.e. 50 % sea water salinity (Miyamoto 1996).
- Their natural distribution area should be along the coastlines of as many arid and semiarid countries as possible.
- The selected species should have a high economic potential, i.e. they should yield good crops (Glenn et al. 1999) and secure sustainable earnings.
- A detailed knowledge about the mechanisms which confer resistance to abiotic stress is necessary. It would also be desirable to gather information about the effects of future atmospheric conditions (e.g. elevated CO_2 or ozone concentrations) or increased temperature on the performance and salt resistance of the species in question.
- It has to be tested if the sustainable use of halophytes is socially compatible, i.e. saline production systems have to fit with the existing agricultural/economic infrastructure (Lieth 2000). If the local population refuses to cultivate certain plants, the project will fail in the end.

Numerous halophytes can be utilised for a large variety of economical and ecological purposes which are listed in Fig. 7.1 (see also Fig. 7.2). Only one aspect of ecological utilisation is to be pointed out in more detail at this point because it has a special connection to global climate change: Every new large plant population which is created for long-lasting use can sequester CO_2 (Güth 2001; Arnalds 2004). Salt marshes and halophytes seem to be particularly suited for a long-term CO_2 sequestration (Caçador et al. 2002), and the cultivation of these plants in order to counteract the greenhouse effect was already proposed by the United Nations Environment Programme (UNEP) in 1993 (UNEP 1993).

The above considerations suggest that the sustainable use of cash crop halophytes can probably ameliorate the negative impacts of global change on environment, agriculture and human nutrition, while it can simultaneously counteract one of its

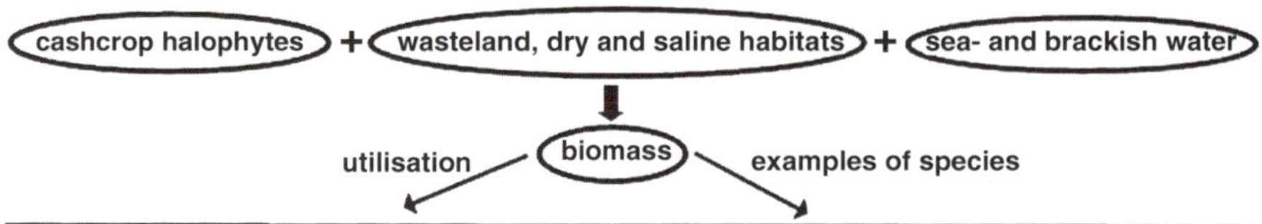

	food: - salad, vegetables - oil (seeds) - fruits	 - *Aster tripolium, Plantago maritima* - *Salicornia* ssp., *Critmum maritimum* - *Avicennia marina*
	fodder	*Atriplex* ssp., *Sporobulus virginicus, Panicum furgidum*
	ornamental plants	*Aster tripolium, Limoniastrum monopatalum, Sesuvium verrucosum, Kosteletzkya virginica*
	wood, fibers	mangroves, *Tamarix* ssp., *Juncus maritimus*
	chemicals (for chemical and pharmaceutical industry, production of clothes or plastics)	*Beta maritima, Juncus acutus, Critmum maritimum*
	renewable resources (e.g. biofuel)	*Spartina alterniflora, Panicum turgidum*
	landscaping, combating soil erosion and desertification (planting of streets, residential areas, dune stabilsation, re-greening)	*Sesuvium portulacastrum, Batis maritima*
	coast protection	mangroves, *Spartina* ssp.
	tertiary treatment of water and soil to reclaim degraded land, soil amelioration (reduction of salinity, oil or heavy metal contamination)	*Suaeda maritima, Salsola kali*
	preservation of biodiversity	salt marsh plants, mangroves
	marine agro-systems	*Salicornia* ssp., mangroves, algae
	CO_2 sequestration	all species

Fig. 7.1 Possibilities for halophyte utilisation. Utilisations of halophytes already existing, utilisation purposes which are investigated and corresponding examples of species. Source of the photo belonging to marine agro-systems: http://www.seawaterfoundation.org/sea_science.html

causes, namely, the rising atmospheric CO_2 concentration. Additionally, the cultivation of halophytes and their manufacturing into foods or industrial products will be cost-effective and economically advantageous. Therefore, this topic is of high sociopolitical relevance.

Fig. 7.2 Utilisable halophytes. (**a**) *Aster tripolium*, (**b**) *Spartina townsendii*, (**c**) *Chenopodium quinoa*, (**d**) *Suaeda maritima*, (**e**) *Salicornia europaea* and (**f**) *Avicennia marina* (mangrove)

4 The Establishment of Saline Production Systems: Necessary Steps

In order to determine the possible contribution of halophytes to the extension of food supplies or to the substitution of conventional crops as renewable resources, detailed studies will be necessary. Some rules and precautions are to be considered while developing saline irrigation systems (Lieth 2000; Lieth and Mochtschenko 2003; for details see Figs. 7.3 and 7.4). First of all, a preselection with the help of existing background information about the foreseeable utilisation potential and environmental demands (climate, soil conditions, water demand) as well as a phytosociological assessment (suitability test) should be conducted. The salt resistance and its threshold as well as the food or fodder quality can be studied by means of a quick check system which can also be used to test the performance of the species under future atmospheric conditions. More details about this system are presented in the following chapter.

In case of positive results, an analysis of yield and sustainability should be carried out in countries which are qualified for cultivation, on local soils, with local saline water resources and suited irrigation systems (for details see Figs. 7.3 and 7.4).

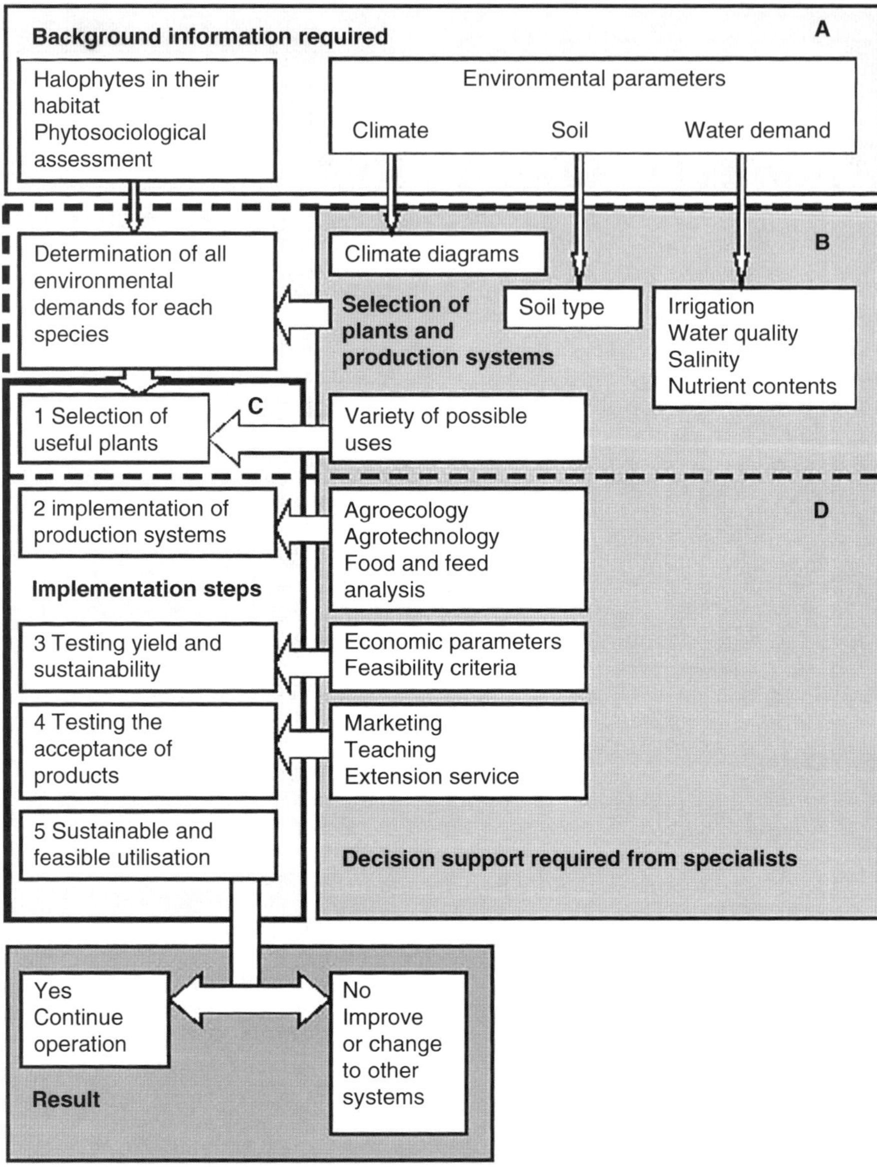

Fig. 7.3 Flowchart of work steps for the implementation of saline production systems (after Lieth, in Lieth and Lohmann 2000)

In doing so, also economic criteria should be considered, including yield and its longevity and the practicability of the culture. The final step will be testing the marketing, i.e. the acceptance and the profitability, of the new product.

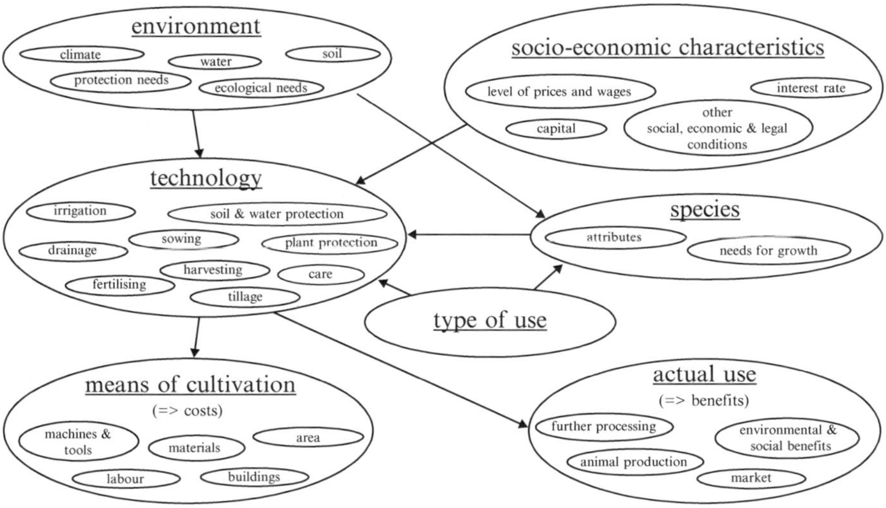

Fig. 7.4 The interdependencies between parameters which have to be considered when establishing saline production systems (after Lieth, in Lieth and Lohmann 2000)

5 Quick Check System: The Basic Step to Select Suited Cash Crop Halophytes

In a so-called quick check system, plants are grown under different salinity levels, but otherwise under identical growth conditions. Several types of cultivation can be used, such as soil cultures, gravel hydroponics with drip irrigation (Fig. 7.5a) or pure hydroponics (Fig. 7.5b). The quick check system is used to analyse the individual (species specific) salt resistance mechanisms and the salt resistance threshold. It can also serve to test the performance of the plants under elevated CO_2 concentration. In this case two identical quick check systems are run in two open-top chambers where the plants are supplied with ambient (just under 400 ppm) and elevated CO_2 concentration (e.g. 540 ppm, simulating the future atmospheric concentration in 40–80 years time; IPCC 2007) (Fig. 7.5b). Owing to these qualities, the quick check system is essential for the selection of promising cash crop halophytes (Flowers and Colmer 2008; Koyro and Lieth 2008). These plants exhibit a broad bandwidth of morphological, physiological and biochemical adaptive mechanisms (Koyro et al. 2006; Huchzermeyer 2011; Debez et al. 2012) and also differ in their response to elevated CO_2. The quick check is well suited to investigate the reaction of plants to the growth constraints of saline habitats (osmotic stress and restriction of CO_2 uptake; ion toxicity and nutrient imbalance; Geissler et al. 2013) and to the influence of elevated CO_2 on salt resistance.

Growth depression under saline conditions can be explained at least for glycophytes such as barley and wheat by a two-phase model (Munns 1993, 2002)

Fig. 7.5 (**a**, **b**) Quick check systems to determine species-specific salt resistance mechanisms and salt resistance threshold. (**a**) *Sporobolus virginicus* (*left/back row*) and *Leptochloa fusca* (*right/front row*), gravel hydroponics with drip irrigation; (**b**) *Aster tripolium*, pure hydroponics culture, permanently aerated, in open-top chamber. (**c**, **d**) Examples of saline production systems. (**a**) *Sporobolus virginicus* cultivated for fodder using sprinkler irrigation in the United Arab Emirates; (**b**) *Aster tripolium* cultivated for food in the Tejo estuary, Portugal (photo: Lieth 2000)

distinguishing between osmotic and ionic stress. It is not proved that halophytes experience these constraints within the same time frame, but they need to react to both of them in any case. Osmotic stress is caused by the reduced osmotic potential of the external medium (physiological drought). Osmotic stress has negative effects on the water balance of plants and causes detrimental changes in cellular components because the biologically active conformation of proteins and biomembranes depends on an intact hydration shell (Schulze et al. 2002). Water loss can be minimised, e.g. by stomatal closure. As a consequence, gas exchange and thus the assimilation rate are inhibited, which in turn may lead to the formation of reactive oxygen species (ROS) which are highly destructive to lipids, nucleic acids, and proteins (Geissler et al. 2010; Ozgur et al. 2013). Compared to salinity, elevated CO_2 concentration often has contrary effects on plants, especially C_3 plants. It leads to a higher CO_2 concentration gradient between the outside air and the intercellular spaces of the leaves, so that the diffusion of CO_2 into the leaves and the pCO_2/pO_2 ratio at the sites of photoreduction are increased (Robredo et al. 2007). Therefore, usually photorespiration and the rates of oxygen activation and ROS formation are reduced due to an increased NADPH utilisation, whereas the net photosynthetic rate and thus the carbon supply are enhanced (Kirschbaum 2004; Long et al. 2004; Ignatova et al. 2005; Pérez-López et al. 2012; Wang et al. 2012). Furthermore, we often find a lower stomatal resistance (Li et al. 2003; Rogers et al. 2004), which, together with higher net assimilation, also leads to a better water use efficiency of photosynthesis (Morgan et al. 2001; Urban 2003). As a consequence of these effects, on the one hand, there might be less need for antioxidants as elevated CO_2 ameliorates oxidative stress (Schwanz et al. 1996). On the other hand, more energy can be provided for energy-dependent stress resistance mechanisms such as the synthesis of osmolytes and antioxidants. Due to both effects mentioned above, elevated CO_2 can increase plant survival under abiotic stress conditions, especially in C_3 plants (Ball and Munns 1992; Rozema 1993; Wullschleger et al. 2002; Geissler et al. 2010).

Ion-specific stress is caused by the accumulation of high amounts of Na^+ and Cl^- ions in the cell wall and in the cytosol, leading to toxic effects on the structure and function of biomembranes and proteins. Moreover, the increased Na^+ uptake can inhibit the uptake of other nutrients and/or disturb their internal distribution within the plant (ion homeostasis). Such an ion imbalance leads to Na^+-induced nutrient deficiencies, particularly to deficiencies of the macronutrients K^+, Mg^{2+} or Ca^{2+} in the plant (Marschner 1995). During this phase of salt stress, one can differentiate between salt-sensitive, less sensitive and salt-resistant species. In general, plants have the ability to prevent high ion concentrations either by ion compartmentalisation (includers) or by ion exclusion (excluders) (Schachtman and Munns 1992). While halophytes often take up high amounts of Na^+ and accumulate them in the leaf vacuoles (inclusion) (Läuchli and Epstein 1984), glycophytes pursue mainly the exclusion strategy (Läuchli 1984). However, this popular classification into includers and excluders is somewhat simplified and undifferentiated because most plants show an individual combination of both inclusion and exclusion mechanisms.

The determination of individual regulation of the growth constraints (which mainly limit the salt resistance of the species in question) is an essential preliminary stage for a successful sustainable use of halophytes. Preferably, the influence of future atmospheric conditions such as elevated CO_2 on salt resistance mechanisms should also be tested in order to get more detailed information about the future potential of the plants as crops. Species which exhibit an enhanced salt resistance under elevated CO_2 may be especially suited for sustainable use.

The above considerations show that the investigations carried out within the frame of the quick check system have to be tightly related to the main constraints of salinity on plant growth (Marschner 1995). Basic parameters characterising plant responses to salinity are determined via this system. If the effect of elevated CO_2 on salt resistance is to be studied, these parameters are measured in plants grown under ambient and elevated CO_2 concentration under otherwise identical conditions. The most important basic parameters to be determined are listed below:

- Life cycle (annual, biennial, perennial), form of asexual reproduction (vegetative multiplication) or sexual reproduction
- Growth parameters of various parts of the plants (e.g. dry weight, ratios leaf area/plant biomass (LAR) and shoot/root)
- Yield parameters (biomass, grain yield, food or fodder quality, etc.)
- Morphological and anatomical adaptations
- Photosynthesis (such as correspondence between electron transport rate and CO_2 fixation), stomatal resistance, chlorophyll content and ratios and antioxidant defence

Table 7.1 Examples of data obtained from quick check systems of different halophytes under varying salinity levels

NaCl (mM)	*A* (μmol/m²/s)	R_s (s/cm)	*E* (mol/m²/s²)	WUE *A*/*E*	ψ (MPa)	FW (g)
Panicum turgidum						
0	10.8±0.15	6.44±0.19	1.55±0.05	6.53±0.32	−0.59±0.05	468.34±30.70
125	14.85±0.17	5.80±0.17	1.95±0.06	7.64±0.24	−0.93±0.05	517.99±41.49
250	8.90±0.05	3.39±0.06	1.38±0.05	6.50±0.19	−1.60±0.04	121.22±6.24
375	6.25±0.06	7.96±0.13	1.38±0.03	4.55±0.11	−1.82±0.01	80.22±3.12
500	5.47±0.03	9.31±0.27	1.20±0.07	4.60±0.27	−3.13±0.02	23.74±0.60
Atriplex nummularia						
0	18.25±0.38	2.08±0.34	5.61±0.14	3.26±0.15	−0.18±0.03	64.10±7.69
125	13.95±1.23	4.04±0.65	2.98±0.24	4.69±0.09	−0.33±0.06	358.97±26.30
250	12.65±0.74	5.62±1.01	2.74±0.31	4.70±0.31	−0.81±0.12	383.97±16.03
500	8.94±0.46	8.03±0.42	1.72±0.19	5.24±0.34	−1.58±0.27	319.23±7.05
750	5.01±0.51	15.95±0.68	0.75±0.10	6.75±0.86	−2.58±0.20	135.90±10.90
Chenopodium quinoa						
0	27.37±5.97	1.10±0.06	6.75±0.44	4.05±0.45	−0.18±0.09	105.81±3.08
125	21.15±1.49	2.70±0.23	2.83±0.22	7.47±2.34	−0.36±0.02	121.20±2.56
375	15.23±1.65	5.48±0.66	0.98±0.26	15.54±1.31	−1.28±0.07	37.26±2.07
500	9.42±1.26	6.55±0.65	0.64±0.11	14.72±1.92	−1.80±0.13	19.32±1.71
Aster tripolium						
0	21.55±6.10	1.65±0.75	4.32±0.87	5.09±1.55	−0.67±0.14	76.87±12.52
125					−1.23±0.22	49.66±4.68
250	11.39±5.96	4.51±2.15	2.84±1.48	4.27±0.87	−1.90±0.54	37.24±3.04
375					−3.50±1.11	28.55±2.26
500	7.57±4.60	9.69±5.42	2.30±1.57	3.50±0.43	−4.95±1.15	17.21±6.00
Aster tripolium, under elevated atmospheric CO_2 concentration (520 ppm)						
0	33.63±6.25	1.15±0.72	6.11±1.45	5.79±1.53	−0.57±0.13	79.37±19.65
125					−1.12±0.37	53.06±12.90
250	20.79±7.79	3.08±2.06	4.00±1.91	5.39±1.06	−1.90±0.50	38.20±4.11
375					−2.48±0.58	28.95±5.46
500	12.96±4.78	6.89±4.90	2.52±1.26	5.62±0.30	−4.09±1.10	19.47±6.33

In case of *A. tripolium*, ambient (380 ppm) and elevated (520 ppm) CO_2 concentration was applied for each salt treatment. Values represent mean±SD values of ≥3 measurements per treatment
A net photosynthetic rate, R_s stomatal resistance, *WUE* water use efficiency of photosynthesis, ψ shoot water potential, *FW* fresh weight

- Water use efficiency (WUE), leaf water potential, osmotic potential and water saturation deficit (WSD)
- Mineral content (indicator of ion deficiency or ion excess), ion selectivity, ion compartmentation and compatible solutes

Some examples of such data obtained from quick check systems of different halophytes (*Panicum turgidum*, *Atriplex nummularia*, *Chenopodium quinoa*, *Aster tripolium*) are presented in Table 7.1. The results show that under ambient CO_2 concentration, all mentioned plant species adjust osmotically to salinity by decreasing their water potential. Under salt stress, the net photosynthetic rates decrease due to a higher stomatal resistance and thus a lower transpiration rate (in case of *P. turgidum*, this trend is transient). However, there are marked differences in the water use efficiencies of photosynthesis (WUE), the most crucial parameter regarding gas exchange. While *A. nummularia* and *C. quinoa* are able to enhance WUE with rising salinity, the WUE in *P. turgidum* shows an optimal value at 125 mM NaCl and decreases at higher salinities, and in *A. tripolium* it continually decreases. These results are in accordance with the data for plant growth which show that *A. nummularia* is the most resistant species (optimal growth at 250 mM NaCl), followed by *C. quinoa*, *P. turgidum* (optimal growth of both species at 125 mM NaCl, but more growth reduction at high salinity in *P. turgidum*) and *A. tripolium* (continual growth depression with increasing salinity). Consequently, *A. nummularia* is probably the most profitable cash crop halophyte among the mentioned species. However, in a CO_2-rich future, the salt resistance of C_3 plants such as *C. quinoa* and *A. tripolium* is likely to be more enhanced than the one of C_4 species, with positive effects on the utilisation potential of C_3 plants. In case of *A. tripolium*, we could show that elevated CO_2 increases water relations (water potential), photosynthesis and WUE in salt-treated plants (Table 7.1). The enhanced energy supply was not employed for producing more biomass (Table 7.1), but for increasing the investment in salt resistance mechanisms, leading to a higher survival rate under saline conditions (Geissler et al. 2009a, b, 2010). Investigations on other plants about the interaction of salinity and elevated CO_2 would be very desirable to assess the future potential of halophytes as crop plants more accurately.

Apart from revealing the most promising future cash crop halophytes, the results of the quick check system often influence and support the selection of suited production or irrigation systems (drip irrigation, ditch irrigation or sprinkler system) and of the type of cultivation (monoculture or mixed cultures) (Khan et al. 2009a).

6 Conclusions and Future Perspective

Due to global climate change which exacerbates freshwater scarcity and salinisation, the last decade has witnessed marked losses of arable land, especially in arid and semiarid regions. At least until the breeding of salt-resistant crops will succeed and be accepted, we need to acquire and test candidate halophyte species, screen germplasm under highly saline conditions and develop management techniques for the productive use of halophytes. The further use of these plants is the only available way for a sustainable utilisation and can ameliorate the water crisis.

While reviewing the cultivation of halophytes in detail relating to NaCl salinity and global change/elevated CO_2 concentration for the first time, the following major conclusions can be drawn: A precondition for the sustainable use of halophytes is a detailed knowledge of individual resistance mechanisms, the food or fodder quality and the performance under new climate change scenarios. Such information has already been obtained for some species by growing them in a so-called quick check system. The results show that many halophytes are high-productive and high-quality crops and that they often exhibit an enhanced salt resistance and/or productivity under elevated CO_2, so they are very promising future crop plants. As their cultivation can sequester CO_2, it can also counteract the causes of global change. However, especially regarding the interaction of salinity with other abiotic factors associated with global change such as elevated CO_2, more knowledge is required to assess the potential of halophytes as future crops. Additionally, saline agriculture needs to be much further developed although several promising pilot projects regarding the sustainable use of halophytes are already in progress worldwide:

- The combination of terrestrial saline cultures with aquacultures of fish or shrimps. This includes the establishment of plantations of food crop halophytes in tropical countries such as Eritrea (Manzanar Project, a combination of fish farming and mangrove cultivation, and Seawater Farms, a combination of fish and shrimp farming, *Salicornia* and mangrove cultivation).
- The cultivation of *Sporobolus virginicus* or *Distichlis spicata* as fodder plants under saline sprinkler irrigation in the United Arab Emirates (Fig. 7.5c).
- The use of halophytes in order to detoxify wastewater (e.g. to decrease heavy metal pollution, Tejo River, Portugal).
- The cultivation of *Aster tripolium* in the Tejo Estuary (Portugal) as food plant (salad or vegetable) (Fig. 7.5d).
- The cultivation of *Salicornia bigelovii* in Jubail (Saudi Arabia) to obtain oil seeds. 12.7–24.6 t of biomass/ha (on average 18.0 t/ha) and 1.39–2.46 t of seeds/ha (on average 2 t/ha), respectively, are produced every year (Glenn et al. 1991).
- The culture of *Spartina alterniflora* (which is flooded regularly) near Dubai (United Arab Emirates). *Spartina* is used as fodder and produces up to 40 t biomass/ha (Odum 1974).
- The cultivation of *Salicornia* for food or of *Tamarix aphylla* for wood production within the frame of the Colorado Delta Project carried out by OASE (Organisation for Agriculture in Saline Environments).
- The cultivation of *Panicum turgidum* for fodder, in mixed cultures with *Suaeda maritima* which is used for wood production (Karachi, Pakistan).
- The cultivation of *Crambe maritima* as food plant on saline soils on the island of Texel (Netherlands).

The current projects were at least ambitious ventures because of the risk of a death spiral in consequence of continuously increasing soil salinity. However, the different combination of salt-resistant plants with controlled flux of salt import (and export), salt accumulators, well-matched composition of species and periodically or permanently flooded areas under cultivation led to sustainable and partially cost-effective systems in nearly all cases (at least for the years of trial period). However, these projects were carried out exclusively with wild plants, so their domestication is given priority (Stanton et al. 1992; Flowers 2004). The domestication of these wild plants enables us to develop halophytic crop plants which will be as competitive and productive as traditional crops like maize or barley. The first step of such a process is the selection of productive genotypes. However, while developing new crop plants, various risks and imponderabilities have to be overcome, such as the optimisation of culture systems, vulnerability to diseases or economic profitability. An example of how such difficulties can be successfully overcome is the cultivation of *Panicum turgidum* in conjunction with *Suaeda fruticosa* at the Sindh/Balochistan coast of Pakistan (Khan et al. 2009b).

Although the utilisation of saline production systems is still in its infancy, the advantages cannot be ignored. Sea water and brackish water are abundant and contain not only Na^+ and Cl^- but many essential macro- and micronutrients. Especially, the irrigation with brackish water would enlarge the application range and the number of suited halophytes and their potential yield, while reducing ecological risks.

Apart from the soil cultures described above, aquacultures, e.g. of seaweeds or microalgae, do also have a high potential for utilisation. Algae contain up to 70 % valuable proteins and minerals in their dry mass and can be used instead of wheat or soya.

Existing potentials should therefore be fully exploited before extending herbicide applications or genetic engineering. Additionally, more research should be done regarding the suitability of salt-resistant plants as crops in a CO_2-rich future. This topic has been very much neglected so far. Only one potential cash crop halophyte (*Aster tripolium*) has been investigated in detail (Geissler et al. 2009a, b, 2010), with promising results, namely, a higher survival rate of salt-treated plants under elevated CO_2. Species such as *A. tripolium*, which seem to be promising crop plants in general, and the salt resistance of which is enhanced under elevated CO_2 concentration, should be given priority for advanced local field studies and the application in saline production systems because they are probably best suited for both counteracting the cause of global climate change and mitigating its consequences.

Considering all the points mentioned above, it has to be emphasised that time is running fast, so if we do not react quickly by carrying out more basic research as well as applied field studies, there will soon be no further necessity to investigate salt resistance because soil salinity will reach levels where not even halophytes can grow.

Acknowledgments The authors would like to thank Mr. Jürgen Franz, Mr. Wolfgang Stein, Mr. Gerhard Mayer, Mrs. Angelika Bölke, Mrs. Gerlinde Lehr and Mrs. Nikol Strasilla for technical assistance regarding the quick check experiments.

References

Arnalds A (2004) Carbon sequestration and the restoration of land health. An example from Iceland. Clim Change 65:333–346

Ball MC, Munns R (1992) Plant responses to salinity under elevated atmospheric concentrations of CO_2. Aust J Bot 40:515–525

Bray EA, Bailey-Serres J, Weretilnyk E (2000) Responses to abiotic stresses. In: Buchanan BB, Gruissem W, Jones RL (eds) Biochemistry and molecular biology of plants. American Society of Plant Physiologists, Rockville, MD, pp 1158–1203

Caçador I, Costa L, Vale C (2002) The importance of halophytes in carbon cycling in salt marshes. In: Xiaojing L, Mengyu L (eds) Halophyte utilization and regional sustainable development of agriculture. Weather Press, Beijing, p 199

Dansereau P (1957) Biogeography: an ecological perspective. Ronald Press, New York
Debez A, Braun HP, Pich A, Taalmalli W, Koyro HW, Abdelly C, Huchzermeyer B (2012) Proteomic and physiological responses of the halophyte *Cakile maritima* to moderate salinity at the germinative and vegetative stages. J Proteomics 75:5667–5694
Deutsche Stiftung Weltbevölkerung (DSW) (2012a) Published on DSW website. http://www.weltbevoelkerung.de/informieren/unsere-themen/bevoelkerungsdynamik/mehr-zum-thema/bevoelkerungsentwicklung.html. Accessed 30 Oct 2012
Deutsche Stiftung Weltbevölkerung (DSW) (2012b) Published on DSW website. http://www.weltbevoelkerung.de/uploads/tx_tspagefileshortcut/FS_Entw_Projekt_web.pdf. Accessed 30 Oct 2012
Evans LT (2005) Is crop improvement still needed? J Crop Improv 14:1–7
Flowers TJ (2004) Improving crop salt tolerance. J Exp Bot 55:307–319
Flowers TJ, Colmer TD (2008) Salinity tolerance in halophytes. New Phytol 179:945–963
Flowers TJ, Galal HK, Bromham L (2010) Evolution of halophytes: multiple origins of salt tolerance in land plants. Funct Plant Biol 37:604–612
Geissler N, Hussin S, Koyro HW (2009a) Interactive effects of NaCl salinity and elevated atmospheric CO_2 concentration on growth, photosynthesis, water relations and chemical composition of the potential cash crop halophyte *Aster tripolium* L. Environ Exp Bot 65:220–231
Geissler N, Hussin S, Koyro HW (2009b) Elevated atmospheric CO_2 concentration ameliorates effects of NaCl salinity on photosynthesis and leaf structure of *Aster tripolium* L. J Exp Bot 60:137–151
Geissler N, Hussin S, Koyro HW (2010) Elevated atmospheric CO_2 concentration enhances salinity tolerance in *Aster tripolium* L. Planta 231:583–594
Geissler N, Huchzermeyer B, Koyro HW (2013) Effect of salt stress on photosynthesis under ambient and elevated atmospheric CO_2 concentration. In: Ahmad P, Azooz MM, Prasad MNV (eds) Salt stress in plants: signalling, omics and adaptations. Springer, New York, pp 377–413
Glenn EP, O'Leary JW, Watson MC, Thompson TL, Kuehl RO (1991) *Salicornia bigelovii* Torr.: an oilseed halophyte for seawater irrigation. Science 251:1065–1067
Glenn EP, Brown JJ, Blumwald E (1999) Salt tolerance and crop potential of halophytes. Crit Rev Plant Sci 18:227–255
Güth M (2001) Halophyte uses in different climates. III. Computer-aided analysis of socioeconomic aspects of the sustainable utilization of halophytes. Progress in biometeorology, vol 15. Backhuys, Leiden
Huchzermeyer B (2011) Salztoleranz bei Halophyten: Grundlagen der Anpassung von Pflanzen an ihre Umwelt. In: Bavarian Academy of Sciences (ed) Leben unter extremen Bedingungen. Rundgespräche der Kommission für Ökologie 39. Pfeil, Munich
Hussin S, Geissler N, Koyro HW (2013) Effect of NaCl salinity on *Atriplex nummularia* (L.) with special emphasis on carbon and nitrogen metabolism. Acta Physiol Plant 35:1025–1038
Ignatova LK, Novichkova NS, Mudrik VA, Lyubimov VY, Ivanov BN, Romanova AK (2005) Growth, photosynthesis, and metabolism of sugar beet at an early stage of exposure to elevated CO_2. Russ J Plant Physiol 52:158–164
IPCC (2007) Climate change 2007. The physical science basis. Contribution of working group I to the fourth assessment report of the Intergovernmental Panel on Climate Change. Cambridge University Press, Cambridge, NY
Khan S, Khan MA, Hanjra MA, Mu J (2009a) Pathways to reduce the environmental footprints of water and energy inputs in food production. Food Policy 34:141–149
Khan MA, Ansari R, Ali H, Gul B, Nielsen BL (2009b) *Panicum turgidum*, a potentially sustainable cattle feed alternative to maize for saline areas. Agr Ecosyst Environ 129:542–546
Kirschbaum MUF (2004) Direct and indirect climate change effects on photosynthesis and transpiration. Plant Biol 6:242–253
Koyro HW, Lieth H (1989) Salinity conversion table, 5th enlarged edn. H. Lieth, Osnabrück
Koyro HW, Lieth H (2008) Global water crisis: the potential of cash crop halophytes to reduce the dilemma. In: Lieth H, Sucre MG, Herzog B (eds) Mangroves and halophytes: restoration and utilization. Tasks for vegetation science, vol 43. Springer, Dordrecht, pp 7–20

Koyro HW, Geissler N, Hussin S, Huchzermeyer B (2006) Mechanisms of cash crop halophytes to maintain yield and reclaim soils in arid areas. In: Khan MA, Weber DJ (eds) Ecophysiology of high salinity tolerant plants. Tasks for vegetation science, vol 40. Springer, Dordrecht, pp 345–366

Koyro HW, Geissler N, Hussin S, Debez A, Huchzermeyer B (2008) Survival at extreme locations: life strategies of halophytes: the long way from system ecology, whole plant physiology, cell biochemistry and molecular aspects back to sustainable utilization at field sites. In: Abdelly C, Öztürk M, Ashraf M, Grignon C (eds) Biosaline agriculture and high salinity tolerance. Birkhäuser, Basel, pp 1–20

Koyro HW, Daoud S, Harrouni MC (2013a) Salt response of some halophytes with potential interest in reclamation of saline soils: gas exchange, water use efficiency and defence mechanism. In: Shahid SA, Abdelfattah MA, Taha FK (eds) Developments in soil salinity assessment and reclamation. Innovative thinking and use of marginal soil and water resources in irrigated agriculture. Springer, Dordrecht, pp 523–542

Koyro HW, Hussain T, Huchzermeyer B, Khan MA (2013b) Photosynthetic and growth responses of a perennial halophytic grass *Panicum turgidum* to increasing NaCl concentrations. Environ Exp Bot 91:22–29

Künnemann TD, Gad G (1997) Überleben zwischen Land und Meer: Salzwiesen. Verlag Isensee, Oldenburg

Läuchli A (1984) Salt exclusion: an adaptation of legumes for crops and pastures under saline conditions. In: Toenniessen GH, Staples RC (eds) Salinity tolerance in plants. Strategies for crop improvement. Wiley, New York, pp 171–187

Läuchli A, Epstein E (1984) Mechanisms of salt tolerance in plants. Calif Agric 38:18–20

Le Houerou HN (1993) Forage halophytes in the Mediterranean basin. In: Choukr-Allah R, Hamdy A (eds) Advanced course on halophyte utilization in agriculture. Hassan II University, Agadir, pp 405–436

Li JH, Dugas WA, Hymus GJ, Johnson DP, Hinkle CR, Drake BG, Hungate BA (2003) Direct and indirect effects of elevated CO_2 on transpiration from *Quercus myrtifolia* in a scrub-oak ecosystem. Glob Chang Biol 9:96–105

Lieth H (1999) Development of crops and other useful plants from halophytes. In: Lieth H, Moschenko M, Lohmann M, Koyro HW, Hamdy A (eds) Halophyte uses in different climates. I. Ecological and ecophysiological studies. Backhuys, Kerkwerve

Lieth H (ed) (2000) Sustainable halophyte utilization in the Mediterranean and subtropical dry regions. Final report. CEC concerted action: sustainable halophyte utilization in the Mediterranean and subtropical dry regions

Lieth H, Lohmann M (eds) (2000) Cash crop halophytes for future halophyte growers. Lieth, Institut für Umweltsystemforschung (IUSF), University of Osnabrück, no. 20. ISSN 09336-3114

Lieth H, Mochtschenko M (eds) (2003) Cash crop halophytes. Recent studies. Kluwer, Dordrecht

Long SP, Ainsworth EA, Rogers A, Ort DR (2004) Rising atmospheric carbon dioxide: plants FACE the future. Annu Rev Plant Biol 55:591–628

Marschner H (1995) Mineral nutrition of higher plants, 2nd edn. Academic, London

Miyamoto S (1996) Salt tolerance, water use and potential irrigation scheduling of halophytes. In: Malcolm CV, Hamdy A, Choukr-Allah R (eds) Halophytes and biosaline agriculture. Marcel Dekker, New York, pp 181–220

Morgan JA, Lecain DR, Mosier AR, Milchunas DG (2001) Elevated CO_2 enhances water relations and productivity and affects gas exchange in C3 and C4 grasses of the Colorado shortgrass steppe. Glob Chang Biol 7:451–466

Munns R (1993) Physiological processes limiting plant growth in saline soils: some dogmas and hypotheses. Plant Cell Environ 16:15–24

Munns R (2002) Comparative physiology of salt and water stress. Plant Cell Environ 25:239–250

Odum EP (1974) Halophytes, energetics and ecosystems. In: Queen WH, Reimold RJ (eds) Ecology of halophytes. Academic, New York, pp 599–602

Ozgur R, Uzilday B, Sekmen A (2013) Reactive oxygen species regulation and antioxidant defence in halophytes. Published on Functional Plant Biology website. http://www.publish.csiro.au/

view/journals/dsp_journals_pip_abstract_Scholar1.cfm?nid=102&pip=FP12389. Accessed 10 Apr 2013 (in press)

Pérez-López U, Robredo A, Lacuesta M, Mena-Petite A, Muñoz-Rueda A (2012) Elevated CO_2 reduces stomatal and metabolic limitations on photosynthesis caused by salinity in *Hordeum vulgare*. Photosynth Res 111:269–283

Rabhi M, Castagna A, Remorini D, Scattino C, Smaoui A, Ranieri A, Abdelly C (2011) Photosynthetic responses to salinity in two obligate halophytes: *Sesuvium portulacastrum* and *Tecticornia indica*. S Afr J Bot 79:39–47

Robredo A, Pérez-López U, de la Maza HS, González-Moro B, Lacuesta M, Mena-Petite A, Muños-Rueda A (2007) Elevated CO_2 alleviates the impact of drought on barley improving water status by lowering stomatal conductance and delaying its effects on photosynthesis. Environ Exp Bot 59:252–263

Rogers A, Allen DJ, Davey PA, Morgan PB, Ainsworth EA, Bernacchi CJ, Cornic G, Dermody O, Dohleman FG, Heaton EA, Mahoney J, Zhu XG, Delucia EH, Ort DR, Long SP (2004) Leaf photosynthesis and carbohydrate dynamics of soybeans grown throughout their life-cycle under free-air carbon dioxide enrichment. Plant Cell Environ 27:449–458

Rozema J (1993) Plant responses to atmospheric carbon dioxide enrichment: interactions with some soil and atmospheric conditions. Vegetatio 104(105):173–190

Rozema J, Flowers TJ (2008) Crops for a salinized world. Science 322:1478–1480

Schachtman DP, Munns R (1992) Sodium accumulation in leaves of *Triticum* species that differ in salt tolerance. Aust J Plant Physiol 19:331–340

Schimper AFW (1898) Pflanzengeographie auf physiologischer Grundlage. Fischer, Jena

Schimper AFW (1903) Plant-geography upon a physiological basis. Clarendon, Oxford

Schubert S (1999) Anpassung von Mais (*Zea mays* L.) an Bodensalinität: Strategien und Konzepte. In: Merbach W, Wittenmayer L, Augustin J (eds) Stoffumsatz im wurzelnahen Raum. 9. Borkheider Seminar zur Ökophysiologie des Wurzelraumes. Teubner, Stuttgart, pp 74–79

Schulze ED, Beck E, Müller-Hohenstein K (2002) Pflanzenökologie. Spektrum Akademischer, Heidelberg

Schwanz P, Picon C, Vivin P, Dreyer D, Guehl J, Polle A (1996) Responses of antioxidative systems to drought stress in pendunculate oak and maritime pine as modulated by elevated CO_2. Plant Physiol 110:393–402

Stanton BR, Perkins AS, Tessarollo L, Sassoon DA, Parada LF (1992) Loss of N-myc function results in embryonic lethality and failure of the epithelial component of the embryo to develop. Genes Dev 6:2235–2247

UNEP (1993) Halophytes for livestock, rehabilitation of degraded land and sequestering atmospheric carbon. United Nations Environment Programme (UNEP), Nairobi

Urban O (2003) Physiological impacts of elevated CO_2 concentration ranging from molecular to whole plant responses. Photosynthetica 41:9–20

Vinocur B, Altman A (2005) Cellular basis of salinity tolerance in plants. Environ Exp Bot 52:113–122

Von Sengbusch P (2003) Halophytes. Published on Botany online. http://www.biologie.uni-hamburg.de/b-online/e56/56d.htm. Accessed 30 Oct 2012

Wang W, Vinocur B, Altman A (2003) Plant responses to drought, salinity and extreme temperatures: towards genetic engineering for stress tolerance. Planta 218:1–14

Wang D, Heckathorn A, Wang X, Philpott SM (2012) A meta-analysis of plant physiological and growth responses to temperature and elevated CO_2. Oecologia 169:1–13

Wullschleger SD, Tschaplinski TJ, Norby RJ (2002) Plant water relations at elevated CO_2: implications for water-limited environments. Plant Cell Environ 25:319–331

Chapter 8
Effects of Heat Stress on Growth and Crop Yield of Wheat (*Triticum aestivum*)

Sonal Mathur and Anjana Jajoo

Contents

1 Introduction

1.1 Global Climatic Changes and Agriculture

More food is required with rising population and it is essential to recognise and enumerate the responses of most of the vital crop plants to varying environmental (or ecological) conditions. Crop yield is very sensitive to changing global climatic

S. Mathur • A. Jajoo (✉)
School of Life Science, Devi Ahilya University, Indore, India
e-mail: anjanajajoo@hotmail.com

P. Ahmad and M.R. Wani (eds.), *Physiological Mechanisms and Adaptation Strategies in Plants Under Changing Environment: Volume 1*, DOI 10.1007/978-1-4614-8591-9_8,

changes like increasing temperature, salinity and water stress. Higher plants, being immobile, have a greater need of protection against stresses. Stress can be defined as 'a condition in which enhanced demands on a plant direct to a preliminary down-regulation/destabilisation of functions, leading to a normalisation and better resistance'. A permanent damage or death is observed in case when the tolerance limits are exceeded and the adaptive ability is overtaxed. The increasing menace of environmental stresses necessitates finding strategies to cope with their ever-increasing adversaries.

World agricultural production and supply patterns are influenced up to a greater extent by the predicted changes in environment especially elevation in atmospheric CO_2, salinity, temperature and precipitation, associated with changes in nitrogen deposition, tropo-stratospheric ozone levels, UV-B radiation, etc. Due to global warming, plants growing in tropical, subtropical and temperate regions will get exposed to high-temperature stress conditions and in turn lead to both ecological and agricultural effects. Global mean temperature will increase 0.3 °C per decade (Jones et al. 1999) attaining approximately 1 and 3 °C above the present value by years 2025 and 2100, respectively, leading to global warming (according to a report from Intergovernmental Panel on Climatic Change [IPCC]) (Wahid et al. 2007, 2012). A comprehensive understanding of the physiological responses of plants to extreme temperatures is essential for future strategies for plant improvement (Silva-Correia et al. 2012).

1.2 Effect of Heat Stress on Plants

An increase in temperature has been directly linked to a decrease in photosynthetic efficiency and finally decreased crop yield (Mathur et al. 2011a). This also leads to a complete loss in productivity with a slight more increase in temperature degrees. Global warming may cause premature ageing in wheat, according to computer modelling studies of the crop's response. Even the changes in climate and extreme weather events are likely to impact agricultural crops. In recent years, erratic weather patterns lead to biotic and abiotic stresses in agricultural sectors resulting in reduced yield potential or crop failure and thereby result in poor economy. High-temperature stress is characterised by the intensity, duration and rate of increase in temperature. As the temperature increases above a threshold level, the magnitude and extent of stress increases rapidly, and as a result, complex acclimation effects occur that depends on temperature and other environmental factors. The degree and extent of possible damage experienced by the plant is decided by developmental phase of the crop exposed to increased temperatures (Slafer and Rawson 1995; Wollenweber et al. 2003). Wheat is particularly sensitive to stress injury in areas where high temperature limits productivity (Modarresi et al. 2010). High day temperature can cause damage to components of leaf photosynthesis, reducing CO_2 assimilation rates during the vegetative stage when compared to environment with optimum temperature. At 46.11 °C temperature, the plant tissue dies. Under certain conditions plant temperature can rise to a critical level. Acute severity in temperatures may lead to premature death of plants. For monocotyledons high day temperature can cause leaf firing leading to the necrosis of the leaf tips. High temperature adversely affects many cellular processes in immature plants. Processes related to plant growth and development, such as tassel initiation and time of flowering, pollen sterility, as well as rate and duration of endosperm cell division, are few of the detrimental effects of high-temperature stress. This stress may result in lipid peroxidation and subsequently membrane injury or damage, protein degradation, enzyme inactivation, pigment bleaching and disruption of DNA strands. A loss in quality as well as yield is known to occur due to high-temperature stress (Stone and Nicholas 1994). As a consequence of the present trend of intensifying temperatures globally, these effects may be expected to increase (Hennessy 1994). Daily mean temperatures up to 30 °C during grain filling increased dough strength, and as the temperature increased above 30 °C, dough strength decreased (Irmak et al. 2008). Heat-associated damage to reproductive tissues in different crops is a major cause for yield loss in agriculture production worldwide (Suzuki et al. 2013).

Programmed cell death is induced in plants under high-temperature stress and as a result may direct to the shredding of leaves, flowers not blooming, little to no production of fruits, or simply the death of plants. The reproductive tissues in plants are particularly sensitive to heat stress. Scalding and scorching of leaves and stems, sunburn on fruits and stems, leaf drop, rapid leaf death and reduction in growth are also results of heat injury. However, susceptibility of species and cultivars to high temperatures may show a discrepancy with the phase of the plant development. Both, vegetative and reproductive stages of plant growth have shown to be affected by heat stress. High temperature affects all phases of crop growth; accelerates floral initiation; reduces the period of spike development; results in shorter spike with lower number of spikelets; and hence influences pollen development (Wahid et al. 2007). In wheat, the impact of high-temperature stress on seedling growth and leaf development has been recognised from the temperature sensitivity of pigmentation and the inhibition of the chloroplast function. If the environmental temperature exceeds the upper threshold of the temperature range to which plants are adapted by about 20 °C within a few hours, there is unambiguous evidence that before the impairment of other cell functions, the photosynthetic apparatus of chloroplasts is reversely damaged (Efeoglu and Terzioglu 2009). Excess heat in plants is dissipated by three major ways: (1) long-wave radiation, (2) heat convection into the air and (3) transpiration.

Plants suffer short-term high temperature during hot summer days and undergo a phenomenon called as midday depression. In it, there is a decline in the values of net photosynthetic rate in the middle of the day, i.e. noon. Midday depression is a result of many ecological, physiological and biochemical factors which are mentioned in Table 8.1. Midday depression of photosynthesis occurs in many plants and significantly affects crop yields. A chain of event in plants is triggered as heat stress becomes more severe, as described in Table 8.2.

Plant species as well as the time of exposure to the new temperature regime decides the temperature sensitivity of photosynthesis. As compared to mature, fully

Table 8.1 Various factors responsible for midday depression

Ecological factors	Sunlight, air temperature, air humidity, soil water stress, CO_2 in the air
Physiological factors	Stomatal closure, enhancement of respiration and/or photorespiration, increase in mesophyll resistance, decrease in leaf water potential, circadian rhythm
Biochemical factors	Photosynthate accumulation, decrease in Rubisco activity, enhanced ABA biosynthesis, decline in PS II photochemical efficiency

Table 8.2 Events taking place in a plant in response to high-temperature stress

1. Decrease in photosynthesis
2. Increased respiration
3. Closing down of photosynthesis—closed stomata stops CO_2 capture and increases photorespiration
4. Major slowdown in transpiration (cooling process loss and internal temperature increase)
5. Cell membrane leakage (signals changes in protein synthesis)
6. Continued physical water loss
7. Growth inhibition
8. Plant starvation through rapid use of food reserves, inefficient food use and inability to call on reserves when and where needed
9. Release of toxins through cell membrane
10. Loss of membrane integrity and protein breakdown

expanded leaves exhibiting a reduced capacity to adjust, young developing leaves of herbaceous plants appear to be more flexible to temperature changes since they are plastic in their photosynthetic response to sudden changes in temperature.

2 Effects of High-Temperature Stress on Various Metabolic Reactions

2.1 Protein Stability

Proteins are an important component of any cellular organisation and form the basic infrastructure of those organisms. Any abiotic and/or biotic stress reflected through plant is indirectly an outcome of change in protein stability. Unfolded protein response (UPR) is initiated by heat stress response (HSR) in both endoplasmic reticulum (ER) and cytosol (Fig. 8.1) (Mittler et al. 2012). In the ER pathway, the transcription of certain genes gets affected by entry of unfolded proteins in the nuclei that causes accretion or accumulation of ER chaperone transcripts. This also shows the capacity to alter cell metabolism in order to withstand the heat stress. UPR also gets activated by specific chemicals and abiotic stresses. Plasma membrane is considered to be more sensitive to heat response as compared to UPR. Under mild heat conditions, few unfolded proteins and HSR chaperones are likely to survive.

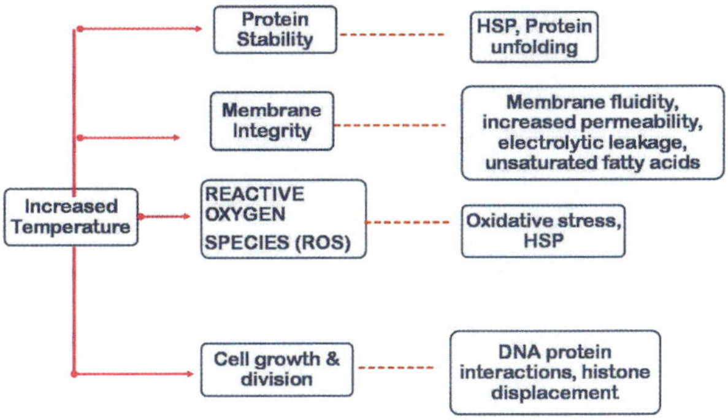

Fig. 8.1 Effect of high-temperature stress on different cellular processes

Abiotic stress develops a cascade of molecular responses in plants (Qu et al. 2013). Few mechanisms of defence such as gene expression which are not expressed under 'normal' conditions are triggered as a result of high temperature as well as other stresses (Morimoto 1993; Feder 2006). Furthermore at molecular level, an increase in the synthesis of protein groups was observed that causes sudden changes in genotypic expression. These groups of proteins were called 'heat-shock proteins' (Hsps), 'stress-induced or stress-activated proteins' or 'stress proteins'. Hsps vary largely in type as well as mode of their expression in plants. In all plants and animal species, Hsps depict the role of molecular chaperones that regulates the folding and accumulation of proteins, their localisation and degradation as well. Hsps as chaperones participate in refolding proteins and also prevent the irreversible aggregation of other proteins during high-temperature stress conditions (Tripp et al. 2009). Hsps are highly conserved in organisms as studied in response to heat stress. Many similarities have been found in diverse eukaryotes in the molecular mechanisms of HSP gene induction. The level of diversification and abundance of the sHsps in a plant reflects the level of adaptation of the plant to heat stress. Molecular pathway leading to the expression of genes to synthesise heat-shock proteins is composed of several mechanisms like sensing temperature which is connected to the mechanism of signal transfer to Hsps where activation of gene expression occurs by binding to the heat-shock element (HSE) in DNA (Larkindale et al. 2005; Khurana et al. 2013; Chauhan et al. 2012). Six nuclear gene families have been observed to encode all plant sHSPs, and each gene family characterises the proteins found in distinct cellular compartments, i.e. cytosol (class I and class II), chloroplast, endoplasmic reticulum, mitochondria and membranes (Waters et al. 1996). More than 20 sHSPs have been found in higher plants and 40 different sHSPs may be found in the same species (Vierling 1991). HSP16.9A, HSP16.9B and HSP16.9C are few HSPs identified in wheat plant (Waters et al. 1996). Changes in enzyme levels, photosynthesis activity, cellular membrane structure and protein metabolism take place as a response of heat shock in plants. Süss and Yordanov (1986) suggested that the interaction of HSP (22 kDa) with the chloroplast external membrane can influence the composition of the membrane and thereby decreases its fluidity, probably increasing the efficiency of the ATP transport (Al-Whaibi 2012). In *Arabidopsis*, heat stress activates the protein kinase MPK6 and in vitro and in vivo evidences have proved that MPK6 specifically targets the major heat stress transcription factor HsfA2 (Evrard et al. 2013). Several heat-shock factors (HSF) have been found responsible for thermotolerance in *Arabidopsis* (Pick et al. 2012; Scharf et al. 2012).

2.2 Membrane Integrity

Heat stress is one of the major factors causing alterations in membrane integrity. Increases in temperature lead to increase in the fluidity of the plasma membrane. Calcium channels become activated by the increased fluidity of the plasma membrane that leads to an influx of Ca^{2+} ions into the cell. However, the actual heat stress sensor in plants cells is still unknown. Over 40 calcium channels have been encoded by *A. thaliana* genome and most of them mark their localisation in the plasma membrane. Ca^{2+} ions are regarded as the significant channels responsible for sensing heat. Apart from activating the calcium channels, lipid signalling may also be triggered by high temperature. When plants subjected to high-temperature stress (Fig. 8.1), the membrane's structure is altered and gradually permeability increases followed by electrolyte leakage and eventually the cell dies (Wang 1988). Cell membranes play a major role in regulating the incoming and outgoing of electrolytes movement of the cell and also provide a stable site for the binding and catalysis of enzymes. A serious injury (in terms of stresses) to cell membrane causes membrane damage, leading to leakage of solutes (Heckman et al. 2002; Salvucci and Crafts-Brandner 2004a) and has been correlated to the inactivation/downregulation of photosynthesis.

A few parameters may prove to be good indicators of free radical damage to cell membranes as well as to assess thermostability of membrane under heat stress. They are measurement of electrolyte leakage (EL) and level of cell membrane lipid peroxidation of unsaturated fatty acids in phospholipids (Wang et al. 2009). In general polyunsaturated fatty acids decrease with increasing high temperature. Also a decrease in trienoic fatty acids including strongly diminished 16:3 produced exclusively within the prokaryotic pathway in the chloroplast has been observed. At high temperature reduced accumulation of trienoic fatty acids is accompanied by an increase in linoleic acid, 18:2, and an increase in 16:0. In addition, it is the degree of fatty acid unsaturation that varies noticeably and not the levels of major leaf lipids themselves. Studies on the time required for the fatty acid composition to adjust to high-temperature growth conditions have shown that major alterations in leaf membranes do not occur rapidly. It is the level of unsaturation of PG that is important for removal and replacement of damaged D1 proteins in plants (Moon et al. 1995). For stabilisation of D1 protein at higher temperature, direct involvement of protein-lipid association is important (Falcone et al. 2004). Membrane lipids that influence membrane trafficking of selected chemicals are precursors of intracellular signalling molecules, involved in the regulation and control of cellular function as well as response to stresses or injury through signal transduction processes (Haucke and Di Paolo 2007; Wang 2004).

In photosynthetic tissues (especially in plastids), glycolipids are believed to be abundantly distributed lipids and are considered to be highly vulnerable to damage by heat stress (Welti et al. 2002). Monogalactosyldiacylglycerol (MGDG) and digalactosyldiacylglycerol (DGDG) are the two most abundant glycolipids in leaves. The ratio of MGDG decreases while that of DGDG increases under high-temperature stress. At higher temperatures the increased ratio of DGDG to MGDG facilitates the maintenance of chloroplast membrane integrity and normal membrane protein function. DGDG forms a more robust bilayer in an aqueous milieu due to the presence of large polar head group, whereas MGDG promotes the formation of a hexagonal phase (Hex II) structure that possibly results in the loss of bilayer integrity due to the presence of smaller head group (Su et al. 2009).

2.3 *Reactive Oxygen Species*

Different metabolic pathways are expected to be enzyme dependent with diverse responsiveness to unnecessary high temperature. High-temperature stress may uncouple some metabolic pathways and lead to the accumulation of unnecessary by-products. Reactive oxygen species (ROS) is one such by-product. Water is produced by total reduction of oxygen while ROS are the product of partial reduction of oxygen including superoxide anion (O^{2-}), hydrogen peroxide (H_2O_2) and the hydroxyl radical ($OH^{\bullet}$). ROS being highly reactive and potentially detrimental to cells results in the oxidation of lipid, proteins and DNA. 'Oxidative stress' is the process under which high levels of ROS production occur. To limit oxidative damage, it is thought that ROS which are the unwanted by-products of photosynthesis should be 'mopped up' from biological systems. ROS are now recognised as important signalling molecules with roles in responses to abiotic and biotic stresses. Organelles possessing high oxidising metabolic activity or intense rate of electron flow, e.g. chloroplasts, mitochondria and microbodies, are the major ROS-producing organelles. Under both stressed and unstressed conditions, chloroplasts are the main sites for generating ROS. It is considered as most vulnerable organelle in leaf tissues of plants to high-temperature stress. A reduced level of trienoic fatty acids and an elevated level of dienoic fatty acids in plant chloroplasts are considered resistant to high temperatures (Murakami et al. 2000; Sohn and Back 2007). The plasma membrane NADPH oxidases, peroxidases, oxalate oxidases and amine oxidases are the other groups of oxidases that produce ROS. ROS plays extensive roles in a number of physiological processes in plants like regulation of photosynthesis and cell wall metabolism and acts as defence against pathogens. However, recent studies have shown that it plays an important role in the control of gene expressions and plant development (Wang et al. 2009).

ROS is known to play a direct role in the stimulation of heat-shock proteins (HSPs). Redox-sensitive transcription factors are activated by stress sensors present in the photosynthetic and respiratory electron transport chains which in turn activate and upregulate the expression of genes encoding HSPs and related proteins, ROS-scavenging enzymes and NADPH oxidases. ROS not only are important as sensors that actively respond to environmental changes but also play a major role in orchestrating plant movement (stomatal closure and tropism) developments. Specific ROS-producing enzymes actively mediate ROS accumulation during heat stress. Studies have suggested that changes in plasma membrane fluidity lead to ROS accumulation. The event is most likely a positive feedback loop since more calcium channels in the plasma membrane open up due to accumulation of ROS resulting in more calcium influx into the cell. Downstream pathways may be activated and redox state of the cell might be altered by ROS accumulation. It is therefore assumed that the heat sensing of plasma membrane is closely interlinked with ROS signalling mechanism. ROS accumulation may lead to programmed cell death.

2.4 Cell Growth and Division

Growth is usually defined as an irreversible increase in size, with 'size' quantified as height, volume or fresh or dry weight. The two primary processes involved in plant growth are cell division and cell growth. Temperature affects morphology through differential effects on cell division and expansion. Higher temperatures are associated with larger specific leaf area (Midmore et al. 1984). Species differ markedly in how temperature affects leaf net photosynthetic rates, and these differences play a major role in adaptation (Björkman et al. 1980). High temperatures cause increased leaf-elongation rates and decreased duration of leaf elongation (Bos et al. 2000). Impact of heat stress on leaf area expansion and dynamics is a comparatively unstated area and needs further research. A noteworthy enhancement in leaf numbers, predominantly when reproductive development was arrested without any decrease in leaf photosynthetic rates, was observed as a consequence of heat stress (Prasad et al. 2006). The amount of solar radiation used by the plant regulates leaf development and duration of crop growth and thereby overall crop yield (Sinclair 1994). Decreased stem growth followed by decreased plant height is a serious heat injury effect (Prasad et al. 2006). Root growth is exceptionally susceptible to water and heat stresses as it decreases with increasing temperature. It is found that, as compared to other growth processes, root growth process has a very narrow optimum temperature range (Porter and Gawith 1999). Root number, root length and root diameter are reduced during heat-stressed conditions. Also root growth decreases when heat stress occur during reproductive development, largely for the reason of decreased carbon partitioning to roots (Batts et al. 1998; Prasad et al. 2008). Severe temperature delays flower initiation and vegetative cycle continues. Other than cell growth, cell division and senescence is also affected by heat stress (Vainonen et al. 2012). Due to a rise in temperature, packaging of DNA leads to a considerable drop in the concentration of certain histones such as H2A.Z-containing nucleosomes that might trigger the changes in transcriptome so that HSR genes are initiated. Less histones in the way allow for more transcriptions of these warming-induced genes. In certain HSP promoters, less H2A.Z occupancy may have an effect on transcription factor and other regulatory proteins' expression as well as DNA binding ability ultimately inducing transcriptome response. It is however controversial that whether the occupancy of the histones is the cause for heat sensing related to acquired thermotolerance.

3 High-Temperature Stress and Photosynthetic Processes

Photosynthesis is considered as one of the most susceptible physiological processes to heat stress. It serves as a global sensor of environmental stress that induces cellular energy imbalance as reflected in the distinct alteration in redox chemistry associated with thylakoid membranes and alteration in cellular sugar status (Biswal et al. 2011). The developmental history of the leaves and the physiological age of the leaf tissue are the deciding factors for the photosynthetic response of plants to changing temperature. At very high temperatures, severe cellular injury and even cell death may occur within minutes, which could be attributed to a catastrophic collapse of cellular organisation (Schöffl et al. 1999). Heat stress causes detrimental effects on different photosynthetic functions involving the photochemical reactions related to PS II as well as the dark reaction (Fig. 8.2). PS II has been considered as one of the most heat-labile part of the photosynthetic apparatus (Srivastava et al. 1997). The two key factors that make PS II electron transport most susceptible to heat stress are:

1. The increase in fluidity of thylakoid membranes at high temperature causes dislodging of PS II light-harvesting complexes from thylakoid membrane.
2. The PS II integrity is dependent on electron dynamics. If high-temperature stress disrupts metabolic processes that either donate or accept electrons from PS II,

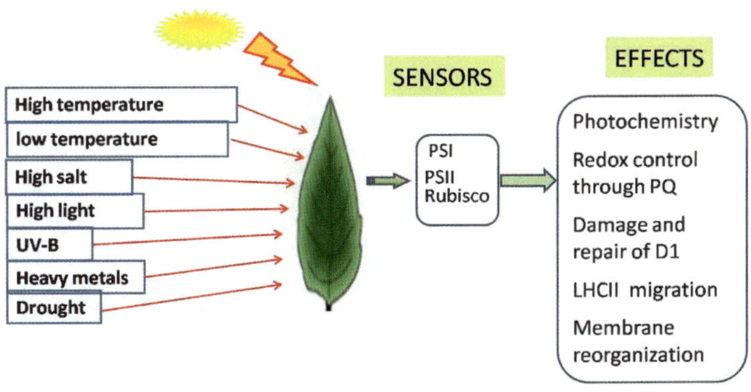

Fig. 8.2 Abiotic stress and its effects on photosynthesis

then the PS II is likely to dislodge from the thylakoid membrane. As a result of heat stress, a loss of oxygen-evolving complex (OEC) activity takes place (Enami et al. 1994; Yamane et al. 1998), an inhibition of electron transfer from primary/secondary electron-accepting plastoquinone of PS II at the acceptor side (Bukhov and Mohanty 1999; Cao and Govindjee 1990) and dissociation of the peripheral antenna complex of PS II from its core complex (Srivastava et al. 1997). According to Gounaris et al. (1983), heat stress possibly induces grana destacking and the formation of cylindrical inverted micelles.

3.1 Effects on Photosystem II: D1 Protein

PS II is a membrane protein complex found in oxygenic photosynthetic organisms (higher plants, green algae and cyanobacteria), which harness light energy to split H_2O into O_2, protons and electrons. A recent crystallographic study of PS II at new resolution (1.9 Å) has been done by Umena group (Umena et al. 2011).

PS II is comprised of more than 25 polypeptides and is surrounded by a number of Chl (*a*)- and Chl (*b*)-binding proteins, known as light-harvesting complexes, that funnel energy into the complex. The (RC) reaction centre core where light energy is converted into electrochemical potential energy and the water-splitting reaction occurs is situated at the heart of PS II complex. Two homologue polypeptides known as D1 and D2 proteins have five transmembrane α-helices, each mark their localisation at the enzymatic heart of the RC complex. The 'special pair', of Chl (*a*) P680, is present between these proteins. The redox-active tyrosine Y_Z and the Q_B plastoquinone are located on D1 protein, while pheophytin and Q_A are present on the D2 protein. As compared to Q_B plastoquinone that moves freely in and out of the D1 protein at the 'Q_B site', the Q_A plastoquinone remains rather fixed in the D2 protein. Closely associated with the D1 and D2 proteins are two Chl-containing proteins (CPs) called CP43 and CP47. Finally, the PS II RC core complex has several extrinsic proteins attached to its lumenal surface, the nature of which varies with different types of organisms, although the PsbO protein is ubiquitous to all types of oxyphototrophs (Bricker and Burnap 2005).

Dephosphorylation of a number of PS II core proteins, namely, D1, D2 and CP43, has been stimulated by heat stress (Rokka et al. 2000; Vener et al. 2001). The regulation of phosphorylation of light-harvesting complex II (LHC II) of the PS II core proteins is quite different (Harrison and Allen 1991). The control of phosphorylation of thylakoid proteins interacts with redox status, while dephosphorylation of LHC II appears to be catalysed by a different phosphatase than dephosphorylation of other thylakoid-associated proteins (Hammer et al. 1997; Vener et al. 1999). Many thylakoid-associated kinases (TAKs) (Snyders and Kohorn 1999, 2001) are present, but still there are few kinases that are not linked to TAKs necessary for state transitions and phosphorylation of LHC II (Depège et al. 2003). Two different regulatory systems work together, i.e. one that regulates and controls thylakoid protein phosphorylation/dephosphorylation of LHC II and state transitions, and the second system shows the regulation of phosphorylation/dephosphorylation of other thylakoid proteins. Heat stress is one of the most efficient ways of modulating the phosphorylation status of many thylakoid proteins (Vener et al. 2001; Hansson and Vener 2003; Sharkey 2005).

3.2 Effect of High-Temperature Stress on Oxygen-Evolving Complex

The lumenally exposed OEC is closely associated with the Mn cluster of PS II, the product of PsbO (33 kDa subunit), PsbP (23 kDa subunit) and PsbQ (16 kDa subunits) found in higher plants and green algae. The 33 kDa subunit plays structural and regulatory role in the optimisation of the oxygen evolution reaction. The 23 kDa subunit enables PS II to evolve oxygen under both Ca^{2+} and Cl^{-} limiting conditions. Water oxidation occurs at the OEC.

One of the most vulnerable sites in PS II apparatus is thought to be on the donor side of PS II, i.e. the OEC, as compared to acceptor side. The dissociation of the Mn-stabilising 33 kDa protein from OEC followed by release of Mn atoms is heat initiated (Enami et al. 1994; Yamane et al. 1998). Dissociation of LHC II occurs due to heat inactivation of PS II (Schreiber et al. 1997). A shift of the redox equilibrium between Q_A and Q_B is an additional heat-induced effect (Havaux et al. 1989). High-temperature stress can damage the OEC and the electron transport at both the donor and the acceptor sides of PS II of the photosynthetic apparatus.

3.3 Effect of High-Temperature Stress on Chlorophyll (a) Fluorescence Transients

Due to its intricate connection with the numerous processes taking place during the energy conversion of light into a stable chemical form, Chl (*a*) fluorescence has proven to be an open window in the heart of the photosynthetic process (Stirbet and Govindjee 2011). The origin of Chl (*a*) fluorescence is mainly from PS II. The OJIP transient has been used for the characterisation of the photochemical quantum yield of PS II photochemistry as well as the electron transport activity (Stirbet and Govindjee 2011; Chen et al. 2012). The OJIP transient originates from at O or F_0 (minimum fluorescence where all Q_A is in oxidised form) and reaches a maximum called P or F_M (where all Q_A is in reduced form) in ~200 ms. The intermediate steps obtained at ~2 and ~30 ms are called J and I. The so-called JIP test is a tool to analyse the polyphasic rise of the Chl (*a*) fluorescence transient and has applications in investigating in vivo the 'vitality' of plants and the response of the photosynthetic apparatus to different stresses (Strasser and Strasser 1995; Tsimilli-Michael et al. 1996; Srivastava and Strasser 1996; Christen et al. 2007) like high temperature

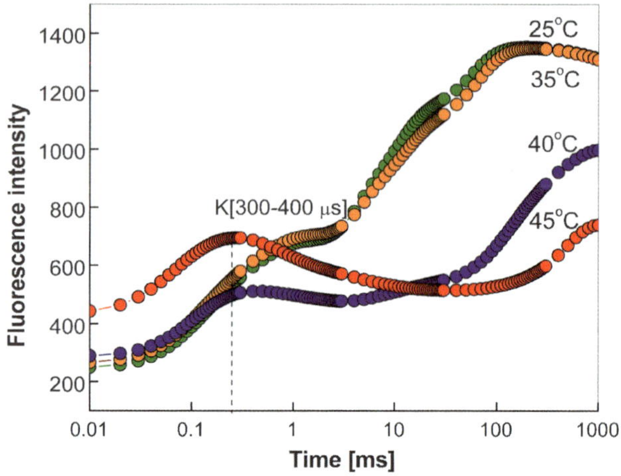

Fig. 8.3 Changes in Chl (*a*) fluorescence induction curves in wheat leaves treated at elevated temperatures. The graph has time axis in log scale

(Mathur et al. 2011a) and salinity stress (Mehta et al. 2010a). Effects of high-temperature stress on Chl (*a*) fluorescence induction kinetics in wheat leaves are shown in Fig. 8.3 where a significant decrease in the relative amplitude of the I-P phase was observed at 40 and 45 °C (Fig. 8.3). An additional K step was observed at 300 μs in response to high-temperature stress (45 °C) in wheat. The change from an OJIP curve to an OKJIP curve seems to be a characteristic response to high-temperature stress. Such changes have not been observed in plants subjected to environmental stresses encountered in various circumstances such as elevated ozone, salt, CO_2, heavy metals, light or water (Lazár et al. 1997; Pospisil et al. 1998). At 45 °C, a dominant K step was observed followed by a dip and later a slight increase to a highly suppressed P step. The origin of a K step may be attributed to an inhibition of the OEC (Guissé et al. 1995; Srivastava et al. 1997; Lazár and Pospisil 1999) and inhibition of electron transport from pheophytin to Q_A (Guissé et al. 1995), or it may reflect changes in the structure of the LHC of PS II. Increase in fluorescence yield at the K-band has also been linked to a partial uncoupling of the OEC (Srivastava et al. 1997). The K step has been shown to originate when electron flow to the acceptor side exceeds electron flow from the donor side thereby leading to oxidation of the RC. Thus, injury to the OEC due to heat stress, for example, induces the K step, by inhibiting efficient electron donation to the RC (Strasser et al. 2004).

The damage in fraction of OEC as calculated by the method of Chen and Cheng (2009) shows no damage till 35 °C (data not shown), but it has been observed that complete inactivation of OEC occurs as the temperature increases to 40 °C as well as 45 °C (showing negative values). This showed that higher temperature causes a complete damage of OEC. The results are in corroboration with the result of K step representing an inhibition of OEC at 45 °C (Mathur et al. 2011a).

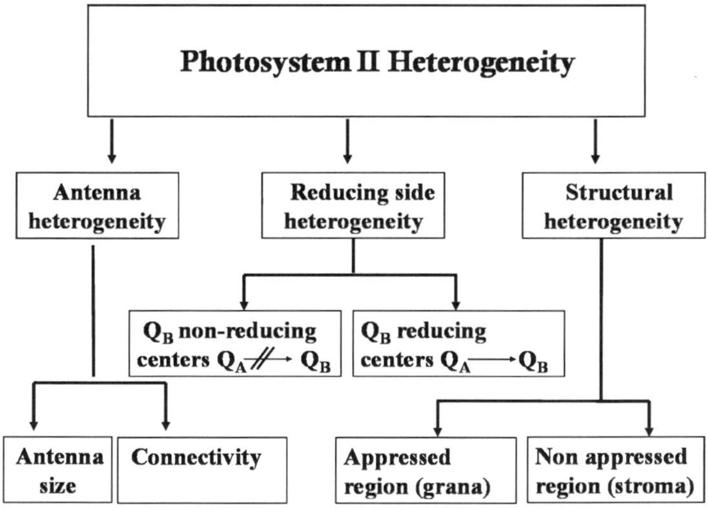

Fig. 8.4 Classification of different types of photosystem II heterogeneity

3.4 Photosystem II Heterogeneity

It is observed that PS II is more heterogeneous than other components such as PS I and Cyt b_6f in several aspects and differs in its structure as well as function (Lavergne and Briantais 1996). This diverse nature of PS II is known as heterogeneity (Fig. 8.4). The reasons behind PS II heterogeneity are as follows: (1) PS II complex comprises of a D1 polypeptide with high turnover rate, (2) there is regulated phosphorylation of some core polypeptide of PS II, and (3) PS II complex is heterogeneously distributed between appressed and nonappressed region of thylakoid membranes (Lavergne and Briantais 1996). Three aspects of PS II heterogeneity have been studied widely, i.e. PS II antenna heterogeneity, PS II reducing side heterogeneity and structural heterogeneity (Fig. 8.4). PS II alpha (α), PS II beta (β) and PS II gamma (γ) centres have been defined on the basis of the differences in the antenna size, while on the basis of acceptor/reducing side function, Q_B-reducing and Q_B-nonreducing centres have been defined, and also on the basis of structural reorganisation and distribution of PS II in appressed and nonappressed region, structural heterogeneity has been defined. Extent and nature of PS II heterogeneity may vary under different physiological conditions, i.e. temperature stress, salinity and osmotic stress (Mathur et al. 2011b; Mehta et al. 2010b; Tomar et al. 2012).

It has been shown that though photochemically competent, a number of PS II centres are unable to transfer electrons efficiently from Q_A^- to Q_B (Lavergne 1982; Melis 1985; Graan and Ort 1986; Guenther et al. 1988). Lavergne termed these centres as PS II Q_B nonreducing (Lavergne 1982). In such centres Q_A^- can be reoxidised only by a back reaction with the donor side of PS II (Schansker and Strasser 2005). Q_B-nonreducing differs from Q_B-reducing centre in not being able to reduce the PQ pool. The Q_B-nonreducing centres are localised in the nonappressed regions while Q_B-reducing centres are located in the appressed regions of the thylakoids (Tyystjarvi and Aro 1990).

Analysis of biphasic data of fluorescence rise obtained in presence of DCMU suggested the presence of three distinct populations of PS II centres in the chloroplast termed as PS II (α), PS II (β) and PS II (γ) (Melis and Duysens 1979; Black et al. 1986). Among the three centres, PS II (α) is the major 'normal' PS II centres localised in grana partition regions (Andersson and Melis 1983) with higher absorption cross-section area, large light-harvesting antenna (~210–250 Chl) and responsible for majority of the water oxidation activity and plastoquinone reduction. PS II β (~130 Chl) and PS II γ (~80 Chl) have smaller light-harvesting antenna due to absence of peripheral part of antenna localised in stroma portion of PS II. These two are considered as the inactive centres.

3.4.1 Effects of High-Temperature Stress on Antenna Heterogeneity

The data obtained from complementary area growth curve method (Fig. 8.5a–d) prompted direct information about the lifetime and the relative proportions of PS II (α), PS II (β) and PS II (γ) in high-temperature-treated wheat leaves. In 25 °C (control), 35 °C, 40 °C and 45 °C leaves (Fig. 8.5), the lifetime (τ) in the control leaves of the fastest α-component, slower β-component and the γ-component being slowest was found to be ~0.260 ms, τ~0.864 ms and τ ~6.606 ms, respectively. Further increase in temperature from 25 to 45 °C led to a decrease in the proportion of α-centres with an increase in the number of β- and γ-centres. The β- and γ-centres probably increased at the cost of α-centres, and thus, these components seem to be interconvertible depending on the environmental conditions. The proportion of α:β:γ centres was 72:25:3 in control, 61:26:14 at 35 °C, 57:32:11 at 40 °C and 45:42:13 at 45 °C (Fig. 8.5) (Mathur et al. 2011b). This study suggested that the PS II (α) centres have been converted into PS II (β) and PS II (γ) centres, and the active centres have been converted into inactive one. It was observed that after high-temperature stress, the changes in PS II (β) centres are more prominent as compared to PS II (γ). Changes in antenna organisation following a heat treatment probably involve the dissociation of PS II (α) into free LHC II and PS II (β) and the latter then migrates from the appressed to the nonappressed thylakoid membranes (Sundby et al. 1986). The fast PS II (γ) component decreased upon heating leaves at various high temperatures (25–45 °C), and the slow PS II (β) component became dominant at the cost of PS II (α) component. At 45 °C it was observed that there were 42 % PS II (β) centres when compared to control leaves (25 %) and 13 % PS II (γ) centres with only 3 % (control) (Mathur et al. 2011b). The α-phase represents a group of PS II reaction centres associated with a large number of pigment molecules. At high-temperature stress, there is a dissociation of the pigment molecules (which are probably bound to proteins) from these PS II (α) centres, leaving behind smaller complexes which are then detected as an increase in the number of PS II (β) and PS II (γ) centres (Sinclair and Spence 1988). The inhibition of OEC and the variations

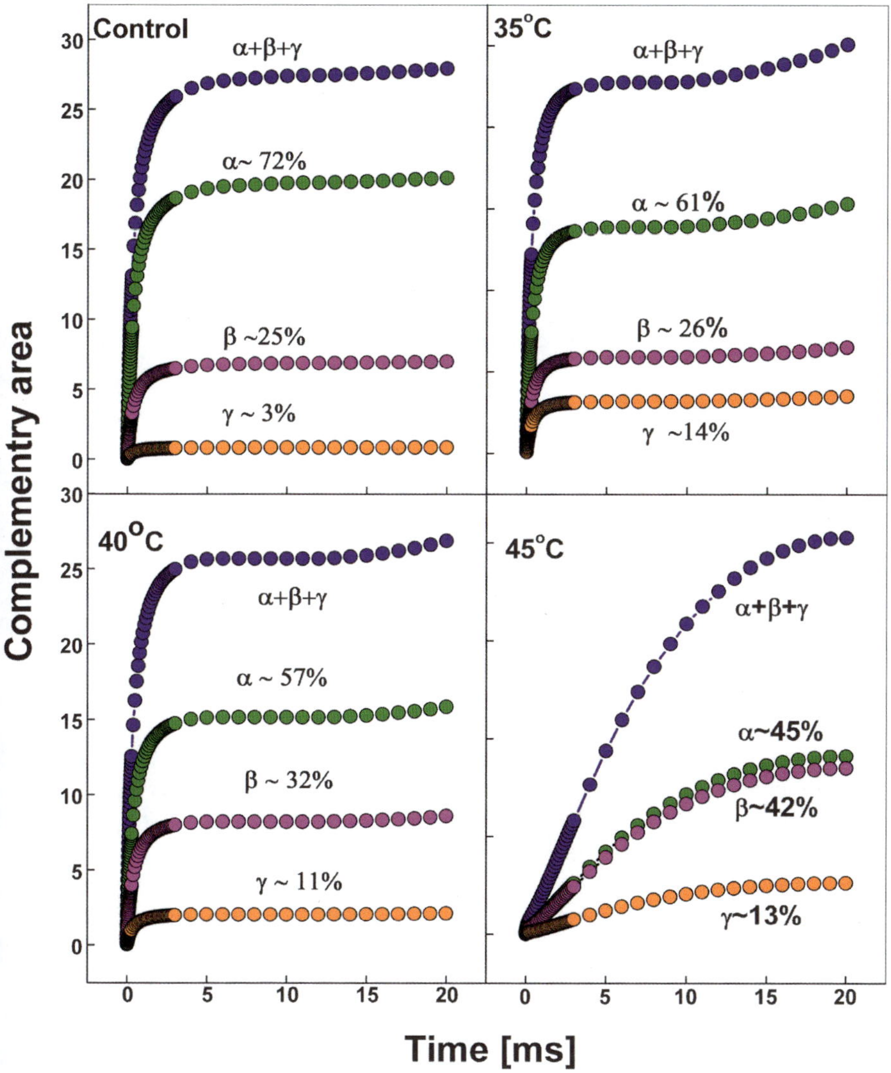

Fig. 8.5 Complementary area growth curve showing percentage of PS II (α), PS II (β) and PS II (γ) centres with increasing temperature in wheat leave

in the complementary areas bring about a partial conversion of PS II (α) to the PS II (β) and PS II (γ) centres (Hsu and Lee 1991).

The antenna heterogeneity can be also calculated on the basis of connectivity (also called grouping) (Fig. 8.6). According to the concept of connectivity, closed PS II reaction centres (RC) may transfer their excitation energy to the open neighbouring PS II units resulting in sigmoidal curve of fluorescence rise rather than exponential curve (Strasser et al. 2004). It was suggested that the three populations

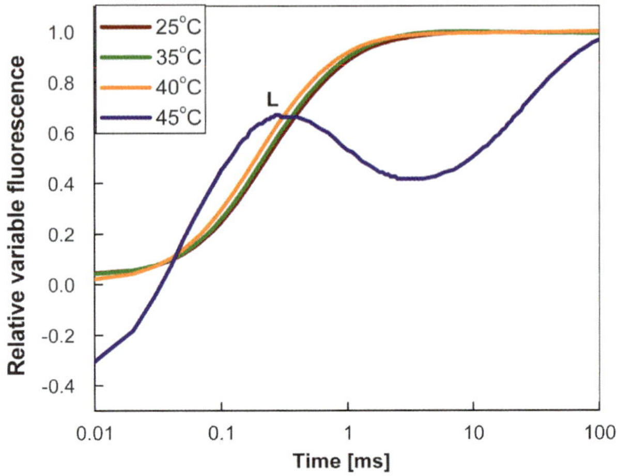

Fig. 8.6 Time course of the DCMU-FR curves in wheat leaves treated with different temperatures. The curves are presented in terms of rFv(t) starting from 10 μs (Fo) and finishing at 10 ms

of PS II units (α, β and γ) are different in their connectivity properties, i.e. the α-centres are regarded to be more grouped (Joliot and Joliot 1964) and can exchange excitation energy with each other, whereas the two others are not. Moreover the trapping efficiency of the γ-centres is thought to be lower. The PS II (β) have been characterised by an exponential rise in the time course of complementary area (CA) against a non-exponential (sigmoidal) rise shown by PS II (α) (Melis and Homann 1976). The exponential shape of this curve for PS II (β) reflects mutual energetic separation of these PS IIs which are not able to exchange excitation energy. On the other hand, the non-exponential fluorescence rise of PS II (α) is believed to reflect energetic connectivity between these PS IIs. The curves are represented by relative variable fluorescence (rFv(t)), which is defined as (F(t) − Fo)/(Fm − Fo), where Fo, Fm and F(t) are the minimal and maximal measured fluorescence intensity at time 't', respectively (Lazár et al. 2001). As compared to 25 °C (control), there is a gradual loss of connectivity in the FR of 40 °C, but at 45 °C there was no grouping observed at all. With an increase in temperature, a decrease in the sigmoidal component of the curve was observed suggesting a decrease in the connectivity between antenna molecules and an increase in the number of inactive centres. An additional positive 'L' step (~150 μs) was observed at 45 °C, i.e. the P_2G phase that demonstrates the overall grouping probability within all PS II antennas (Strasser et al. 2007). The presence of this L step at 45 °C indicates that the PS II units in high-temperature-treated samples are less grouped and less energy was being exchanged between independent PS II units (Chen and Cheng 2009). Loosing cooperativity (ungrouping) indicates that the PS II units of high-temperature-treated samples have lost stability and have become more fragile. The appearance of K step (Fig. 8.3) also reflects the changes in the energetic connectivity between the PS II units. A loss in connectivity also indicates an increase in the fraction of closed RCs, i.e. Q_B-nonreducing centres (Strasser and Tsimilli-Michael 1998).

Table 8.3 Relative amounts (normalised) or number of Q_B-reducing and Q_B-nonreducing centres as a result of high-temperature treatment in detached wheat leaves

Temp. (°C)	Relative amount of Q_B-reducing centres	Relative amount of Q_B-nonreducing centres	Chl conc. (mg/mL)	% Chl in 10 K pellet
25	82±2	18±3	2.09±0.03	70
35	80±3	20±2	1.82±0.12	60
40	61±2	39±2	1.39±0.10	46
45	50±3	50±3	1.26±0.14	42

Table also shows degree of stacking of thylakoid membranes by digitonin fractionation method. The values show amount of Chl in 10 K pellet which indicates degree of stacking of thylakoid membranes, after treatment at high temperatures

3.4.2 Effects of High-Temperature Stress on Reducing Side Heterogeneity

The relative population of Q_B-nonreducing centres was quantified by the method of Strasser and Tsimilli-Michael (1998) shown in Table 8.3. The relative fractions of Q_B-nonreducing PS II centres are not affected up to 35 °C, but as the temperature increased to 40 °C, a change in the fractions of Q_B-nonreducing PS II centres was observed, while at temperature greater than 40 °C, i.e. at 45 °C, the fraction of Q_B-nonreducing increased drastically up to 50 % (Table 8.3). At high temperatures (45 °C), the fractions of Q_B-nonreducing centres increased implying that these centres were not able to reduce PQ pool and that the active Q_B-reducing centres convert to inactive Q_B-nonreducing centres. The decrease in the fraction of fully active RCs (Q_A^- and Q_B-reducing centres) is supported by an increase in the fraction of heat sink centres, i.e. Q_B-nonreducing centres or heat sink centres which are regarded as a downregulation mechanism in order to dissipate the excess of absorbed light. The PS II at higher temperature thus switches from a process that converts light energy into biochemical energy storage which in turn converts absorbed light energy into heat (Bussotti et al. 2007).

3.4.3 Effects of High-Temperature Stress on Structural Heterogeneity

High-temperature stress may show its effects by causing structural alterations (stacking/destacking) of the thylakoid membranes. Digitonin fractionation is generally used to determine the structural heterogeneity and gives the information about changes in the structural organisation of the thylakoid membranes. Percentage Chl in 10 K pellet gives an estimate of the relative amount of grana (stacked or appressed region) and stroma (unstacked or nonappressed region). High percent Chl in 10 K pellet suggests more concentration of grana in the samples indicating presence of more stacked membranes. As shown in Table 8.3, with increasing temperature (from 35 to 45 °C), the membranes become more unstacked in wheat leaves and are expected to abolish differences between PS II centres located in grana and stroma thylakoids (Pospíšil and Tyystjärvi 1999).

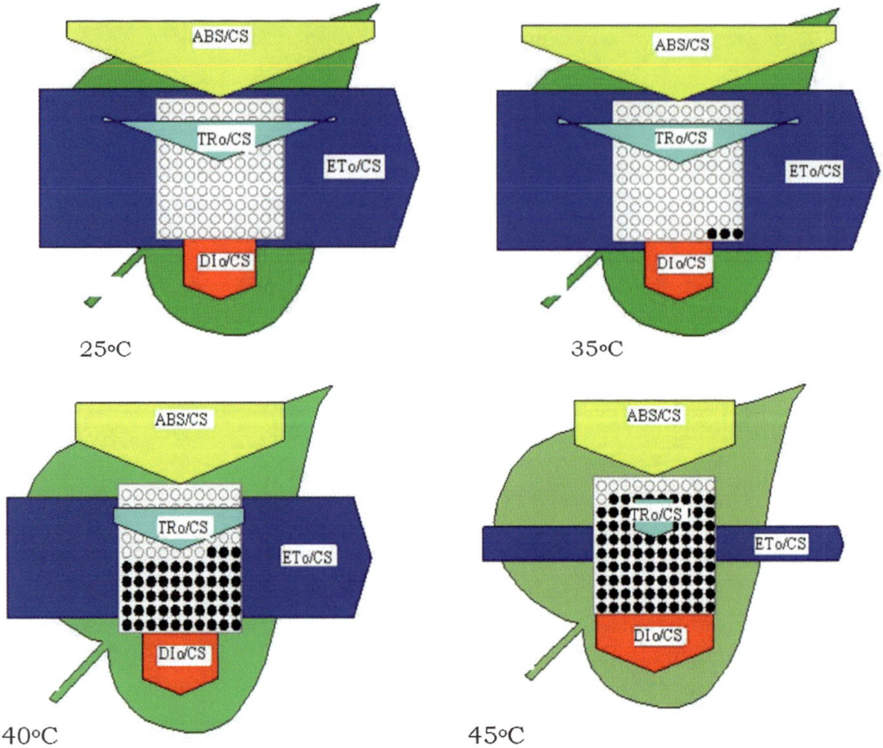

Fig. 8.7 Energy pipeline leaf model of wheat leaves treated at high temperatures

3.4.4 Energy Pipeline Model Under High-Temperature Stress

The energy pipeline leaf model in response to high-temperature treatment has been proposed (Fig. 8.7). This model gives information about the efficiency of flow of energy from antenna to the electron transport chain components through the reaction centre (RC) of PS II. The area of the arrows for various parameters such as ABS/CSo, ETo/CSo and DIo/CSo indicates the efficiency of light absorption, trapping and electron transport and dissipation per cross section of PS II, respectively. Under high-temperature stress, ABS/CSo and ETo/CSo have shown to decrease in wheat leaves where CSo arbitrarily equals to Fo. ABS/CSo gives information about the number of photons absorbed by antenna molecules of active and inactive PS II RCs over the excited cross section of the tested sample represented by the dark-adapted Fo. A decrease in ABS/CSo at high temperature indicates a decrease in the energy absorbed per excited cross section. ETo/CSo describes electron transport in a PS II cross section that gives information about the rate of reoxidation of reduced Q_A via electron transport over a cross section of active RCs (Force et al. 2003). A decrease in this ratio shows that inactivation of RC complexes and the OEC has taken place that inhibits the donor side of PS II. An increase in the total dissipation measured over the cross section (DIo/CSo) indicates a decrease in the density of active RCs (indicated as open circles) and an increase in the density of inactive RCs (indicated as filled circles) in response to high-temperature treatment. Dissipation of absorbed energy may occur as heat, fluorescence and energy transfer to other systems. An increase in energy dissipation at high temperature suggests that energy available for photochemistry is reduced under stress conditions (Strasser et al. 1996, 2000; Kruger et al. 1997).

3.5 *Effects on CO_2: Fixation/Rubisco Activity*

The properties of Rubisco activase and other components of the photosynthetic machinery are affected by heat-shock proteins, changes in the chloroplast milieu and other factors. In turn, these factors undoubtedly affect the precise response of photosynthesis and Rubisco activation to temperature and also the degree to which a given plant species can acclimatise to temperatures outside its optimal range (Law and Crafts-Brandner 1999). Photosynthesis is negatively affected by high temperatures (>40 °C), while CO_2 assimilation is significantly inhibited by moderately high-temperature stress (Feller et al. 1998; Salvucci and Crafts-Brandner 2004a, b). The two substrates of Rubisco, carbon dioxide and oxygen, regulate the response of photosynthesis to heat stress. The inhibition of activation of Rubisco via direct effect on Rubisco activase is associated with decreased CO_2 assimilation (Feller et al. 1998; Haldimann and Feller 2004; Salvucci and Crafts-Brandner 2004a, b; Tang et al. 2007). At high temperatures, the solubility of oxygen is decreased to a lesser extent than solubility of CO_2, resulting in an increased rate of photorespiration and lower rates of photosynthesis. High-temperature stress basically inhibits Rubisco by inhibiting the enzyme Rubisco activase (Crafts-Brandner and Salvucci 2000), and the mechanism responsible for its inactivation is related to inability of Rubisco activase to overcome the inherently faster rates of Rubisco inactivation (Prasad et al. 2008). A balance between sequestration of Rubisco active sites in a closed, inactive conformation and the reactivation of these sites by conformational changes induced by Rubisco activase represents the active state of Rubisco (Andrews 1996; Spreitzer and Salvucci 2002; Portis 2003). A biochemical study of inactivation of Rubisco under heat stress has shown that this balance shifts to a lower activation state at higher temperature because of faster rates of Rubisco inactivation and slower rates of activase activity (Crafts-Brandner and Salvucci 2000; Salvucci and Crafts-Brandner 2004b). However, the activase activity in vitro is sufficient for Rubisco activation at optimal temperatures but at the same time is insufficient to keep pace with the faster rates of Rubisco inactivation at high temperatures (Crafts-Brandner and Salvucci 2000). Thus, the activation state of Rubisco decreases under heat stress because activase activity cannot overcome the faster rates of Rubisco inactivation. The poor performance of activase at high temperature is probably because of its relatively low temperature optimum for catalysis, caused in part by thermal instability (Salvucci et al. 2001; Rokka et al. 2001), as well as to other unspecified causes (Sharkey 2000).

3.6 Effects of Heat Stress on Grain Filling and Crop Yield

Grain- or seed-filling duration is the time from seed set to physiological maturity, and at this stage, high temperature may influence both the quantity and quality of the yield. It is thought that extreme temperatures are more important than average temperatures in determining plant responses. Crop yields are affected by net primary productivity and also by the phenology of crop development. Increased temperature can speed phenological development, reducing the grain-filling period for crops, change in proteomics and hence yield (Majoul-Haddad et al. 2013). Crop yields were greater under elevated CO_2, but warmer temperatures reduced the duration of crop growth and, hence, the yield of determinate crops such as winter wheat and onion (Wheeler et al. 2000). High temperature causes a reduction in grain filling after anthesis (Veisz et al. 2008), more rapid apoptosis and attainment of earlier harvest maturity (Altenbach et al. 2003). A reduction in the starch content, which makes up more than 65 % of the dry weight of cereals, leads to severe yield losses (Rakszegi et al. 2006; Barnabás et al. 2008; Yan et al. 2008). In the early phases of grain filling, reductions in grain weight in response to stress could be due to a lower number of endosperm cells (Nicolas et al. 1985). Furthermore limited supplies of grain assimilates may lead to a decrease in starch synthesis during the later phases of grain filling (Blum 1998) or by direct effects on the process of synthesis in the grain (Yang et al. 2004). Reduction in 1,000-kernel weight, diameter and starch content was observed in grains exposed to stress (Labuschange et al. 2009). Stem elongation seems to be the most sensitive phase, while booting and anthesis were moderately sensitive and the phase between heading and anthesis was found to be least sensitive (Ugarte et al. 2007). High temperature not only causes delay in flowering but also affects the fertilisation and therefore shows delayed seed-set time. Heat stress decreases the seed-filling duration resulting in smaller seed size. The increase in seed-filling rate does not compensate for loss of duration thereby produces smaller seed size and seed yields. Various developmental stages of crop are differentially sensitive to stress conditions. In peanut (Prasad et al. 1999), wheat (Saini and Aspinall 1981), rice (Matsui et al. 2001) and maize (Claassen and Shaw 1970), stress just before anthesis and at anthesis causes significant increase in floral abortion and lower seed numbers. At high temperatures, plants can be damaged at seedling emergence, reproductive development, stem elongation, heading and flowering. Photosynthesis slows down and plant respiration increases when wheat leaves are exposed to high temperatures and finally results in overall loss of plant growth. Hot weather that occurs during the conversion from vegetative to reproductive stages can cause small grain plants with lesser ability to carry out respiration and supply materials for new growth. High temperatures during flowering have been shown to result in head sterility.

4 Protective Mechanisms in Plants/Recovery

All aspects of plant development, growth, reproduction and yield are adversely affected by heat stress. Being immobile, plants are forced to invest valuable resources in modifying their metabolism to prevent damage caused by heat, through a process generally referred to as acclimatisation. Alternatively, plants can activate programmed cell death in specific cells or tissues, a process that can lead to the shedding of leaves, the abortion of flower or fruit formation or even death of the entire plant. Plants reprogramme their transcriptome, protome, metabolome and lipidome in response to the changes of surrounding temperature. This action helps the plants to reset a new metabolic balance so that the cells can survive and function as normal even at high temperature, basically by inducing changes in the composition of certain transcripts, metabolites and lipids. However, when the temperature comes back to normal, plant cells can reverse the reprogramme process and get back to the original metabolic balance. In addition, plants can also programme cell death in response to the heat stress resulting in leaves shedding.

Plants have developed mechanisms for stress adaptation, defence and repair (Biswal et al. 2011). The biological processes in plants that are involved in adaptation to stress are complex. Confronted to changes in temperatures, plants readjust their biochemical make-up to adapt and survive (Ruelland and Zachowski 2010). In case of high-temperature stress, adaptive mechanisms include change in the composition and degree of saturation of fatty acids (Quinn and Williams 1985), changes in the levels of protective antioxidants (Almeselmani et al. 2006; Snider et al. 2010) and trienoic fatty acids (Murakami et al. 2000), changes in membrane permeability and increased cyclic electron flow (Zhang and Sharkey 2009; Sharkey and Zhang 2010). In addition to these, synthesis of heat-shock proteins (Burke 2001), activation of synthesis of new form of Rubisco activase (Kurek et al. 2007; Salvucci and Crafts-Brandner 2004a; Allakhverdiev et al. 2008) and rapid dephosphorylation of PS II facilitating repair of PS II (Rokka et al. 2000) play very important roles to enable plants to tolerate heat stress. PS II hetergeneity changes seem to be an adaptive mechanism of plants to tolerate stress conditions (Mathur et al. 2011b; Mehta et al. 2010b).

Recent reports have suggested that certain environmental stresses like high-temperature stress promote interconversions of PS IIα centres into PS IIβ and PS IIγ and the active Q_B-reducing centres into inactive Q_B-nonreducing centres (Mathur et al. 2011b). The plant responds to high-temperature stress by transient interconversions of PS II heterogeneity up to temperature of 40 °C. Some reversible structural rearrangements in PS II may provide an adaptive mechanism. However, at 45 °C an irreversible loss or permanent damage occurs to PS II, and no recovery was observed in both types of heterogeneity.

5 Conclusions and Future Prospective

In plants, changes in ambient temperature are sensed via a complex network of sensors located in different parts of the plant. The impacts of changing climate on wheat can be counter-intuitive and that the severity of the impact depends strongly on cultivar characteristics. Many questions remain unanswered, despite recent advances in our understanding of the biophysical mechanisms involved in high-temperature stress in plants. Research programmes should be developed which are dedicated and aimed at enhancing the tolerance to combinations of different abiotic stresses and particularly those related to global changes (e.g. elevated carbon dioxide and ultraviolet-B radiation, drought, high temperature). The mechanisms to tolerate heat may be different; therefore, an integrated approach needs to be taken for development of cultivar. As our knowledge is improved and is able to quantify the impacts of both short-term and long-term effects of high-temperature stress on growth, development, yield, quality of crops as well as various physiological functions, the chances of incorporation of these effects into crop models will be better. Improving knowledge on the physiological and genetic nature of tolerance can lead to increased grain yield and improved quality of crops. Improved models will affect performance of the crop to future climates and also help to identify traits that can potentially be improved to obtain higher and stable crop yields under stress environments.

References

Allakhverdiev SI, Kreslavski V, Klimov V, Los D, Carpentier R, Mohanty P (2008) Heat stress: an overview of molecular responses in photosynthesis. Photosynth Res 98:541–550

Almeselmani M, Deshmukh PS, Sairam RK, Kushwaha SR, Singh TP (2006) Protective role of antioxidant enzymes under high temperature stress. Plant Sci 171:382–388

Altenbach S, DuPont B, Kothari FM, Chan KM, Johnson R, Lieu EL (2003) Temperature, water and fertilizer influence the timing of key events during grain development in a US spring wheat. J Cereal Sci 37:9–20

Al-Whaibi MH (2012) Plant heat-shock proteins: a mini review. J King Saud Univ 23(2):139–150 doi:10.1016/j.jksus.2010.06.022

Andersson JM, Melis A (1983) Localization of different photosystems in separate regions of chloroplast membranes. Proc Natl Acad Sci U S A 80:745–749

Andrews TJ (1996) The bait in the Rubisco mousetrap. Nat Struct Biol 3:3–7

Barnabás B, Jäger K, Fehér A (2008) The effect of drought and heat stress on reproductive processes in cereals. Plant Cell Environ 31:11–38

Batts GR, Ellis RH, Morison JIL, Nkemka PN, Gregory PJ, Hadley P (1998) Yield and partitioning of crops of contrasting cultivars of winter wheat in response to CO_2 and temperature in field studies using temperature gradient tunnels. J Agric Sci 130:17–27

Biswal B, Joshi PN, Raval MK, Biswal UC (2011) Photosynthesis, a global sensor of environmental stress in green plants: stress signalling and adaptation. Curr Sci 101:47–56

Björkman O, Badger MR, Armong PA (1980) Response and adaptation of photosynthesis to high temperatures. In: Turner NC, Kramer PJ (eds) Adaptation of plants to water and high temperature stress. Wiley, New York, pp 233–249

Black MT, Brearley TH, Horton P (1986) Heterogeneity in chloroplast photosystem II. Photosynth Res 89:193–207

Blum A (1998) Improving wheat grain filling under stress by stem reserve mobilisation. Euphytica 100:77–83

Bos HJ, Tijani-Eniola T, Struik PC (2000) Morphological analysis of leaf growth of maize: responses to temperature and light intensity. Neth J Agric Sci 48:181–198

Bricker TM, Burnap RL (2005) The extrinsic proteins of photosystem II. In: Wydrzynski TJ, Satoh K (eds) Photosystem II: the light-driven water: plastoquinone, oxidoreductase. Advances in photosynthesis and respiration. Springer, Dordrecht, The Netherlands, pp 95–120

Bukhov NG, Mohanty P (1999) Elevated temperature stress effects on photosystems: characterization and evaluation of the nature of heat induced impairments. In: Singhal GS, Renger G, Sopory SK, Irrgang KD, Govindjee (eds) Concepts in photobiology: photosynthesis and photomorphogenesis. Narosa Publishers, New Delhi, pp 617–648

Burke JJ (2001) Identification of genetic diversity and maturation in higher plant acquired thermotolerance. Physiol Plant 112:167–170

Bussotti F, Desotgiu R, Cascio C, Strasser RJ, Gerosa G, Marzuoli R (2007) Photosynthesis responses to ozone in young tress of three species with different sensitivities in a 2 year open top chamber experiment (Curno, Italy). Plant Physiol 130:122–135

Cao J, Govindjee (1990) Chlorophyll *a* fluorescence transients as an indicator of active and inactive photosystem II in thylakoid membranes. Biochim Biophys Acta 1015:180–188

Chauhan H, Khurana N, Nijhavan A, Khurana JP, Khurana P (2012) The wheat chloroplastic small heat shock protein (sHSP26) is involved in seed maturation and germination and imparts tolerance to heat stress. Plant Cell Environ 35:1912–1931. doi:10.1111/j.1365-3040.2012.02525.x

Chen LS, Cheng L (2009) Photosystem 2 is more tolerant to high temperature in apple (*Malus domestica* Borkh) leaves than in fruit peel. Photosynthetica 47:112–120

Chen WR, Zheng JS, Li YQ, Guo WD (2012) Effects of high temperature on photosynthesis, chlorophyll fluorescence, chloroplast ultrastructure, and antioxidant activities in fingered citron. Russ J Plant Physiol 59:732–740

Christen D, Schőnmanna S, Jermini M, Strasser RJ, Defago G (2007) Characterization and early detection of grapevine (*Vitis vinifera*) stress responses to esca disease by *in situ* chlorophyll fluorescence and comparison with drought stress. Environ Exp Bot 60:504–514

Claassen MM, Shaw RH (1970) Water deficit effects on corn. II. Grain components. Agron J 62:652–655

Crafts-Brandner SJ, Salvucci ME (2000) Rubisco activase constrains the photosynthetic potential of leaves at high temperature and CO_2. Proc Natl Acad Sci U S A 97:13430–13435

Depège N, Bellafiore S, Rochaix JD (2003) Role of chloroplast protein kinase Stt7 in LHCII phosphorylation and state transition in *Chlamydomonas*. Science 299:1572–1575

Efeoglu B, Terzioglu S (2009) Photosynthetic responses of two wheat varieties to high temperature. EurAsia J Biosci 3:97–106

Enami I, Kitamura M, Tomo T, Isokawa Y, Ohta H, Katoh S (1994) Is the primary cause of thermal inactivation of oxygen evolution in spinach PS II membranes release of the extrinsic 33 kDa protein or of Mn? Biochim Biophys Acta 1186:52–58

Evrard A, Kumar M, Lecourieux D, Lucks J, von Koskull-Doring P, Hirt H (2013) Regulation of the heat stress response in Arabidopsis by MPK6-targeted phosphorylation of the heat stress factor HsfA2. Peer J 1:e59. doi:10.7717/peerj.59

Falcone DL, Ogas JP, Somerville CR (2004) Regulation of membrane fatty acid composition by temperature in mutants of Arabidopsis with alterations in membrane lipid composition. BMC Plant Biol 17:4–17

Feder ME (2006) Integrative biology of stress: molecular actors, the ecological theatre, and the evolutionary play. In: International symposium on environmental factors, cellular stress and evolution, Varanasi, India, 13–15 Oct, p 21

Feller U, Crafts-Brandner SJ, Salvucci ME (1998) Moderately high temperatures inhibit ribulose-1,5-bisphosphate carboxylase/oxygenase (Rubisco) activase-mediated activation of Rubisco. Plant Physiol 116:539–546

Force L, Critchley C, van Rensen JJS (2003) New fluorescence parameters for monitoring photosynthesis in plants 1. The effect of illumination on the fluorescence parameters of the JIP-test. Photosynth Res 78:17–33

Gounaris K, Brain ARR, Quinn PJ, Williams WP (1983) Structural and functional changes associated with heat-induced phase-separations of non-bilayer lipids in chloroplast thylakoid membranes. FEBS Lett 153:47–52

Graan T, Ort DR (1986) Detection of oxygen-evolving photosystem II centres inactive in plastoquinone reduction. Biochim Biophys Acta 852:320–330

Guenther JE, Nemson JA, Melis A (1988) Photosystem stoichiometry and chlorophyll antenna size in *Dunaliella salina* (green algae). Biochim Biophys Acta 934:108–117

Guissé B, Srivastava A, Strasser RJ (1995) The polyphasic rise of the chlorophyll *a* fluorescence (O-K-J-I-P) in heat-stressed leaves. Arch Sci Geneva 48:147–160

Haldimann P, Feller U (2004) Inhibition of photosynthesis by high temperature in oak (*Quercus pubescens* L.) leaves grown under natural conditions closely correlates with a reversible heat-dependent reduction of the activation state of ribulose-1,5-bisphosphate carboxylase/oxygenase. Plant Cell Environ 27:1169–1183

Hammer MF, Markwell J, Sarath G (1997) Purification of a protein phosphatase from chloroplast stroma capable of dephosphorylating the light-harvesting complex-II. Plant Physiol 113:227–233

Hansson M, Vener AV (2003) Identification of three previously unknown *in vivo* protein phosphorylation sites in thylakoid membranes of *Arabidopsis thaliana*. Mol Cell Proteomics 2:550–559

Harrison MA, Allen JF (1991) Light-dependent phosphorylation of photosystem II polypeptides maintains electron transport at high light intensity: separation from effects of phosphorylation of LHC-II. Biochim Biophys Acta 1058:289–296

Haucke V, Di Paolo G (2007) Lipids and lipid modifications in the regulation of membrane traffic. Curr Opin Cell Biol 19:426–435

Havaux M, Lorrain L, Leblanc RM (1989) *In vivo* measurement of spectroscopic and photochemical properties of intact leaves using the 'mirage effect'. FEBS Lett 250:395–399

Heckman NL, Horst GL, Gausson RE, Tavener BT (2002) Trinexapac-ethyl influence on cell membrane thermostability of Kentucky bluegrass leaf tissue. Sci Hortic 92:83–186

Hennessy KJ (1994) Climate change impact on wheat. In: Coombs B (ed) Australian grains. Morescope Publications, Melbourne, pp 213–219

Hsu BD, Lee JY (1991) A study on the fluorescence induction curve of the DCMU-poisoned chloroplast. Biochim Biophys Acta 1056:285–292

Irmak S, Naeem HA, Lookhart GL, MacRitchie F (2008) Effect of heat stress on wheat proteins during kernel development in wheat near-isogenic lines differing at Glu-D1. J Cereal Sci 48:513–516

Joliot A, Joliot MP (1964) Etude cinetique de la reaction photochimique liberant l oxygeneau cours de la photosynthese. C R Acad Sci Paris 258:4265–4622

Jones PD, New M, Parker DE, Mortin S, Rigor IG (1999) Surface area temperature and its change over the past 150 years. Rev Geophys 37:173–199

Khurana N, Chauhan H, Khurana P (2013) Wheat chloroplast targeted sHSP26 promoter confers heat and abiotic stress inducible expression in transgenic *Arabidopsis* plants. PLoS One 8(1):e54418. doi:10.1371/journal.pone.0054418

Kruger GHJ, Tsimilli-Michael M, Strasser RJ (1997) Light stress provokes plastic and elastic modifications in structure and function of photosystem II in camellia leaves. Physiol Plant 101:265–277

Kurek I, Chang TK, Bertain SM, Madrigal A, Liu L, Lassner MW, Zhu G (2007) Enhanced thermostability of Arabidopsis Rubisco activase improves photosynthesis and growth rates under moderate heat stress. Plant Cell 19:3230–3241

Labuschange MT, Elago O, Koen E (2009) The influence of temperature extremes on some quality and starch characteristics in bread, biscuit and durum wheat. J Cereal Sci 49:184–189

Larkindale J, Hall JD, Knight MR, Vierling E (2005) Heat stress phenotypes of *Arabidopsis* mutants implicate multiple signaling pathways in the acquisition of thermotolerance. Plant Physiol 138:882–897

Lavergne J (1982) Two types of primary acceptors in chloroplasts photosystem II. Photobiochem Photobiophys 3:257–285

Lavergne J, Briantais JM (1996) Photosystem II heterogeneity. In: Ort RD, Yocum CF (eds) Oxygeneic photosynthesis: the light reactions. Kluwer, Dordrecht, The Netherlands, pp 265–287

Law R, Crafts-Brandner SJ (1999) Inhibition and acclimation of photosynthesis to heat stress is closely correlated with activation of ribulose-1,5-bisphosphate carboxylase/oxygenase. Plant Physiol 120:173–182

Lazár D, Pospisil P (1999) Mathematical simulation of chlorophyll *a* fluorescence rise measured with 3-[3′,4′-dichlorophenyl]-1,1-dimethylurea-treated barley leaves at room and high temperature. Eur Biophys J 28:468–477

Lazár D, Iliek P, Naus I (1997) An appearance of K peak in fluorescence induction depends on the acclimation of barley leaves to higher temperature. J Lumin 72:595–596

Lazár D, Tomek P, Ilík P, Naus J (2001) Determination of the antenna heterogeneity of PS II by direct simultaneous fitting of several fluorescence rise curves measured with DCMU at different light intensities. Photosynth Res 68:247–257

Majoul-Haddad T, Bancel E, Martre P, Triboi E, Branlard G (2013) Effect of short heat shocks applied during grain development on wheat (Triticum aestivum L.) grain proteome. J Cereal Sci 57(3):486–495. doi:10.1016/j.jcs.2013.02.003

Mathur S, Allakhverdiev SI, Jajoo A (2011a) Analysis of high temperature stress on the dynamics of antenna size and reducing side heterogeneity of photosystem II in wheat (*Triticum aestivum*). Biochim Biophys Acta 1807:22–29

Mathur S, Jajoo A, Mehta P, Bharti S (2011b) Analysis of elevated temperature-induced inhibition of photosystem II by using chlorophyll *a* fluorescence induction kinetics in wheat leaves (*Triticum aestivum*). Plant Biol 13:1–6

Matsui T, Omasa K, Horie T (2001) The difference in sterility due to high temperature during flowering period among japonica rice varieties. Plant Prod Sci 4:90–93

Mehta P, Allakhverdiev SI, Jajoo A (2010a) Characterization of photosystem II heterogeneity in response to high salt stress in wheat leaves (*Triticum aestivum*). Photosynth Res 105: 249–255

Mehta P, Jajoo A, Mathur S, Bharti S (2010b) Chlorophyll *a* fluorescence study revealing effects of high salt stress on photosystem II in wheat leaves. Plant Physiol Biochem 48:16–20

Melis A (1985) Functional properties of PS IIβ in spinach chloroplasts. Biochim Biophys Acta 808:334–342

Melis A, Duysens LMN (1979) Biphasic energy conversion kinetics and absorbance difference spectra of photosystem II of chloroplasts. Evidence for two different photosystem II reaction centres. Photochem Photobiol 29:373–382

Melis A, Homann PH (1976) Heterogeneity of the photochemical centres in system II of chloroplasts. Photochem Photobiol 23:343–350

Midmore DJ, Cartwright PM, Fischer RA (1984) Wheat in tropical environments II. Crop growth and grain yield. Field Crop Res 8:207–227

Mittler R, Finka A, Goloubinoff P (2012) How do plants feel the heat? Trends Biochem Sci 37(3):118–125

Modarresi M, Mohammadi V, Zali A, Mardi M (2010) Response of wheat yield and yield related traits to high temperature. Cereal Res Commun 38:23–31

Moon BY, Higashi S, Gombos Z, Murata N (1995) Unsaturation of the membrane lipids of chloroplasts stabilizes the photosynthetic machinery against low-temperature photoinhibition in transgenic tobacco plants. Proc Natl Acad Sci U S A 92(14):6219–6223

Morimoto RI (1993) Cells in stress: the transcriptional activation of heat shock genes. Science 259:1409–1410

Murakami Y, Tsuyama M, Kobayashi Y, Kodama H, Iba K (2000) Trienoic fatty acids and plant tolerance of high temperature. Science 287:476–479

Nicolas ME, Gleadow RM, Dalling MJ (1985) Effect of post-anthesis drought on cell-division and starch accumulation in developing wheat grains. Ann Bot 55:433–444

Pick T, Jaskiewicz M, Peterhänsel C, Conrath U (2012) Heat shock factor HsfB1 primes gene transcription and systemic acquired resistance in Arabidopsis. Plant Physiol 159:52–55

Porter JR, Gawith M (1999) Temperature and the growth and development of wheat: a review. Eur J Agron 10:23–36
Portis AR Jr (2003) Rubisco activase: Rubisco's catalytic chaperone. Photosynth Res 75:11–27
Pospíšil P, Tyystjärvi E (1999) Molecular mechanism of high temperature-induced inhibition of acceptor side of photosystem II. Photosynth Res 62:55–66
Pospisil P, Skotnica J, Naus J (1998) Low and high temperature dependence of minimum Fo and maximum Fm chlorophyll fluorescence in vivo. Biochim Biophys Acta 1363:95–99
Prasad PVV, Craufurd PQ, Summerfield RJ (1999) Fruit number in relation to pollen production and viability in groundnut exposed to short episodes of heat stress. Ann Bot 84:381–386
Prasad PVV, Boote KJ, Allen LH Jr (2006) Adverse high temperature effects on pollen viability, seed-set, seed yield and harvest index of grain sorghum (*Sorghum bicolor* (L.) Moench) are more severe at elevated carbon dioxide due to higher tissue temperatures. Agric For Meteorol 139:237–251
Prasad PVV, Staggenborg SA, Ristic Z (2008) Impacts of drought and/or heat stress on physiological, developmental, growth, and yield processes of crop plants. In: Advances in agricultural systems modeling series 1, Madison, WI, pp 301–355
Qu A-L, Ding Y-F, Jiang Q, Zhu C (2013) Molecular mechanisms of the plant heat stress response. Biochem Biophys Res Commun 432(2):203–207. doi:10.1016/j.bbrc.2013.01.104
Quinn PG, Williams WP (1985) Environmentally induced changes in chloroplast membranes and their effects on photosynthetic function. In: Barber J, Baker NR (eds) Photosynthetic mechanisms and the environment. Elsevier, Amsterdam, pp 1–47
Rakszegi M, Láng L, Bedi Z (2006) Importance of starch properties in quality oriented wheat breeding. Cereal Res Commun 34:637–640
Rokka A, Aro EM, Herrmann RG, Andersson B, Vener AV (2000) Dephosphorylation of photosynthetic membranes as an immediate response to abrupt elevation of temperature. Plant Physiol 123:1525–1535
Rokka A, Zhang L, Aro EM (2001) Rubisco activase: an enzyme with a temperature-dependent dual function? Plant J 25:463–471
Ruelland E, Zachowski A (2010) How plants sense temperature. Environ Exp Bot 69:225–232
Saini HS, Aspinall D (1981) Effect of water deficit on sporogenesis in wheat (*Triticum aestivum* L.). Ann Bot 48:623–633
Salvucci ME, Crafts-Brandner SJ (2004a) Relationship between the heat tolerance of photosynthesis and the thermal stability of rubisco activase in plants from contrasting thermal environments. Plant Physiol 134:1460–1470
Salvucci ME, Crafts-Brandner SJ (2004b) Inhibition of photosynthesis by heat stress: the activation state of Rubisco as a limiting factor in photosynthesis. Physiol Plant 120:179–186
Salvucci ME, Osteryoung KW, Crafts-Brandner SJ, Vierling E (2001) Exceptional sensitivity of Rubisco activase to thermal denaturation in vitro and in vivo. Plant Physiol 127:1053–1064
Schansker G, Strasser RJ (2005) Quantification of non-Q_B-reducing centres in leaves using a far-red pre-illumination. Photosynth Res 84:145–151
Scharf KD, Berberich T, Ebersberger I, Nover L (2012) The plant heat stress transcription factor (Hsf) family: structure, function and evolution. Biochim Biophys Acta 1819(2):104–119
Schöffl F, Prandl R, Reindl A (1999) Molecular responses to heat stress. In: Shinozaki K Yamaguchi-Shinozaki K (eds) Molecular responses to cold, drought, heat and salt stress in higher plants. RG Landes, Austin, TX, pp 81–98
Schreiber U, Gademan R, Ralph PJ, Larkum AWD (1997) Assessment of photosynthetic performance of prochloron in *Lissoclinum patellain hospite* by chlorophyll fluorescence measurement. Plant Cell Physiol 38:945–951
Sharkey TD (2000) Some like it hot. Science 287:435–437
Sharkey TD (2005) Effects of moderate heat stress on photosynthesis: importance of thylakoid reactions, rubisco deactivation, reactive oxygen species, and thermotolerance provided by isoprene. Plant Cell Environ 28:269–277
Sharkey TD, Zhang R (2010) High temperature effects on electron and proton circuits of photosynthesis. J Integr Plant Biol 52:712–722

Silva-Correia J, Azevedo H, Lino-Neto T, Tavares RM (2012) Understanding heat stress tolerance of suspended cells in the model plant Populus euphratica. In: Papers in scientific journals, ISRN Forestry, pp 1–5. doi:10.5402/2012/243694

Sinclair TR (1994) Limits to crop yield. In: Boote KJ et al (eds) Physiology and determination of yield. ASA, Madison, WI, pp 509–532

Sinclair J, Spence SM (1988) The analysis of fluorescence induction transients from dicholoro-phenyldimethylurea-poisoned chloroplasts. Biochim Biophys Acta 935:184–194

Slafer GA, Rawson HM (1995) Base and optimum temperatures vary with genotype and stage of development in wheat. Plant Cell Environ 18:671–679

Snider JL, Oosterhuis DM, Kawakami EM (2010) Genotypic differences in thermotolerance are dependent upon prestress capacity for antioxidant protection of the photosynthetic apparatus in *Gossypium hirsutum*. Physiol Plant 138:268–277

Snyders S, Kohorn BD (1999) TAKs, thylakoid membrane protein kinases associated with energy transduction. J Biol Chem 274:9137–9140

Snyders S, Kohorn BD (2001) Disruption of thylakoid-associated kinase 1 leads to alteration of light harvesting in Arabidopsis. J Biol Chem 276:32169–32176

Sohn SO, Back K (2007) Transgenic rice tolerant to high temperature with elevated contents of dienoic fatty acids. Biol Plant 51:340–342

Spreitzer RJ, Salvucci ME (2002) Rubisco: interactions, associations and the possibilities of a better enzyme. Annu Rev Plant Biol 53:449–475

Srivastava A, Strasser RJ (1996) Stress and stress management of land plants during a regular day. J Plant Physiol 148:445–455

Srivastava A, Guissé B, Greppin H, Strasser RJ (1997) Regulation of antenna structure and electron transport in photosystem II of *Pisum sativum* under elevated temperature probed by the fast polyphasic chlorophyll *a* fluorescence transient: OKJIP. Biochim Biophys Acta 1320:95–106

Stirbet A, Govindjee (2011) On the relation between the Kautsky effect (chlorophyll *a* fluorescence induction) and photosystem II: basics and applications of the OJIP fluorescence transient. J Photochem Photobiol B 104:236–257

Stone PJ, Nicholas ME (1994) Wheat cultivars vary widely in their responses of grain yield and quality to short periods of post-anthesis heat stress. Aust J Plant Physiol 21:887–900

Strasser BJ, Strasser RJ (1995) Measuring fast fluorescence transients to address environmental questions: the JIP-test. In: Mathis P (ed) Photosynthesis: from light to biosphere. Kluwer Academic, Dordrecht, pp 977–980

Strasser RJ, Tsimilli-Michael M (1998) Activity and heterogeneity of PS II probed in vivo by the chlorophyll *a* fluorescence rise O-[K]-J-I-P. In: Garab G (ed) Photosynthesis: mechanisms and Effects. Kluwer Academic, Dordrecht, The Netherlands, pp 4321–4324

Strasser RJ, Eggenberg P, Strasser BJ (1996) How to work without stress but with fluorescence. Bull Soc R Sci Liege 65:330–349

Strasser RJ, Srivastava A, Tsimilli-Michael M (2000) The fluorescence transient as a tool to characterize and screen photosynthetic samples. In: Yunus M, Pathre U, Mohanty P (eds) Probing photosynthesis: mechanisms, regulation and adaptation. Taylor & Francis, London, pp 443–483

Strasser RJ, Tsimilli-Michael M, Srivastava A (2004) Analysis of chlorophyll *a* fluorescence transient. In: Papageorgiou G, Govindjee (eds) Advances in photosynthesis and respiration: chlorophyll *a* fluorescence: a signature of photosynthesis. Springer, Dordrecht, The Netherlands, pp 321–362

Strasser RJ, Tsimilli-Michael M, Dangre D, Rai M (2007) Biophysical phenomics reveals functional building blocks of plants system biology: a case study for the evaluation of the impact of mycorrhization with *P. indica*. In: Varma A, Oelmuller R (eds) Soil biology, advanced techniques in soil microbiology. Springer, Berlin, pp 319–341

Su K, Bremer DJ, Jeannotte R, Welti R, Yang C (2009) Membrane lipid composition and heat tolerance in cool-season Turf grasses, including a Hybrid Bluegrass. J Amer Soc Hort Sci 134(5):511–520

Sundby C, Melis A, Mäenpää P, Andersson B (1986) Temperature-dependent changes in the antenna size of photosystem II. Reversible conversion of photosystem IIα to photosystem IIβ. Biochim Biophys Acta 851:475–483
Süss KH, Yordanov I (1986) Biosynthetic cause of vivo acquired thermotolerance of photosynthetic light reactions and metabolic responses of chloroplasts to heat stress. Plant Physiol 81:192–199
Suzuki N, Miller G, Sejima H, Harper J, Mittler R (2013) Enhanced seed production under prolonged heat stress conditions in Arabidopsis thaliana plants deficient in cytosolic ascorbate peroxidase 2. J Exp Bot 64(1):253–263. doi:10.1093/jxb/ers335
Tang Y, Wen X, Lu Q, Yang Z, Cheng Z, Lu C (2007) Heat stress induces an aggregation of the light-harvesting complex of photosystem II in spinach plants. Plant Physiol 143:629–638
Tomar RS, Mathur S, Allakhverdiev SI, Jajoo A (2012) Changes in PS II heterogeneity in response to osmotic and ionic stress in wheat leaves (*Triticum aestivum*). J Bioenerg Biomembr 44:411–419
Tripp J, Mishra SK, Scharf KD (2009) Functional dissection of the cytosolic chaperone network in tomato mesophyll protoplasts. Plant Cell Environ 32:123–133
Tsimilli-Michael M, Kruger GHJ, Strasser RJ (1996) About the perpetual state changes in plants approaching harmony with their environment. Arch Sci Geneva 49:173–203
Tyystjarvi E, Aro EM (1990) Temperature-dependent changes in PS II heterogeneity support a cycle of PS II during photoinhibition. Photosynth Res 26:109–117
Ugarte C, Calderini DF, Slafer GA (2007) Grain weight and grain number responsiveness to pre-anthesis temperature in wheat, barley and triticale. Field Crop Res 100:240–248
Umena Y, Kawakami K, Shen JR, Kamiya N (2011) Crystal structure of oxygen-evolving photosystem II at a resolution of 1.9 Å. Nature 475:55–60
Vainonen JP, Jaspers P, Wrzaczek M, Lamminmaki A, Reddy RA et al (2012) RCD1-DREB2A interaction in leaf senescence and stress responses in *Arabidopsis* thaliana. Biochem J 442:573–581. doi:10.1042/BJ20111739
Veisz O, Bencze SZ, Balla K, Vida GY (2008) Change in water stress resistance of cereals due to atmospheric CO_2 enrichment. Cereal Res Commun 36:1095–1098
Vener AV, Rokka A, Fulgosi H, Andersson B, Herrmann RG (1999) A cyclophilin-regulated PP2A-like protein phosphatase in thylakoid membranes of plant chloroplasts. Biochemistry 38:14955–14965
Vener AV, Harms A, Sussman MR, Vierstra RD (2001) Mass spectrometric resolution of reversible protein phosphorylation in photosynthetic membranes of *Arabidopsis thaliana*. J Biol Chem 276:6959–6966
Vierling E (1991) The role of heat shock proteins in plants. Annu Rev Plant Phys 42:579–620
Wahid A, Gelani S, Ashraf M, Foolad MR (2007) Heat tolerance in plants: an overview. Environ Exp Bot 61:199–223
Wahid A, Farooq M, Hussain I, Rasheed R, Galani S (2012) Responses and management of heat stress in plants. In: Ahmad P, Prasad MNV (eds) Environmental adaptations and stress tolerance of plants in the era of climate change. Springer, New York
Wang B (1988) Biological free radicals and membrane damage of plant. Plant Physiol Commun 2:12–16
Wang X (2004) Lipid signaling. Curr Opin Plant Biol 7:329–336
Wang JZ, Cui LJ, Wang Y, Li JL (2009) Growth, lipid peroxidation and photosynthesis in two tall fescue cultivars differing in heat tolerance. Biol Plant 53(2):237–242
Waters ER, Lee GJ, Vierling E (1996) Evolution, structure and function of the small heat shock proteins in plants. J Exp Bot 47:325–338
Welti R, Li W, Li M, Sang Y, Biesida H, Zhou H, Rajashekar CB, Williams TD, Wang X (2002 Profiling membrane lipids in plant stress responses. J Biol Chem 277:31994–32002
Wheeler TR, Craufurd PQ, Ellis RH, Porter JR, Prasad PV (2000) Temperature variability and the yield of annual crops. Agric Ecosyst Environ 82:159–167
Wollenweber B, Porter JR, Schellberg J (2003) Lack of interaction between extreme high temperature events at vegetative and reproductive growth stages in wheat. J Agron Crop Sc 189:142–150

Yamane Y, Kashino Y, Koike H, Satoh K (1998) Effects of high temperatures on the photosynthetic systems in spinach: oxygen-evolving activities fluorescence characteristics and the denaturation process. Photosynth Res 57:51–59

Yan SH, Yin YP, Li WY, Li Y, Liang TB, Wu YH, Geng QH, Wang ZL (2008) Effect of high temperature after anthesis on starch formation of two wheat cultivars differing in heat tolerance. Acta Ecol Sin 28:6138–6147

Yang JC, Zhang JH, Wang ZQ, Xu GW, Zhu QS (2004) Activities of key enzymes in sucrose-to-starch conversion in wheat grains subjected to water deficit during grain filling. Plant Physiol 135:1621–1629

Zhang R, Sharkey TD (2009) Photosynthetic electron transport and proton flux under moderate heat stress. Photosynth Res 100:29–43

Chapter 9
Low Temperature Stress in Plants: An Overview of Roles of Cryoprotectants in Defense

Kalpna Bhandari and Harsh Nayyar

Contents

1 Introduction

Cold stress has very wide and far-reaching consequences on various economically important crops (Beck et al. 2007) that are sensitive to temperatures below 10 °C (Ouellet 2007). Low temperature stress may be further categorized into chilling stress (0–15 °C) and freezing stress (<0 °C). The crop plants susceptible to chilling stress include mainly those of tropical and subtropical origin, e.g., rice (*Oryza sativa*), maize (*Zea mays*), and chickpea (*Cicer* sp.). Other crops being negatively affected by low temperature conditions include soybean (*Glycine max* L.), lima bean (*Phaseolus lunatus* L.), cucurbits (*Cucurbita* sp.), tomato (*Lycopersicon esculentum* Mill.), pepper (*Capsicum annuum* L.), eggplant (*Solanum melongena* L.), and various cereal crops (Thakur et al. 2010).

K. Bhandari • H. Nayyar (✉)
Department of Botany, Panjab University, Chandigarh, India
e-mail: harshnayyar@hotmail.com

P. Ahmad and M.R. Wani (eds.), *Physiological Mechanisms and Adaptation Strategies in Plants Under Changing Environment: Volume 1*, DOI 10.1007/978-1-4614-8591-9_9,

Cold stress affects the normal functioning of plants at various levels, viz., morphological, physiological, and cellular levels. Morphologically, the plants may show various effects such as anthocyanin accumulation, stunted growth, wilting, reduced and deformed leaves (Rymen et al. 2007), chlorosis (Yoshida et al. 1996), and even necrosis. At genetic level, cold-responsive genes are upregulated resulting in increased expression of various enzymatic and structural proteins, thus helping the plant to mitigate cold and chilling stress. Prolonged exposure to chilling temperatures at reproductive stage of development is bound to ultimately affect the yield of cold-sensitive crops manifested in the form of reduced number of flowers, reduced fruit set ultimately resulting in lower quality and quantity of yield (Kaur et al. 2008; Kumar et al. 2011), or even complete crop failure in some cases (Hudak and Salaj 1999). Reproductive stage being directly related to the yield and its products being the major source of food, cold stress during reproductive phase has extensive socioeconomic impacts (Thakur et al. 2010).

2 Cryoprotectants

To counter the cold stress and maintaining their survival, plants adopt a variety of cellular mechanisms that include the accumulation of diverse types of cryoprotectants, each of which is unique with reference to its structure and function. A variety of biomolecules may act as cryoprotectants which include amino acids such as proline, some betaines, various sugars, and sugar alcohols. Here, we present an updated knowledge about their emerging roles especially with respect to cold stress.

2.1 Proline

Proline (Pro) is a nonessential imino acid, very well known to have osmoprotective functions. It acts as an important plant osmoprotectant under various abiotic stress conditions (Delauney and Verma 1993; Hare et al. 1999) especially chilling stress (Burbulis et al. 2011; Zhang et al. 2011; Jonytiene et al. 2012; Yang et al. 2012). Proline is not universally accumulated in all the plants, e.g., it is naturally accumulated in significant amounts in plants such as tobacco, barley, wheat, and *Arabidopsis* whereas plants like sugarcane (Naik and Joshi 1983) and pigeon pea (Joshi 1984) do not accumulate proline.

2.1.1 Structure, Biosynthesis, and Metabolism

Proline is a unique nonessential imino acid in having an imino group (–NH) instead of usual amino group ($-NH_2$) and the imino group being fixed to a pyrrolidine ring as shown in Fig. 9.1.

Fig. 9.1 Structure of proline

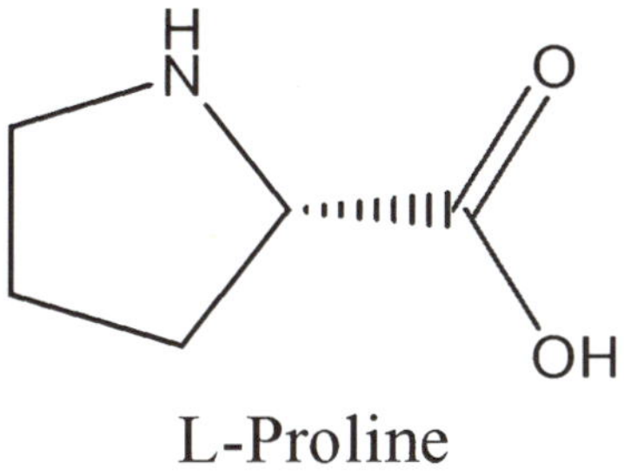

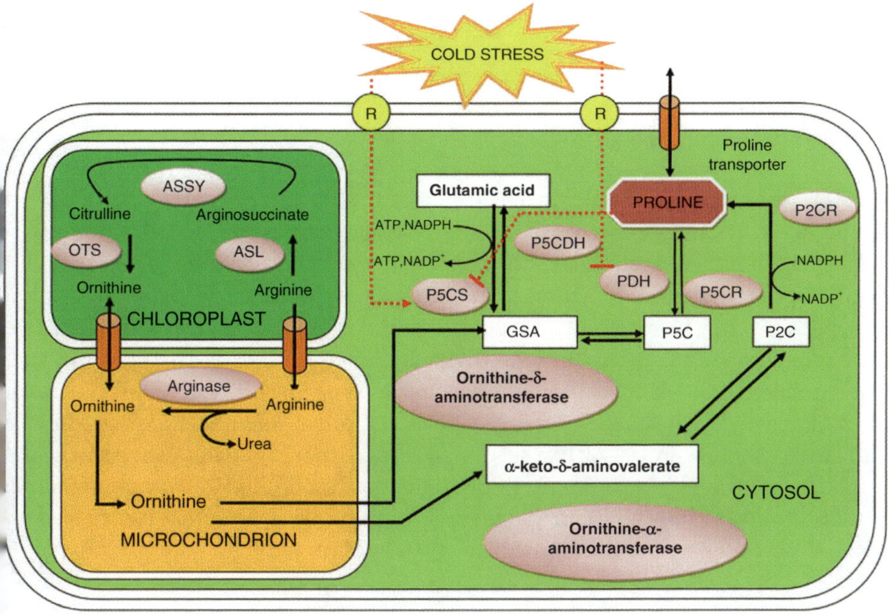

Fig. 9.2 Proline biosynthesis and metabolism, signaling pathway under cold stress (Delauney and Verma 1993; Hare et al. 1999; Funck et al. 2008; Szabados and Savoure 2009). *R* putative receptor, *P5C* pyrroline-5-carboxylate, *GSA* glutamic semialdehyde, *P5CS* pyrroline 5 carboxylate synthetase, *P5CR* pyrroline-5-carboxylate reductase, *PDH* proline dehydrogenase, *P2C* pyrroline-2-carboxylate, *Imt1* myoinositol-*O*-methyltransferase, *P2CR* pyrroline-5-carboxylate reductase, *ASL* argininosuccinate lyase, *ASSY* argininosuccinate synthetase, *OTC* ornithine transcarbamylase, *P5CDH* P5C dehydrogenase

Proline biosynthesis occurs in cytosol and the main precursor of proline biosynthesis is glutamic acid which is acted upon by enzyme pyrroline-5-carboxylate synthetase (P5CS) into glutamate semialdehyde (GSA) which is then spontaneously converted into pyrroline-5-carboxylate (P5C) (Hu et al. 1992; Savoure et al. 1995) as shown in Fig. 9.2. P5C thus produced is finally reduced to proline by P5C reductase (P5CR) (Verbruggen et al. 1993). Alternatively, arginine and ornithine pathways may also lead to the accumulation of proline in the cytosol. Arginine pathway occurs in the chloroplast and involves the formation of argininosuccinate from

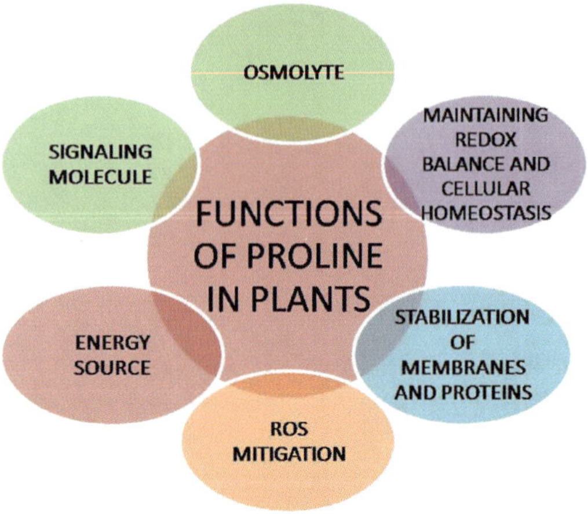

Fig. 9.3 Summary of various functions of proline in plants growing under cold-stressed conditions

arginine in the presence of enzyme ASL (argininosuccinate lyase) which is then further acted upon by enzyme ASSY (argininosuccinate synthetase) to form citrulline. The latter then finally yields ornithine through OTC (ornithine transcarbamylase) enzyme-catalyzed reaction. Ornithine when transported to mitochondria generates a series of reactions that constitute the ornithine pathway. Alternatively, arginine may be transported into mitochondria where it is acted upon by enzyme arginase leading to the formation of ornithine accompanied by the removal of urea. Ornithine pathway has been found to play an important role under various stresses and developmental processes like seedling growth (Roosens et al. 1998; Armengaud et al. 2004; Xue et al. 2009). Ornithine thus produced undergoes transamination reaction catalyzed by ornithine-delta-transaminase to yield proline by a series of reactions involving various intermediates such as GSA and P5C. Alternatively, ornithine in the presence of enzyme ornithine-alpha-aminotransferase may form P2C (pyrroline-2-carboxylate) which on reduction by P2C reductase also forms proline, thus further increasing the proline levels in the plant system. Under cold stress, genes controlling the activity of P5CS are upregulated, whereas genes controlling the activity of PDH (proline dehydrogenase) are downregulated, thus resulting in the accumulation of increased levels of proline and minimizing its reversal back into enzyme (Hare et al. 1999). Proline has been found to regulate its own production beyond the required limits by feedback inhibition by controlling the activity of P5CS genes (Hu et al. 1992; Thiery et al. 2004). Some plasma membrane proline transporters have also been proposed to bring about active cell to cell or cell to organ transport of proline (Rentsch et al. 1996; Ueda et al. 2001; Grallath et al. 2005).

2.1.2 Functions in the Plants

Proline is undoubtedly a multifunctional molecule (Szabados and Savoure 2009) that helps in mitigating various abiotic stresses and protects the plants in multiple ways (summarized in Fig. 9.3). Proline plays a plethora of protective functions in the plant systems and is universally accepted as a compatible solute as it can accumulate up to very high levels under stress conditions (Hare and Cress 1997; Kavi Kishor et al. 2005; Verbruggen and Hermans 2008) since it is water soluble and does not bear any charge at neutral pH. Thus, it maintains the osmotic potential of the cell and hence saves the plant from chilling, salt, or drought-induced water deficit. Accumulation of proline may be due to either increased biosynthesis or decreased degradation (Delauney and Verma 1993; Yoshiba et al. 1995). Li et al. (2011) studied the role of *OsTPS1* gene in abiotic stress tolerance in rice and observed that accumulated proline levels in transgenics were significantly higher than the wild type. Under normal conditions, two different high proline expression lines had accumulated 21.76 and 20.37 μg $(g\ FW)^{-1}$ proline as compared to 18.29 μg $(g\ FW)^{-1}$ proline in wild type. When subjected to cold treatment (4 °C for 5 days), proline levels in transgenic lines were raised by four- to fivefold to 110.85 and 110.88 μg $(g\ FW)^{-1}$, whereas wild-type plants had accumulated 81.02 μg $(g\ FW)^{-1}$. Xin and Browse (1998) reported that wild-type *Arabidopsis* plants required 48 h cold treatment (4 °C) to increase the activity of *AtP5CS1* by three times and proline levels by ten times, while in *OsP5CS*-overexpressing transgenic rice (Igarashi et al. 1997), the activities of proline-synthesizing enzymes were found to have been enhanced within 2 h exposure to 4 °C. Kaur et al. (2011) studied the effect of exogenous 10 μM proline application on chilling-stressed chickpea plants and reported that plants supplied with proline had accumulated 310 μmol g^{-1} DW proline after 4 days of chilling treatment, whereas in untreated plants, the proline level was 230 μmol g^{-1} DW. Additionally, the levels of proline were found to have decreased drastically to 28 μmol g^{-1} DW after 7 days in proline-untreated plants, whereas in proline-treated plants, it remained significantly higher. Proline levels in plants have been found to be useful in maintaining redox potential and cellular homeostasis as well. Krackhardt and Guerrier (1995) during their studies on soybean seeds found that proline maintains the cytosolic pH by directing excess protons (H^+) produced under stress towards proline biosynthesis. Proline regulates NAD^+/NADH ratio as NAD^+ is generated during proline synthesis, whereas NADH is produced during proline degradation; thus, an overall balance is maintained (Verbruggen and Hermans 2008; Aghaee et al. 2011). Since proline and glutamate are interconvertible, cytosolic oxidation of NADPH may be coupled to mitochondrial electron transport system by facilitating the electron transfer and thereby the conversion of NADPH to NAD^+ and by maintaining a balance between energy absorbed and energy utilized (Hare and Cress 1997). This results in the increased activity of oxidative pentose phosphate pathway, thus providing the starting material for the generation of various secondary metabolites and ATP.

Proline also acts as a protein-stabilizing hydrotrope for stabilization of enzymes and biomembranes (Rajendrakumar et al. 1994; Zhang et al. 2011). Proline is proposed to maintain the structure and conformation of proteins by acting as chaperone and thus protecting the enzymes and other proteins from denaturation under temperature extremes (Rajendrakumar et al. 1994) and salt stress (Ozturk and Demir 2002). Rajendrakumar et al. (1994) reported that proline exhibited protective effect on enzyme M4 lactate dehydrogenase. The protective quality of proline was attributed to its ability to form hydrophilic colloids with water and having a hydrophobic backbone for interacting with proteins, thereby maintaining the structure and conformation of proteins indispensable for their functional integrity. Jonytiene et al. (2012) studied the effect of exogenous proline supply to rapeseed shoots and found that proline-treated shoots were cold tolerant and the electrolyte leakage content, which is an indicator of membrane injury, was lowered considerably by an amount of 4.2–17.88 %. Transforming *Arabidopsis* plants with *Imtl* (mannitol-1-phosphate dehydrogenase gene) resulted in higher proline and soluble sugar contents, and the transgenics thus obtained had lower membrane leakage and reduced malondialdehyde (MDA) content indicating lower membrane damage (Zhu et al. 2012). Presence of proline was found to protect isolated thylakoid membranes of *Brassica juncea* from photoinhibition (Alia et al. 1991), thus proving the membrane-stabilizing properties of proline. Similarly, Van Rensburg et al. (1993) showed that maintenance of chloroplast ultrastructure and membrane integrity was positively correlated with proline levels in the various tobacco cultivars. Alternatively, some proline-degrading enzymes can also be downregulated under stress to increase proline levels. Nanjo et al. (1999) observed that anti-ProDH (proline dehydrogenase) *Arabidopsis thaliana* transgenics recorded 59 % ion leakage as compared to 90–95 % leakage in wild-type plants. The transformation was thus found to increase the survival significantly. Heber et al. (1973) proved that 100 mM proline levels were quite effective in alleviating the freezing induced membrane inactivation in vivo and in vitro.

Proline has also been found to be quite effective in scavenging reactive oxygen species (ROS) (Szabados and Savoure 2009; Aghaee et al. 2011; Zhang et al. 2011) especially by acting as singlet oxygen quencher (Matysik et al. 2002). The possible mechanism of detoxifying various ROS is by forming stable compounds with ROS (Floyd and Nagy 1984; Naidu et al. 1991). Alia et al. (1991) worked on isolated photoinhibited chloroplast thylakoids from *Brassica juncea* and found that normal electron transport was restored and lipid peroxidation was considerably reduced when supplied with 1 M proline. Thus, proline is effective in protecting chloroplasts from oxidative damage especially PSII (Alia and Mohanty 1997). Kaur et al. (2011) studied the oxidative metabolism in proline-treated and proline-untreated chickpea plants. Oxidative stress, evaluated as level of thiobarbituric acid-reactive substances (TBARS), showed that levels in chilling stress and proline-untreated plants were 423 μmol g^{-1} DW, but in proline-treated ones, it was recorded to be 310 μmol g^{-1} DW. Also, the activities of various antioxidative enzymes such as SOD (superoxide dismutase), CAT (catalase), and crucial nonenzymatic antioxidants such as ascorbic acid and glutathione were lower in proline-supplied plants, but the activity of APO (ascorbate peroxidase) had increased. Since biosynthetic pathway of proline utilizes a large amount of reducing power and oxidation of a single molecule of proline yields about 30 ATP equivalents (Atkinson 1977), thus, proline is a potential energy source during the stress as well as during recovery from the stress. It brings about the stress mitigation by conservation of energy and generation of ATP under stress by providing reducing equivalents to mitochondrial electron transport chain (ETC) (Hare and Cress 1997) and amino acids for post-stress growth. The role of proline in oxidative respiration has also been proposed by various workers as evident from the fact that proline degradation and glutamate dehydrogenase enzyme occur inside mitochondria; thus, proline has also been postulated to be contributing carbon to TCA cycle. Skubatz et al. (1989) suggested that proline is the main energy source for heat production in voodoo lilies (*Sauromatum guttatum*).

Furthermore, the role of proline in nitrogen fixation has also been proved by the studies carried out by Kohl et al. (1988) on bacteroides of ureide-producing legumes which showed higher levels of proline oxidation. In short, proline acts as a source of nitrogen, carbon, and reducing power under stress (Nanjo et al. 1999; Szabados and Savoure 2009). Proline synthesis has been suggested to play an important signal in response to various abiotic stresses, thus acting as a potential secondary messenger that is capable of bringing about various physiological responses at cellular level (Hare and Cress 1997; Mattioli et al. 2009; Szabados and Savoure 2009; Hayat et al. 2012). Hence, proline has several cellular functions in various organelles and organs of the plant, and its accumulation during stress could be of vital significance for maintaining the cellular homeostasis. The functions of proline are summarized in Fig. 9.3.

2.1.3 Endogenous Levels in Cold-Stressed Plants

Proline is one of the most important compatible solutes that have been proposed to accumulate under low and freezing temperatures (Van Swaaij et al. 1985; Naidu et al. 1991). The increase in endogenous proline levels depends on the type of plant as well as the severity and duration of stress experienced by the plant (Kavi Kishor et al. 2005). Proline levels have been proved to go up in various crops such as chickpea (Kumar et al. 2010), *Solanum* spp. (Shi et al. 2011), wheat (Aghaee et al. 2011), *B. napus* (Jonytiene et al. 2012), *Passiflora edulis* (Chen et al. 2012), and *B. carinata* (Klima et al. 2012) when subjected to cold shock. Cold-acclimated in vitro raised winter rapeseed shoots showed considerably enhanced endogenous levels of proline and soluble sugars, thereby conferring the frost tolerance to the shoots (Burbulis et al. 2011). Soybean (*Glycine max*) plants when acclimated by subjecting the plants to nonlethal temperature of 4 °C showed increased levels of proline. The non-acclimated plants also recovered slowly as compared to the acclimated ones. Proline levels varied during various phases, i.e., increased in the acclimation phase, decreased in the chilling phase, and then again showed increase in the recovery phase (Yadeghari et al. 2008). Though there are many reports of proline as an osmoprotectant as well as an effective cryoprotectant, some workers have challenged the positive correlation between proline accumulation and stress tolerance (Delauney and Verma 1993). Many workers have attributed the proline accumulation to the water deficits at low temperatures, but studies have shown that proline levels rise even in cold-hardened plants with adequate water supply (Van Swaaij et al. 1985; Naidu et al. 1991). Higher accumulation of proline under cold-stressed conditions has been found to be related to cold tolerance in chickpea (Kumar et al. 2010) and *Medicago* species (Zhang et al. 2011).

2.1.4 Exogenous Application for Cold Tolerance

In case of proline non-accumulating plants or inadequately accumulating plants, exogenous application of proline has been proved to be a good alternative to increase the cellular levels of proline. There are numerous reports where exogenously given proline has resulted in improved cold tolerance. Duncan and Widholm (1987) reported that maize calli supplied with proline-acquired cold tolerance performed better under cold conditions and survived longer cold treatments. Exogenous application of 1 M proline enhanced the photochemical electron transport system in the isolated thylakoids of *Brassica juncea* (Alia et al. 1991), whereas in some other cases, exogenously applied proline has also resulted in inhibited root and shoot growth of rice seedlings and *Brassica napus* (Garcia et al. 1993; Chen and Kao 1995). Exogenous application in case of wheat and barley genotypes increased the foliar proline and imparted frost resistance (Petcu et al. 1998). In chickpea plants treated with proline at reproductive stage, better pollen viability, pollen germination, pollen tube growth, better flower, and pod retention were observed (Kumar et al. 2011). Also, the level of other compatible solutes such as trehalose and sucrose was elevated. Proline when exogenously applied to tobacco culture cells resulted in decreased lipid peroxidation but increased SOD and catalase activities (Islam et al. 2009). Jonytiene et al. (2012) studied the effect of exogenously applied proline to in vitro cultured *B. napus* (rapeseed) shoots. Rapeseed shoots were first cold acclimated at 4 °C for 14 days and then de-acclimated at 16/18 °C for 1, 3, 5, and 7 days. After 7 days of de-acclimation, the shoots growing on medium supplied with exogenous proline showed elevated endogenous proline levels, and the electrolyte leakage content had decreased by 4.2–17.88 %, thus rendering the shoots to be more cold tolerant.

2.1.5 Proline Toxicity

Proline may not be compatible for cells at its supraoptimal levels as has been indicated by some studies where exogenous proline application led to toxicity Exogenously applied proline also resulted in inhibited root and shoot growth of rice seedlings and *Brassica napus* (Garcia et al. 1993; Chen and Kao 1995). Proline was observed to be toxic under non-stressed conditions, but the toxicity has been found to have reduced in case of glucose and salt supply. Thus, proline degradation and not its accumulation have been proposed to be the cause of toxicity. Degradation of proline yields two products—P5C and Glc (glucose); the former, i.e., P5C, has been proved to be the toxic one by the detailed studies of genes involved in proline toxicity and proline signal transduction (Hellmann et al. 2000). Deuschle et al. (2004) reported that exogenous application of both proline and P5C has been found to be toxic to the plants as it led to reduced growth, activation of various stress-responsive genes, and increased levels of ROS, thus inflicting oxidative injury. Also, they found that plants overexpressing P5CDH were tolerant to proline toxicity, whereas P5CDH knockout mutants were highly sensitive and showed much higher mortality rates.

2.1.6 Transgenic Approach

One of the alternative ways to prove the role of proline in cold tolerance involves incorporating or overexpressing genes in low proline producers. Consequently, several reports have appeared on this aspect where many plants, viz., *Arabidopsis*, tobacco, rice, rapeseed, and tomato, have been genetically engineered with proline-enhancing genes taken from other sources such as maize, rice, lettuce, and ice plant leading to acquisition of cold tolerance in the transformed plants (details in Table 9.1), e.g., Zhu et al. (2012) raised transgenic *Arabidopsis* lines overexpressing *Imtl*, myoinositol-*O*-methyltransferase gene, derived from ice plant (*Mesembryanthemum crystallinum*). Transformed plants showed increased accumulation of compatible solutes such as proline and soluble sugars than wild type and resulted in increased cold tolerance as evident from enhanced MDA content and decreased levels of electrolyte leakage. Pan et al. (2012) transformed *Arabidopsis* with *ZmMPK* gene from maize, and the transgenics thus obtained showed higher levels of proline and other compatible solutes such as soluble sugars. The plants were cold tolerant and showed better germination rate under low temperature conditions. Since accumulation of any molecule can be brought about either by upregulating its synthesis or by downregulating its degradation, thus to enhance proline accumulation, some antisense transgenics may also be produced which accumulate higher levels of proline than wild types, e.g., Nanjo et al. (1999) generated antisense *Arabidopsis* transgenic lines transformed with AtProDH cDNA encoding proline dehydrogenase (ProDH), the enzyme controlling proline degradation. Anti-ProDH plants thus obtained were cold and salinity tolerant. These examples strongly suggest a role of proline in conferring cold tolerance. Future studies need to involve proline mutants in cold stress-related studies as well to find out the precise functions of this cryoprotectant.

2.2 *Glycine Betaine*

Betaines are quaternary ammonium compounds, and in plants, glycine betaine *N*,*N*,*N*-trimethylglycine (GB) form is the most common one besides proline betaine, alanine betaine, etc. (McNeil et al. 2000). Glycine betaine (GB) is an important osmolyte, and various naturally betaine accumulating plants have been found to accumulate its higher levels when subjected to salt, drought, and cold conditions (Ashraf and Foolad 2007; Chen and Murata 2011; Giri 2011). Besides plants, the osmoprotective functions of glycine betaine have been proved in lower organisms as well, e.g., in *Bacillus subtilis* (Hoffmann and Bremer 2011) and *Pseudomonas syringae* (Li et al. 2013).

2.2.1 Structure and Biosynthesis

Being a quaternary ammonium compound, all the four H atoms bound to N atom are replaced by methyl ($-CH_3$) groups, i.e., nitrogen atom is fully methylated (Fig. 9.4). Glycine betaine is thus *N*,*N*,*N*-trimethylglycine ($C_5H_{11}NO_2$).

Table 9.1 Various proline overproducing transgenics and their stress responses along with respective references

Transgenic plant	Gene	Source	Stress response	References
Arabidopsis thaliana	*AtProDH*	*Arabidopsis thaliana*	Altered levels of proline dehydrogenase Cold and salt tolerance	Nanjo et al. (1999)
Nicotiana tabacum	*Atp5cs*	*Arabidopsis thaliana*	Enhanced proline levels and hence increased freezing tolerance	Konstantinova et al. (2002)
Malus sp.	*Osmyb4*	*Oryza sativa*	Increased cold tolerance Increased proline and soluble sugar levels	Mattana et al. (2005)
Oryza sativa	*ICE1*	*Arabidopsis thaliana*	Lowered mortality Increased proline levels Increased cold tolerance	Xiang et al. (2008)
Nicotiana tabacum *Arabidopsis thaliana*	*OsSPX1*	*Oryza sativa*	Cold stress tolerant Better survival	Zhao et al. (2009)
Arabidopsis thaliana	*Osmyb4*	*Oryza sativa*	Increased cold tolerance Increased proline and soluble sugar levels	Mattana et al. (2005)
Oryza sativa	*OsMYB3R-2*	*Oryza sativa*	Increased cold tolerance	Ma et al. (2009)
Cucumis sativus	*ICE1*	*Arabidopsis thaliana*	Enhanced cold tolerance Accumulation of proline, soluble sugars, and low MDA content	Liu et al. (2010)
Oryza sativa	*OVP1*	*Oryza sativa*	Increased cold tolerance Increased proline levels	Zhang et al. (2011)
Citrus sinensis	*TERF1*	*Nicotiana tabacum*	Cold, drought and Citrus canker resistance	Qin et al. (2011)
Lycopersicon sp.	*AtCBF1*	*Arabidopsis thaliana*	Increased cold tolerance	Singh et al. (2011)
Arabidopsis thaliana	*ZmMKK4*	*Zea mays*	Increased cold tolerance Higher proline and soluble sugars content Increased germination	Kong et al. (2011)
Nicotiana tabacum	*CBF3*	*Capsicum annuum*	Chilling tolerant Higher proline and soluble sugar contents	Yang et al. (2011)
Oryza sativa japonica	*OsTPS1*	*Oryza sativa indica*	Cold stress mitigation Increased levels of proline and trehalose	Li et al. (2011)
Oryza sativa	*Osmyb4*	*Oryza sativa*	Increased cold tolerance Higher proline and soluble sugars content	Yang et al. (2012)

(continued)

Table 9.1 (continued)

Transgenic plant	Gene	Source	Stress response	References
Brassica napus	*Osmyb4*	*Oryza sativa*	Increased cold tolerance Higher proline and soluble sugars content	Gomaa et al. (2012)
Nicotiana tabacum	*ZmMPK17*	*Zea mays*	Increased cold tolerance Higher proline and soluble sugars content Increased germination	Pan et al. (2012)
Arabidopsis thaliana	*Imt1*	*Mesembryanthemum crystallinum*	Elevated cold tolerance Higher proline and sugars content	Zhu et al. (2012)

mtlD, mannitol-1-phosphate dehydrogenase gene, *ZmMAPK*, *Zea mays* MAPK gene, *OsMYB 4*, *Oryza sativa* myeloblasts transcription factor gene 4, *LsICE 1*, *Lactuca sativa* inducer of CBF (C-repeat binding factor) expression gene 1, *AtCOR15*, *Arabidopsis thaliana* cold-responsive gene, *ZmMKK*, *Zea mays* mitogen-activated protein kinase kinase gene, *TERF1*, tomato ethylene-responsive factor gene 1, *OVP1*, vacuolar proton translocating inorganic pyrophosphate (VPPase) gene 1, *OsSPX1*, *Oryza sativa* SPX domain (SYG1 (suppressor of yeast gpal)/Pho81 (CDK inhibitor in yeast PHO pathway)/XPR1 (xenotropic and polytropic retrovirus receptor)) protein gene, *GmDREB3*, *Glycine max* drought-responsive element binding protein gene 3, *AtProDH*, *Arabidopsis thaliana* proline dehydrogenase, *P5CS*, pyrroline-5-carboxylate synthetase gene

Fig. 9.4 Structure of betaine

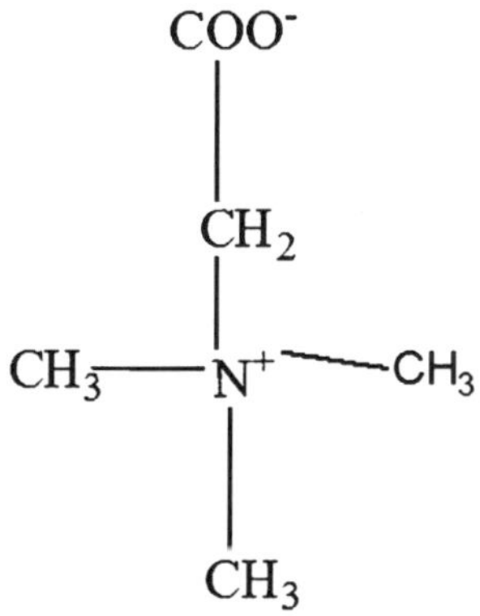

Betaine

In most organisms, it is synthesized from either choline or glycine via two distinct pathways: (a) dehydrogenation of choline and (b) N-methylation of glycine. In higher plants, the synthesis occurs mainly in the chloroplast (Hanson et al. 1985) by two-step oxidation/reduction reaction (Fig. 9.5). Choline is first oxidized by enzyme choline monooxygenase (CMO) into unstable intermediate betaine aldehyde and is later oxidized in the presence of NAD^+-dependent betaine aldehyde dehydrogenase (BADH) into betaine (Hanson et al. 1985; Takabe et al. 2006). In *E. coli*, the biosynthesis differs only in the first step where oxidation is being carried out by CDH (choline dehydrogenase) encoded by *bet A* gene, while the second

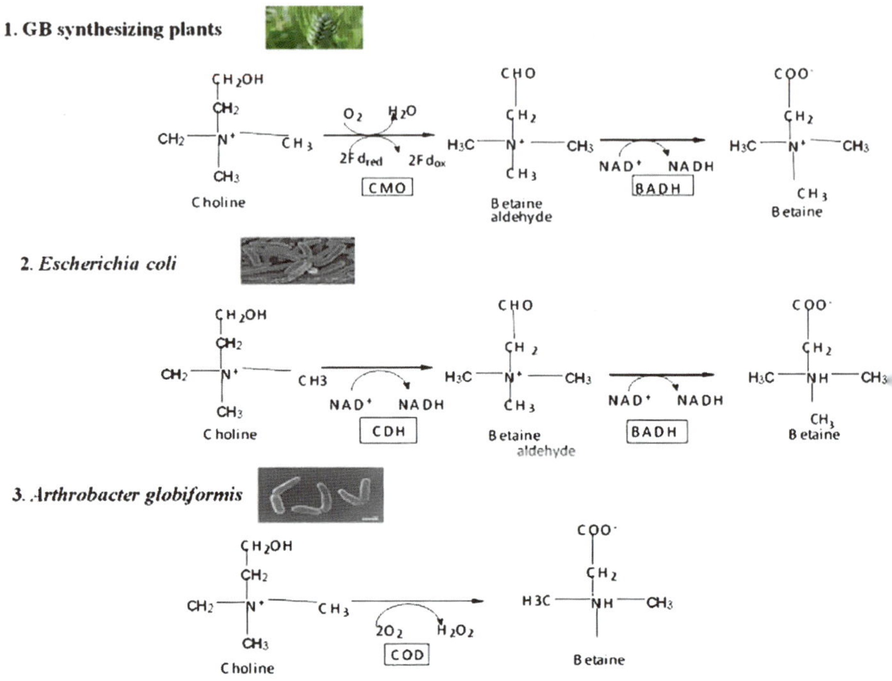

Fig. 9.5 Various pathways leading to the formation of betaines. *BADH* betaine aldehyde dehydrogenase, *CDH* choline dehydrogenase, *CMO* choline monooxygenase, *COD* choline oxidase

step is similar to that of higher plants. In *Arthrobacter globiformis*, a single-step reaction carried out by COD (choline oxidase) encoded by *Cod A* changes choline into betaine. Thus, the pathways for betaines formation differ in the terms of number and types of enzymes involved in various living systems.

2.2.2 Functions in the Plants

Glycine betaine (GB) has diverse roles in cold-stressed plants, which have been proven evidently using exogenous application of GB or transgenics producing this molecule in higher quantity. The cryoprotective and osmoprotective functions of GB (summarized in Fig. 9.6) have been proved by extensive studies on some of the important GB-accumulating higher plants such as maize, spinach, and barley (Rhodes and Hanson 1993; Chen and Murata 2008). Allard et al. (1998) examined the GB content of two different cultivars of wheat (*Triticum aestivum* L. cv. Glenlea and cv. Fredrik) under unstressed and stressed conditions. Unstressed wheat leaves had accumulated 6.5 μmol g^{-1} FW and 8.56 μmol g^{-1} FW GB in Glenlea and Fredrik, respectively, but when subjected to cold treatment (6 °C/2 °C), the GB levels increased significantly to 15.3 μmol g^{-1} FW and 21.3 μmol g^{-1}, respectively, in both

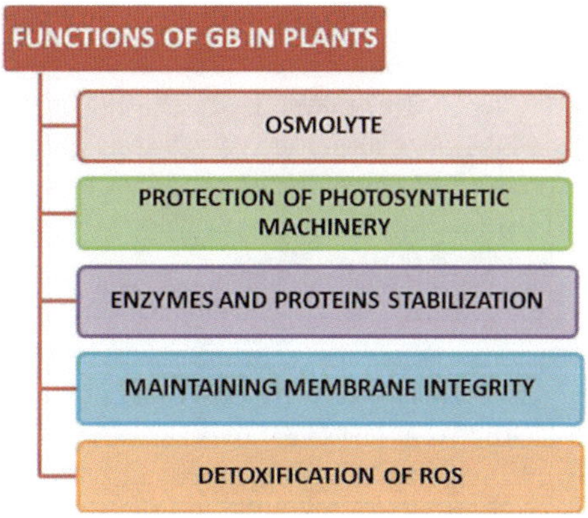

Fig. 9.6 Various functions of glycine betaine in plants growing under cold-stressed conditions

the cultivars. Quan et al. (2004) transformed maize plants with *betA*, and the transformed plants thus obtained were cold tolerant and reported 5.7 μmol g^{-1} FW GB in leaves, whereas wild-type plants had accumulated only 1.6 μmol g^{-1} FW GB, showing a strong positive correlation between GB accumulation and cold tolerance. It has been observed in some studies that the levels of GB accumulated under stress or in GB-overproducing transgenics are not sufficient enough for protecting the plants by acting as an osmoprotectants only (Holmstrom et al. 2000; Park et al. 2004); thus, many other functions are also attributed to GB.

GB has been proved to be crucial for preventing the photoinhibition-induced inactivation and disassociation of various important components of photosynthetic machinery (Holmstrom et al. 2000; Sakamoto and Murata 2000; Chen and Murata 2011; Giri 2011). Hayashi et al. (1997) evaluated the efficacy of GB in mitigating chilling-induced photoinhibition of PSII in *codA* transformed *Arabidopsis* plants and found that photoinhibition in wild plants was as high as 80 %, whereas for transformed lines it was only 45 %. The mechanism of protection is attributed most probably to the stabilization of Mn (manganese) cluster (Mohanty et al. 1993) and the protection of various components of PSII reaction center (D1/D2/Cytb559 complex), thus maintaining the oxygen-evolving machinery (Papageorgiou and Murata 1995; Allakhverdiev et al. 1996). Additionally, GB is also proposed to preserve the de novo protein synthesis required for repairing photodamaged PSII which is otherwise suppressed by various ROS. In various transgenics such as COD-transgenic rice (Sakamoto et al. 1998), *Arabidopsis* (Sakamoto et al. 2000), and tomato (Park et al. 2004), GB gets localized inside chloroplast and results in enhanced cold tolerance, thereby clearly indicating a positive correlation between GB and PSII activity.

Also, GB has the ability to stabilize quaternary structure of proteins and enzymes against various abiotic stresses (Sakamoto and Murata 2000, 2002; Chen and Murata 2011). The protective effect of GB is credited to the presence of both hydrophilic and hydrophobic groups in it which facilitates its interaction with various enzymes and proteins (Sakamoto and Murata 2002). GB molecules have been proposed to act either by preferentially getting replaced from the surface of charged proteins so as to maintain the hydration envelope around proteins or its hydrophobic part may interact with hydrophobic parts of protein, thus exposing the hydrophobic domains of proteins to water and preventing dehydration-induced denaturation (Papageorgiou and Murata 1995). GB protects numerous indispensable enzymes such as RUBISCO by acting as molecular chaperones, thus maintaining the structure and form and if required can even bring about refolding to original conformation (Bourot et al. 2000). Diamant et al. (2001) have also proposed the chaperone-regulated protein disaggregation by glycine betaines.

Moreover, GB has been proved to be quite effective in maintaining the structural integrity of various biomembranes at low temperatures (Gorham 1995). GB maintains the structural and functional properties of various organellar membranes such as chloroplast and mitochondria, thereby contributing towards the proper functioning of various essential processes such as ETC inside these organelles (Hamilton and Heckathorn 2001). Chilling-induced lipid peroxidation is also prevented by GB supply, thus protecting the various membranes from oxidative damage (Chen et al. 2000). Similar protective effects of GB were also reported in case of CDH-overexpressing maize transgenics as well (Quan et al. 2004). GB has also been suggested to directly check the ROS-induced K^+ ions efflux by maintaining the membrane integrity or by blocking the K^+ ion channel (Chen and Murata 2011). As chilling disrupts the normal functioning and integrity of various membranes, brings about the denaturation of various enzymes, and interrupts the electron transport system in chloroplast and mitochondria, all these lead to the production of ROS. GB, when available in the plant system, is found to bring about the mitigation of chilling-induced oxidative stress by activating the action of various antioxidative enzymes (Giri 2011). Einset et al. (2007a) have reported the upregulation of several genes encoding for membrane trafficking *RabA4c* GTPase, root-specific NADPH-dependent ferric reductase (FRO_2) (Fig. 9.7) in GB-treated *Arabidopsis* plants. *RabA4c* thus activated further brings about ROS detoxification, thus corroborating the proposed interrelationship between GB accumulation and ROS mitigation. The levels of H_2O_2 and catalase were recorded to go up in chilling-stressed tomato plants when exogenously supplied with GB (Park et al. 2004) resulting in upregulation of H_2O_2-mediated ROS detoxification involving catalase. GB was found effective against methylviologen (MV)-induced ROS generation as well. Alia et al. (1999) studied the activities of various antioxidative enzymes in *codA* *Arabidopsis* transgenics and observed higher CAT and APX activities in transformed plants.

2.2.3 Endogenous GB Levels in Cold-Stressed Plants

The endogenous levels of GB have been proved to go up under low and chilling temperatures (Wang et al. 2010; Chen and Murata 2011; Giri 2011). Accumulation of betaines has been confirmed in the leaves of winter-type barley grown at 5 °C

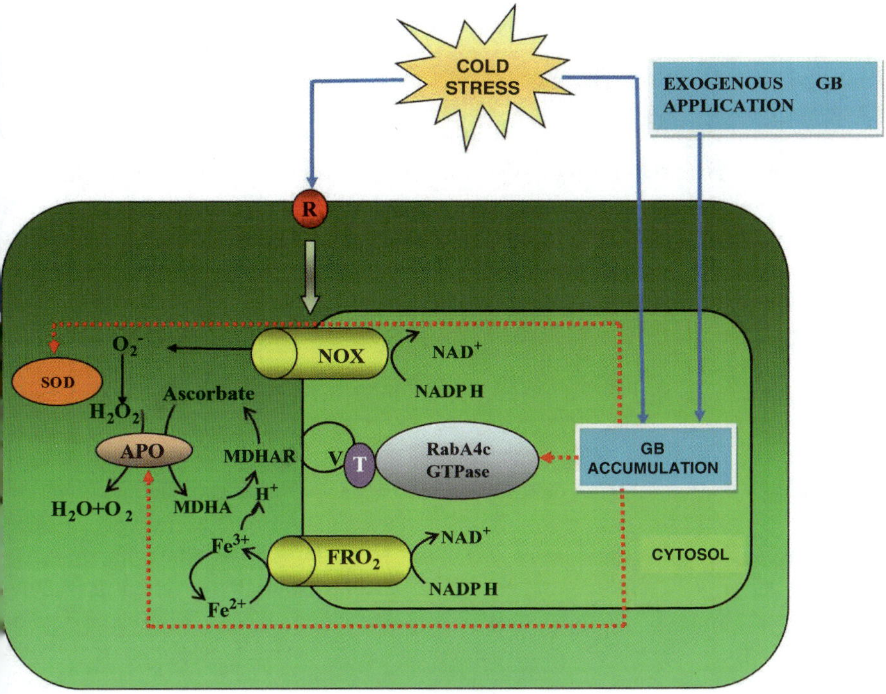

Fig. 9.7 Mechanism of action of various GB-upregulated genes under stress (Einset and Connolly 2009). *R* putative receptor, *PM* plasma membrane, *NOX* NADPH oxidase activity, *APO* ascorbate oxidase, *SOD* superoxide dismutase, *MDHAR* monodehydroascorbate reductase, *MDHAR* monodehydroascorbate, *T* tethering protein, *V* vesicle, FRO_2 ferric reductase

The results thus obtained indicated significant correlation between higher levels of betaines and ability to overcome harmful effects of freezing in winter (Kishitani et al. 1994). Rajashekar et al. (1999) when subjected strawberry (*Fragaria* × *ananassa* Duch.) to 4-week cold treatment reported about twofold increase in GB levels, i.e., from 1.8 to 3.7 μmol g^{-1} FW, and cold tolerance increased drastically from −5.5 to −17 °C. Similarly, GB has been recorded to accumulate under cold temperatures in carrot (Fallon and Phillips 1989), spinach (McCue and Hanson 1990), Puma rye (Koster and Lynch 1992), sorghum (Wood et al. 1996), and wheat (Naidu et al. 1991; Allard et al. 1998). Various important plants such as rice, tobacco, potato, and *Arabidopsis* do not produce glycine betaine, whereas crops like wheat, maize, and barley are not able to accumulate significant amounts under various abiotic stresses; thus in such cases the growth and yield of stressed plants can be subsequently enhanced by exogenous application as well as by genetically engineering GB-overexpressing plants (Nayyar et al. 2005; Chen and Murata 2008; Wang et al. 2010).

2.2.4 Exogenous Application for Cold Tolerance

Exogenous application of GB has been found to be effective in mitigating harmful effects of various abiotic stresses, thereby conferring a positive effect on the growth and yield of the particular plant (Chen et al. 2000; Yang and Lu 2005; Chen and Murata 2008). Experiments with wheat seedlings (Allard et al. 1998) demonstrated that cold-acclimated wheat plants when supplied with 250 mM exogenous GB application showed additive role in improving freezing tolerance of the plants. Under cold stress, Nayyar et al. (2005) reported the protective effect on reproductive stage of chickpea. It not only increased the pollen viability, pollen tube growth, stigma receptivity, and ovule viability at bud stage but also increased the plant biomass and yield. Similar cryoprotective effects of exogenously applied GB were also confirmed when applied to *Medicago* seedlings (Zhao et al. 1992), potato (Somersalo et al. 1996), strawberry (Rajashekar et al. 1999), maize (Chen et al. 2000; Farooq et al. 2008), *Arabidopsis* (Xing and Rajashekar 2001), and tomato (Park et al. 2004). Exogenously applied GB can be taken up by leaves as well as roots from where it is transported to other organs resulting in increased growth and yield by increasing abiotic stress tolerance (Mäkelä et al. 1998; Chen and Murata 2008). In particular, reproductive organs, i.e., flowers and fruits, accumulate higher levels of GB under stressed conditions (Chen and Murata 2008). Exogenously applied glycine betaine improves cold resistance in various plants, e.g., 2 mM GB application to cold-stressed *N. tabacum* roots increased photosynthetic efficiency by protecting chloroplast at ultrastructural level (Wang et al. 2008a). When supplied with exogenous glycine betaine, cold-stressed cucumber plants showed better survival, enhanced photosynthetic efficiency, and reduced MDA content and ROS (Li et al. 2004).

2.2.5 GB Toxicity

The prolonged exposure of plants to high GB levels has been reported to be toxic to the plants, and the levels above 100 mM have been found to inflict leaf burns and leaf tip necrosis in tomato and wheat (Allard et al. 1998; Mäkelä et al. 1998). GB may also act as suitable substrate for the growth of some fungi such as *Fusarium graminearum* (Rhodes and Hanson 1993); thus, many factors have to be taken care of during its exogenous application. Gibon et al. (1997) also reported the negative effects of GB supply on higher non-accumulating plants such as *Brassica napus*. Since betaine aldehydes are generally toxic to plants, plants overexpressing BADH enzyme have been developed, e.g., BADH-overexpressing tobacco lines developed by Rathinasabapathi et al. (1994) that efficiently metabolize betaine aldehydes, but generally, the transgenics have slow growth than the control plants. However, unlike transgenics overexpressing other compatible solutes that show various phenotypic variations such as deformed leaf size and shape in trehalose-overexpressing plants, GB-overproducing transgenics have not yet been reported to show any such phenotypic alterations (Giri 2011).

2.2.6 GB Action Under Cold Stress

The plant upon experiencing cold stress activates plasma membrane NADPH oxidase activity (NOX) (Fig. 9.7) leading to the production of superoxide radicals in the cell wall, thereby initiating a series of ROS signaling (Einset et al. 2007b). Under these conditions, GB, if present, starts playing its indirect antioxidative role by upregulating the activity of antioxidative enzymes like ascorbate oxidase and monodehydroascorbate reductase (MDHAR), which then carry out their respective antioxidative functions. Additionally, enzyme RabA4c GTPase (enzyme bound to plasma membrane by a tethering protein and involved in membrane trafficking) is also upregulated which facilitates the movement of ascorbate oxidase and MDHAR from cytosol to cell wall, thus increasing the availability of these enzymes. Some studies have proposed the role of various ferric reductases (FRO_2) during the mitigation of chilling-induced ROS by GB (Robinson et al. 1999; Jeong et al. 2008). The activity of plasma ferric reductase ($PMFRO_2$) is also upregulated under cold stress which brings about the reduction of ferric ions (Fe^{3+}) to ferrous ions (Fe^{2+}) and generates reducing potential (H^+). During the entire process of ferrous ion-mediated oxidative stress mitigation, concentrations of ferrous ions are kept low to accentuate the ROS production.

2.2.7 Transgenic Approach

Many economically important crops such as rice, potato, and tomato do not accumulate GB; thus, transgenics accumulating GB in amounts higher than their corresponding wild relatives have been developed for various cereals, vegetables, fruits, and other crops of agricultural and horticultural importance (McCue and Hanson 1990) from different sources ranging from microorganisms such as *E. coli* and *Arthrobacter* to higher plants such as rice and spinach (listed in Table 9.2). Out of various genes transferred, BADH (coding for BADH enzyme that catalyzes the production of betaines from their respective aldehydes in higher plants), *cod A* (coding for COD that catalyzes production of betaines directly from choline in *Arthrobacter globiformis*), and *Osmyb4* (*Oryza sativa* myeloblast transcription factor gene 4) were the most important ones. Pioneers in this field were Hayashi et al. (1997) who developed the first ever GB-overproducing transgenic *Arabidopsis* transformed with *codA* gene from *Arthrobacter globiformis*. The transgenic plants thus obtained were reported to accumulate approximately 1 μg GB per gram fresh weight in the chloroplasts. The transgenics thus obtained clearly outperformed the wild ones under various stresses in terms of higher biomass, better germination, and survival of transgenic seeds. Similarly, Park et al. (2004) transformed tomato plants with *codA* gene from *Arthrobacter globiformis* which were quite tolerant to cold than wild ones. Increased levels of GB were accumulated in leaves and reproductive parts of tomato plants reaching up to 0.3–1.2 μg GB per gram fresh weight of the plant. Transgenic lines thus obtained exhibited better seed germination, plant survival, and enhanced yield (about 10–30 % increase) than their wild counterparts.

Table 9.2 Glycine betaine-overexpressing transgenic plants and their stress responses along with respective references

Transgenic plant	Gene transferred	Source	Stress response	References
Arabidopsis	*Cod A*	*Arthrobacter globiformis*	Freezing tolerance Cold and salt resistant Protective effect on PSII	Alia et al. (1991)
Arabidopsis	*Cod A*	*Arthrobacter globiformis*	Freezing tolerance Cold and salt resistant Protective effect on PSII	Hayashi et al. (1997)
Oryza sativa	*Cod A*	*Arthrobacter globiformis*	Cold and salt resistant Protective effect on PS II	Sakamoto et al. (1998)
Arabidopsis	*Cod A*	*Arthrobacter globiformis*	Freezing tolerance Cold and salt resistant Protective effect on PSII	Sakamoto et al. (2000)
Oryza sativa	*Cod A*	*Arthrobacter globiformis*	Freezing tolerance	Konstantinova et al. (2002)
Lycopersicon esculentum	*Cod A*	*Arthrobacter globiformis*	Chilling stress resistant Detoxification of ROS	Park et al. (2004)
Zea mays	*Cod A*	*E. coli*	Cold and drought tolerance Protection of photosynthetic machinery and detoxification of ROS	Quan et al. (2004)
Arabidopsis	*Osmyb4*	*Oryza sativa*	Accumulation of glycine betaine, proline, glucose, fructose, and sucrose	Mattana et al. (2005)
Oryza sativa	*Cod A*	*Spinacia oleracea*	Cold, salt, and drought tolerance Detoxification of ROS	Shirasawa et al. (2006)
Arabidopsis thaliana	*RabA4c*	*Arabidopsis thaliana*	Increased scavenging of ROS Increased chilling stress tolerance	Einset et al. (2007a)
Apple (*Malus pumila* Mill.)	*Osmyb4*	*Oryza sativa*	Cold and drought tolerant Accumulation of glycine betaine, proline, and soluble sugars	Pasquali et al. (2008)
Oryza sativa	*Cod A*	*Spinacia oleracea*	Cold, salt, and drought tolerance Detoxification of ROS Membrane protection	Kathuria et al. (2009)
B. campestris L. spp. *chinensis*	*Cod A*	*Arthrobacter globiformis*	Increased levels of glycine betaine Protection of photosynthetic machinery	Wang et al. (2010)
Triticum aestivum	*BADH*	*Atriplex hortensis*	Glycine betaine overproduction Increased antioxidant enzymes Increased compatible solutes	Zhang et al. (2010)
Ipomoea batatas	*SoBADH*	*Spinacia oleracea*	GB accumulation ROS detoxification Cold, salt, and oxidative stress tolerance	Fan et al. (2012)

BADH, betaine aldehyde dehydrogenase gene, *cod A*, choline oxygenase gene, *OsMYB 4*, *Oryza sativa* myeloblast transcription factor gene 4, *RabA4c*, *Arabidopsis thaliana* Rab GTPase

2.3 *Sugars*

Sugars are multifunctional biomolecules that play a crucial role as substrates for various indispensable energy-providing reactions or may be stored as reserve food material (Ramon et al. 2008). Levels of various sugars have been widely demonstrated to go up not only when the plant experiences cold stress (Burbulis et al. 2012) but also under natural conditions when winters commence (Janská et al. 2010) and the levels go down when favorable season approaches (Yuanyuan et al. 2010). Acclimation to cold occurs when the plants experience brief and low intensity of cold stress and sugars play an important role during this process. Among all, the most important sugar is sucrose (Guy et al. 1980), but in addition to sucrose, other sugars also accumulate, e.g., glucose, fructose, fructans, raffinose, and stachyose.

2.3.1 Structure, Biosynthesis, and Lysis

Sucrose is a disaccharide composed of glucose and fructose subunits as shown in Fig. 9.8. It may be represented as *O*-β-D-fructofuranosyl-(2, 1)-α-D-glucopyranoside. It is a nonreducing sugar as it contains no free anomeric carbon atom and can be easily hydrolyzed back into its constituents.

Sucrose biosynthesis in higher plants may be catalyzed by either sucrose or synthase or pathway involving sucrose phosphate synthase (SPS) and sucrose phosphatase (Fig. 9.8). Sucrose phosphate first acts on UDP-glucose and fructose-6-phosphate (F6P) resulting in the removal of UDP and formation of sucrose-6-phosphate. Sucrose phosphatase then comes into play and removes phosphate from sucrose-6-phosphate, thus finally forming sucrose (Hawker 1977). Sucrose thus formed may be hydrolyzed irreversibly by invertase into glucose and fructose or hydrolyzed reversibly by sucrose synthase into UDP-glucose or ADP-glucose and fructose (Zrenner et al. 1995).

2.3.2 Functions in the Plants

Sugar metabolism is profoundly affected by cold temperatures (Perras and Sarhan 1984), and endogenous levels of sugars have been reported to have elevated under cold stress in members of Compositae (Frehner et al. 1984) and Gramineae families (Volenec and Nelson 1984). Cold tolerance has been found to be directly correlated with the sugar levels in many cases (Magel et al. 2001; Sasaki et al. 2001; Gudleifsson 2010). Tissue sugar content is considerably affected by seasonal variations especially in temperate perennials in which severalfold increase in free sugar content has been observed during autumn and winter (Yelenosky and Guy 1977). Under low temperature conditions, the active growth is usually slowed down or suspended resulting in decreased metabolic utilization processes due to which need for the

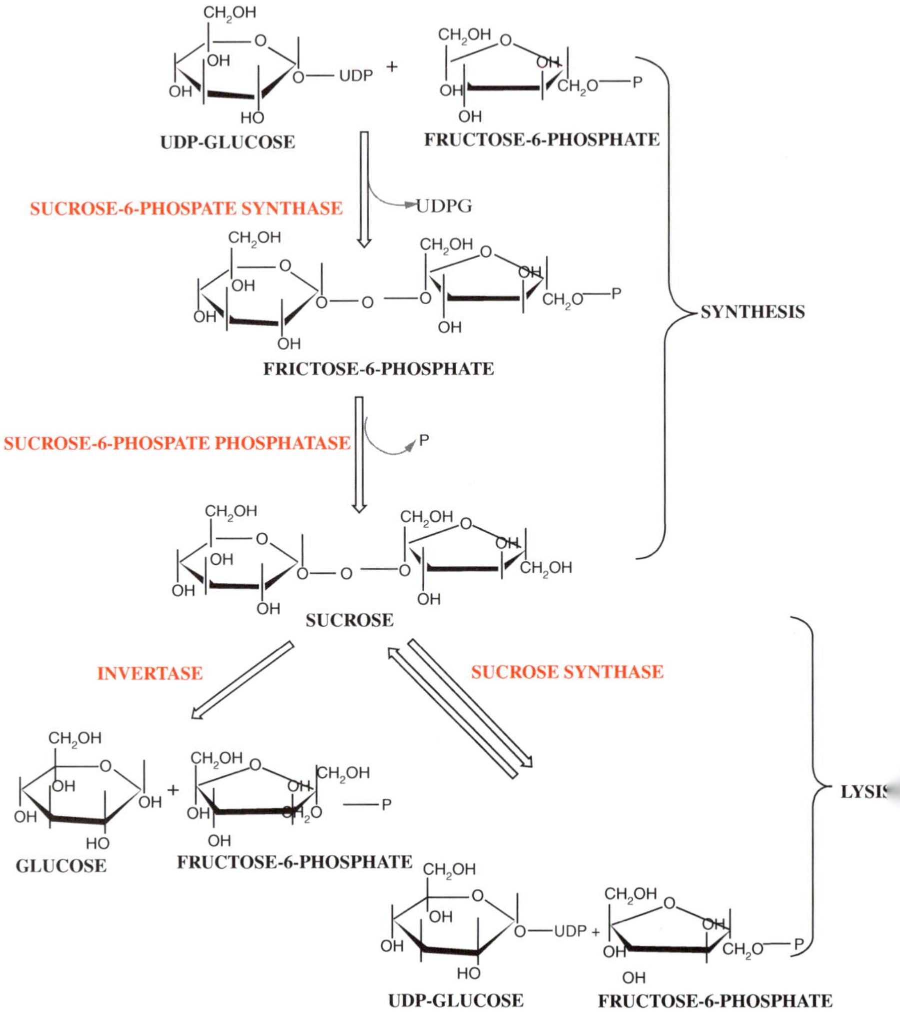

Fig. 9.8 Biosynthesis and lysis pathway of sucrose

products of photosynthesis is much lower than its production. Usually, the surplus photosynthates are stored in the form of starch but, surprisingly during low temperature exposure, decrease in starch content, but increase in free sugars was observed. Low temperature also leads to increased levels of enzymes like SPS, SS (sucrose synthase), INV (invertase), and RUBISCO (ribulose-1,5-bisphosphate carboxylase or oxygenase) (Rapacz 2002) aimed at increasing the levels of sucrose (Ramon et al. 2008) as sucrose is the main source of energy for various biochemical processes. Thus, sugars have also been found to be playing a nutritive role under low temperature conditions and also facilitate revival from freezing stress (Eagles et al. 1993).

Decreasing temperature results in the loss of water in terms of formation of ice crystals in apoplast which grow at the cost of cytoplasmic water. Thus, to avoid water stress, plants accumulate various compatible solutes called cryoprotectants such as proline, glycine betaine, and sugars (Burbulis et al. 2012). The role of sugars as an osmolyte and as a cryoprotectant is well supported by various studies carried out worldwide (Wanner and Junttila 1999; Burbulis et al. 2012). These not only increase the osmotic potential of the cells but also bring about the depression in freezing point (Ouellet 2007). Sugars play an important role in the stabilization of structure and conformations of various enzymes and proteins, thereby preventing denaturation and hence maintaining their function. Sugars can be replaced by water molecules and prevent collapsing of cell due to dehydration by filling intermembrane spaces (Crowe et al. 1988). Also, under low temperatures, sugars do not get crystallized rather get vitrified into its glass-like form which prevents the coagulation of already denatured proteins (Crowe et al. 1988; Anchordoguy et al. 1987).

Koster and Lynch (1992) in their studies on winter rye leaves found that cold-induced accumulation of sugars occurs not in the vacuole but in the cytoplasm, thereby protecting various membrane systems. Uemura and Steponkus (2003) reported that high levels of exogenous sucrose concentrations (up to 400 mM) were capable of protecting membranes from various chilling-induced deformities. Zhang et al. (2011) found *Medicago falcata* to be freezing tolerant than *Medicago truncatula* due to higher accumulation of osmolytes such as soluble sugars and proline and the enhanced levels of sugars were effective in protecting membranes from chilling injury. The protective properties of sugars have been attributed to their ability to interact with lipid bilayer (Shalaev and Steponkus 2001; Shao et al. 2006), thus maintaining their fluidity even at low temperatures. Sugars have also been proposed and proved to be the important antioxidants in various plant systems by numerous workers world over (Stoyanova et al. 2011; Hernandez-Marin and Martínez 2012; Keunen et al. 2013; Peshev et al. 2013). Sugars not only protect the plants from cold-induced oxidative stress indirectly by triggering the protective mechanism of other antioxidants but may also directly act as ROS scavengers (Peshev and Van den Ende 2013). Alternatively, sugars also have been shown to facilitate regeneration of various antioxidants, e.g., ascorbate biosynthesis and regeneration in harvested broccoli florets (Nishikawa et al. 2005). Besides being the primary source of energy and their function as osmolytes, sugars also act as signaling molecules under cold stress (Wang et al. 2008b; Xiao et al. 2009), thus controlling various transcriptional, posttranscriptional, and posttranslational processes (Rolland et al. 2006; Muller et al. 2011). Therefore, sugars play a pivotal role in controlling and coordinating various processes related to plant metabolism, growth, and development (Yuanyuan et al. 2010 as summarized in Fig. 9.9).

9.3.3 Endogenous Levels in Cold-Stressed Plants

Low temperature conditions have been proved to trigger accumulation of total soluble sugar content in various plant species, e.g., cabbage (Sasaki et al. 1996), *Lolium*

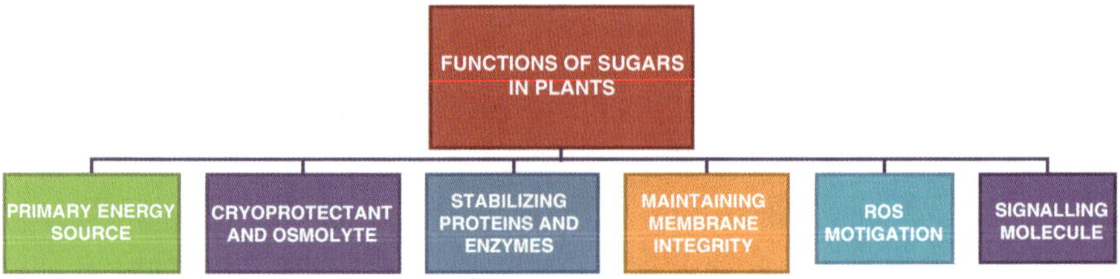

Fig. 9.9 Various functions of sugars in plants growing under cold-stressed conditions

(Bhowmik et al. 2006), alfalfa (Mo et al. 2011), *Arabidopsis thaliana* (Sicher 2011), cassava (An et al. 2012), wheat (Zeng et al. 2011; Janmohammadi et al. 2012), and *Brassica* sp. (Burbulis et al. 2012). Cold-tolerant chickpea genotypes are reported to have higher activities of enzymes involved in carbohydrate metabolism (Kaur et al. 2009; Kumar et al. 2011). Wang et al. (2007) observed increased accumulation of soluble sugar content in rice in relation to cold tolerance. Similar trend was observed in case of cold-treated *Arabidopsis* plants (Cook et al. 2004). Puma rye (*Secale cereale* L. cv. Puma) was cold acclimated and the total soluble sugar content was found to have nearly raised twofold mainly contributed by sucrose and raffinose (Koster and Lynch 1992). Soluble sugars and proline contents were also found to have increased significantly in cotton (*Gossypium hirsutum* L.) when subjected to chilling treatment of 15 and 5 °C (Azymi et al. 2012). Wheat (*Triticum aestivum*) plants subjected to cold shock treatment (2–4 °C) showed five- to sixfold increase in sucrose synthase activity of leaves and hence increased sucrose accumulation (Crespi et al. 1991). Burbulis et al. (2012) observed a positive correlation between cold acclimation and endogenous levels of soluble sugars and proline. Rapacz and Hura (2002) studied the cold acclimation and de-acclimation processes in rapeseed (*Brassica napus* L.) and found activities of RUBISCO (ribulose-1,5-bisphosphate carboxylase or oxygenase), sucrose synthase (SS), and SPS to go up during acclimation but decreased during de-acclimation. The observations recorded by Sasaki et al. (2001) in case of cabbage seedlings corroborated the above findings.

2.3.4 Exogenous Application to Increase Cold Tolerance

There have been various studies where sugars have been exogenously supplied in certain plant species to show the involvement of sugars in cold and freezing tolerance. There are many reports of cold tolerance induced by exogenously supplied sugars to the whole plants or cells grown in vitro in culture media, e.g., eucalyptus (Travert et al. 1997) and barley cells (Tabaei-Aghdaei et al. 2003). Some studies have indicated the differential response of whole plant and cell cultures with sugar feeding under warm and cold conditions. Exogenous sugar supply has been found to be effective in inducing freezing tolerance in cell cultures under both cold as well as warm conditions (Tabaei-Aghdaei et al. 2003), whereas whole plants were responsive under cold conditions only (Tabaei-Aghdaei et al. 2003). *Arabidopsis* seedlings when supplied with 10–35 and 30–400 mM sucrose concentrations conferred freezing tolerance to the seedlings and protected the membranes from chilling-induced deformations (Uemura and Steponkus 2003). In strawberry plants raised in vitro, Lukoseviciute et al. (2009) demonstrated the positive effects of exogenously applied sucrose, raffinose, and proline on cold acclimation. Sucrose and proline were found to improve the survival, whereas raffinose had a negative effect. Palonen and Junttila (1999) studied the effect of exogenously applied sucrose in raspberry (*Rubus idaeus* L.) in relation to cold tolerance. The levels of sucrose, glucose, and fructose were enhanced, and the cold hardiness acquired was positively correlated with sucrose levels. Rekarte-Cowie et al. (2008) proved that exogenously applied sucrose regulates the cold acclimation in *Arabidopsis* by inducing the overproduction cold-inducible gene expression resulting in increased freezing tolerance and the accumulation of soluble carbohydrates (Gilmour et al. 2000). Pretreatment of various plant cells with sugars has been found to be effective in increasing the survival after cryopreservation of various higher plant species, e.g., preincubation with glucose in cabbage leaf cells (Jitsuyama et al. 1997) and sucrose incubation for Jerusalem artichoke (Harris et al. 2004). Some researchers have pointed out the role of soluble sugars in the induction of senescence (Paul and Pellny 2003). *Arabidopsis* plants when supplied with 2 % glucose accompanied by low nitrogen supply showed leaf senescence (Pourtau et al. 2004; Wingler et al. 2004).

2.3.5 Sugar Signaling

Sugars not only act as energy sources and as osmolytes under stress but also act as primary messengers in signal transduction (Rolland et al. 2002) and interact with other signal pathways working inside a cell, thereby making sugar sensing and signaling a complex pathway. A typical plant cell has various receptors and transporters on cell wall and cell membrane for sensing sugars such as hexose transporters for glucose (Glu), fructose (Fru), and sucrose (Suc) (Rolland et al. 2002) (Fig. 9.10). Sugars thus transported are sensed through either HXK (hexokinase)-dependent pathway involving phosphorylation or HXK-independent pathway without phosphorylation (Rosa et al. 2009). Glucose is acted upon by glucokinases forming glucose-6-phosphate (G6P) leading to the formation of trehalose by a series of enzyme-catalyzed reactions, whereas sucrose (Suc) gets broken into fructose and glucose; each subunit further gets phosphorylated into F6P and G6P by the action of fructokinases and glucokinases, respectively. KIN 10/11 (*Arabidopsis* protein kinase, catalytic subunit of SnRK (sucrose-non-fermenting-1-related protein kinase)) is also involved in sugar sensing and signaling (Rosa et al. 2009). It is proposed to regulate the phosphorylation of various class II trehalose phosphate synthase (TPS) enzymes, thereby controlling trehalose synthesis. KIN is further regulated by sugar phosphatases such as G6P. Extracellular glucose, if present, may be influxed inside the cell by G-protein-coupled receptor RGS1 (regulator of G-protein) present on cell membrane and GPA1 (G-protein-α-subunit). The influxed

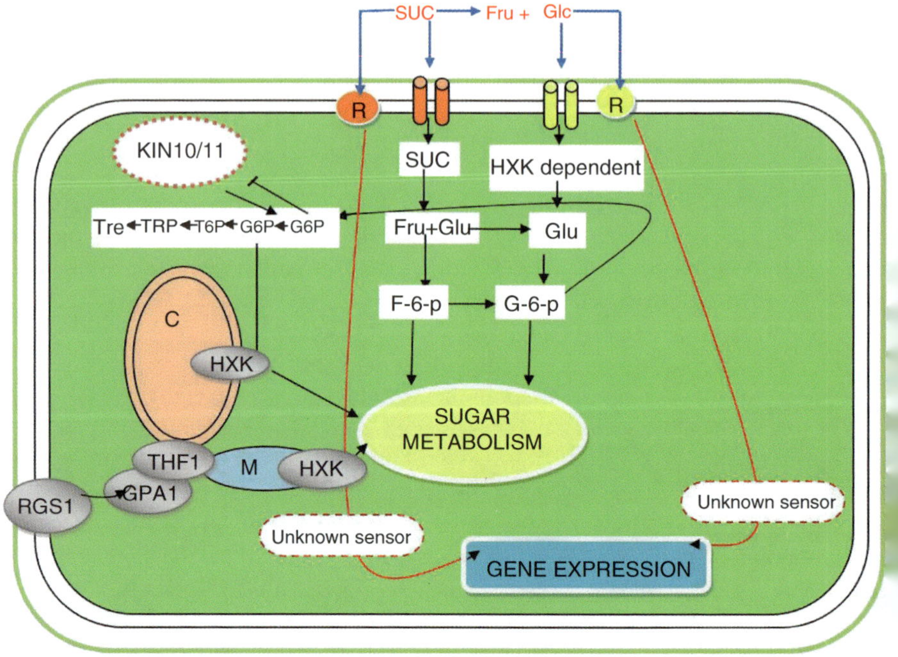

Fig. 9.10 Sugar sensing and signaling in plants (Rolland et al. 2002; Gupta and Kaur 2005; Ramon et al. 2008; Thakur et al. 2010). *R* receptor, *C* chloroplast, *M* mitochondria, *HXK* hexokinase, *Glu* glucose, *Suc* sucrose, *Fru* fructose, *G6P* glucose-6-phosphate, *F6P* fructose-6-phosphate, *RGS* regulator of G-protein, *THF* thylakoid formation protein, *KIN Arabidopsis* protein kinase, *GPA* regulator of G-protein

sugar then carries out its further signaling via THF (thylakoid formation) proteins present on the plastids (Ramon et al. 2008). Alternatively, glucose and sucrose may also bring about gene expression through some unknown sensor as well (Gupta and Kaur 2005).

2.3.6 Transgenic Approach

There are some examples where some plant species have been genetically manipulated to enhance the accumulation of endogenous sugar levels in order to improve their cold tolerance (listed in Table 9.3). Hurry et al. (2000) studied the behavior of transgenic *Arabidopsis* plants having inhibited sugar-hydrolyzing enzymes like CBFPase (cytosolic fructose-1,6-bisphosphate) and SPS or overexpressing maize SPS; all this resulted in increasing the sugar levels in the plant and was positively correlated to cold tolerance. *Eskimo1* mutants and *sfr* (sensitive to freezing) mutants of *Arabidopsis* also supported this positive correlation. The *sfr4* mutants having sugar levels lower than wild type were quite sensitive to freezing (McKown

Table 9.3 Sugar-overexpressing transgenic plants and their stress responses along with respective references

Transgenic plant	Gene transferred	Source	Stress response	References
Arabidopsis thaliana	*Osmyb4*	*Oryza sativa*	Increased levels of glucose, fructose, sucrose, proline, and glycine betaine Enhanced cold tolerance	Mattana et al. (2005)
Arabidopsis thaliana	*TaSnRK2.4*	*Triticum aestivum*	Freezing, salt, and drought tolerance	Mao et al. (2005)
Nicotiana tabacum	*1-SST*	*Lactuca sativa*	Freezing resistance	Li et al. (2007)
Malus pumila	*Osmyb4*	*Oryza sativa*	Cold and drought tolerance Metabolite accumulation	Pasquali et al. (2008)
Nicotiana tabacum	*OsSPX*	*Oryza sativa*	Better survival	Zhao et al. (2009)
Arabidopsis thaliana			Cold and freezing tolerance	
Nicotiana tabacum	*CBF3*	*Capsicum annuum*	Chilling tolerance Higher proline and soluble sugar contents	Yang et al. (2011)
Arabidopsis thaliana	*ZmMKK4*	*Zea mays*	Increased cold tolerance Higher proline and soluble sugar content Increased germination	Kong et al. (2011)
Nicotiana tabacum	*SpERD15*	*Solanum pennellii*	Accumulation of soluble sugars and proline Multiple stress tolerance	Ziaf et al. (2011)
Oryza sativa	*Osmyb4*	*Oryza sativa*	Increased cold tolerance Higher soluble sugars and proline content	Yang et al. (2012)
Brassica rapa	*Osmyb4*	*Oryza sativa*	Accumulation of sugars (sucrose, fructose, glucose), proline, and anthocyanins Increased cold tolerance	Gomaa et al. (2012)

Osmyb4, *Oryza sativa* myeloblast transcription factor gene 4, *SpERD15*, *Solanum pennellii* early responsive-to-dehydration gene, *ICE1*, inducer of CBF (C-repeat binding factor) expression gene1, *AtCOR*, *Arabidopsis thaliana* cold-responsive genes, *ZmMKK*, *Zea mays* mitogen-activated protein kinase kinase gene, *1-SST*, sucrose:sucrose 1-fructosyltransferase, *OsSPX*, *Oryza sativa* SPX domain (SYG1 (suppressor of yeast gpa1)/Pho81 (CDK inhibitor in yeast PHO pathway)/XPR1 (xenotropic and polytropic retrovirus receptor) protein gene, *TaSnRK*, *Triticum aestivum* sucrose-non-fermenting (Snf)-related protein kinase gene

et al. 1996; Warren et al. 1996). Thus, these mutants clearly give a good insight into the relation between sugar levels and freezing tolerance in plants. Zhao et al. (2009) transformed tobacco with rice gene *OsSPX* and subjected both wild-type and transgenic seedlings to freezing stress (−2 °C for 20 h) and then allowed to recover under normal conditions. Transgenic plants performed much better than wild types

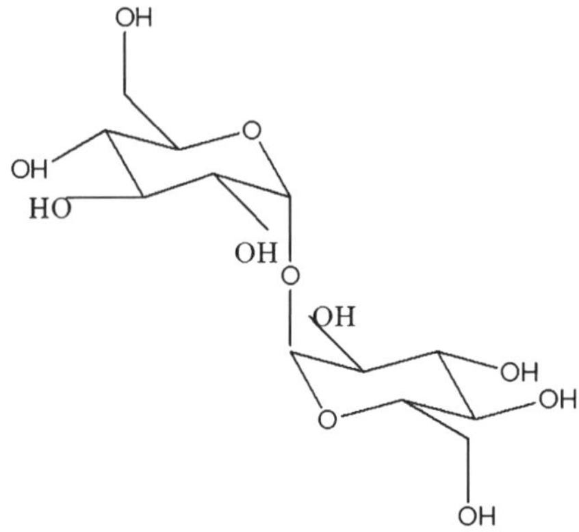

Fig. 9.11 Structure of trehalose

under cold-stressed conditions in terms of better survival, accumulation of compatible solutes such as soluble sugars, proline, and decreased membrane injury as evident from reduced electrolyte leakage content. Similarly, transgenic rice overexpressing *OsMYB* genes was tolerant to cold, salt, and drought stress than wild type. Transgenics had accumulated higher levels of soluble sugars and showed higher activities of various stress-regulated genes and antioxidative enzymes, e.g., CAT, SOD, and APX (Yang et al. 2012).

2.4 *Trehalose*

Trehalose is a common sugar in both prokaryotes and eukaryotes such as bacteria, fungi, and invertebrates (Elbein et al. 2003; Baroja-Fernandez et al. 2009) where it serves as an energy source or stress protectant (Muller et al. 1999). With the notable exception of the desiccation-tolerant "resurrection plants," including the ferns *Selaginella lepidophylla* (Adams et al. 1990) and *S. sartorii* (Iturriaga et al. 2009), trehalose is not accumulated in detectable levels in most plants (Garg et al. 2002); rather most of the plant species accumulate sucrose as a major transport sugar (Wingler et al. 2000). Although present in trace amounts in higher plants, trehalose biosynthesis genes are present universally in all the plants and have been discovered recently in many plants (Leyman et al. 2001; Ramon and Rolland 2007; Paul et al 2008; Schluepmann and Paul 2009).

2.4.1 Structure, Biosynthesis, and Lysis

Trehalose is a natural alpha-linked disaccharide formed by an α, α-1, 1-glucoside bond between two glucose units (Fig. 9.11). Trehalose is structurally α-D glucopyranosyl-[1, 1]-α-D-glucopyranoside, a nonreducing disaccharide since it is formed by the bonding of two reducing groups thereby unable to reduce other compounds (Muller et al. 1999).

Trehalose is synthesized in two steps (Goddjin and Van 1999; Elbein et al. 2003) from G6P and uridine diphosphoglucose (UDP-glucose).

Step 1: G6P and UDP-glucose in the presence of enzyme TPS react to form trehalose-6-phosphate and UDP.

$$Glucose\ \ 6\ \ PO_4 + UDP\ \ Glucose \xrightarrow{TPS} Trehalose\ \ 6\ \ PO_4 + UDP$$

TPS proteins so far identified in various living systems have been categorized into class I TPS and class II TPS based on their homology to yeast TPS1.

Step 2: In this step, the products of step 1 in the presence of enzyme trehalose-6-phosphate phosphatase (TPP) react to form the ultimate product, i.e., trehalose.

$$Trehalose\ \ 6\ \ PO_4 + UDP \xrightarrow{TPP} Trehalose + Pi$$

Trehalose thus produced may be hydrolyzed into constituent glucose subunits by the activity of enzyme trehalase (Tre). Trehalase is found to be present in almost every plant (Muller et al. 2001) and has been identified in soybean as *GMTRE1* and in *Arabidopsis* as *AtTRE1*.

$$Trehalose \xrightarrow{Trehalose} 2Glucose$$

2.4.2 Functions in the Plants

Trehalose plays an important role as a protectant under various abiotic stresses (Iturriaga et al. 2009; Josine et al. 2011; Li et al. 2011), e.g., trehalose levels were found to be elevated manyfold in response to drought, salt stress (Yeo et al. 2000; Cortina and Culianez-Macia 2005; Karim et al. 2007), and the protective effect of trehalose under low temperature conditions (summarized in Fig. 9.12), i.e., cold (Iordachescu and Imai 2011) and freezing (Lopez et al. 2008) have been widely studied and accepted. Trehalose may accumulate in certain organelles under stress-induced water-deficit conditions, thus playing an important role as osmoregulator (Mahajan and Tuteja 2005; Karim et al. 2007). Many reports of enhanced trehalose levels in cold-stressed plants confirm its protective role, e.g., rice (Pramanik and Imai 2005) and *Arabidopsis* (Kaplan et al. 2004). Trehalose also prevents the stress-induced denaturation of proteins. It replaces water and itself participates in H bonding to polar heads, thus maintaining hydration envelope even under cold-induced water-deficit conditions (Clegg 1985). Trehalose also prevents the aggregation of already denatured proteins by undergoing vitrification instead of crystallization under low temperatures, thus transforming into an amorphous glass structure which prevents the movement of proteins (Elbein et al. 2003) as it has the ability to remain in its amorphous form even under very low temperatures and

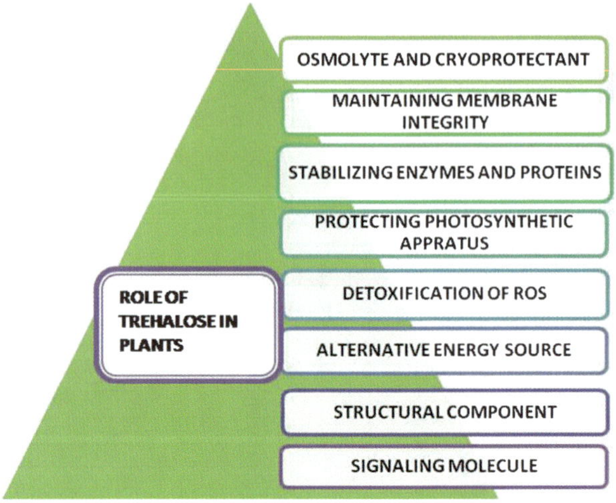

Fig. 9.12 Various functions of trehalose in plants growing under cold-stressed conditions

completely dehydrated conditions, thus forming amorphous matrix (Sun et al. 1996; Sundaramurthi et al. 2010).

Trehalose has also been found effective in stabilizing and maintaining the membrane integrity and protein structure. Since the functioning of membranes is controlled by its fluidity, trehalose protects various biomembranes by depressing the phase transition temperature and thereby preventing membrane rigidification and maintaining fluidity even at low temperatures (Crowe and Crowe 1990). The membrane-stabilizing property of trehalose is attributed to its stereochemistry; the structural conformation of two glucose subunits is such that it makes trehalose suitable to fit within polar heads of lipid bilayer, thus preventing their fusion and phase transitions (Rudolph et al. 1990). Trehalose therefore not only protects the cell membranes but also protects the various cellular organelles at ultrastructural levels (Elbein et al. 2003). It was suggested to prevent photooxidation of chloroplasts by protecting PSII (Garg et al. 2002; Jang et al. 2003) to maintain the proper function of photosynthetic machinery under stress.

As cold stress is generally accompanied by oxidative stress (Peng et al. 2013), hence, the cryoprotectants are expected to play an important role in mitigating oxidative stress as well. Trehalose potentially brings about the chemical detoxification of various ROS, thus preventing oxidative stress (Benaroudj et al. 2001; Nery et al. 2008) by upregulating the activities of various antioxidative enzymes such CAT, SOD, and APX. Transgenic plants overexpressing trehalose have been found to show better survival by mitigating chilling-induced oxidative injury by stimulating the action of various antioxidative enzymes (Eun et al. 2007; Yang et al. 2011). Additionally, trehalose may also act as reserve food material and alternative energy source in some lower organisms such as fungi by hydrolyzing into constituent glucose subunits during early germination (Thevelein 1984). Besides performing its function as an important cryoprotectant and osmolyte and regulating various physiological processing occurring in a plant cell, it may also form an integral part of cell wall in many mycobacteria and corynebacteria (Belisle et al. 1997; Chatterjee 1997). T6P, an intermediate of trehalose biosynthesis pathway, is an important signaling molecule and has been found to regulate cytosolic and plastidic carbohydrate metabolism which ultimately affects plant growth and development (Eastmond et al. 2003). T6P regulates starch synthesis and photosynthesis in the chloroplast (Lunn et al. 2006). It negatively regulates the activity of SnRK1 (bZIP11 activator) which controls various energy requirement-related processes (Delatte et al. 2011) and also may upregulate ABI4 which induces ABA signaling and photosynthesis (Ponnu et al. 2011).

2.4.3 Endogenous Levels in Cold-Stressed Plants

Under stress conditions, levels of trehalose have been found to go up in various plant organs (Garg et al. 2002; El-Bashiti et al. 2005), e.g., trehalose was found to be accumulated in *Arabidopsis thaliana* shoots when subjected to temperature extremes, i.e., high and low temperatures (Kaplan et al. 2004), whereas rice plants experiencing chilling exhibited accumulation of trehalose in its shoots as well as roots due to increased induction of *OTSPP1* and *OTSPP2* in these organs (Pramanik and Imai 2005; Shima et al. 2007). *Arabidopsis thaliana sweetie* mutants were found to accumulate four times or more trehalose levels compared to their wild types, and similar to other trehalose-overproducing transgenics, these mutants showed higher constitutive expression of abiotic stress-related genes (Veyres et al. 2008). Activity of various trehalose-metabolizing enzymes also followed a regular trend whereby activity of trehalose-6-phosphate (TPS) showed increase (Kolbe et al. 2005), whereas that of trehalase decreased under stress. Increase in TPS activity was found to be parallel with trehalose accumulation under stress (El-Bashiti et al. 2005). In *Arabidopsis thaliana*, cold stress upregulates the activity of trehalose biosynthesis genes as evident from eightfold increase in trehalose levels after 4 days of 4 °C cold treatment (Kaplan et al. 2004). In rice seedlings, also the trehalose levels increased due to simultaneous activation of two TPP genes (OsTPP1 and OsTPP2) when exposed to chilling, drought, and abscisic acid application (Pramanik and Imai 2005). Thus, many studies have put forward the interdependence of OsTTP1 and OsTPP2 expression and ABA (Pramanik and Imai 2005; Shima et al. 2007).

2.4.4 Exogenous Application for Cold Tolerance

Exogenous application of trehalose has been found to be helpful in mitigating various stresses including cold in some plants. Exogenous supply to *Arabidopsis thaliana* seedlings for 12 h (Hanhong et al. 2005) significantly increased the sugar levels. Levels of sucrose and trehalose were raised by 3.2-fold and 145-fold, respectively. Various genes related to stress response were triggered, leading to the formation of respective stress proteins such as cytosolic dehydroascorbate reductase 1 (DHAR1) and *S*-adenosylmethionine synthetase 2 (SAMS2), Phi glutathione *S*-transferase 2 (AtGSTF2), and flavin mononucleotide-binding flavodoxin-like quinone reductase 1 (FQR1). All these proteins participate in defense mechanisms such as detoxification of ROS and bring about the stress response. Muller et al. (2000) reported the expression of a fructan-synthesizing enzyme sucrose: fructan-6-fructosyl-transferase in excised barley leaves upon trehalose feeding. Presoaking of baker's yeast in trehalose solution has also been found to impart freezing tolerance (Hirasawa et al. 2001). Similarly, both increased endogenous and exogenous levels of trehalose considerably enhanced the freezing tolerance of *Cryptococcus laurentii* and *Rhodotorula glutinis* (Li et al. 2008). All these effects of exogenous application are attributed to increased T6P content rather than accumulation of trehalose as osmolyte (Schluepmann et al. 2003).

2.4.5 Trehalose Toxicity

In spite of ubiquitous occurrence of trehalose, an important signaling molecule and even as energy source in many cases, it has been reported to adversely affect plant growth and even proved to be toxic for *Arabidopsis thaliana* (Schluepmann et al. 2004; Ramon et al. 2008; Schluepmann and Paul 2009) and seedlings of dodder vine (*Cuscuta reflexa*) (Veluthambi et al. 1982a, b). The seedlings of *Arabidopsis growing* with 30 mM trehalose showed inhibition of root and shoot growth. Fritzius et al. (2001) reported up to 80 % decrease in root length upon trehalose exposure in *Arabidopsis*. The toxicity of trehalose is due to the accumulation of T6P which downregulates SnRK1, a HXK required for maintaining the carbon supply during various growth and development processes (Schluepmann et al. 2004; Zhang et al. 2009), thus impairing plant growth and development. The negative effects could be reversed to some extent by increasing sucrose concentrations and not trehalose (Schluepmann et al. 2004; Zhang et al. 2009).

2.4.6 Trehalose Signaling

When plant cell experiences cold stress, sucrose uptake is increased and sucrose when lysed makes glucose and fructose available to the cell. These two molecules further perform various functions including trehalose biosynthesis through a series of enzyme-catalyzed reactions (Fig. 9.13). Trehalose may directly enter a cell or via some unknown transporter, thus resulting in increased endogenous trehalose levels. Trehalose itself does not act as signaling molecule rather its phosphorylated form i.e., T6P acts as signaling molecule (Schluepmann et al. 2003, 2004). The levels of T6P have been found to go up when sucrose concentration goes up and not glucose thus indicating a link between sucrose and T6P. The latter is present in the cytosol and is proposed to be transported inside chloroplast by some unknown transporter (shown as "?"). Inside the chloroplast, T6P brings about the thioredoxin-mediated

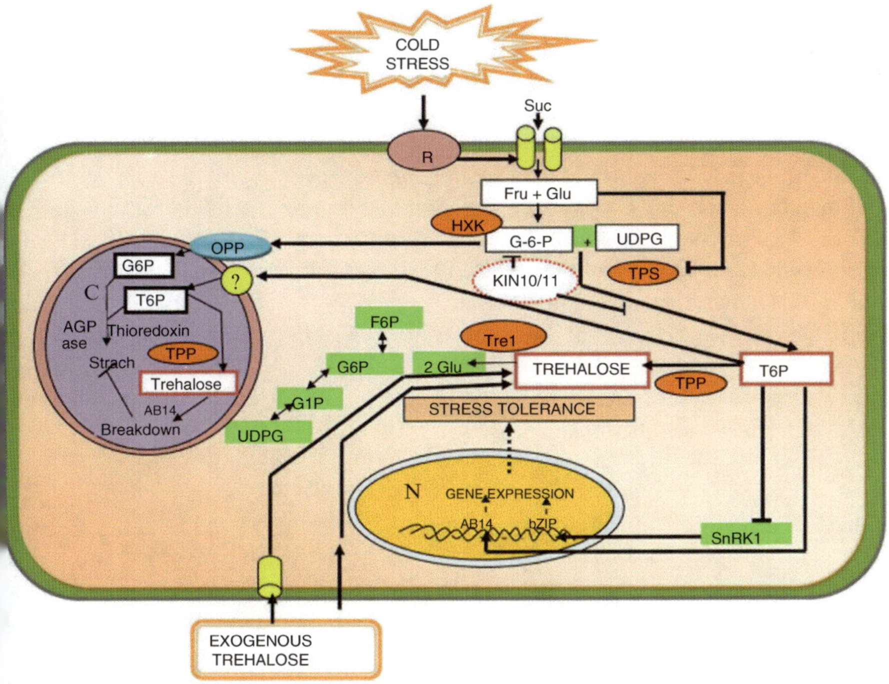

Fig. 9.13 Trehalose biosynthesis and signal transduction under stress (Paul et al. 2008; Iturriaga et al. 2009; Delatte et al. 2011; Ponnu et al. 2011). *R* putative receptor, *HXK* hexokinase, *C* chloroplast, *N* nucleus, *Suc* sucrose, *Fru* fructose, *Glu* glucose, *G6P* glucose-6-phosphate, *G1P* glucose-1-phosphate, *T6P* trehalose-6-phosphate, *UDPG* uridine diphosphate glucose, *TPS* trehalose-6-phosphate synthase, *TPP* trehalose-6-phosphate phosphatases, *Tre 1* trehalase 1, *KIN* Arabidopsis protein kinase, *OPP* oxidative pentose phosphate pathway, *AGPase* ADP-glucose pyrophosphorylase, *ABI4* ABA (abscisic acid) insensitive transcription factor gene 4, *bZIP* basic region leucine zipper transcription factor, *SnRK* sucrose-non-fermenting (Snf)-related protein kinase

activation of AGPase (ADP-glucose pyrophosphorylase), an important starch-synthesizing enzyme, thus leading to starch accumulation (Kolbe et al. 2005). Trehalose can be lysed by enzyme Tre (trehalase) into its constituent glucose subunits; the latter then may get phosphorylated to form G6P which further may be converted into G1P (glucose-1-phosphate) and F6P. Muller et al. (2001) during their studies on *Arabidopsis* reported that application of fungicide validamycin A, a potent trehalase inhibitor, led to increased levels of trehalose, while levels of sucrose and starch went down, thereby clearly indicating a cytosolic and plastidial carbohydrate metabolism nexus. T6P thus acts as a positive regulator of both photosynthesis and starch synthesis in plastids (Wingler et al. 2000; Lunn et al. 2006). G6P present in the cytosol is also transported inside chloroplast by means of OPP (oxidative pentose phosphate pathway), which also participates in thioredoxin activation of AGPase. T6P negatively regulates SnRK1, which is bZIP11 (basic region leucine zipper transcription factor 11) activator and performs an important role in responding to the energy requirements of the cell, thereby regulating various processes related to growth and development metabolism (Hardie 2007; Polge and Thomas 2007; Halford and Hey 2009). When levels of sucrose decrease, i.e., under nutrient stress, T6P levels also go down, thereby resulting in the upregulation of SnRK1, which is an important HXK responsible for maintaining the carbon supply for various growth and metabolic activities, as a result bringing about upregulation of various carbon utilization processes and enhancing the carbon supply to growing cells (Delatte et al. 2011). ABI4 has been proposed to be directly involved in the expression of starch lysis genes, thus regulating the starch synthesis inside chloroplasts (Acevedo-Hernandez et al. 2005). T6P may upregulate ABI4 transcription factor which induces ABA signaling and photosynthesis (Ponnu et al. 2011). KIN10/11, which is the catalytic subunit of Snf-related protein kinase (SnRK), may also regulate trehalose synthesis at various steps.

2.4.7 Transgenic Approach

Keeping in view the crucial role of trehalose, various trehalose-overproducing transgenics have been developed by targeting its biosynthetic pathway. The first report was by Garg et al. (2002), who raised transgenic rice lines by transferring genes *OTS A-OTS B* from *E. coli*. Transgenic rice plants thus obtained were reported to accumulate 200 times more trehalose than wild type and were tolerant to cold, drought, and salt stress. Similarly, many trehalose-overproducing transgenics have been developed so far (Table 9.4) for plants like rice, *Arabidopsis*, tobacco, and barley having cold and other abiotic stress tolerance. The transgenics produced by transformation with *TPS1* genes from *E. coli* and yeast showed some undesirable phenotypic variations such as lancelet-shaped leaves, stunted root and shoot growth, reduced yield, and even sterility in some cases (Yeo et al. 2000; Garg et al. 2002). The possible reason was attributed to the accumulation of T6P. The problem was overcome by constructing a transgene with fused coding regions of both TPS and TPP regions called TPSP. Additionally, a stress-inducible promoter can be used to maintain required trehalose levels even in unstressed conditions, and the resulting transgenics were free from any morphological abnormalities (Garg et al. 2002; Jang et al. 2003). Following is the list of some transgenics overexpressing trehalose showing their response to abiotic stresses.

2.5 Sugar Alcohols (Polyols)

Sugar alcohols constitute an important class of compatible solutes besides a number of metabolites such as proline, polyamines, glycine betaine, fructans, and sugars. There are several naturally occurring sugar alcohols, viz., sorbitol, D-mannitol, galactitol, erythritol, D-arabitol, ribitol, rhamnitol, and xylitol, found in a wide

Table 9.4 Trehalose-overexpressing transgenic plants and their stress responses along with respective references

Transgenic plant	Gene transferred	Gene source	Stress response	References
Oryza sativa	*OTS A-OTS B*	*E. coli*	Cold, drought, and salinity tolerance	Garg et al. (2002)
Nicotiana tabacum	*AtTPS1*	*Arabidopsis thaliana*	Cold, drought, and salinity tolerance	Almeida et al. (2005)
Arabidopsis thaliana	*CaXTH3*	*Capsicum annuum*	Cold, drought, and salinity tolerance	Seok et al. (2006)
Arabidopsis thaliana	*ScTPS1-ScTPS2*	*Saccharomyces cerevisiae*	Freezing drought, salt, and heat tolerance	Miranda et al. (2007)
Arabidopsis thaliana	*OsUGE-1*	*Oryza sativa*	Freezing, salt, and drought tolerance	Liu et al. (2007)
Oryza sativa	*AtNDPK2*	*Oryza sativa*	Chilling and antioxidant signaling	Eun et al. (2007)
Medicago sativa	*ScTPS1-ScTPS2*	*Saccharomyces cerevisiae*	Freezing, heat drought, and salinity tolerance	Saurez et al. (2008)
Oryza sativa	*TPP1*	*Oryza sativa*	Cold and salt tolerance	Ge et al. (2008)
Oryza sativa	*TaSTRG*	*Triticum aestivum*	Cold, drought, and salt tolerance	Zhou et al. (2009)
Arabidopsis thaliana	*ZmCOl6.1*	*Zea mays*	Cold and salt tolerance	Guerra-Peraza et al. (2009)
Nicotiana tabacum	*CaBi-1*	*Capsicum annuum*	Low and high temperature, drought, and salinity tolerance	Isbat et al. (2009)
Lolium temulentum	*eFF-1a, ACT11, UBQ 5*	*Lolium temulentum*	Cold, drought, salt, and heat tolerance	Dombrowski and Martin (2009)
Xenopus laevis	*GhTIP1*	*Gossypium hirsutum*	Freezing tolerance, heavy metal stress tolerance	Li et al. (2009)
Oryza sativa	*miRNAs*	*Oryza sativa*	Cold, drought, and salt tolerance	Jian et al. (2010)
Oryza sativa	*OsABF2*	*Oryza sativa*	Cold stress tolerance	Hossain et al. (2010)
Oryza sativa	*OsAP2LP*	*Oryza sativa*	Cold, drought, and salt tolerance	Ma et al. (2010)
Nicotiana tabacum	*CRT/DRE binding factor 1 (CBF1)*	*Arabidopsis thaliana*	Increased low temperature tolerance of plants tolerance	Yang et al. (2010)

(continued)

Table 9.4 (continued)

Transgenic plant	Gene transferred	Gene source	Stress response	References
Malus domestica	*cyMDH*	*M. domestica callus*, *Solanum lycopersicum*	Cold and salt tolerance	Yao et al. (2011)
Citrus unshiu	*CitERF*	*Citrus unshiu*	Low temperature tolerance	Yang et al. (2011)
Hordeum vulgare	*HvTIP1;2*, *HvTIP2;1*, *HvTIP2;2*, *HvTIP2;3*, and *HvTIP4;1*	*Hordeum vulgare*	Cold and osmotic stress tolerance	Ligaba et al. (2011)
Oryza sativa	*OsTPS1*	*Oryza sativa*	Low temperature tolerance	Li et al. (2011)
Arabidopsis thaliana	*ftl1-1D/ddf1*	*Arabidopsis thaliana*	Cold, drought, heat, and salinity tolerance	Kang et al. (2011)

CitERF, citrus ethylene-responsive element binding factor (ERF), *HvTIP*, *Hordeum vulgare* tonoplast intrinsic protein, *ftl1-1D/ddf1*, freezing tolerant line1 (ftl1-1D)/dwarf and delayed flowering 1 (DDF1), *OsTPS*, *Oryza sativa* trehalose-6-phosphate synthase, *cyMDH*, cytosolic malate dehydrogenase, *CRT*, C-repeat, *DRE*, dehydration-responsive element, *OsAP2PL*, *Oryza sativa* AP2 (APETALA2)/ERF (ethylene-responsive element binding factor) like protein, *OsaABF2*, *Oryza sativa* ABA-responsive element binding factor 2, *miRNA*, microRNA, *GhTIP1*, *Gossypium hirsutum* tonoplast intrinsic protein 1, *CaBi-1*, *Capsicum annuum BAX* (BCL2-associated X protein) inhibitor 1, *TaSTRG*, *Triticum aestivum* salt tolerance-related genes, *AtNDPK2*, *Arabidopsis thaliana* nucleoside diphosphate kinase 2, *OsUGE*, *Oryza sativa* UDP-glucose 4-epimerase, *ScTPS1*, *Saccharomyces cerevisiae* trehalose-6-phosphate synthase 1, *ScTPS2*, *Saccharomyces cerevisiae* trehalose-6-phosphate synthase 2, *CaXTH3*, *Capsicum annuum* xyloglucan endotransglucosylase/hydrolase, *OTS A*, *E. coli* gene encoding TPS1, *OTS B*, *E. coli* gene encoding TPP

variety of organisms ranging from bacteria to algae, fungi, and even higher plants (Bieleski 1982; Stoop et al. 1996). These various polyols are translocated through phloem along with sucrose (Zimmerman and Zeigler 1975). Mannitol, the most widely distributed sugar alcohol (Stoop et al. 1996), occurs in various plant exudates (Jennings et al. 1998) and sorbitol in various berries and higher plants (Ahmad et al. 1979), while erythritol occurs in trace amounts in fungi (mushrooms) (Yoshida and Hashimoto 1986) and some fruits (Shindou et al. 1989). It may be approximated that out of total carbon fixed, 30 % is in the form of polyols (Bieleski 1982).

2.5.1 Structure and Biosynthesis

Structurally sugar alcohols are acyclic polyols containing three or more hydroxy groups as shown in Fig. 9.14. Being chemically more reduced than their corresponding sugars, from which they have been derived, sugar alcohols contain higher energy.

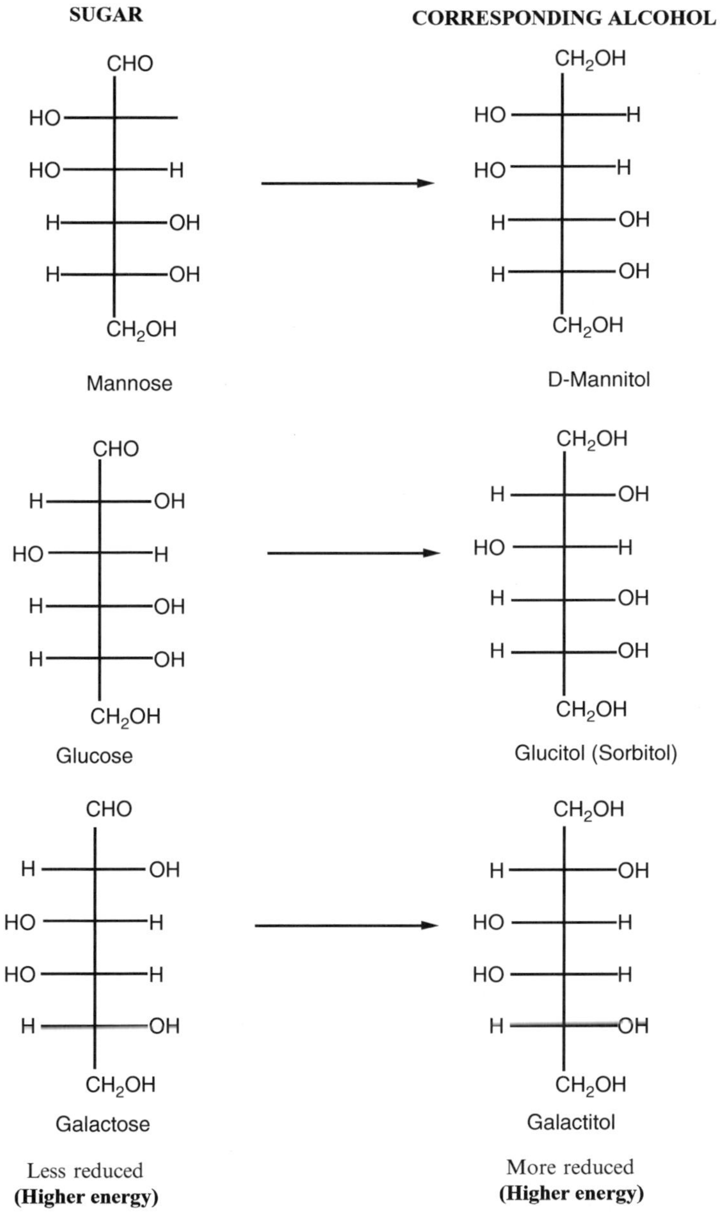

ig. 9.14 Various sugars and corresponding sugar alcohols derived from them

DHAP (dihydroxyacetone phosphate) generated during glycolysis generates F6P hrough a series of reactions (Fig. 9.15). F6P thus may get isomerized into G6P by he action of enzyme G6P isomerase (represented as I). Enzyme aldose-6-phosphate eductase (represented as II) then acts on to yield sorbitol-6-phosphate which then

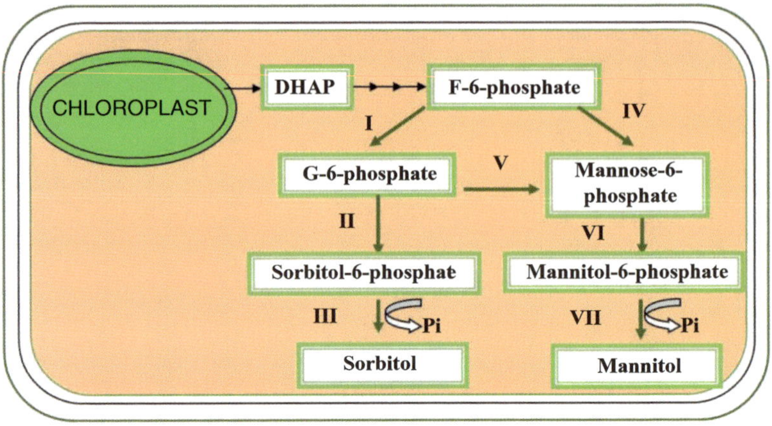

Fig. 9.15 Biosynthesis pathway of sugar alcohols (Rumpho et al. 1983). I–VII are the various enzymes catalyzing the respective reactions. I: glucose-6-phosphate isomerase, II: aldose-6-phosphate reductase, III: sorbitol-6-phosphate phosphatase, IV: mannose-6-phosphate isomerase, V: mannose-6-phosphate epimerase, VI: mannose-6-phosphate reductase, VII: mannose-6-phosphate phosphatase

undergoes dephosphorylation in the presence of enzyme sorbitol-6-phosphate phosphatase (represented as III) to finally generate sorbitol. Alternatively, F-6-phosphate may be converted into mannose-6-phosphate by the action of enzyme mannose-6-phosphate isomerase (represented as IV). Likewise, mannitol-6-phosphate also yields mannitol after undergoing a series of reduction and dephosphorylation reactions catalyzed by enzymes mannose-6-phosphate reductase (represented as VI) and mannose-6-phosphate phosphatase (represented as VII), respectively. Similarly, other sugar alcohols like cyclitol myoinositol are also produced from G6P (Loewus and Loewus 1983).

2.5.2 Functions in the Plants

Sugar alcohols have been found to protect plants from multiple abiotic stresses such as cold, salt, and drought (Rathinasabapathi 2000) via various protective strategies (summarized in Fig. 9.16). As well-established compatible solutes, these are involved in osmotic stress adaptation (Brown 1990; McCue and Hanson 1992); hence, the plants overproducing polyols do have an edge over nonproducing plants with respect to stress tolerance. It may be pointed out that besides sucrose and various nonreducing sugars, sugar alcohols are the only types of carbohydrates to be translocated via phloem and thus assimilated in higher plants. Sugar alcohols are excellent osmolytes and the osmoprotective function of sugar alcohols or polyols is attributed to the presence of water like hydroxyl groups which enable polyols to mimic the structure of water, thereby assembling around the macromolecule

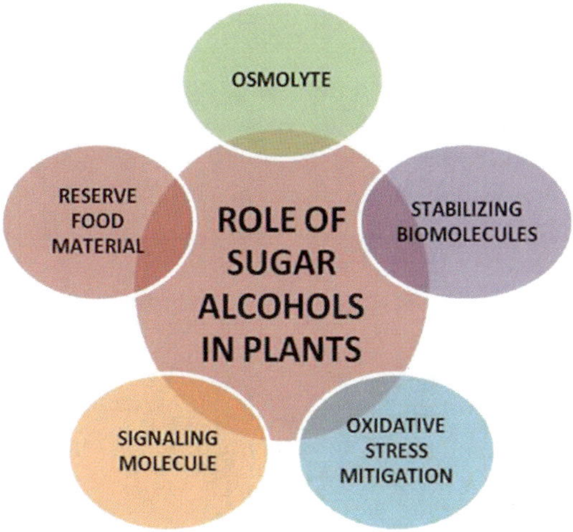

Fig. 9.16 Various functions of sugar alcohols in plants growing under cold-stressed conditions

resulting in the formation of an artificial "hydration envelope" (Yancey et al. 1982; Bohnert et al. 1995; Stoop et al. 1996). Thus, sugar alcohols may also act as membrane stabilizers and proteins-protecting chaperones (Bohnert and Jensen 1996).

Further, sugar alcohols also help in mitigating oxidative stress by scavenging the ROS, thereby preventing cell damage (Bohnert et al. 1995; Shen et al. 1997b; Wang et al. 2003). Shen et al. (1997a) targeted mannitol biosynthesis to tobacco chloroplasts and proposed the role of mannitol in hydroxyl scavenging as done by other workers (Adams et al. 1993; Kavi Kishor et al. 1995). Thus, mannitol was proposed to protect various thiol (SH) enzymes such as PRK (phosphoribulokinase, a Calvin cycle enzyme) whose activity was otherwise reported to be drastically reduced by about 65 % due to photooxidation-induced hydroxyl radicals. Additionally, other thiol regulating chloroplast components such as ferredoxin (Fd), thioredoxin, and glutathione are also protected by mannitol. Nishizawa et al. (2008) reported that galactinol synthase-overexpressing *Arabidopsis* plants were tolerant to chilling and salt stress-induced oxidative damage by detoxifying hydroxyl radicals. Khare et al. (2010) reported SOD, CAT (catalase), and MDA levels to have increased in transgenic tomato overproducing mannitol, thus protecting the plants from chilling-induced oxidative stress. The possible mode of action of hydroxyl (OH^-) scavenging has been proposed to be the formation of mannitol radical by donating its H atom to hydroxyl (OH^-), thus forming a water molecule; mannitol radicals then may be oxidized into mannose (Franzini et al. 1994; Shen et al. 1997b). Sugar alcohols have been proposed to act as low molecular weight chaperones, thereby maintaining the structure and function of biomembranes and proteins (Bohnert and Jensen 1996).

Under low temperature conditions, carbon supplies have been observed to be partly directed towards sugar alcohol synthesis (Bohnert et al. 1995). Sugar alcohols besides being chemically inert and water soluble like sucrose may also serve the function of stored form of reduced carbon (Nadwodnik and Lohaus 2008).

Myoinositol may generate various RFOs (raffinose family oligosaccharides) such as raffinose, stachyose, and verbascose. In celery (*Apium graveolens*), sugar alcohols are accumulated in high quantities approximately equal to that of sucrose. In addition to these functions, sugar alcohols are also suggested to play a role in the regulation of coenzymes and storage of reduced carbon and energy (Loescher 1987). Myoinositol (non-phosphorylated form of inositol) is a crucial sugar alcohol that further generates many biologically active phosphorylated forms which constitute an integral part of various signaling pathways, membrane synthesis, and stress mitigation processes (Loewus and Murthy 1999). Inositol in its phosphorylated form, i.e., IP_3 (inositol 1,4,5-trisphosphate), mediates Ca^{2+} signaling and maintains Ca^{2+} homeostasis in the cell by phosphorylation into IP_4 (inositol 1,3,4,5-tetrakisphosphate) (Communi et al. 1995). It also mediates calcium release and absorption (Hill et al. 1988) by mobilizing calcium from its intracellular reserves such as endoplasmic reticulum (Berridge 1997). IP_3 and IP_4 both are important secondary messengers and thus help relay the external changes in the environment or the stresses that the particular cell is experiencing to the inside of the cell.

2.5.3 Endogenous Levels in Cold-Stressed Plants

Sugar alcohols accumulate naturally in various crops, e.g., in celery (*Apium graveolens*), mannitol is formed in amounts equivalent to those of sucrose (Loester et al. 1992). Maruyama et al. (2009) carried out the metabolite profiling of *Arabidopsis* plants exposed to cold and dehydration stress and observed upregulation of various stress-responsive genes and increased levels of sugars and sugar alcohols such as myoinositol and galactinol. The accumulated endogenous levels of polyols are not high enough to account for significant osmotic adjustments; thus, it was proposed that sugar alcohols do not protect the plants just by acting as osmolytes only; rather they adopt some other mechanisms also to reinforce their protective strategies such as ROS scavenging (Shen et al. 1997a, b), protecting membranes and proteins (Bohnert and Jensen 1996). Relationship between sugar alcohols and stress mitigation has been further confirmed by various sugar alcohol-overproducing transgenics which have been proved to survive better under a variety of stresses. Prabhavathi et al. (2002) transformed eggplant (*Solanum melongena*) with bacterial gene mtlD resulting in higher accumulation of sugar alcohols. Consequently, these were found to grow well under chilling, salt, and drought-induced osmotic stress accompanied by higher chlorophyll content in some of the plants. In transgenic tobacco 1–7 μg mol^{-1} FW mannitol was accumulated (Shen et al. 1997a), whereas transgenic *Arabidopsis* accumulated 0.05–12 μg mol^{-1} FW mannitol (Thomas et al. 1995).

2.5.4 Exogenous Application for Cold Tolerance

There are only few reports of exogenous application of sugar alcohols for stress mitigation. Maize cultures when supplied with mannitol resulted in increased proline levels, thus imparting cold tolerance to them. Free proline accumulation in the callus was positively correlated to mannitol supply (Duncan and Widholm 1987). Smirnoff and Cumbes (1989) carried out in vitro experiments to establish the relation between sugars and stress mitigation and found that mannitol, sorbitol, and proline at low concentration (approx. 20 mM) exerted protective effects on various enzymes such as phosphoribulokinase (Shen et al. 1997a, b).

2.5.5 Toxicity

Higher levels of mannitol and other sugar alcohols or polyols have been found to negatively affect the plant growth, e.g., plants accumulating sorbitol up to 2–3 μmol g^{-1} fresh weight exhibited no phenotypic variations, but with its further increase many abnormalities like necrotic lesions begun to appear in sorbitol-overproducing transgenic tobacco and even higher concentrations, i.e., exceeding 15–20 μmol g^{-1} fresh weight rendered the plants infertile and even resulted in loss of rooting capability in primary regenerants in culture or soil (Sheveleva et al. 1998). These abnormalities were attributed to the depleted myoinositol levels and could be possibly eliminated by carrying out further transformations.

2.5.6 Transgenic Approach

Keeping in view their widely accepted and confirmed role of polyols in various stresses, various abiotic stress-tolerant transgenic plants have been raised. Plants like tobacco and *Arabidopsis* which naturally do not accumulate polyols can be genetically transformed to mitigate various abiotic stresses such as cold, drought, and salinity by accumulating different sugar alcohols such as sorbitol (Tao et al. 1995; Sheveleva et al. 1998), inositol (Smart and Flores 1997), mannitol (Thomas et al. 1995; Karakas et al. 1997; Shen et al. 1997a), and D-inositol (Vernon et al. 1993; Sheveleva et al. 1997). Numerous genes encoding various sugar alcohols have been identified and isolated, but so far *mltD gene* has been found to be the most important one. *mltD* encodes for mannitol 1 phosphate dehydrogenase, the enzyme responsible for catalyzing reversible reaction converting F6P into mannitol-1-phosphate. Various nonspecific phosphatases then act on mannitol-1-phosphate to convert it into mannitol (Thomas et al. 1995). Khare et al. (2010) generated transgenic tomato lines by transforming with bacterial gene mtlD encoding mannitol-1-phosphate dehydrogenase. Plants thus obtained proved to be cold tolerant and survived at temperature which otherwise proved lethal to non-transformed ones as shown in Table 9.5.

2.6 *Fructans*

Fructans are the polymers of fructose which may be linear or branched (Cairns 2003), synthesized from sucrose by fructosyltransferases (FTs) and found mostly in

Table 9.5 Various sugar alcohol-overexpressing transgenics and their stress response along with respective references

Transgenic plant	Gene	Gene source	Traits conferred	References
Nicotiana tabacum	*mltD*	*E. coli*	Oxidative stress	Shen et al. (1997a)
Solanum melongena	*mtlD*	*E. coli*	Chilling, drought, and salt tolerance	Prabhavathi et al. (2002)
Petunia hybrida	*mtlD*	*E. coli*	Chilling tolerance	Chiang et al. (2005)
Nicotiana tabacum	*AtIpk-2β*	*Arabidopsis*	Freezing, drought, and salt tolerance	Yang et al. (2008)
Lycopersicon esculentum	*mtlD*	*E. coli*	Cold, drought, and salt tolerance	Khare et al. (2010)

mtlD, mannitol-1-phosphate dehydrogenase gene, *AtIpk-2β*, *Arabidopsis* inositol polyphosphate 6-/3-kinase

temperate and cold zone grasses (French and Waterhouse 1993). Fructans reportedly occur in 12–15 % of higher plants (Hendry 1987) as well as in a wide range of bacteria and fungi (Van Hijum et al. 2003; Martínez-Fleites et al. 2005) besides primary storage carbohydrates, i.e., sucrose and starch. Storage of fructans is not restricted to any particular plant organ; rather it may be present in any vegetative organ such as stem or leaves or reproductive parts as well. Fructans are synthesized under the conditions when the rate of photosynthesis exceeds the metabolic demands of the plant under stress especially when sucrose level in the sink organs has reached its critical level (Pollock et al. 1988). Hence, fructans accumulate during autumn and winters in the leaves, stem, and various underground parts of different perennial forage grasses and cereals (Meier and Reid 1982; Pontis and De Campillo 1985) and then made available to the plant during periods of early growth in spring, regrowth, and during grain filling (Blacklow et al. 1984; Pollock 1984). Natural accumulators of fructans include Asteraceae members such as dahlia, members of Liliaceae family such as asparagus, and Poaceae such as wheat, barley, and ryegrass (Shiomi et al. 1997; Van den Ende and Van Laere 2007).

2.6.1 Structure

Depending upon the source or starting sugar, different fructans may differ on the basis of degree of polymerization (DP), the presence or absence of branches, and the basis of linkage type between fructosyl residues and the position of glucose residue. Thus, fructans may be broadly classified into following basic types:

1. *Inulin type*: Those having β-(2,1) linkage between fructosyl residue and glucose residue and are mostly found in the dicots (Bonnett et al. 1994; Koops and Jonker 1996). Simplest inulin type of fructans is a trisaccharide 1-kestose (Fig. 9.17) also known as isokestose (Valluru and Van den Ende 2008).

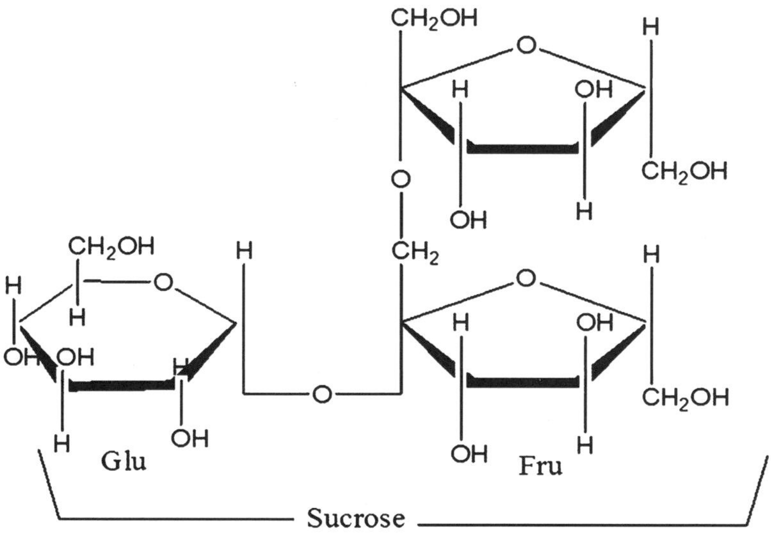

Fig. 9.17 Structure of 1-kestose

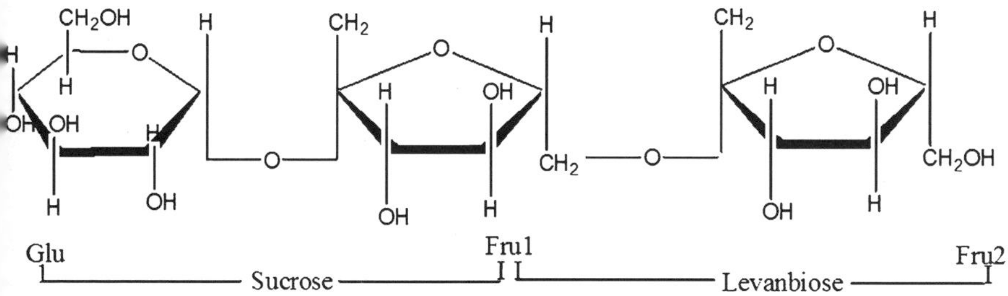

Fig. 9.18 Structure of 6-kestose

2. *Levan type*: They are the linear fructans having β-(2,6) linkage between fructosyl residue and glucose residue. They are found mainly in monocots, e.g., some grasses such as *Dactylis glomerata* (Bonnett et al. 1994). The smallest levan-type fructan is a trisaccharide 6-kestose (Fig. 9.18) and having the following structure (Valluru and Van den Ende 2008).
3. *Graminan type*: They are the branched fructans having both β-(2,6) and β-(2,1) linkage between fructosyl residue and glucose residue (Fig. 9.19). Levan and graminan types generally occur in bacteria and monocots.

9.6.2 Biosynthesis and Metabolism

Fructans are synthesized in the plants during normal development, i.e., during winters and autumn and when the plant is exposed to unfavorable environmental

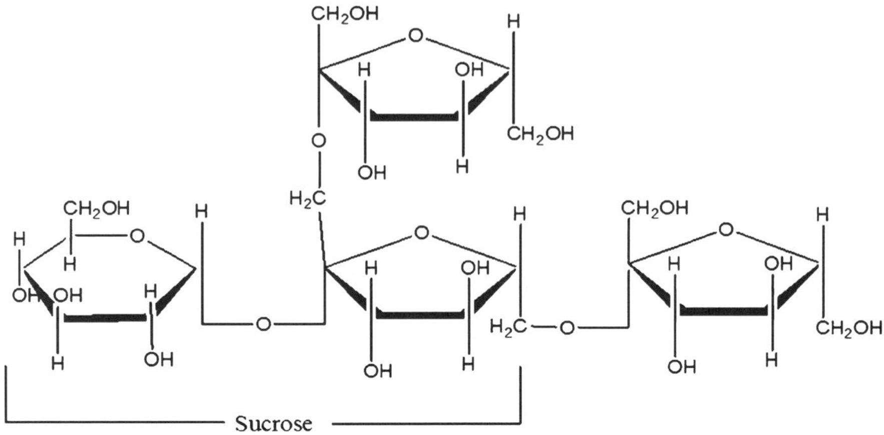

Fig. 9.19 Structure of bifurcose

conditions such as cold, drought, and nitrogen deficiency (Kerepesi et al. 1998; Wei et al. 2002). Unlike sucrose which is synthesized in the cytoplasm, fructans are synthesized in the vacuoles. Specific enzymes (fructosyltransferases) synthesize fructans by bringing about the transfer of fructose from sucrose to the growing fructan chain (Vijn and Smeekens 1999). Fructose breakdown in plants is brought about by fructan exohydrolases (FEHs), releasing terminal fructosyl units using water as an acceptor (Fig. 9.20). Sucrose, the starting biomolecule for fructan synthesis, in the presence of enzyme 1-SST, forms 1-kestose which is the key fructan generating various other fructans. Alternatively, sucrose can also form 6-kestose and then further form levan type of fructans in the presence of 6-SFT. 1-Kestose in the presence of sucrose and 6G-FFT forms neokestose which may then form levan and inulin neo-series. 1-Kestose may also form bifurcose which is then acted upon by 6-SFT and 1-FFT to generate both levan type and mixed levan type of fructans. Inulin types of fructans are also generated from 1-kestose by enzyme 1-FFT.

2.6.3 Functions in the Plants

Fructans play a variety of protective role in plants under cold stress (summarized in Fig. 9.21). Being soluble and localized inside vacuole, the osmoregulatory compartment of the cell, fructans are assumed to a play role in osmotic adjustment. Fructans comprise the only kind of polysaccharides resembling disaccharides and RFOs in stabilizing lipid bilayer by interacting with lipid head groups (Hincha et al. 2000) In case of the plants subjected to low temperature, fructans are preferred over starch for mitigating chilling stress due to these reasons:

1. Higher solubility of fructans in water.
2. Protective effect of fructans on membranes by preventing chilling-induced crystallization.

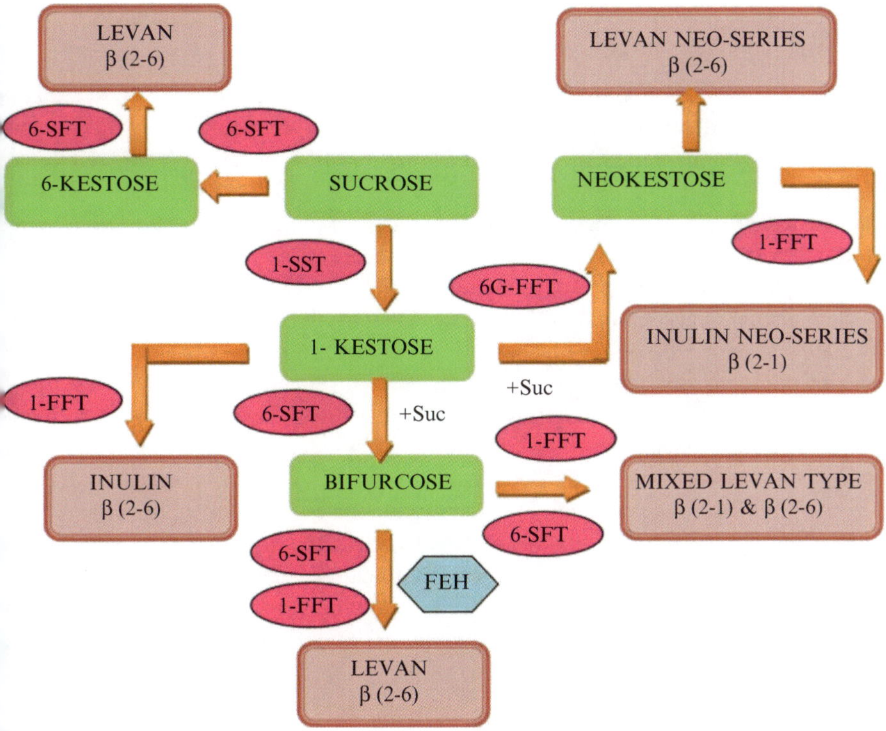

Fig. 9.20 Fructan biosynthesis in plants (Wiemken et al. 1995). *Ellipsoids*—different fructosyltransferases. *FEH* fructan exohydrolase, *1-FFT* fructan:fructan 1-fructosyltransferase, *1-SST* Suc:Suc 1-fructosyltransferase, *6G-FFT* fructan:fructan 6G-fructosyltransferase, *6-SFT* Suc:fructan 6-fructosyltransferase

Fig. 9.21 Various functions of fructans in plants growing under cold-stressed conditions

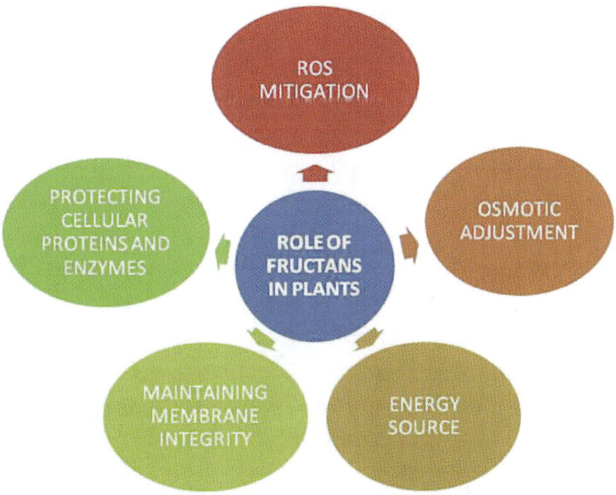

3. Even under low temperature conditions, biosynthesis of fructans proceeds normally (Vijn et al. 1997; Vijn and Smeekens 1999).

Since fructans are mainly stored in the vacuole (Wagner et al. 1983; Wiemken et al. 1986) besides some reports of their presence in the apoplast (Livingston and Henson 1998) and phloem (Wang and Nobel 1998), they clearly play a role in osmotic adjustment under stress. Van den Ende et al. (2004) reported the levels of fructose hydrolyzing enzyme 6-kestose exohydrolase (6-KEH) to have been upregulated under cold in wheat sink tissues, thereby making simple hexoses available to the plant. Additionally, its presence in the vacuole depresses the freezing temperature of the plant cell, and this makes the plant cold resistant (Van den Ende et al. 2004). Fructans act as main source of energy in overwintering plants for survival under freezing temperatures until the commencement of favorable season (Yukawa and Watanabe 1991). Fructans thus directly help in stress mitigation by accumulating under stress, acting as energy and carbon source by making hexoses available upon hydrolysis as evident from increased fructan hydrolases at low temperatures (Ji et al. 2007), or indirectly by bringing about the production of other osmoprotectants.

Fructans similar to other sugars are capable of maintaining structural integrity of various enzymes and proteins, thus protecting against stress-induced denaturation. Fructans thus help in maintaining various physiological processes under adverse conditions by maintaining the normal functioning of enzymes associated with them. Similar to other sugars, fructans help maintaining a hydration envelope by hydrogen bonding to phosphate and choline group of various cellular biomembranes under freeze-drying conditions (Hincha et al. 2000), thus minimizing the water loss and maintaining their integrity and fluidity. Inulin-type fructans have variable conformations, thus imparting it access to deeper regions of membranes (Valluru and Van den Ende 2008). Fructans under freezing temperature undergo vitrification, as seen in other sugars, and it immobilizes lipid head groups which prevents the water-deficit induced fusion of various organellar membranes and cell membrane as well (Vereyken et al. 2003), e.g., fructans of inulin type (from chicory roots) and levan type (from *Bacillus subtilis*) protected liposomes during freezing-drying or air-drying (Hincha et al. 2000; Vereyken et al. 2003). The transport of fructans from site of synthesis, i.e., vacuole to site of action, i.e., apoplast where fructans and FEHs are present, seems to be yet unclear, but vesicle-mediated tonoplast-derived fructan transport has been proposed to be the most likely pathway (Valluru et al. 2008). Fructans protect the plants under chilling stress from oxidative injury also by detoxifying ROS or by upregulating other antioxidative processes (Parvanova et al. 2004; Bolouri-Moghaddam et al. 2010). Li et al. (2007) observed lowered lipid peroxidation in fructan-overproducing transgenic tobacco plants.

However, some studies contradict the positive correlation between sugars concentration and freezing tolerance (Suzuki and Nass 1988; Spollen and Nelson 1994). In two *Lolium perenne* cultivars, freezing and soluble sugar content was not found to be positively related (Pollock et al. 1988). Extensive study on 23 oat genotypes subjected to freezing under cold conditions showed that fructans with low degree of polymerization (DP<6) only exhibited protective effect (Livingston and Henson

1998). However, in wheat, triticale, and several rye cultivars, the opposite was found to be true (Suzuki and Nass 1988). All these studies indicate that role of fructans under abiotic stress is clearly much more complex and is governed by various factors like source, size, structure, and localization of fructan molecules (Livingston et al. 2007).

2.6.4 Endogenous Levels in Cold-Stressed Plants

Fructans have been reported to play crucial role during various stress, and fructan levels have been found to go up whenever plants are experiencing unfavorable conditions such as drought and cold (Yoshida et al. 1998; Vijn and Smeekens 1999; Kawakami and Yoshida 2005). This mechanism of cold tolerance, i.e., the accumulation of fructans, is a crucial survival mechanism adopted by various plants such as wheat, barley, and ryegrass against cold stress (Chalmers et al. 2005). Levels of various compatible solutes such as sucrose, glucose, fructose, and fructans were found to have elevated during cold hardening in winter oat (*Avena sativa*) by Livingston and Henson (1998). Many varieties of wheat were found to accumulate fructans in addition to other sugars during winter hardening and the freezing tolerance thus acquired that increased positively with the increasing levels of fructans (Yoshida et al. 1998; Vágújfalvi et al. 1999). This accumulation of fructans was attributed to the increased activities of various FT (fructosyltransferase) genes (Kawakami and Yoshida 2002). Similar observations were recorded in chicory (Van Laere and Van den Ende 2002) and grasses (Wei and Chatterton 2001; Wei et al. 2002). Wheat has been found to accumulate fructans in large proportion under low temperature conditions (Santoiani et al. 1993; Kawakami and Yoshida 2005). Kawakami and Yoshida (2005) reported that during cold hardening of winter wheat, both 6-SFF and 1-FFT participate in the biosynthesis of graminan type of fructans. Fructan levels in wheat were found to constitute about 10 % or even more of total fresh weight of winter wheat crown (Yoshida et al. 1998). In wheat calli, cold hardening was accompanied by higher sugar and fructan levels in frost-tolerant genotypes than sensitive ones (Kerepesi et al. 2004). Livingston and Henson (1998) while studying the behavior of various apoplastic sugars in oats during second phase of cold hardening found fructan levels to have increased in apoplastic fluid indicating a clear role of fructans in freezing tolerance. Dependence on sucrose for fructan accumulation during cold conditions was shown in tall fescue calli where fructans were synthesized in cold without any sucrose supply, whereas for the calli raised under controlled conditions, sucrose supply was required (Damiani et al. 2012).

2.6.5 Transgenic Approach

So far many transgenics showing increased fructan biosynthesis under abiotic stress have been produced utilizing bacteria, plants, and fungi as the gene sources, viz., *Solanum tuberosum*, *Nicotiana tabacum*, *Nicotiana plumbaginifolia*, *Pichia*

Table 9.6 Fructan-overproducing transgenics and their stress response along with respective references

Transgenic plant	Gene	Source	Stress response	References
Nicotiana tabacum	*Sac B*	*Bacillus subtilis*	Freezing tolerance	Konstantinova et al. (2002)
Nicotiana tabacum	*Sac B*	*Bacillus subtilis*	Freezing tolerance	Parvanova et al. (2004)
Lolium perenne	*wft2*, *wft1*	*Triticum aestivum*	Freezing tolerance	Hisano et al. (2004)
Nicotiana tabacum	*Ls1-SST*	*Lactuca sativa*	Freezing tolerance	Li et al. (2007)
Oryza sativa	*wft2*, *wft1*	*Triticum aestivum*	Chilling tolerance	Kawakami et al. (2008)
Nicotiana tabacum	*Bp6-SFT*	*Bromus pictus*	Freezing tolerance	del Viso et al. (2011)

Bp6-SFT, *Bromus pictus* sucrose:fructan 6-fructosyltransferase, *wft2*, wheat sucrose:sucrose 1-fructosyltransferase, *wft1*, wheat sucrose:fructan 6-fructosyltransferase, *Ls1-SST*, *Lactuca sativa* sucrose:sucrose 1-fructosyltransferase, *Sac B*, *Bacillus subtilis* levansucrase

pastoris, *Beta vulgaris*, *Zea mays*, and *Oryza sativa*. Sprenger et al. (1995) were the pioneers in cloning 6-SFT gene (fructan biosynthesis gene) from barley (*Hordeum vulgare*). Subsequently, various fructan biosynthesis genes have been identified, isolated, and cloned from various plants such as chicory, *Cichorium intybus* (de Halleux and Van Cutsem 1997), *Allium cepa* (Vijn et al. 1998), *Festuca arundinacea* (Lüscher et al. 2000), *Agropyron cristatum* (Wei and Chatterton 2001), *Poa secunda* (Wei et al. 2002), *Triticum aestivum* (Kawakami and Yoshida 2002), *Lolium perenne* (Lidgett et al. 2002; Chalmers et al. 2003), and *H. vulgare* (Nagaraj et al. 2004). These genes when introduced into various fructan non-accumulators resulted in fructan accumulation and tolerance against various abiotic stresses such as cold, drought, and freezing (Pilon-Smits et al. 1998; Konstantinova et al. 2002; Hisano et al. 2004).

2.6.6 Abnormalities in the Transgenics

Plants transformed with bacterial fructan synthesis genes were found to have some morphological alterations such as stunted growth, chlorosis, and necrosis (Cairns 2003), but those with plant fructan synthesis genes were free from any such aberrations (Hisano et al. 2004) possibly because of the fact that accumulation of fructans in any other cellular compartment than vacuole increases the chances of affecting cellular metabolism (Ebskamp et al. 1994; Van der Meer et al. 1994). Table 9.6 indicates some examples of fructan-overproducing transgenics and the consequent effects on cold tolerance.

2.7 *RFOs (Raffinose Family Oligosaccharides)*

RFOs refer to the galactosyl derivatives of sucrose and include raffinose, stachyose, and verbascose. RFOs may be represented as (alpha-1,6-galactosyl)$_n$-Suc, where n represents the number of galactosyl branches attached to sucrose and $n \leq$ approximately 7, but RFOs with degree of polymerization even up to 9 have been reported. These constitute the most abundant carbohydrate type after sucrose. RFOs have been reported to be synthesized in almost every plant during the course of its life cycle at some stage or other, may be herbaceous (Bachmann et al. 1994; Taji et al. 2002) or woody (Stushnoff et al. 1993). Some of them may not accumulate RFOs at all, whereas in others the stored reserves may even constitute up to 20–25 % of total dry weight of the plant. RFOs may be stored in any part of the plant, e.g., in tubers of *Stachys sieboldii*, seeds of soybean, or in the mesophyll tissue of green photosynthetic leaves of *Ajuga patens* (Bachmann et al. 1994; Bachmann and Keller 1995). Accumulation of RFOs has been reported in many plant species during cold acclimation, thereby imparting frost tolerance (Bachmann et al. 1994; Castonguay and Nadeau 1998; Gilmour et al. 2000).

2.7.1 Structure

By definition, RFOs are the α-galactosyl derivatives of sucrose. The simplest type of RFO, which consists of galactosyl attached to sucrose moiety, is a trisaccharide raffinose consisting of glucose, fructose, and galactosyl residue (Fig. 9.22). Likewise, tetrasaccharide stachyose consists of two galactosyl residues attached to sucrose (Fig. 9.23). Thus, in this way, the basic carbon skeleton remains the same, and more and more galactosyl residues may be added and further higher polysaccharides may be generated, e.g., pentasaccharide verbascose (Fig. 9.24) and so on.

2.7.2 Biosynthesis

RFO synthesis occurs by the action of a series of α-galactosyltransferases (Kandler and Hopf 1984; Dey 1985).

$$UDP \quad Galactose + myoinositol \xrightarrow{Galactinol\ synthase} Galactinol + UDP$$

$$Galactinol + Sucrose \xrightarrow{Raffinose\ synthase} Raffinose + myoinositol$$

$$Galactinol + Raffinose \xrightarrow{Stachyose\ synthase} Stachyose + myoinositol$$

Galactinol donates galactosyl to sucrose, thus forming raffinose (DP = 3), having three galactosyl residues attached to sucrose. Similarly, stachyose may be formed

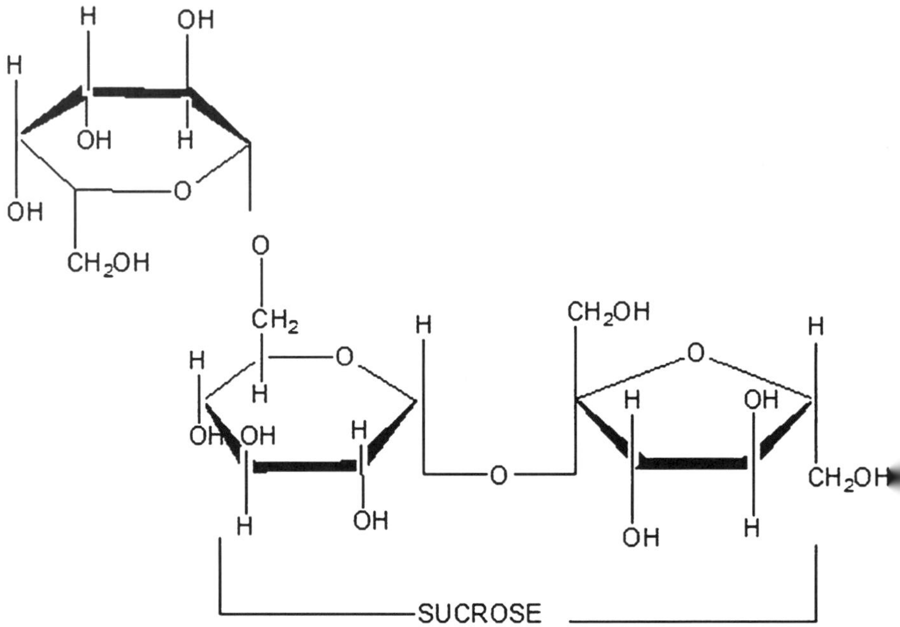

Fig. 9.22 Structure of raffinose

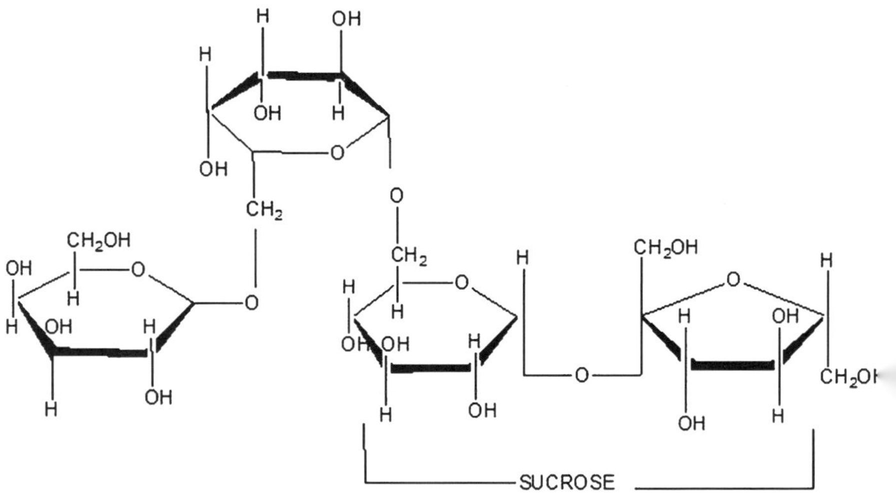

Fig. 9.23 Structure of stachyose

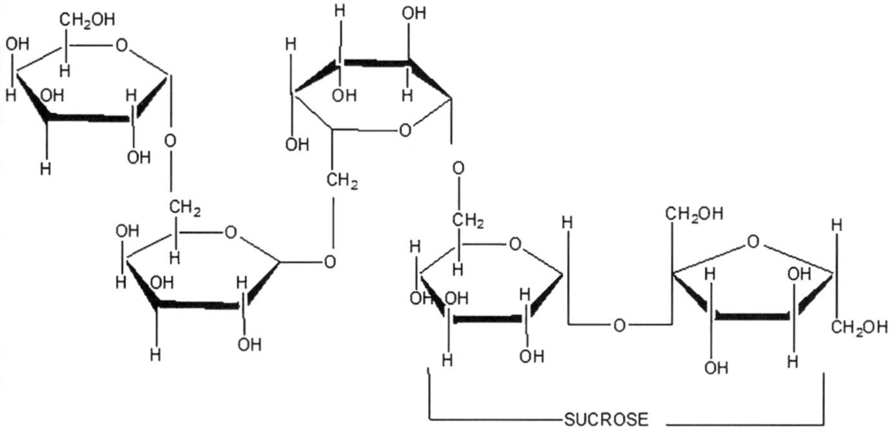

Fig. 9.24 Structure of verbascose

by attaching one more galactosyl residue, i.e., DP=4, to raffinose. Reactions are catalyzed by enzymes raffinose synthase and stachyose synthase, respectively (Kandler and Hopf 1982; Keller and Pharr 1996). Further higher oligosaccharides with increased DP (DP>4) can also be formed by galactinol-independent galactosyl transfer reactions utilizing raffinose, stachyose, and verbascose as substrates and attaching more galactosyl residues catalyzed by enzyme GGT, i.e., galactan: galactan galactosyl transferase, functional analog of fructan: fructan fructosyltransferase found in the fructan synthetic pathway (Bachmann et al. 1994; Gilbert et al. 1997).

9.7.3 Functions in the Plants

RFOs such as raffinose and galactinol have been proved to be helpful in mitigating various types of abiotic stresses such as freezing (Pennycooke et al. 2003), cold, and drought (Taji et al. 2002) in several ways (summarized in Fig. 9.25). Like other sugars, RFO does also play a role as excellent osmolytes, thus maintaining the osmotic potential of the cells under stress. RFOs such as raffinose do not change to crystalline form during cold rather change into glassy solid or vitrified state (Sun et al. 1994), thus imparting a protective effect. RFOs help in maintaining the hydration envelope around lipid heads of membrane and hence maintain its fluidity (Hincha et al. 2003). In addition to synthesis and accumulation of conventional carbon reserves such as starch and sucrose, RFOs serve as very good alternatives for this purpose in a variety of plants (Kandler and Hopf 1982; Keller and Pharr 1996). RFOs are reported to perform both these functions, i.e., storage and transport, very well in *Ajuga patens* (Bachmann et al. 1994). Thus, similar to other sugars, RFOs may also serve the dual function by acting as carbon reserves and also participating in carbon translocation through phloem as some RFOs such as raffinose and stachyose are phloem mobile (Sprenger and Keller 2000; Ayre et al. 2003).

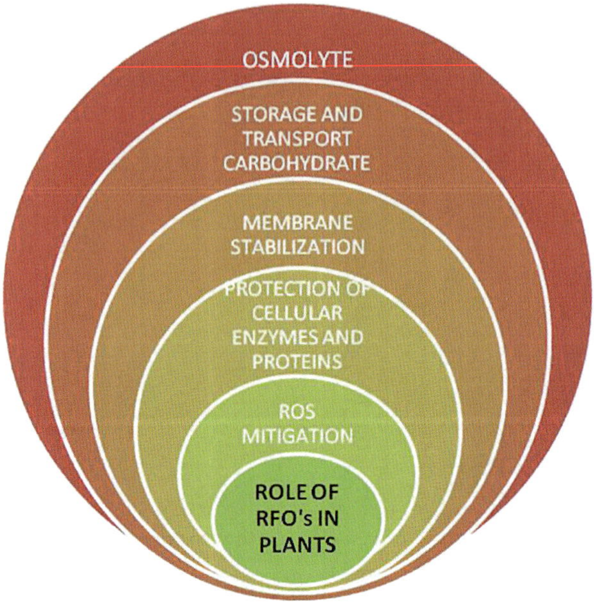

Fig. 9.25 Various functions of RFOs in plants growing under cold-stressed conditions

Further, RFOs are also suggested to stabilize cellular membranes and proteins. Pioneer studies in this regard were carried out by Santarius (1973) using isolated chloroplast membranes of spinach (*Spinacia oleracea*) and reported that raffinose protected the photosynthetic machinery by maintaining the normal activities of electron and cyclic photophosphorylation under adverse climatic conditions such as freezing, drought, and high temperature stress. Santarius and Milde (1977) also corroborated the above findings. The structure and molecular size of protecting molecules are also important factors and raffinose being a trisaccharide was suggested to be more effective in stress mitigation than disaccharides (such as sucrose) or monosaccharides (such as glucose). Likewise, studies carried out by Lineberger and Steponkus (1980) on isolated thylakoid membranes and by Hincha et al. (2003) on liposomes supported the protective effect of raffinose most probably by interacting with protein and lipid bilayer (Hoekstra et al. 2001). Raffinose also helps in vitrification of sucrose rather than its crystallization at low temperatures, thus imparting a protective effect (Caffrey et al. 1988; Sun et al. 1994).

Nishizawa et al. (2008) reported high levels of raffinose and galactinol in transgenic *Arabidopsis* plants. Consequently, the plants were suggested to detoxify hydroxyl radical that protected them from oxidative stress injury due to environmental stresses such as salinity or chilling or on the application of chemicals like methylviologen (MV) which aggravate the production of hydroxyl radical under controlled conditions. RFOs thus may protect the photosynthetic machinery from oxidative damage.

2.7.4 Endogenous Levels in Cold-Stressed Plants

RFO levels have been reported to go up under various abiotic stresses such as cold, drought, and salinity, e.g., in soybeans (Blackman et al. 1992) and maize seeds (Brenac et al. 1997). Taji et al. (2002) while studying the effect of salinity and cold on the endogenous levels of various osmolytes found galactinol and raffinose to have accumulated in addition to sucrose. Bachmann et al. (1994) in *Ajuga reptans*, Zuther et al. (2004) in *Arabidopsis*, and Stushnoff et al. (1998) in pansies (*Viola wittrockiana*) proposed that RFO accumulation could be induced by low temperature treatments. The activity of enzyme galactinol synthase (GolS), key enzyme in RFO synthesis, was found to go up in kidney beans, thus accumulating higher levels of RFOs, and in case of *Arabidopsis*, the expression of *GolS* genes was cold inducible (Liu et al. 1998). Similarly, the activities of genes GolS (galactinol synthase) and RS (raffinose synthase) were reported to be upregulated under unfavorable climatic conditions such as cold and desiccation in *Arabidopsis* and tomato seedlings (Downie et al. 2003; Zuther et al. 2004). Sprenger and Keller (2000) while studying the allocation of RFOs in *Ajuga reptans* proposed that there were at least two isoforms of *GolS*, i.e., *GolS1* and *GolS2*. *GolS1* was mainly found to be involved in the synthesis of storage RFOs, whereas *GolS2* was credited with the synthesis of transport RFOs. However, Taji et al. (2002) have identified seven *GolS* related genes from *Arabidopsis thaliana* named as *AtGolS* 1,2,3,4,5,6,7. The levels of RFOs have been observed to be positively correlated with cold tolerance in various cold-hardy species (Hinesley et al. 1992; Wiemken and Ineichen 1993). RFOs have been reported to have accumulated to very high levels in members of Cucurbitaceae, Lamiaceae, Oleaceae, and Scrophulariaceae and up to 100-fold accumulation in cucurbits. Bachmann et al. (1994) while working with metabolism of RFOs in *Ajuga reptans* leaves found that with the commencement of low temperatures in the autumn, starch hydrolysis was accompanied by increased levels of sucrose and RFO in the shoots (200 mg g^{-1} fresh weight), whereas RFO levels were lowest during summers (75 mg g^{-1} fresh weight). Plants when exposed to cold treatment (14 day at 10/3 °C as day and night) recorded about tenfold increase in nonstructural carbohydrates contributed mainly by RFO. Klotke et al. (2004) reported that *Arabidopsis* ecotype C24 was more tolerant to freezing than Columbia ecotype as the former was found to have accumulated higher levels of raffinose under cold stress.

When exposed to cold stress, leaves of *Ajuga patens*, *Arabidopsis thaliana*, and *Spinacia oleracea* were found to accumulate raffinose in the chloroplast, but the raffinose synthesis enzymes, galactinol synthase and raffinose synthase, are cytosolic, and thus raffinose is proposed to be transported into chloroplast actively through a transporter in chloroplast envelope (Schneider and Keller 2009).

2.7.5 Transgenic Approach

As already discussed, galactinol synthase catalyzes the crucial step involving the formation of galactinol; thus, it is the principal enzyme in the biosynthesis which regulates the carbon partitioning between sucrose and RFOs (Saravitz et al. 1987). The gene coding this enzyme (*GolS* gene) can thus be manipulated to control RFO levels in the plants and hence finds application in genetically modifying plants to raise RFO-overproducing plants capable of mitigating various abiotic stresses (Taji et al. 2002; Nishizawa et al. 2008). Not only just *GolS* but plants may be transformed with genes such as *UGE* coding for enzyme UDP-glucose 4-epimerase, which brings about inter conversion of UDP-glucose and UDP-galactose. The resultant mutants were found to be tolerant to various abiotic stresses, and the tolerance was mainly attributed to enhanced levels of raffinose in the plant, e.g., *Oryza sativa-derived OsUGE-1* gene-overexpressing *Arabidopsis* plants accumulated raffinose and were tolerant to abiotic stresses such as cold, drought, and salinity (Liu et al. 2007). Since galactosidase is the enzyme involved in RFO hydrolysis, thus, alternatively the activity of this enzyme can also be altered to control raffinose accumulation and degradation. Pennycooke et al. (2003) studied the activity of α-galactosidase enzyme, responsible for the hydrolysis of α-1,6 linkage of RFO during de-acclimation, and raised transgenic *Arabidopsis* plants by transferring galactosidase-encoding enzyme *Lea-Gal* gene from tomato, and the transgenic plants were then cold acclimated and de-acclimated. Raffinose levels were elevated by a significant amount; about 12–22-fold increase was recorded in non-acclimated ones and 22–53-fold increase in acclimated plants, thus increasing the freezing tolerance of the plant. Downregulated activities of α-galactosidase rendered the plants freezing tolerant, whereas upregulation had the reverse impact. However, Zuther et al. (2004) proposed that cold tolerance and raffinose were not universally corelated. They raised GS (galactinol synthase) overexpressing as well as RS (raffinose synthase) knockout mutants of *Arabidopsis*. GS-overexpressing lines accumulated about 20 times increased levels of raffinose, whereas RS mutants did not accumulate any. However, there was not much difference in the freezing tolerance levels of both types of transgenics, suggesting that increased raffinose levels did not significantly alter the freezing tolerance. Table 9.7 mentions some transgenic with altered expression of genes coding for biosynthetic genes of RFOs, which acquired stress tolerance.

3 Conclusion and Future Perspectives

The above information on cryoprotectants indicates their diverse functions (summarized in Fig. 9.26) under cold stress that effectively defend the plants from cold injury. All the cryoprotectants, described above, are excellent osmoprotectants, maintain membrane integrity, stabilize cellular membranes and proteins, and also bring about the detoxification of ROS either directly or indirectly by activating othe

Table 9.7 Some RFO-overproducing transgenics and their stress response along with respective references

Transgenic plant	Gene	Source	Stress response	References
Arabidopsis thaliana	*AtGolS2*	*Arabidopsis thaliana*	Raffinose accumulation Cold and drought tolerance	Taji et al. (2002)
Petunia hybrida	*Lea-Gal gene*	*Lycopersicon esculentum*	Increased raffinose levels Freezing tolerance	Pennycooke et al. (2003)
Arabidopsis thaliana	*GolS*	*Cucumis sativus*	20 times more raffinose accumulation No stress mitigation	Zuther et al. (2004)
Arabidopsis thaliana	*OsUGE-1*	*Oryza sativa*	Raffinose accumulation Cold and drought tolerance	Liu et al. (2007)
Arabidopsis thaliana	*AtGolS1*, *AtGolS2*	*Arabidopsis thaliana*	Mitigate oxidative stress induced chilling, salt, or by MV treatment	Nishizawa et al. (2008)
Medicago falcata	*MfGolS1*	*Medicago falcata*	Raffinose accumulation Cold and drought tolerance	Zhuo et al. (2012)

AtGolS2, *Arabidopsis thaliana* galactinol synthase, *MfGolS1*, *Medicago falcata* galactinol synthase, *Lea-Gal gene*, *Lycopersicon esculentum* α-galactosidase, *Os UGE*, *Oryza sativa* UDP-glucose 4-epimerase

antioxidants. Stress-induced signaling in plants has been reported to be carried out efficiently by proline, glycine betaine, sucrose, trehalose, and various sugar alcohols. Additionally, proline and glycine betaine both help in maintaining the NAD/NADH ratio in the plant cell. The normal functioning of chloroplasts even under cold stress is maintained by cryoprotectants such as glycine betaine and trehalose by imparting protective effect on PSII. Since these biomolecules play crucial role in stress tolerance but these are not accumulated universally in all plants and even in accumulators, the levels are not sufficiently high to be able to mitigate the stress. Thus, to increase their availability, plants can be exogenously fed upon these cryoprotectants or plants may be genetically modified to overexpress enzymes catalyzing their synthesis. Considerable progress has been made in this regard in the recent times, but the transgenics in many cases have failed to deliver the desired results under field conditions. Hence, persistent efforts are required to specify the key strategies that plants exercise to combat a complex of stresses that they are being exposed to under field conditions.

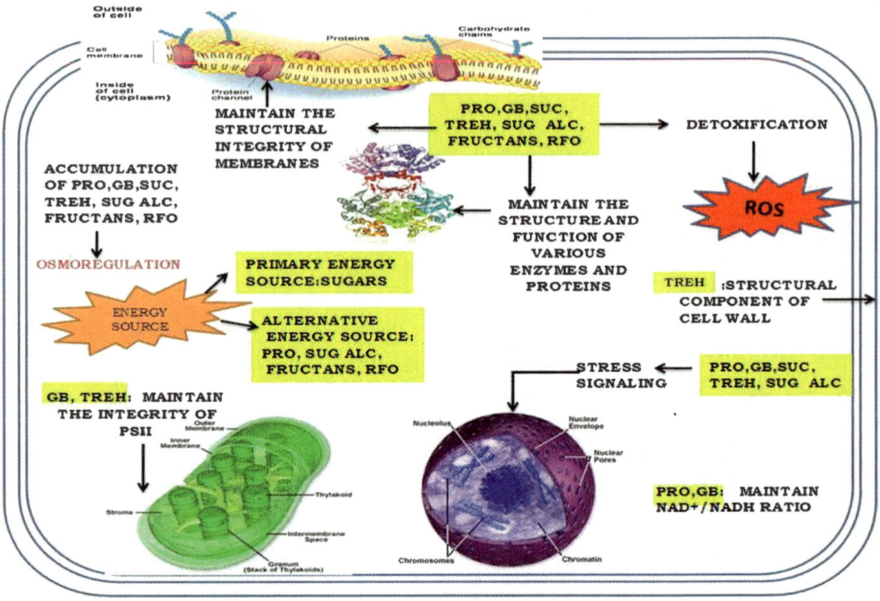

Fig. 9.26 Summary of protective roles of various cryoprotectants in a plant cell under cold-stressed conditions. *Pro* proline, *GB* glycine betaine, *Suc* sucrose, *Treh* trehalose, *Sug Alc*: sugar alcohol, *RFO* raffinose family oligosaccharides

References

Acevedo-Hernandez GJ, Leon P, Herrera-Estrella LR (2005) Sugar and ABA responsiveness of a minimal RBCS light responsive unit is mediated by direct binding of ABI4. Plant J 43:506–519

Adams R, Kendall E, Kartha KK (1990) Comparison of free sugars in growing and desiccated plants of *Selaginella lepidophylla*. Biochem Syst Ecol 18:107–110

Adams P, Zegeer A, Bohnert HJ, Jensen RG (1993) Anion exchange separation and amperometric detection of inositols from flower petals. Anal Biochem 214:321–324

Aghaee A, Moradi F, Zare-Maivan H, Zarinkamar F, Pour Irandoost H, Sharifi P (2011) Physiological responses of two rice (*Oryza sativa* L.) genotypes to chilling stress at seedling stage. Afr J Biotechnol 10:7617–7621

Ahmad I, Larher F, Stewart GR (1979) Sorbitol, a compatible osmotic solute in Plantago maritima. New Phytol 82:671–678

Alia PS, Mohanty P (1997) Involvement of proline in protecting thylakoid membranes against free radical-induced photodamage. J Photochem Photobiol B 38:253–257

Alia A, Pardha Saradhi P, Mohanty P (1991) Proline enhances primary photochemical activities in isolated thylakoid membranes of *Brassica juncea* by arresting photo-inhibitory damage. Biochem Biophys Res Commun 181:1238–1244

Alia KY, Sakamoto A, Nonaka H, Hayashi H, Saradhi PP, Chen THH, Murata N (1999) Enhanced tolerance to light stress of transgenic *Arabidopsis* plants that express the *codA* gene for a bacterial choline oxidase. Plant Mol Biol 40:279–288

Allakhverdiev SI, Feyziev Ya M, Ahmed A, Hayashi H, Aliev Ja A, Klimov VV, Murata N, Carpentier R (1996) Stabilization of oxygen evolution and primary electron transport reactions in photosystem II against heat stress with glycine betaine and sucrose. J Photochem Photobiol B 34:149–157

Allard F, Houde M, Krol M, Ivanov A, Huner NPA, Sarhan F (1998) Betaine improves freezing tolerance in wheat. Plant Cell Physiol 39:1194–1202

Almeida AM, Villalobos E, Araújo SS, Leyman B, van Dijk P, Alfaro-Cardoso L, Fevereiro PS, Torné JM, Santos DM (2005) Transformation of tobacco with an *Arabidopsis thaliana* gene involved in trehalose bio-synthesis increases tolerance to several abiotic stresses. Euphytica 146:165–176

An D, Yang J, Zhang P (2012) Transcriptome profiling of low temperature-treated cassava apical shoots showed dynamic responses of tropical plant to cold stress. BMC Genomics 13:64

Anchordoguy TJ, Rdolph AS, Carpenter JF, Crowe JH (1987) Modes of interaction of cryoprotectants with membrane phospholipids during freezing. Cryobiology 24:324–331

Armengaud P, Thiery L, Buhot N, Grenier-De March G, Savoure A (2004) Transcriptional regulation of proline biosynthesis in *Medicago truncatula* reveals developmental and environmental specific features. Physiol Plant 120:442–450

Ashraf M, Foolad MR (2007) Roles of glycine betaine and proline in improving plant abiotic stress resistance. Environ Exp Bot 59:206–216

Atkinson DE (1977) Cellular energy metabolism and its regulation. Academic, New York

Ayre BG, Keller F, Turgeon R (2003) Symplastic continuity between companion cells and the translocation stream. Long distance transport is controlled by retention and retrieval mechanisms in the phloem. Plant Physiol 131:1518–1528

Azymi S, Sofalian O, Jahanbakhsh G, Khomari S (2012) Effect of chilling stress on soluble protein, sugar and proline accumulation in cotton (*Gossypium hirsutum* L.) genotypes. Int J Agric Crop Sci 4:825–830

Bachmann M, Keller F (1995) Metabolism of the raffinose family oligosaccharides in leaves of *Ajuga reptans* L. inter and intracellular compartmentation. Plant Physiol 109:991–998

Bachmann M, Matile P, Keller F (1994) Metabolism of the raffinose family oligosaccharides in leaves of *Ajuga reptans* L. Cold acclimation, translocation, and sink to source transition: discovery of a chain elongation enzyme. Plant Physiol 105:1335–1345

Baroja-Fernandez E, Munoz FJ, Montero M, Etxeberria E, Sesma MT, Ovecka M, Bahaji A, Ezquer I, Li J, Prat S, Pozueta-Romero J (2009) Enhancing sucrose synthase activity in transgenic potato (*Solanum tuberosum* L.) tubers results in increased levels of starch, ADP glucose and UDP glucose and total yield. Plant Cell Physiol 50:1651–1662

Beck EH, Fettig S, Knake C, Hartig K, Bhattarai T (2007) Specific and unspecific responses of plants to cold and drought stress. J Biosci 32:501–510

Belisle JT, Vissa VD, Sievert T, Takayama K, Brennan PJ, Bersa GS (1997) Role of the major antigen of Mycobacterium tuberculosis in cell wall biogenesis. Science 276:1420–1422

Benaroudj N, Lee DH, Goldberg AL (2001) Trehalose accumulation during cellular stress protects cells and cellular proteins from damage by oxygen radicals. J Biol Chem 276:24261–24267

Berridge MJ (1997) Elementary and global aspects of calcium signaling. J Physiol 499:291–306

Bhowmik PK, Tamura K, Sanada Y, Tase K, Yamada T (2006) Sucrose metabolism of perennial ryegrass in relation to cold acclimation. Z Naturforsch C 61:99–104

Bieleski RL (1982) Sugar alcohols. In: Loewus A, Tanner W (eds) Encyclopedia of plant physiology, plant carbohydrates, intracellular carbohydrates. Springer, Berlin, pp 158–192

Blacklow WM, Darbyshire B, Pheloung P (1984) Fructans polymerised and depolymerised in the internodes of winter wheat as grain-filling progressed. Plant Sci Lett 36:213–218

Blackman SA, Obendorf RL, Leopold AC (1992) Maturation proteins and sugars in desiccation tolerance of developing soybean seeds. Plant Physiol 100:225–230

Bohnert HJ, Jensen RG (1996) Strategies for engineering water stress tolerance in plants. Trends Biotechnol 14:89–97

Bohnert HJ, Nelson DE, Jensen RG (1995) Adaptations to environmental stresses. Plant Cell 7:1099–1111

Bolouri-Moghaddam MR, Le Roy K, Xiang L, Rolland F, Van den Ende W (2010) Sugar signalling and antioxidant network connections in plant cells. FEBS J 277:2022–2037

Bonnett GD, Sims IM, John JAS, Simpson RJ (1994) Purification and characterization of fructans with b-2,1- and b-2,6-glycosidic linkages suitable for enzyme studies. New Phytol 127:261–269

Bourot S, Sire O, Trautwetter A, Touze T, Wu F, Blanco C, Bernard T (2000) Glycine betaine-assisted protein folding in alysA mutant of *Escherichia coli*. J Biol Chem 275:1050–1056

Brenac P, Horbowicz M, Downer SM, Dickerman AM, Smith ME, Obendorf RL (1997) Raffinose accumulation related to desiccation tolerance during maize (*Zea mays* L.) seed development and maturation. J Plant Physiol 150:481–488

Brown AD (1990) Microbial water stress physiology, principles and perspectives. Wiley, New York

Burbulis N, Jonytiene V, Kupriene R, Blinstrubiene A (2011) Changes in proline and soluble sugars content during cold acclimation of winter rape-seed shoots in vitro. J Food Agric Environ 9:371–374

Burbulis N, Jonytienė V, Kupriene R, Blinstrubienė A, Vytautas Liakas V (2012) Biochemical and physiological factors related to cold de-acclimation and by trehalose-6-phosphate. Plant Physiol 149:1860–1871

Caffrey M, Fonseca V, Leopold AC (1988) Lipid-sugar interactions. Plant Physiol 86:754–758

Cairns AJ (2003) Fructan biosynthesis in transgenic plants. J Exp Bot 54:549–567

Castonguay Y, Nadeau P (1998) Enzymatic control of soluble carbohydrate accumulation in cold-acclimated crowns of alfalfa. Crop Sci 38:1183–1189

Chalmers J, Johnson X, Lidgett A, Spangenberg G (2003) Isolation and characterisation of a sucrose: sucrose 1-fructosyltransferase gene from perennial ryegrass (*Lolium perenne*). J Plant Physiol 160:1385–1391

Chalmers J, Lidgett A, Cummings N, Cao Y, Forster J, Spangenberg G (2005) Molecular genetics of fructan metabolism in perennial ryegrass. Plant Biotechnol J 3:459–474

Chatterjee D (1997) The mycobacterial cell wall: structure, biosynthesis and sites of drug action. Curr Opin Chem Biol 1:578–588

Chen SL, Kao CH (1995) Cd induced changes in proline level and peroxidase activity in roots of rice seedlings. Plant Growth Regul 17:67–71

Chen THH, Murata N (2008) Glycine betaine: an effective protectant against abiotic stress in plants. Trends Plant Sci 13:499–505

Chen THH, Murata N (2011) Glycine betaine protects plants against abiotic stress: mechanisms and biotechnological applications. Plant Cell Environ 34:1–20

Chen WP, Li PH, Chen THH (2000) Glycine betaine increases chilling tolerance and reduces chilling-induced lipid peroxidation in *Zea mays* L. Plant Cell Environ 23:609–661

Chen Y, Chen X, Wang N, Hu F, Cao FL (2012) Physiological response to cold stress and cell ultrastructure changes in *Passiflora edulis* leaves. Acta Bot Bor Occid Sin 3:019

Chiang YJ, Stushnoff C, McSay AE (2005) Overexpression of mannitol-1-phosphate dehydrogenase increases mannitol accumulation and adds protection against chilling injury in petunia J Am Soc Hortic Sci 130:605–610

Clegg JS (1985) The physical properties and metabolic status of *Artemia* cysts at low water contents: "The water replacement hypothesis". In: Leopold AC (ed) Membranes, metabolism and dry organisms. Cornell University Press, Ithaca, NY, pp 169–187

Communi D, Lecocq R, Vanweyenberg V, Erneux C (1995) Active site labeling of inositol 1,4,5-trisphosphate 3-kinase A by phenylglyoxal. Biochem J 310:109–115

Cook D, Fowler S, Fiehn O, Thomashow MF (2004) A prominent role for the CBF cold response pathway in configuring the low temperature metabolome of *Arabidopsis*. Proc Natl Acad Sc U S A 101:15243–15248

Cortina C, Culianez-Macia FA (2005) Tomato abiotic stress enhanced tolerance by trehalose biosynthesis. Plant Sci 169:75–82

Crespi MD, Zabaleta EJ, Pontis HG, Salerno GL (1991) Sucrose synthase expression during cold acclimation in wheat. Plant Physiol 96:887–891

Crowe JH, Crowe LM (1990) Lyotropic effects of water on phospholipids. In: Franks F (ed) Water science reviews. Cambridge University Press, Cambridge, England, pp 1–23

Crowe JH, Crowe LM, Carpenter JF, Rudolph AS, Wistrom CA, Spargo BJ, Anchordoguy TJ (1988) Interactions of sugars with membranes. Biochem Biophys Acta 947:367–384

Damiani CR, Pistelli L, Guglielminetti L (2012) Fructan metabolism in tall fescue calli under different environmental condition. Afr J Biotechnol 11:10953–10961

de Halleux S, Van Cutsem P (1997) Cloning and sequencing of the 1-SST cDNA from chicory root (accession no. U81520) (PGR97-036). Plant Physiol 113:1003

del Viso F, Casabuono AC, Couto AS, Hoppa HE, Puebla AF, Heinz RA (2011) Functional characterization of a sucrose:fructan 6-fructosyltransferase of the cold-resistant grass *Bromus pictus* by heterologous expression in *Pichia pastoris* and *Nicotiana tabacum* and its involvement in freezing tolerance. J Plant Physiol 168:493–499

Delatte TL, Sedijani P, Kondou Y, Matsui M, de Jong GJ, Govert W, Somsen GW, Wiese-Klinkenberg A, Primavesi LF, Paul MJ, Schluepmann H (2011) Growth arrest by trehalose-6-phosphate: an astonishing case of primary metabolite control over growth by way of the SnRK1 signaling pathway. Plant Physiol 157:160–174

Delauney AJ, Verma DPS (1993) Proline biosynthesis and osmoregulation in plants. Plant J 4:215–223

Deuschle K, Funck D, Forlani G, Stransky H, Biehl A, Leister D, van der Graaff E, Kunzee R, Frommer WB (2004) The role of delta(1)-pyrroline-5-carboxylate dehydrogenase in proline degradation. Plant Cell 16:3413–3425

Dey PM (1985) D-galactose containing oligosaccharides. In: Dey PM, Dixon RA (eds) The biochemistry of storage carbohydrates in green plants. Academic, London, pp 53–129

Diamant S, Eliahu N, Rosenthal D, Goloubinoff P (2001) Chemical chaperones regulate molecular chaperones in vitro and in cells under combined salt and heat stresses. J Biol Chem 276:39586–39591

Dombrowski JE, Martin RC (2009) Evaluation of reference genes for quantitative RT-PCR in *Lolium temulentum* under abiotic stress. Plant Sci 176:390–396

Downie B, Gurusinghe S, Dahal P, Thacker RR, Snyder JC, Nonogaki H, Yim K, Fukanaga K, Alvarado V, Bradford KJ (2003) Expression of a GALACTINOL SYNTHASE gene in tomato seeds is up-regulated before maturation desiccation and again after imbibition whenever radicle protrusion is prevented. Plant Physiol 131:1347–1359

Duncan DR, Widholm JM (1987) Proline accumulation and its implication in cold tolerance of regenerable maize. Plant Physiol 83:703–708

Eagles CF, Williams J, Louis DV (1993) Recovery after freezing in *Avena sativa* L., *Lolium perenne* L. and *L. multiflorum* Lam. New Phytol 123:477–483

Eastmond PJ, Li Y, Graham IA (2003) Trehalose-6-phosphate a regulator of sugar metabolism in plants? J Exp Bot 54:533–537

Ebskamp MJM, van der Meer IM, Spronk BA, Weisbeek PJ, Smeekens SCM (1994) Accumulation of fructose polymers in transgenic tobacco. Biotechnology 12:272–275

Einset J, Connolly EL (2009) Glycine betaine enhances extracellular processes blocking ROS signaling during stress. Plant Signal Behav 4:197–199

Einset J, Nielsen E, Connolly EL, Bones A, Sparstad T, Winged P, Zhue JK (2007a) Membrane-trafficking RabA4c involved in the effect of glycine betaine on recovery from chilling stress in *Arabidopsis*. Physiol Plant 130:511–518

Einset J, Winge P, Bones A (2007b) ROS signaling pathways in chilling stress. Plant Signal Behav 2:365–367

El-Bashiti T-E, Hamamci H, Oktem HA, Yucel M (2005) Biochemical analysis of trehalose and its metabolising enzymes in wheat under abiotic stress conditions. Plant Sci 169:47–54

Elbein AD, Pan YT, Pastuszak I, Carrol D (2003) New insights on trehalose: a multifunctional molecule. Glycobiology 13:17–27

Eun SS, Jia G, Young HK, Joon-Hyeong C, Chun KL, Jang HH, Myeong-Hyeon W (2007) Regulations of marker genes involved in biotic and abiotic stress by overexpression of the AtNDPK2 gene in rice. Biochem Biophys Res Commun 363:126–132

Fallon KM, Phillips R (1989) Responses to water stress in adapted carrot cell suspension cultures. J Exp Bot 40:681–687
Fan W, Zhang M, Zhang H, Zhang P (2012) Improved tolerance to various abiotic stresses in transgenic sweet potato (*Ipomoea batatas*) expressing spinach betaine aldehyde dehydrogenase. PLoS One 7:e37344
Farooq M, Aziz T, Hussain M, Rehman H, Jabran K, Khan MB (2008) Glycine betaine improves chilling tolerance in hybrid maize. J Agron Crop Sci 194:152–160
Floyd RA, Nagy ZS (1984) Formation of long-lived hydroxyl free radical adducts of proline and hydroxyproline in a Fenton reaction. Biochim Biophys Acta 790:94–97
Franzini E, Sellak H, Hakim J, Pasquier C (1994) Comparative sugar degradation by (OH^-) produced by the iron-driven Fenton reaction and gamma radiolysis. Arch Biochem Biophys 309:261–265
Frehner M, Keller F, Wiemken A (1984) Localization of fructan metabolism in the vacuoles isolated from protoplasts of Jerusalem artichoke. J Plant Physiol 116:197–208
French AD, Waterhouse AL (1993) Chemical structure and characteristics. In: Suzuki M, Chatterton NJ (eds) Science and technology of fructans. CRC Press, Boca Raton, pp 41–81
Fritzius T, Aeschbacher R, Wiemken A, Wingler A (2001) Induction of ApL3 expression by trehalose complements the starch-deficient *Arabidopsis* mutant adg2-1 lacking ApL1, the large subunit of ADP-glucose pyrophosphorylase. Plant Physiol 126:883–889
Funck D, Stadelhofer B, Wolfgang Koch W (2008) Ornithine-δ aminotransferase is essential for arginine catabolism but not for proline biosynthesis. BMC Plant Biol 8:40
Garcia AB, Engler AJ, Van Montagu M, Caplan A (1993) Effect of NaCl and osmoprotectants on growth inhibition, ion accumulation, and gene expression in rice (*Oryza sativa*). In: Close TJ, Bray EA (eds) Plant responses to cellular dehydration during environmental stress. ASPP, Washington, DC, pp 272–273
Garg AK, Kim J-K, Owens TG, Ranwala AP, Choi YD, Kochian LV, Wu RJ (2002) Trehalose accumulation in rice plants confers high tolerance levels to different abiotic stresses. Proc Natl Acad Sci U S A 99:15898–15903
Ge LF, Chao DY, Shi M, Zhu MZ, Gao JP, Lin HX (2008) Over-expression of the trehalose-6-phosphate phosphatase gene OsTPP1 confers stress tolerance in rice and results inactivation of stress responsive genes. Planta 228:191–201
Gibon Y, Bessieres MA, Larher F (1997) Is glycine betaines a non-compatible solute in higher plants that do not accumulate it? Plant Cell Environ 20:329–340
Gilbert GA, Wilson C, Madore MA (1997) Root-zone salinity alters raffinose oligosaccharide metabolism and transport in *Coleus*. Plant Physiol 115:1267–1276
Gilmour SJ, Sebolt AM, Salazar MP, Everard JD, Thomashow MF (2000) Over-expression of the *Arabidopsis* CBF3 transcriptional activator mimics multiple biochemical changes associated with cold acclimation. Plant Physiol 124:1854–1865
Giri J (2011) Glycine betaine and abiotic stress tolerance in plants. Plant Signal Behav 6:1746–1751
Goddjin OJM, Van DK (1999) Trehalose metabolism in plants. Trends Plant Sci 4:315–319
Gomaa AM, Raldugina GN, Burmistrova NA, Radionov NV, Kuznetsov VV (2012) Response of transgenic rape plants bearing the Osmyb4 gene from rice encoding a trans-factor to low above-zero temperature. Russ J Plant Physiol 59:105–114
Gorham J (1995) Betaines in higher plants: biosynthesis and role in stress metabolism. In: Wallsgrove RM (ed) Amino acids and their derivatives in higher plants. Cambridge University Press, Cambridge, pp 171–203
Grallath S, Weimar T, Meyer A, Gumy C, Suter-Grotemeyer M, Neuhaus JM, Rentsch D (2005) The AtProT family compatible solute transporters with similar substrate specificity but differential expression patterns. Plant Physiol 137:117–126
Gudleifsson BE (2010) Ice tolerance and metabolite accumulation of herbage crops in Iceland and impact of climate change. Icel Agric Sci 23:111–122
Guerra-Peraza O, Nguyen HT, Stamp P, Jörg L (2009) ZmCOI6.1, *a* novel, alternatively spliced maize gene, whose transcript level changes under abiotic stress. Plant Sci 176:783–791

Gupta AK, Kaur N (2005) Sugar signaling and gene expression in relation to carbohydrate metabolism under abiotic stresses in plants. J Biosci 30:761–776
Guy CL, Yelenosky G, Sweet HC (1980) Light exposure and soluble sugars in Citrus frost hardiness. Fla Sci 43:265–268
Halford NG, Hey SJ (2009) Snf1-related protein kinases (SnRKs) act within an intricate network that links metabolic and stress signalling in plants. Biochem J 419:247–259
Hamilton EW, Heckathorn SA (2001) Mitochondrial adaptations to NaCl complex I is protected by antioxidants and small heat shock proteins, whereas complex II is protected by proline and betaine. Plant Physiol 126:1266–1274
Hanhong B, Eliot H, Richard S (2005) Exogenous trehalose promotes non-structural carbohydrate accumulation and induces chemical detoxification and stress response proteins in *Arabidopsis* thaliana grown in liquid culture. Plant Sci 168:1293–1301
Hanson AD, May AM, Grumet R, Bode J, Jamieson GC, Rhodes D (1985) Betaine synthesis in chenopods: localization in chloroplasts. Proc Natl Acad Sci U S A 82:3678–3682
Hardie DG (2007) AMP-activated/SNF1 protein kinases: conserved guardians of cellular energy. Nat Rev Mol Cell Biol 8:774–785
Hare PD, Cress WA (1997) Metabolic implications of stress-induced proline accumulation in plants. Plant Growth Regul 21:79–102
Hare PD, Cress WA, van Staden J (1999) Proline synthesis and degradation: a model system for elucidating stress-related signal transduction. J Exp Bot 50:413–434
Harris W, Lynch PT, Hargreavs AJ, Bonner PLR (2004) Cryopreservation of *Helianthus tuberosus* cell suspension cultures: the effect of pre-culture treatments on cytoskeletal proteins and trans glutaminase activity. Cryo Lett 25:213–217
Hawker JS (1977) Enzymes concerned with sucrose synthesis and transformations in seeds of maize, broad bean and castor bean. Phytochemistry 10:2313–2322
Hayashi H, Alia KY, Mustardy L, Deshnium P, Ida M, Murata N (1997) Transformation of *Arabidopsis thaliana* with the *cod A* gene for choline oxidase; accumulation of glycine betaine and enhanced tolerance to salt and cold stress. Plant J 12:133–142
Hayat S, Hayat Q, Alyemeni MN, Wani AS, Pichtel J (2012) Role of proline under changing environments: a review. Plant Signal Behav 7:1456–1466
Heber U, Tyankova L, Santarius KA (1973) Effects of freezing on biological membranes in vivo and in vitro. Biochim Biophys Acta 291:23–37
Hellmann H, Funck D, Rentsch D, Frommer WB (2000) Hypersensitivity of an *Arabidopsis* sugar signaling mutant toward exogenous proline application. Plant Physiol 122:357–367
Hendry G (1987) The ecological significance of fructan in contemporary flora. New Phytol 106:201–216
Hernandez-Marin E, Martínez A (2012) Carbohydrates and their free radical scavenging capability: a theoretical study. J Phys Chem 116:9668–9675
Hill TD, Dean NM, Boyton AL (1988) Inositol 1, 3, 4, 5-tetrakisphosphate induces Ca^{2+} sequestration in rat liver cells. Science 242:1176–1178
Hincha DK, Hellwege EM, Heyer AG, Crowe JH (2000) Plant fructans stabilize phosphatidylcholine liposomes during freeze-drying. Eur J Biochem 267:535–540
Hincha DK, Zuther E, Heyer AG (2003) The preservation of liposomes by raffinose family oligosaccharides during drying is mediated by effects on fusion and lipid phase transitions. Biochim Biophys Acta 1612:172–177
Hinesley LE, Pharr DM, Snelling LK, Funderburk SR (1992) Foliar raffinose and sucrose in four conifer species: relationship to seasonal temperature. J Am Soc Hortic Sci 117:852–855
Hirasawa R, Yokoigawa K, Isobe Y, Kawai H (2001) Improving the freeze tolerance of baker's yeast by loading with trehalose. Biosci Biotechnol Biochem 65:522–526
Hisano H, Kanazawa A, Kawakami A, Yoshida M, Shimamoto Y, Yamada T (2004) Transgenic perennial ryegrass plants expressing wheat fructosyltransferase genes accumulate increased amounts of fructan and acquire increased tolerance on a cellular level to freezing. Plant Sci 167:861–868

Hoekstra FA, Golovina EA, Buitink J (2001) Mechanisms of plant desiccation tolerance. Trends Plant Sci 6:431–438
Hoffmann T, Bremer E (2011) Protection of *Bacillus subtilis* against cold stress via compatible-solute acquisition. J Bacteriol 193:1552–1562
Holmstrom K, Somersalo S, Mandal A, Palva TE, Welin B (2000) Improved tolerance to salinity and low temperature in transgenic tobacco producing glycine betaine. J Exp Bot 51:177–185
Hossain MA, Cho JI, Han M, Ahn CH, Jeon JS, An G, Park PB (2010) The *ABRE*-binding bZIP transcription factor OsABF2 is a positive regulator of abiotic stress and ABA signaling in rice. J Plant Physiol 167:1512–1520
Hu CAA, Delauney AJ, Verma DPS (1992) A bifunctional enzyme (Δ^1-pyrroline-5-carboxylate synthetase) catalyzes the first two steps in proline biosynthesis in plants. Proc Natl Acad Sci U S A 89:9354–9358
Hudak J, Salaj J (1999) Effect of low temperatures on the structure of plant cells. In: Pessarakli M (ed) Handbook of plant and crop stress. Marcel Dekker, New York, pp 441–464
Hurry V, Strand Å, Furbank R, Stitt M (2000) The role of inorganic phosphate in the development of freezing tolerance and the acclimatization of photosynthesis to low temperature is revealed by the pho mutants of *Arabidopsis thaliana*. Plant J 24:383–396
Igarashi Y, Yoshiba Y, Sanada Y, Yamaguchi-Shinozaki K, Wada K, Shinozaki K (1997) Characterisation of the gene for D1-pyrroline-5-carboxylate synthetase and correlation between the expression of the gene and salt tolerance in *Oryza sativa* L. Plant Mol Biol 33:857–865
Iordachescu M, Imai R (2011) Trehalose and abiotic stress in biological systems. In: Shanker AK, Venkateswarlu B (eds) Abiotic stress in plants: mechanisms and adaptations. InTech, Croatia, pp 215–234
Isbat M, Zeba N, Kim SR, Hong CB (2009) A BAX inhibitor-1 gene in *Capsicum annuum* is induced under various abiotic stresses and endows multi-tolerance in transgenic tobacco. J Plant Physiol 166:1685–1693
Islam MM, Hoque MA, Okuma E, Banu MNA, Shimoishi Y, Nakamura Y, Murata Y (2009) Exogenous proline and glycinebetaine increase antioxidant enzyme activities and confer tolerance to cadmium stress in cultured tobacco cells. J Plant Physiol 166:1587–1597
Iturriaga G, Suárez R, Nova-Franco B (2009) Trehalose metabolism: from osmoprotection to signaling. Int J Mol Sci 10:3793–3810
Jang IC, Oh SJ, Seo JS, Choi WB, Song SI, Kim CH, Kim YS, Seo HS, Choi YD, Nahm BH, Kim JK (2003) Expression of a bifunctional fusion of the *Escherichia coli* genes for trehalose-6-phosphate synthase and trehalose-6-phosphate phosphatase in transgenic rice plants increases trehalose accumulation and abiotic stress tolerance without stunting growth. Plant Physiol 131:516–524
Janmohammadi M, Enayati V, Sabaghnia N (2012) Impact of cold acclimation, de-acclimation and re-acclimation on carbohydrate content and antioxidant enzyme activities in spring and winter wheat. Icel Agric Sci 25:3–11
Janská A, Mars P, Zelenková S, Ovesna J (2010) Cold stress and acclimation-what is important for metabolic adjustment? Plant Biol 12:395–405
Jennings DB, Ehrenshaft M, Pharr DM, Williamson JD (1998) Roles for mannitol and mannitol dehydrogenase in active oxygen mediated plant defense. Proc Natl Acad Sci U S A 95 15129–15133
Jeong J, Cohu C, Kerkeb L, Pilon M, Connolly EL, Guerinot ML (2008) Chloroplast Fe(III) chelate reductase activity is essential for seedling viability under iron limiting conditions. Proc Natl Acad Sci U S A 105:10619–10624
Ji XM, Van den Ende W, Schroeven L, Clerens S, Geuten K, Cheng S, Bennett J (2007) The rice genome encodes two vacuolar invertases with fructan exohydrolase activity but lacks the related fructan biosynthesis genes of the Pooideae. New Phytol 173:50–62
Jian X, Zhang L, Li G, Zhang L, Wang X, Cao X, Fang X, Chen F (2010) Identification of novel stress-regulated microRNAs from *Oryza sativa* L. Genomics 95:47–55
Jitsuyama Y, Suzuki T, Harada T, Fujikawa S (1997) Ultrastructural study on mechanism of increased freezing tolerance due to extracellular glucose in cabbage cells. Cryo Lett 18:33–44

Jonytiene V, Burbulis N, Kupriene R, Blinstrubiene A (2012) Effect of exogenous proline and de-acclimation treatment on cold tolerance in *Brassica napus* shoots cultured in vitro. J Food Agric Environ 10:327–330

Joshi SS (1984) Effect of salinity stress on organic and mineral constituents in the leaves of pigeon pea (*Cajanus cajan* L. Var. C-11). Plant Soil 82:69–76

Josine TL, Ji J, Wang G, Guan CF (2011) Advances in genetic engineering for plants abiotic stress control. Afr J Biotechnol 10:5402–5413

Kandler O, Hopf H (1982) Oligosaccharides based on sucrose-sucrosyl oligosaccharides. In: Person A, Zimmerman MM (eds) Encyclopedia of plant physiology-new series. Springer, New York, pp 348–383

Kandler O, Hopf H (1984) Biosynthesis of oligosaccharides in vascular plants. In: Lewis DH (ed) Storage carbohydrates in vascular plants. Cambridge University Press, Cambridge, UK, pp 115–131

Kang HG, Kim J, Kim B, Jeong H, Choi SH, Kim EK, Lee HY, Lim PO (2011) Over-expression of FTL1/DDF1, an AP2 transcription factor, enhances tolerance to cold, drought, and heat stresses in *Arabidopsis thaliana*. Plant Sci 180:634–641

Kaplan F, Kopka J, Haskel DW, Zhao W, Schiller KC, Gatzke N, Sung DY, Guy CL (2004) Exploring the temperature-stress metabolome of *Arabidopsis*. Plant Physiol 136:4159–4168

Karakas B, Ozias-Akins P, Stushnoff C, Suefferheld M, Rieger M (1997) Salinity and drought tolerance in mannitol-accumulating transgenic tobacco. Plant Cell Environ 20:609–616

Karim S, Aronsson H, Ericson H, Pirhonem M, Leyman B, Welin B, Mantyla E, Palva ET, Van Dijck P, Holmstrom KO (2007) Improved drought tolerance without undesired side effects in transgenic plants producing trehalose. Plant Mol Biol 64:371–386

Kathuria H, Giri J, Nataraja KN, Murata N, Udayakumar M, Tyagi AK (2009) Glycine betaine-induced water-stress tolerance in *cod A*-expressing transgenic indica rice is associated with up-regulation of several stress responsive genes. Plant Biotechnol J 7:512–526

Kaur G, Kumar S, Nayyar H, Upadhyaya HD (2008) Cold stress injury during the pod-filling phase in chickpea (*Cicer arietinum* L.): effects on quantitative and qualitative components of seeds. J Agron Crop Sci 194:457–464

Kaur S, Gupta AK, Kaur N, Sandhu JS, Gupta SK (2009) Antioxidative enzymes and sucrose synthase contribute to cold stress tolerance in chickpea. J Agron Crop Sci 195:393–397

Kaur G, Kumar S, Thakur P, Malik JA, Bhandhari K, Sharma KD, Nayyar H (2011) Involvement of proline in response of chickpea (*Cicer arietinum* L.) to chilling stress at reproductive stage. Sci Hortic 128:174–181

Kavi Kishor PB, Hong Z, Miao GH, Hu CAA, Verma DPS (1995) Over-expression of [delta]-pyrroline-5-carboxylate synthetase increases proline production and confers osmotolerance in transgenic plants. Plant Physiol 108:1387–1394

Kavi Kishor PB, Sangam S, Amrutha RN, Laxmi PS, Naidu KR, Rao KRSS, Rao S, Reddy KJ, Theriappan P, Sreenivasulu N (2005) Regulation of proline biosynthesis, degradation, uptake and transport in higher plants: its implications in plant growth and abiotic stress tolerance. Curr Sci 88:424–438

Kawakami A, Yoshida M (2002) Molecular characterization of sucrose: sucrose 1-fructosyltransferase and sucrose:fructan 6-fructosyltransferase associated with fructan accumulation in winter wheat during cold hardening. Biosci Biotechnol Biochem 66:2297–2305

Kawakami A, Yoshida M (2005) Fructan: fructan 1-fructosyltransferase, a key enzyme for biosynthesis of graminan oligomers in hardened wheat. Planta 223:90–104

Kawakami A, Sato Y, Yoshida M (2008) Genetic engineering of rice capable of synthesizing fructans and enhancing chilling tolerance. J Exp Bot 59:793–802

Keller F, Pharr DM (1996) Metabolism of carbohydrates in sinks and sources: galactosyl-sucrose oligosaccharides. In: Zamski E, Schaffer AA (eds) Photoassimilate distribution in plants and crops. Marcel Dekker, New York, pp 115–184

Kerepesi I, Galiba G, Bányai-Stefanovits E (1998) Osmotic and salt stresses induced differential alteration in water-soluble carbohydrate content in wheat seedlings. J Agric Food Chem 46: 5347–5354

Kerepesi I, Bányai-Stefanovits E, Galiba G (2004) Cold acclimation and abscisic acid induced alterations in carbohydrate content in calli of wheat genotypes differing in frost tolerance. J Plant Physiol 161:131–133

Keunen E, Peshev D, Vangronsveld J, Van den Ende W, Cuypers A (2013) Alternative respiration as a primary defence during cadmium-induced mitochondrial oxidative challenge in *Arabidopsis thaliana*. Environ Exp Bot 91:63–73

Khare N, Goyary D, Singh NK, Shah P, Rathore M, Anandhan S, Sharma D, Arif M, Ahmed Z (2010) Transgenic tomato cv. Pusa Uphar expressing a bacterial mannitol-1-phosphate dehydrogenase gene confers abiotic stress tolerance. Plant Cell Tissue Organ Cult 103:267–277

Kishitani S, Watanabe K, Yasuda S, Arakawa K, Takabe T (1994) Accumulation of glycine betaine during cold acclimation and freezing tolerance in leaves of winter and spring barley plants. Plant Cell Environ 17:89–95

Klima M, Vitamvas P, Zelenkova S, Vyvadilova M, Prasil IT (2012) Dehydrin and proline content in *Brassica napus* and *B. carinata* under cold stress at two irradiances. Biol Plant 56:157–161

Klotke J, Kopka J, Gatzke N, Heyer AG (2004) Impact of soluble sugar concentrations on the acquisition of freezing tolerance in accessions of *Arabidopsis thaliana* with contrasting cold adaptation—evidence for a role of raffinose in cold acclimation. Plant Cell Environ 27: 1395–1404

Kohl DH, Schubert KR, Carter MB, Hagedorn CH, Shearer G (1988) Proline metabolism in N_2-fixing root nodules: energy transfer and regulation of purine synthesis. Proc Natl Acad Sci U S A 85:2036–2040

Kolbe A, Tiesse A, Schluepmann H, Paul M, Ulrich S, Geigenberger P (2005) Trehalose 6-phosphate regulates starch synthesis via post-translational redox activation of ADP-glucose pyrophosphorylase. Proc Natl Acad Sci U S A 102:11118–11123

Kong X, Pan J, Zhang M, Xing X, Zhou Y, Liu Y, Li D (2011) ZmMKK4, a novel group C mitogen-activated protein kinase kinase in maize (*Zea mays*), confers salt and cold tolerance in transgenic *Arabidopsis*. Plant Cell Environ 34:1291–1303

Konstantinova T, Parvanova D, Atanassov A, Djilianov D (2002) Freezing tolerant tobacco, transformed to accumulate osmo-protectants. Plant Sci 163:157–164

Koops AJ, Jonker HH (1996) Purification and characterization of the enzymes of fructan biosynthesis in tubers of *Helianthus tuberosus* (Colombia). Purification of sucrose-sucrose 1-fructosyl transferase and reconstitution of fructan synthesis *in vitro* with purified sucrose-sucrose 1-fructosyl transferase and fructan fructan 1-fructosyl transferase. Plant Physiol 110: 1167–1175

Koster KL, Lynch DV (1992) Solute accumulation and compartmentation during the cold accumulation of Puma Rye. Plant Physiol 98:108–113

Krackhardt M, Guerrier G (1995) Effect of osmotic and ionic stresses on proline and organic acid contents during imbibition and germination of soybean seeds. J Plant Physiol 146:725–730

Kumar S, Nayyar H, Bhanwara RK, Upadhyaya HD (2010) Chilling stress effects on reproductive biology of chickpea. J SAT Agric Res 8:1–14

Kumar S, Malik J, Thakur P, Kaistha S, Sharma KD, Upadhyaya HD, Berger JD, Nayyar H (2011) Growth and metabolic responses of contrasting chickpea (*Cicer arietinum* L.) genotypes to chilling stress at reproductive phase. Acta Physiol Plant 33:779–787

Leyman B, Van Dijck P, Thevelein JM (2001) An unexpected plethora of trehalose biosynthesis genes in *Arabidopsis thaliana*. Trends Plant Sci 6:510–513

Li YY, Liang GJ, Li YH (2004) Effect of exogenous betaine on cold resistance of cucumber seedlings. Plant Physiol Commun 40:673–676

Li HJ, Yang AF, Zhang XC, Gao F, Zhang JR (2007) Improving freezing tolerance of transgenic tobacco expressing sucrose:sucrose1-fructosyl transferase gene from *Lactuca sativa*. Plant Cell Tissue Organ Cult 89:37–48

Li BQ, Zhou ZW, Tian SP (2008) Combined effects of endo- and exogenous trehalose on stress tolerance and biocontrol efficacy of two antagonistic yeasts. Biol Control 46:187–193

Li DD, Tai FJ, Zhang ZT, Li Y, Zheng Y, Wu YF, Li XB (2009) cotton gene encodes a tonoplast aquaporin that is involved in cell tolerance to cold stress. Gene 438:26–32

Li HW, Zang BS, Deng XW, Wang XP (2011) Over-expression of the trehalose-6-phosphate synthase gene OsTPS1 enhances abiotic stress tolerance in rice. Planta 234:1007–1018

Li S, Yu X, Beattie GA (2013) Glycine betaine catabolism contributes to *Pseudomonas syringae* tolerance to hyperosmotic stress by relieving betaine-mediated suppression of compatible solute synthesis. J Bacteriol 195(10):2415–2423. doi:10.1128/JB.00094-13

Lidgett A, Jennings K, Johnson X, Guthridge K, Jones E, Spangenberg G (2002) Isolation and characterisation of a fructosyltransferase gene from perennial ryegrass (*Lolium perenne*). J Plant Physiol 159:1037–1043

Ligaba A, Katsuhara M, Shibasaka M, Djira G (2011) Abiotic stresses modulate expression of major intrinsic proteins in barley (*Hordeum vulgare*). C R Biol 334:127–139

Lineberger D, Steponkus PL (1980) Cryoprotection by glucose, sucrose, and raffinose to chloroplast thylakoids. Plant Physiol 65:298–304

Liu JJ, Krenz DC, Galvez AF, de Lumen BO (1998) Galactinol synthase (GS): increased enzyme activity and levels of mRNA due to cold and desiccation. Plant Sci 134:11–20

Liu H, Dai X, Xua Y, Chong K (2007) Over-expression of OsUGE-1 altered raffinose level and tolerance to abiotic stress but not morphology in *Arabidopsis*. J Plant Physiol 164:1384–1390

Liu L, Duan L, Zhang J, Zhang Z, Mi G, Ren H (2010) Cucumber (*Cucumis sativus* L.) overexpressing cold-induced transcriptome regulator ICE1 exhibits changed morphological characters and enhances chilling tolerance. Sci Hortic 124:29–33

Livingston DP, Henson CA (1998) Apoplastic sugars, fructans, fructan exohydrolase, and invertase in winter oat: responses to second-phase cold hardening. Plant Physiol 116:403–408

Livingston DP III, Hincha DK, Heyer AG (2007) The relationship of fructan to abiotic stress tolerance in plants. In: Shiomi N, Benkeblia N, Onodera S (eds) Recent advances in fructooligosaccharides research. Research Signpost, Trivandrum, India, pp 181–199

Loescher WH (1987) Physiology and metabolism of sugar alcohols in higher plants. Physiol Plant 70:553–557

Loester WH, Tyson RH, Everard JD, Redgwell RJ, Bieleski RL (1992) Mannitol synthesis in higher plants: evidence for the role and characterization of a NADPH-dependent mannose-6-phosphate reductase. Plant Physiol 98:1396–1402

Loewus FA, Loewus MW (1983) *Myo*-inositol: its biosynthesis and metabolism. Annu Rev Plant Physiol 34:137–161

Loewus FA, Murthy PPN (1999) Myo-inositol metabolism in plants. Plant Sci 150:119

Lopez M, Tejera NA, Iribarne C, Lluch C, Herrera-Cervera JA (2008) Trehalose and trehalase in root nodules of *Medicago truncatula* and *Phaseolus vulgaris* in response to salt stress. Physiol Plant 134:575–582

Lukoseviciute V, Rugienius R, Sasnauskas A, Stanys V, Bobinas C (2009) Impact of exogenous sucrose, raffinose and proline on cold acclimation of strawberry *in vitro*. Acta Hortic 839: 203–208

Lunn JE, Feil R, Hendriks JHM, Gibon Y, Morcuende R, Osuna D, Scheible WR, Carillo P, Hajirezaeis MR, Stitt M (2006) Sugar-induced increases in trehalose 6-phosphate are correlated with redox activation of ADP glucose pyrophosphorylase and higher rates of starch synthesis in *Arabidopsis thaliana*. Biochem J 397:139–148

Lüscher M, Hochstrasser U, Vogel G, Aeschbacher R, Galati V, Nelson CJ, Boller T, Wiemken A (2000) Cloning and functional analysis of sucrose:sucrose 1-fructosyltransferase from tall fescue. Plant Physiol 124:1217–1227

Ma Q, Bin Dai X, Xu Y, Guo J, Liu Y, Chen N, Xiao J, Zhang D, Xu Z, Zhang X, Chong K (2009) Enhanced tolerance to chilling stress in OsMYB3R-2 transgenic rice is mediated by alteration in cell cycle and ectopic expression of stress genes. Plant Physiol 150:244–256

Ma H, Zhou H, Zhang H, Zhao J (2010) Cloning and expression analysis of an AP2/ERF gene and its responses to phytohormones and abiotic stresses in rice. Rice Sci 17:1–9

Magel E, Abdel-Latif A, Hampp R (2001) Non-structural carbohydrates and catalytic activities of sucrose metabolizing enzymes in trunks of two *Juglans* species and their role in heartwood formation. Holzforschung 55:135–145

Mahajan S, Tuteja N (2005) Cold, salinity and drought stresses: an overview. Arch Biochem Biophys 444:139–158

Mäkelä P, Jokinen K, Kontturi M, Peltonen-Sainio P, Pehu E, Somersalo S (1998) Foliar application of glycine betaine—a novel product from sugar beet-as an approach to increase tomato yield. Ind Crop Prod 7:139–148

Mao T, Jin L, Li H, Liu B, Yuan M (2005) Two microtubule-associated proteins of the *Arabidopsis* MAP65 family function differently on microtubules. Plant Physiol 138:654–662

Martínez-Fleites C, Ortíz-Lombardía M, Pons T, Tarbouriech N, Taylor EJ, Arrieta JG, Hernández L, Davies GJ (2005) Crystal structure of levansucrase from the Gram negative bacterium *Gluconacetobacter diazotrophicus*. Biochem J 390:19–27

Maruyama K, Takeda M, Kidokoro S, Yamada K, Sakuma Y, Urano K, Fujita M, Yoshiwara K, Matsukura S, Morishita Y, Sasaki R, Suzuki H, Saito K, Shibata D, Shinozaki K, Yamaguchi-Shinozaki K (2009) Metabolic pathways involved in cold acclimation identified by integrated analysis of metabolites and transcripts regulated by DREB1A and DREB2A. Plant Physiol 150:1972–1980

Mattana M, Biazzi E, Consonni R, Locatelli F, Vannini C, Provera S, Coraggio I (2005) Overexpression of Osmyb4 enhances compatible solute accumulation and increases stress tolerance of *Arabidopsis thaliana*. Physiol Plant 125:212–223

Mattioli R, Costantino P, Trovato M (2009) Proline accumulation in plants. Not only stress. Plant Signal Behav 4:1016–1018

Matysik J, Alia BB, Mohanty P (2002) Molecular mechanism of quenching of reactive oxygen species by proline under stress in plants. Curr Sci 82:525–532

McCue KF, Hanson AD (1990) Drought and salt tolerance: towards understanding and application. Trends Biotechnol 8:358–362

McCue KF, Hanson AD (1992) Effects of soil salinity on the expression of betaine aldehyde dehydrogenase in leaves: investigation on hydraulic, ionic and biochemical signals. Aust J Plant Physiol 19:555–564

McKown R, Kuroki G, Warren G (1996) Cold responses of *Arabidopsis* mutants impaired in freezing tolerance. J Exp Bot 47:1919–1925

McNeil SC, Rhodes D, Russell BL, Nuccio ML, Shachar-Hill Y, Hanson AD (2000) Metabolic modeling identifies key constraints on an engineered glycine betaine synthesis pathway in tobacco. Plant Physiol 124:153–162

Meier H, Reid JSG (1982) Reserve polysaccharides other than starch in higher plants. In: Loewus FA, Tanner W (eds) Encyclopedia of plant physiology. Plant carbohydrates I: intracellular carbohydrates. Springer, Berlin, pp 418–471

Miranda JA, Avonce N, Suarez R, Thevelein JM, Dijck P, Iturriaga G (2007) A bifunctional TPS-TPP enzyme from yeast confers tolerance to multiple and extreme abiotic-stress conditions in transgenic *Arabidopsis*. Planta 226:1411–1421

Mo Y, Liang G, Shi W, Xie J (2011) Metabolic responses of alfalfa (*Medicago sativa* L.) leaves to low and high temperature induced stresses. Afr J Biotechnol 10:1117–1124

Mohanty PS, Hayashi H, Papageorgiou GC, Murata N (1993) Stabilization of the Mn-cluster of the oxygen-evolving complex by glycine betaine. Biochim Biophys Acta 1144:92–96

Muller J, Wiemken A, Aeschbacher R (1999) Trehalose metabolism in sugar sensing and plant development. Plant Sci 147:37–47

Muller J, Aeschbacher RA, Sprenger N, Boller T, Wiemken A (2000) Disaccharide mediated regulation of sucrose: fructan-6-fructosyltransferase, a key enzyme of fructan synthesis in barley leaves. Plant Physiol 123:265–274

Muller J, Aeschbacher RA, Wingler A, Boller T, Wiemken A (2001) Trehalose and trehalase in *Arabidopsis*. Plant Physiol 125:1086–1093

Muller B, Pantin F, Génard M, Turc O, Freixes S, Piques M, Gibon Y (2011) Water deficits uncouple growth from photosynthesis, increase C content, and modify the relationships between C and growth in sink organs. J Exp Bot 62:1715–1729

Nadwodnik J, Lohaus G (2008) Subcellular concentrations of sugar alcohols and sugars in relation to phloem translocation in *Plantago major*, *Plantago maritima*, *Prunus persica*, and *Apium graveolens*. Planta 227:1079–1089

Nagaraj VJ, Altenbach D, Galati V, Lüscher M, Meyer AD, Boller T, Wiemken A (2004) Distinct regulation of sucrose: sucrose-1-fructosyltransferase (1-SST) and sucrose: fructan-6-fructosyltransferase (6-SFT), the key enzymes of fructan synthesis in barley leaves: 1-SST as the pacemaker. New Phytol 161:735–748

Naidu BP, Paleg LG, Aspinall D, Jennings AC, Jones GP (1991) Amino acid and glycine betaine accumulation in cold-stressed wheat seedlings. Phytochemistry 30:407–409

Naik CR, Joshi CL (1983) Ineffectual role of proline metabolism in salt stressed sugarcane leaves. Proc Natl Acad Sci U S A 92:265–269

Nanjo T, Kobayashia M, Yoshiba Y, Kakubaric Y, Yamaguchi-Shinozakid K, Shinozakia K (1999) Antisense suppression of proline degradation improves tolerance to freezing and salinity in *Arabidopsis thaliana*. FEBS Lett 461:205–210

Nayyar H, Chander K, Kumar S, Bains T (2005) Glycine betaine mitigates cold stress damage in chickpea. Agron Sustain Dev 25:381–388

Nery DCM, da Silva CG, Mariani D, Fernandes PN, Pereira MD, Panek AD, Eleutherio ECA (2008) The role of trehalose and its transporter in protection against reactive oxygen species. Biochim Biophys Acta 1780:1408–1411

Nishikawa F, Kato M, Hyodo H, Ikoma Y, Sugiura M, Yano M (2005) Effect of sucrose on ascorbate level and expression of genes involved in the ascorbate biosynthesis and recycling pathway in harvested broccoli florets. J Exp Bot 56:65–72

Nishizawa A, Yabuta Y, Shigeoka S (2008) Galactinol and raffinose constitute a novel function to protect plants from oxidative damage. Plant Physiol 147:1251–1263

Ouellet F (2007) Cold acclimation and freezing tolerance in plants. Encyclopedia of life sciences. Wiley, London, pp 1–6

Ozturk L, Demir Y (2002) *In vivo* and *in vitro* protective role of proline. Plant Growth Regul 38:259–264

Palonen P, Junttila O (1999) Cold hardening of raspberry plants *in vitro* is enhanced by increasing sucrose in the culture medium. Physiol Plant 106:386–392

Pan J, Zhang M, Kong X, Xing X, Liu Y, Zhou Y, Liu Y, Sun L, Li D (2012) ZmMPK17, a novel maize group D MAP kinase gene, is involved in multiple stress responses. Planta 235: 661–676

Papageorgiou GC, Murata N (1995) The unusually strong stabilizing effects of glycine betaine on the structure and function of the oxygen evolving photosystem II complex. Photosynth Res 44:243–252

Park EJ, Jeknić Z, Sakamoto A, DeNoma J, Yuwansiri R, Murata N, Chen THH (2004) Genetic engineering of glycinebetaine synthesis in tomato protects seeds, plants, and flowers from chilling damage. Plant J 40:474–487

Parvanova D, Ivanov S, Konstantinova T, Karanov E, Atanassov A, Tsvetkov T, Alexieva V, Djilianov D (2004) Transgenic tobacco plants accumulating osmolytes show reduced oxidative damage under freezing stress. Plant Physiol Biochem 42:57–63

Pasquali G, Biricolti S, Locatelli F, Baldoni E, Mattana M (2008) Osmyb4 expression improves adaptive responses to drought and cold stress in transgenic apples. Plant Cell Rep 27: 1677–1686

Paul MJ, Pellny TK (2003) Carbon metabolite feedback regulation of leaf photosynthesis and development. J Exp Bot 54:539–547

Paul MJ, Lucia F, Primavesi LF, Jhurreea D, Zhang Y (2008) Trehalose metabolism and signaling. Annu Rev Plant Biol 59:417–441

Peng D, Jiang Y, Yi Y, Li FZ, Bai LP, Ma XG, Ma H, Zhong M, Guo ZF (2013) Cold resistance in wheat and its related species. J Chem Biol Phys Sci 3:1132–1141

Pennycooke JC, Jones ML, Stushnoff C (2003) Down-regulating a-galactosidase enhances freezing tolerance in transgenic petunia. Plant Physiol 133:901–909

Perras M, Sarhan F (1984) Energy state of spring and winter wheat during cold hardening: soluble sugars and adenine nucleotides. Physiol Plant 60:129–132

Peshev D, Van den Ende W (2013) Sugars as antioxidants in plants. In: Tuteja N, Gill SS (eds) Crop improvement under adverse conditions. Springer, Berlin, pp 285–308

Peshev D, Vergauwen R, Moglia A, Hideg E, Van den Ende W (2013) Towards understanding vacuolar antioxidant mechanisms: a role for fructans? J Exp Bot 64(4):1025–1038. doi:10.1093/jxb/ers377
Petcu E, Hagima I, Perbea M, Ittu G (1998) Evaluation of frost tolerance in winter wheat. Rom Agric Res 9:37–41
Pilon-Smits EAH, Terry N, Sears T, Kim H, Zayed A, Hwang S, van Dun K, Voogd E, Verwoerd TC, Krutwagen RWHH, Goddijn OJM (1998) Trehalose producing transgenic tobacco plants show improved growth performance under drought stress. J Plant Physiol 152:525–553
Polge C, Thomas M (2007) SNF1/AMPK/SnRK1 kinases, global regulators at the heart of energy control? Trends Plant Sci 12:20–28
Pollock CJ (1984) Physiology and metabolism of sucrosylfructans. In: Lewis DH (ed) Storage carbohydrates in vascular plants. Cambridge University Press, Cambridge, pp 97–113
Pollock CJ, Eagles CF, Sims IM (1988) Effect of photoperiod and irradiance changes upon development of freezing tolerance and accumulation of soluble carbohydrate in seedlings of *Lolium perenne* grown at 2°C. Ann Bot 62:95–100
Ponnu J, Wahl V, Schmid M (2011) Trehalose6phosphate: connecting plant metabolism and development. Front Plant Sci 2:70
Pontis HG, De Campillo E (1985) Fructans. In: Dixon RA (ed) Biochemistry of storage carbohydrates in green plants. Academic, London, pp 205–227
Pourtau N, Marès M, Purdy S, Quentinm N, Ruël A, Wingler A (2004) Interactions of abscisic acid and sugar signaling in the regulation of leaf senescence. Planta 219:765–772
Prabhavathi VR, Yadav JS, Kumar PA, Rajam MV (2002) Abiotic stress tolerance in transgenic eggplant (*Solanum melongena* L.) by introduction of bacterial mannitol phosphodehydrogenase gene. Mol Breed 9:137–147
Pramanik MHR, Imai R (2005) Functional identification of a trehalose 6-phosphate phosphatase gene that is involved in transient induction of trehalose biosynthesis during chilling stress in rice. Plant Mol Biol 58:751–762
Qin L, Luo SL, Deng ZN, Yan JW, Li N, Yuan FR (2011) Enhanced resistance to canker disease and tolerances to biotic and abiotic stresses in terf1 transgenic sweet orange. Acta Hortic 892:165–172
Quan R, Shang M, Zhang H, Zhao Y, Zhang J (2004) Improved chilling tolerance by transformation with betA gene for the enhancement of glycinebetaine synthesis in maize. Plant Sci 166: 141–149
Rajashekar CB, Zhou H, Marcum KB, Prakash O (1999) Glycine betaine accumulation and induction of cold tolerance in strawberry (*Fragaria*×*ananassa* Duch.) plants. Plant Sci 148: 175–183
Rajendrakumar CSV, Reddy BVD, Reddy AR (1994) Proline-protein interactions: protection of structural and functional integrity of M4lactate dehydrogenase. Biochem Biophys Res Commun 201:957–963
Ramon M, Rolland F (2007) Plant development: introducing trehalose metabolism. Trends Plant Sci 12:185–188
Ramon M, Rolland F, Sheen J (2008) Sugar sensing and signaling in *Arabidopsis*. In: The *Arabidopsis* Book. ASPB, Rockville, MD. doi:10.1199/tab.0117
Rapacz M (2002) Cold-deacclimation of oilseed rape (*Brassica napus* var. oleifera) in response to fluctuating temperatures and photoperiod. Ann Bot 89:543–549
Rapacz M, Hura K (2002) The pattern of changes in photosynthetic apparatus in response to cold acclimation and de-acclimation in two contrasting cultivars of oilseed rape. Photosynthetica 40:63–69
Rathinasabapathi B (2000) Metabolic engineering for stress tolerance: installing osmoprotectant synthesis pathways. Ann Bot 86:709–716
Rathinasabapathi B, McCue KF, Gage DA, Hanson AD (1994) Metabolic engineering of glycine betaine synthesis: plant betaine aldehyde dehydrogenases lacking typical transit peptides are targeted to tobacco chloroplasts where they confer betaine aldehyde resistance. Planta 193: 155–162

Rekarte-Cowie I, Ebshish OS, Mohamed KS, Pearce RS (2008) Sucrose helps regulate cold acclimation of *Arabidopsis thaliana*. J Exp Bot 59:4205–4217

Rentsch D, Hirner B, Schmelzer E, Frommer WB (1996) Salt stress-induced proline transporters and salt stress-repressed broad specificity amino acid permeases identified by suppression of a yeast amino acid permease-targeting mutant. Plant Cell 8:1437–1446

Rhodes D, Hanson AD (1993) Quaternary ammonium and tertiary sulfonium compounds in higher plants. Annu Rev Plant Physiol Plant Mol Biol 44:357–384

Robinson NJ, Procter CM, Connolly EL, Guerinot ML (1999) A ferric-chelate reductase for iron uptake from soils. Nature 397:694–697

Rolland F, Winderickx J, Thevelein JM (2002) Glucose-sensing and signalling mechanisms in yeast. FEMS Yeast Res 2:183–201

Rolland F, Baena-Gonzales E, Sheen J (2006) Sugar sensing and signalling in plants: conserved and novel mechanisms. Annu Rev Plant Biol 57:675–709

Roosens NHCJ, Thu TT, Iskandar HM, Jacobs M (1998) Isolation of the ornithine d-aminotransferase cDNA and effect of salt stress on its expression in *Arabidopsis thaliana*. Plant Physiol 117: 263–271

Rosa M, Carolina Prado C, Podazza G, Interdonato R, González JA, Hila M, Prado FE (2009) Soluble sugars—metabolism, sensing and abiotic stress. A complex network in the life of plants. Plant Signal Behav 4:388–393

Rudolph AS, Chandrasekhar E, Gaber BP, Nagumo M (1990) Molecular modeling of saccharide-lipid interactions. Chem Phys Lipids 53:243–261

Rumpho ME, Edwards GE, Loescher WH (1983) A pathway for photosynthetic carbon flow to mannitol in celery leaves. Plant Physiol 73:869–873

Rymen B, Fiorani F, Kartal F, Vandepoele K, Inze D et al (2007) Cold nights impair leaf growth and cell cycle progression in maize through transcriptional changes of cell cycle genes. Plant Physiol 143:1429–1438

Sakamoto A, Murata N (2000) Genetic engineering of glycine betaine synthesis in plants: current status and implications for enhancement of stress tolerance. J Exp Bot 51:81–88

Sakamoto A, Murata N (2002) The role of glycine betaine in the protection of plants from stress: clues from transgenic plants. Plant Cell Environ 25:163–171

Sakamoto A, Alia H, Murata N (1998) Metabolic engineering of rice leading to biosynthesis of glycine betaine and tolerance to salt and cold. Plant Mol Biol 38:1011–1019

Sakamoto A, Valverde R, Alia H, Chen THH, Murata N (2000) Transformation of *Arabidopsis* with the *cod A* gene for choline oxidase enhances freezing tolerance of plants. Plant J 22:449–453

Santarius KA (1973) The protective effect of sugars on chloroplast membranes during temperature and water stress and its relationship to frost, desiccation and heat resistance. Planta 113: 105–114

Santarius KA, Milde H (1977) Sugar compartmentation in frost-hardy and partially dehardened cabbage leaf cells. Planta 136:163–166

Santoiani CS, Tognetti JA, Pontis HG, Salerno GL (1993) Sucrose and fructan metabolism in wheat roots at chilling temperatures. Physiol Plant 87:84–88

Saravitz DM, Pharr DM, Carter TE (1987) Galactinol synthase activity and soluble sugars in developing seeds of four soybean genotypes. Plant Physiol 83:85–189

Sasaki H, Ichimura K, Oda M (1996) Changes in sugar content during cold acclimation and deacclimation of cabbage seedlings. Ann Bot 78:365–369

Sasaki H, Ichimura K, Imada S, Yamaki S (2001) Sucrose synthase and sucrose phosphate synthase, but not acid invertase, are regulated by cold acclimation and deacclimation in cabbage seedlings. J Plant Physiol 158:847–852

Saurez R, Calderon C, Iturriaga G (2008) Enhanced tolerance to multiple abiotic stresses in transgenic alfalfa accumulating trehalose. Crop Sci 49:1791–1799

Savoure A, Jaoua S, Hua X-J, Ardiles W, Montagu MV, Verbruggen N (1995) Isolation, characterization, and chromosomal location of a gene encoding the delta1-pyrroline-5-carboxylate synthetase in Arabidopsis thaliana. FEBS Lett 372:13–19

Schluepmann H, Paul MJ (2009) Trehalose metabolites in Arabidopsis: elusive, active and central. Arabidopsis Book 7:e0122. doi:10.1199/tab.0122

Schluepmann H, Pellny T, van Dijken A, Smeekens S, Paul M (2003) Trehalose 6-phosphate is indispensable for carbohydrate utilization and growth in *Arabidopsis thaliana*. Proc Natl Acad Sci U S A 100:6849–6854

Schluepmann H, van Dijken A, Aghdasi M, Wobbes B, Paul M, Smeekens S (2004) Trehalose mediated growth inhibition of *Arabidopsis* seedlings is due to trehalose-6-phosphate accumulation. Plant Physiol 135:879–890

Schneider T, Keller F (2009) Raffinose in chloroplasts is synthesized in the cytosol and transported across the chloroplast envelope. Plant Cell Physiol 50:2174–2182

Seok KC, Jee EK, Jong AP, Tae JE, Woo TK (2006) Constitutive expression of abiotic stress-inducible hot pepper CaXTH3, which encodes a xyloglucan endotransglucosylase/hydrolase homolog, improves drought and salt tolerance in transgenic *Arabidopsis* plants. FEBS Lett 580:3136–3144

Shalaev EY, Steponkus PL (2001) Phase behavior and glass transition of 1, 2-dioleoylphosphadity lethanolamine (DOPE) dehydrated in the presence of sucrose. Biochem Biophys Acta Biomembr 1514:100–116

Shao HB, Liang ZS, Shao MA (2006) Osmotic regulation of 10 wheat (*Triticum aestivum L.*) genotypes at soil water deficits. Biointerfaces 47:132–139

Shen B, Jensen RG, Bohnert HJ (1997a) Increased resistance to oxidative stress in transgenic plants by targeting mannitol biosynthesis to chloroplasts. Plant Physiol 113:1177–1183

Shen B, Jensen RG, Bohnert HJ (1997b) Mannitol protects against oxidation by hydroxyl radicals. Plant Physiol 115:527–532

Sheveleva E, Chmara W, Bohnert HJ, Jensen RC (1997) Increased salt and drought tolerance by D-ononitol production in transgenic *Nicotiana tabacum*. Plant Physiol 115:1211–1219

Sheveleva EV, Marquez S, Chmara W, Zegeer A, Jensen RG, Bohnert HJ (1998) Sorbitol-6-phosphate dehydrogenase expression in transgenic tobacco. High amounts of sorbitol lead to necrotic lesions. Plant Physiol 117:831–839

Shi Y, Zhang J, Ning J, Jie L (2011) Physiological mechanism of chilling tolerance in eggplant seedling. Acta Bot Bor Occid Sin 31:2498–2502

Shima S, Matsui H, Tahara S, Imai R (2007) Biochemical characterization of rice trehalose-6-phosphate phosphatases supports distinctive functions of these plant enzymes. FEBS J 274:1192–1201

Shindou T, Sasaki Y, Eguchi T, Hagiwara K, Ichikawa T (1989) Identification of erythritol by HPLC and GC-MS and quantitative measurement in pulps of various fruits. J Agric Food Chem 37:1474–1476

Shiomi N, Onodera S, Sakai H (1997) Fructooligosaccharide content and fructosyltransferase activity during growth of onion bulbs. New Phytol 136:105–113

Shirasawa K, Takabe T, Kishitani S (2006) Accumulation of glycine betaine in rice plants that over-express choline monooxygenase from spinach and evaluation of their tolerance to abiotic stress. Ann Bot 98:565–571

Sicher R (2011) Carbon partitioning and the impact of starch deficiency on the initial response of *Arabidopsis* to chilling temperatures. Plant Sci 181:167–176

Singh S, Rathore M, Goyary D, Singh RK, Anandhan S, Sharma DK, Ahmed Z (2011) Induced ectopic expression of At-CBF1 in marker-free transgenic tomatoes confers enhanced chilling tolerance. Plant Cell Rep 30:1019–1028

Skubatz H, Meeuse BJD, Bendich AJ (1989) Oxidation of proline and glutamate by mitochondria of the inflorescence of voodoo lily (*Sauromatum guttatum*). Plant Physiol 91:530–535

Smart CC, Flores S (1997) Over-expression of D-myo-inositol-3-phosphate synthase leads to elevated levels of inositol in *Arabidopsis*. Plant Mol Biol 33:811–820

Smirnoff N, Cumbes QJ (1989) Hydroxyl radical scavenging activity of compatible solutes. Phytochemistry 28:1057–1060

Somersalo S, Kyei-Boahen S, Pehu E (1996) Exogenous glycine betaine application as a possibility to increase low temperature tolerance of crop plants. Nord Jordbruk 78:10

Spollen WG, Nelson CJ (1994) Response of fructan to water deficit in growing leaves of tall fescue. Plant Physiol 106:329–336

Sprenger N, Keller F (2000) Allocation of raffinose family oligosaccharides to transport and storage pools in *Ajuga reptans*: the roles of two distinct galactinol synthases. Plant J 21:249–258

Sprenger N, Bortlik K, Brandt A, Boller T, Wiemken A (1995) Purification, cloning, and functional expression of sucrose:fructan 6-fructosyltransferase, a key enzyme of fructan synthesis in barley. Proc Natl Acad Sci U S A 92:11652–11656

Stoop JMH, Williamson JD, Pharr DM (1996) Mannitol metabolism in plants: a method for coping with stress. Trends Plant Sci 1:139–144

Stoyanova S, Geuns J, Hideg E, Van den Ende W (2011) The food additives inulin and stevioside counteract oxidative stress. Int J Food Sci Nutr 62:207–214

Stushnoff C, Rammele RL Jr, Essensee V, McNeil M (1993) Low temperature induced biochemical mechanisms, implications for cold acclimation and de-acclimation. NATO ASI Ser I 16:647–657

Stushnoff C, Seufferheld MJ, Creegan T (1998) Oligosaccharides as endogenous cryoprotectants in woody plants. In: Li PH, Chen THH (eds) Plant cold hardiness: molecular biology, biochemistry and physiology. Plenum Press, New York, pp 301–309

Sun WQ, Irving TC, Leopold AC (1994) The role of sugar, vitrification and membrane phase transition in seed desiccation tolerance. Physiol Plant 90:621–628

Sun WQ, Leopold AC, Crowe LM, Crowe JH (1996) Stability of dry liposomes in sugar glasses. Biophys J 70:1769–1776

Sundaramurthi P, Patapoff TW, Suryanarayanan R (2010) Crystallization of trehalose in frozen solutions and its phase behavior during drying. Pharm Res 27:2374–2383

Suzuki M, Nass HG (1988) Fructan in winter wheat, triticale, and fall rye cultivars of varying cold hardiness. Can J Bot 66:1723–1728

Szabados L, Savoure A (2009) Proline: a multifunctional amino acid. Trends Plant Sci 15:89–97

Tabaei-Aghdaei SR, Pearce RS, Harrison P (2003) Sugars regulate cold-induced gene expression and freezing-tolerance in barley cell cultures. J Exp Bot 54:1565–1575

Taji T, Ohsumi C, Iuchi S, Seki M, Kasuga M, Kobayashi M, Yamaguchi-Shinozaki K, Shinozaki K (2002) Important roles of drought- and cold-inducible genes for galactinol synthase in stress tolerance in *Arabidopsis thaliana*. Plant J 29:417–426

Takabe T, Rai V, Hibino T (2006) Metabolic engineering of glycinebetaine. In: Rai A, Takabe T (eds) Abiotic stress tolerance in plants: toward the improvement of global environment and food. Springer, Dordrecht, The Netherlands, pp 137–151

Tao R, Uratsu SL, Dandekar AM (1995) Sorbitol synthesis in transgenic tobacco with apple cDNA encoding NADP-dependent sorbitol-6-phosphate dehydrogenase. Plant Cell Physiol 366: 525–532

Thakur P, Kumar S, Malik JA, Berger JD, Nayyar H (2010) Cold stress effects on reproductive development in grain crops: an overview. Environ Exp Bot 67:429–443

Thevelein JM (1984) Regulation of trehalose mobilization in fungi. Microbiol Rev 48:42–59

Thiery L, Leprince AS, Lefebvre D, Ghars MA, Debarbieux E, Savouré A (2004) Phospholipase D is a negative regulator of proline biosynthesis in *Arabidopsis thaliana*. J Biol Chem 279:14812–14818

Thomas JC, Sepahi M, Arendall B, Bohnert HJ (1995) Enhancement of seed germination in high salinity by engineering mannitol expression in *Arabidopsis thaliana*. Plant Cell Environ 18:801–806

Travert S, Valerio L, Fouraste I, Boudet AM, Teulieres C (1997) Enrichment in specific soluble sugars of two eucalyptus cell-suspension cultures by various treatments enhances their frost tolerance via a noncolligative mechanism. Plant Physiol 114:1433–1442

Ueda A, Shi W, Sanmiya K, Shono M, Takabe T (2001) Functional analysis of salt-inducible proline transporter of barley roots. Plant Cell Physiol 42:1282–1289

Uemura M, Steponkus PL (2003) Modification of the intracellular sugar content alters the incidence of freeze-induced membrane lesions of protoplasts isolated from *Arabidopsis thaliana* leaves. Plant Cell Environ 26:1083–1096

Vágújfalvi A, Kerepesi I, Galiba G, Tischner T, Sutka J (1999) Frost hardiness depending on carbohydrate changes during cold acclimation in wheat. Plant Sci 144:85–92
Valluru R, Van den Ende W (2008) Plant fructans in stress environments: emerging concepts and future prospects. J Exp Bot 59:2905–2916
Valluru R, Lammens W, Claupein W, Van den Ende W (2008) Freezing tolerance by vesicle mediated fructan transport. Trends Plant Sci 13:409–414
Van den Ende W, Van Laere A (2007) Fructans in dicotyledonous plants: occurrence and metabolism. In: Shiomi N, Benkeblia N, Onodera S (eds) Recent advances in fructooligosaccharides research. Research Signpost, Kerala, India, pp 1–14
Van den Ende W, De Coninck B, Van Laere A (2004) Plant fructan exohydrolases: a role in signaling and defense? Trends Plant Sci 9:523–528
Van der Meer IM, Koops AJ, Hakkert JC, Van Tunen AJ (1994) Cloning of the fructan biosynthesis pathway of Jerusalem artichoke. Plant J 15:489–500
Van Hijum SAFT, van der Maarel MJEC, Dijkhuizen L (2003) Kinetic properties of an inulosucrase from *Lactobacillus reuteri*. FEBS Lett 534:207–210
Van Laere A, Van den Ende W (2002) Inulin metabolism in dicots: chicory as a model system. Plant Cell Environ 25:803–813
Van Rensburg L, Krüger GHJ, Krüger H (1993) Proline accumulation as drought tolerance selection criterion: its relationship to membrane integrity and chloroplast ultrastructure in *Nicotiana tabacum* L. J Plant Physiol 141:188–194
Van Swaaij AC, Jacobsen E, Feenstra WJ (1985) Effect of cold hardening, wilting and exogenously applied proline on leaf proline content and frost tolerance of several genotypes of *Solanum*. Physiol Plant 64:230–236
Veluthambi K, Mahadevan S, Maheshwari R (1982a) Trehalose toxicity in *Cuscuta reflexa*: sucrose content decreases in shoot tips upon trehalose feeding. Plant Physiol 69:1247–1251
Veluthambi K, Mahadevan S, Maheshwari R (1982b) Trehalose toxicity in *Cuscuta reflexa*: cell wall synthesis is inhibited upon trehalose feeding. Plant Physiol 70:686–688
Verbruggen N, Hermans C (2008) Proline accumulation in plants: a review. Amino Acids 35:753–759
Verbruggen N, Villarroel R, Van M (1993) Osmoregulation of a pyrroline-5 carboxylate reductase gene in Arabidopsis thaliana. Plant Physiol 103:771–781
Vereyken IJ, Chupin V, Islamov A, Kuklin A, Hincha DK, de Kruijff B (2003) The effect of fructan on the phospholipid organization in the dry state. Biophys J 85:3058–3065
Vernon DM, Tarczynski MC, Jensen RG, Bohnert HJ (1993) Cyclitol production in transgenic tobacco. Plant J 4:199–205
Veyres N, Aono M, Sangwan-Norreel BS, Sangwan RS (2008) The *Arabidopsis* sweetie mutant is affected in carbohydrate metabolism and defective in the control of growth, development and senescence. Plant J 55:665–686
Vijn I, Smeekens S (1999) Fructan: more than a reserve carbohydrate? Plant Physiol 120:351–359
Vijn I, van Dijken A, Sprenger N, van Dun K, Weisbeek P, Wiemken A, Smeekens S (1997) Fructan of the inulin neoseries is synthesized in transgenic chicory plants (*Cichorium intybus* L.) harbouring onion (*Allium cepa* L.) fructan:fructan 6G-fructosyltransferase. Plant J 11:387–398
Vijn I, Van Dijken A, Lüscher M, Bos A, Smeets E, Weisbeek P, Wiemken A, Smeekens S (1998) Cloning of sucrose:sucrose 1-fructosyltransferases from onion and synthesis of structurally defined fructan molecules from sucrose. Plant Physiol 117:1507–1513
Volenec JJ, Nelson CJ (1984) Carbohydrate metabolism in leaf meristems of Tall Fescue: II Relationship to leaf elongation rates modified by nitrogen fertilization. Plant Physiol 74 595–600
Wagner W, Keller F, Wiemken A (1983) Fructan metabolism in cereals: induction in leaves and compartmentation in protoplasts and vacuoles. Z Pflanzenphysiol 112:359–372
Wang N, Nobel PS (1998) Phloem transport of fructans in the crassulacean acid metabolism species *Agave deserti*. Plant Physiol 116:709–714

Wang WX, Vinocur B, Altman A (2003) Plant responses to drought, salinity and extreme temperatures: towards genetic engineering for stress tolerance. Planta 219:1–14

Wang G, Li G, Zhen F (2007) Responses of enzyme activities in carbon metabolism to chilling stress in rice with different chilling tolerance. Acta Agron Sin 33:1197–1200

Wang C, Ma XL, Hui Z, Wang W (2008a) Glycine betaine improves thylakoid membrane function of tobacco leaves under low-temperature stress. Photosynthetica 46:400–409

Wang ZY, Li FM, Xiong YC, Xu BC (2008b) Soil-water threshold range of chemical signals and drought tolerance was mediated by ROS homeostasis in winter wheat during progressive soil drying. J Plant Growth Regul 27:309–319

Wang QB, Xu W, Xue QZ, Su WA (2010) Transgenic *Brassica chinensis* plants expressing a bacterial codA gene exhibit enhanced tolerance to extreme temperature and high salinity. J Zhejiang Univ Sci B 11:851–861

Wanner LA, Junttila O (1999) Cold-induced freezing tolerance in *Arabidopsis*. Plant Physiol 120:391–399

Warren G, McKown R, Marin A, Teutonico R (1996) Isolation of mutations affecting the development of freezing tolerance in *Arabidopsis thaliana* (L) Heynh. Plant Physiol 111:1011–1019

Wei JZ, Chatterton NJ (2001) Fructan biosynthesis and fructosyltransferase evolution: the expression of the 6-SFT (sucrose: fructan 6-fructosyltransferase) gene in crested wheatgrass (*Agropyron cristatum*). J Plant Physiol 158:1203–1213

Wei JZ, Chatterton NJ, Harrison PA, Wang RR, Larson SR (2002) Characterization of fructan biosynthesis in big bluegrass (*Poa secunda*). J Plant Physiol 159:705–715

Wiemken V, Ineichen K (1993) Effect of temperature and photoperiod on the raffinose content of spruce roots. Planta 190:387–392

Wiemken A, Frehner M, Keller F, Wagner W (1986) Fructan metabolism, enzymology and compartmentation. Curr Top Plant Biochem Physiol 5:17–37

Wiemken A, Sprenger N, Boller T (1995) Fructan-an extension of sucrose by sucrose. In: Pontis HG, Salerno GL, Echeverria EJ (eds) International symposium on sucrose metabolism. American Society of Plant Physiology Press, Rockville, MD, pp 179–188

Wingler A, Fritzius T, Wiemken A, Boller T, Aeschbacher RA (2000) Trehalose induces the ADP-glucose pyrophosphorylase gene, ApL3, and starch synthesis in *Arabidopsis*. Plant Physiol 124:105–114

Wingler A, Marès M, Pourtau N (2004) Spatial patterns and metabolic regulation of photosynthetic parameters during leaf. New Phytol 161:781–789

Wood AJ, Saneoka H, Rhodes D, Joly RJ, Goldsbrough PB (1996) Betaine aldehyde dehydrogenase in sorghum molecular cloning and expression of two related genes. Plant Physiol 110:1301–1308

Xiang DJ, Hu XY, Zhang Y, Yin KD (2008) Over-expression of *ICE1* gene in transgenic rice improves cold tolerance. Rice Sci 15:173–178

Xiao BZ, Chen X, Xiang CB, Tang N, Zhang QF, Xiong LZ (2009) Evaluation of seven function-known candidate genes for their effects on improving drought resistance of transgenic rice under field conditions. Mol Plant 2:73–83

Xin Z, Browse J (1998) *Eskimo1* mutants of *Arabidopsis* are constitutively freezing tolerant. Plant Physiol 103:607–613

Xing W, Rajashekar CB (2001) Glycine betaine involvement in freezing tolerance and water stress in *Arabidopsis thaliana*. Environ Exp Bot 46:21–28

Xue X, Liu A, Hua X (2009) Proline accumulation and transcriptional regulation of proline biosynthesis and degradation in Brassica napus. BMB Rep 42:28–34

Yadeghari LZ, Heidari R, Carapetian J (2008) The influence of cold acclimation on proline, malondialdehyde (MDA), total protein and pigments contents in soybean (*Glycine max*) seedlings. Res J Biol Sci 3:74–79

Yancey P, Clark M, Hand S, Bowlus R, Somero G (1982) Living with water stress: evolution of osmolyte systems. Science 217:1214–1222

Yang X, Lu C (2005) Photosynthesis is improved by exogenous glycine betaine in salt-stressed maize plants. Physiol Plant 124:343–352

Yang L, Tang R, Zhu J, Liu H, Mueller-Roeber B, Xia H, Zhang H (2008) Enhancement of stress tolerance in transgenic tobacco plants constitutively expressing AtIpk2*b*, an inositol polyphosphate 6-/3-kinase from Arabidopsis thaliana. Plant Mol Biol 66:329–343

Yang JS, Wang R, Meng JJ, Bi YP, Xu PL, Guo F, Wan SB, He QH, Li XG (2010) Over-expression of *Arabidopsis* CBF1 gene in transgenic tobacco alleviates photo-inhibition of PSII and PSI during chilling stress under low irradiance. J Plant Physiol 167:534–539

Yang XY, Xie JX, Lu XP, Liu YZ, Peng SA (2011) Isolation of a citrus ethylene-responsive element binding factor gene and its expression in response to abiotic stress, girdling and shading. Sci Hortic 127:275–281

Yang A, Dai XY, Zhang WH (2012) A R2R3-type MYB gene, OsMYB2, is involved in salt, cold, and dehydration tolerance in rice. J Exp Bot 63:2541–2556

Yao YX, Dong QL, Zhai H, You CX, Hao YJ (2011) The functions of an apple cytosolic malate dehydrogenase gene in growth and tolerance to cold and salt stresses. Plant Physiol Biochem 49:257–264

Yelenosky G, Guy CL (1977) Carbohydrate accumulation in leaves and stems of 'Valencia' orange at progressively colder temperatures. Bot Gaz 138:13–17

Yeo ET, Kwon HB, Han SE, Lee JT, Ryu JC, Byun MO (2000) Genetic engineering of drought resistant potato plants by introduction of the trehalose-6-phosphate synthase (TPS1) gene from *Saccharomyces cerevisiae*. Mol Cells 10:263–268

Yoshiba Y, Kiyosue T, Katagiri T, Ueda H, Mizoguchi T, Yamaguchi-Shinozaki K, Wada K, Harada Y, Shinozaki K (1995) Correlation between the induction of a gene for delta1-pyrroline-5-carboxylate synthetase and the accumulation of proline in Arabidopsis thaliana under osmotic stress. Plant J 7:751–760

Yoshida M, Hashimoto K (1986) Assessment of the pathogenicity of yeasts used in the production of single cell protein. Agric Biol Chem 50:2117–2118

Yoshida R, Kanno A, Kameya T (1996) Cool temperature-induced chlorosis in rice plants. II. Effects of cool temperature on the expression of plastid-encoded genes during shoot growth in darkness. Plant Physiol 112:585–590

Yoshida M, Abe J, Moriyama M, Kuwabara T (1998) Carbohydrate levels among winter wheat cultivars varying in freezing tolerance and snow mold resistance during autumn and winter. Physiol Plant 103:8–16

Yuanyuan M, Zhang Y, Lu J, Shao H (2010) Roles of plant soluble sugars and their responses to plant cold stress. Afr J Biotechnol 8:2004–2010

Yukawa T, Watanabe Y (1991) Studies on fructan accumulation in wheat (*Triticum aestivum* L.). I. Relationship between fructan concentration and overwintering ability from aspect on the pedigree. Jpn J Crop Sci 60:385–391

Zeng Y, Yu J, Cang J, Liu L, Mu Y, Wang J, Zhang D (2011) Detection of sugar accumulation and expression levels of correlative key enzymes in winter wheat (*Triticum aestivum*) at low temperatures. Biosci Biotechnol Biochem 75:681–687

Zhang Y, Primavesi LF, Jhurreea D, Andralojc PJ, Mitchell RA, Powers SJ, Schluepmann H Delatte T, Wingler A, Paul MJ (2009) Inhibition of SNF1-related protein kinase1 activity and regulation of metabolic pathways by trehalose-6-phosphate. Plant Physiol 149 1860–1871

Zhang XY, Liang C, Wang GP, Luo Y, Wang W (2010) The protection of wheat plasma membrane under cold stress by glycine betaine overproduction. Biol Plant 54:83–88

Zhang LL, Zhao MG, Tian QY, Zhang WH (2011) Comparative studies on tolerance of *Medicago truncatula* and *Medicago falcata* to freezing. Planta 234:445–457

Zhao Y, Aspinall D, Paleg LG (1992) Protection of membrane integrity in *Medicago sativa* L. by glycine betaine against the effects of freezing. J Plant Physiol 140:541–543

Zhao L, Liu F, Xu W, Di C, Zhou S, Xue Y, Yu J, Su Z (2009) Increased expression of OsSPX enhances cold/subfreezing tolerance in tobacco and *Arabidopsis thaliana*. Plant Biotechno J 7:550–561

Zhou W, Li Y, Zhao B, Ge RC, Shen YZ, Wang G, Huang ZJ (2009) Over-expression of *TaSTR* gene improves salt and drought tolerance in rice. J Plant Physiol 166:1660–1671

Zhu B, Peng RH, Xiong AS, Xu J, Fu XY, Zhao W, Jin XF, Meng XR, Gao JJ, Cai R, Yao QH (2012) Transformation with a gene for myo-inositol *O*-methyltransferase enhances the cold tolerance of *Arabidopsis thaliana*. Biol Plant 56:135–139

Zhuo C, Wang T, Lu S, Zhao Y, Li X, Guo Z (2013) A cold responsive galactinol synthase gene from *Medicago falcata (MfGolS1)* is induced by myo-inositol and confers multiple tolerances to abiotic stresses. Physiol Plant 149:67–78

Ziaf K, Loukehaich R, Gong P, Liu H, Han Q, Wang T, Li H, Ye Z (2011) A multiple stress responsive gene ERD15 from *Solanum pennellii* confers stress tolerance in tobacco. Plant Cell Physiol 52:1055–1067

Zimmerman MH, Zeigler H (1975) List of sugars and sugar alcohols in sieve-tube exudates. In: Zimmerman MH, Milburn JA (eds) Encyclopedia of plant physiology new series. Springer, New York, pp 480–503

Zrenner R, Salanoubat M, Willmitzer L, Sonnewald U (1995) Evidence of the crucial role of sucrose synthase for sink strength using transgenic potato plants (*Solanum tuberosum* L.). Plant J 7:97–107

Zuther E, Buchel K, Hundertmark M, Stitt M, Hincha DK, Heyer AG (2004) The role of raffinose in the cold acclimation response of *Arabidopsis thaliana*. FEBS Lett 576:169–173

Chapter 10
Lignins and Abiotic Stress: An Overview

Renu Bhardwaj, Neha Handa, Resham Sharma, Harpreet Kaur, Sukhmeen Kohli, Vinod Kumar, and Parminder Kaur

Contents

1 Introduction

Lignins are aromatic 3-dimentional heterogeneous polymers that comprise phenylpropane units (Sjöström 1993). They are the second most abundant biopolymers constituting biosphere's 30 % of organic carbon (Ralph et al. 2007). According to fossil records, the process of lignification in plants appeared approximately 450 million years ago (Stewart and Rothwell 1993). Lignin deposition in the cell walls helped the plants to adapt in the terrestrial habitat during the course of evolution. The transition from aquatic to terrestrial habitat presented the plants with new

R. Bhardwaj (✉) • N. Handa • R. Sharma • H. Kaur
S. Kohli • V. Kumar • P. Kaur
Department of Botanical and Environmental Sciences,
Guru Nanak Dev University, Amritsar 143005, Punjab, India
e-mail: dr.renubhardwaj@yahoo.in; renu_bhardwaj@rediffmail.com

P. Ahmad and M.R. Wani (eds.), *Physiological Mechanisms and Adaptation Strategies in Plants Under Changing Environment: Volume 1*, DOI 10.1007/978-1-4614-8591-9_10,

challenges of radiation, lack of mechanical support, desiccation, pathogens and herbivores (Raven 1984). For acclimatisation to these changes, evolution of various metabolic pathways occurred, and one of them was phenylpropanoid pathway which led to lignin formation. The initial land plants, however, were small and lacked the need for mechanical support, but with the advent of tracheophytes, the plants developed the mechanism to deposit phenylpropanoid units in the cell wall and form the polymer lignin (Weng and Chapple 2010).

Physiologically, lignins play an important role in the life of plants. The process of lignification helps in binding the plant fibres together and thereby providing mechanical support to the plant. They also provide structural integrity to plant cell walls and are important for plant development (Boerjan et al. 2003). Studies conducted by Jones et al. (2001) on *irregular xylem4* (irx4) mutants of *Arabidopsis thaliana* showed that mutant plants had 50 % less lignins than the normal plants. These changes in the content of lignins had severe effects on morphology and architecture of the cell walls thereby leading to changed physical properties of stems. Lignified cell wall of xylem becomes less permeable to water and therefore helps in the transport of water and nutrients. It also provides disease resistance in plants. Lignin and lignin-like phenolic compounds accumulate as a defence response in a variety of plant–microbe interactions (Vance et al. 1980; Nicholson and Hammerschmidt 1992). During penetration by fungal appressoria, lignification provides the plant cell wall more resistance against the mechanical pressure and water, thus makes it more difficult to be accessed by cell wall degrading enzymes (Vance et al. 1980; Nicholson and Hammerschmidt 1992; Zeyen et al. 2002).

Recent studies have revealed that various abiotic stresses faced by plants have impacts on lignin content. These stresses affect the biosynthesis and deposition of lignin units in the cell wall thereby establishing a mechanism of shielding the plants from their adverse effects. The present chapter lays emphasis on the multiple means of lignin composition, content and accumulation and rearrangements in response to a number of abiotic stresses in the plant kingdom.

1.1 Biosynthesis of Lignins

Lignins are complex molecules which are hydrophobic in nature and are composed of a network of phenylpropanoid units. These phenylpropanoids are, namely, types of hydroxycinnamyl alcohol precursors which polymerise oxidatively to form lignins (Higuchi 1985). The three precursors of hydroxycinnamyl alcohol are *p-coumaryl* alcohol, coniferyl alcohol and sinapyl alcohol and polymerisation takes place by radical coupling (Boerjan et al. 2003). The lignins which are derivatives of these precursors are commonly known as guaiacyl (G), hydroxyphenyl (H) and syringyl (S) lignins, respectively (Weng and Chapple 2010). The lignin composition can vary according to the plant family and morphological region (Lewis and Yamamoto 1990).

Figure 10.1 shows the most accepted pathway of the formation of H, S and G lignins. The formation starts from phenylalanine which undergoes a series of reactions like deamination, hydroxylation, transacylation, methylation and reduction that

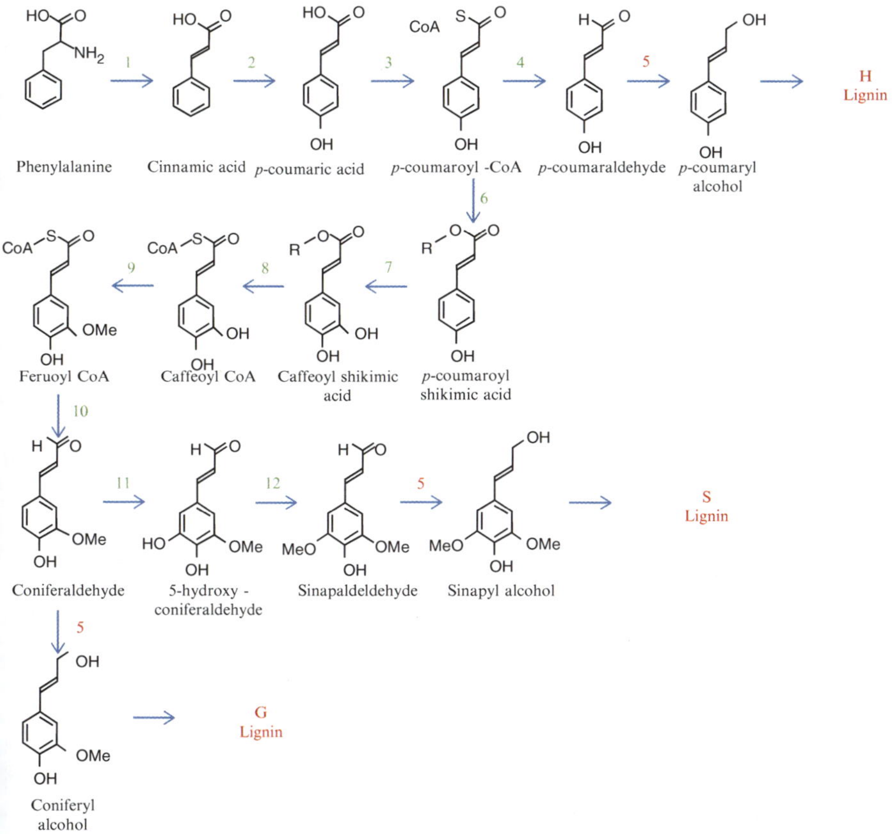

Fig. 10.1 Pathway of lignin synthesis: (1) phenylalanine ammonia-lyase; (2) cinamate-4-hydroxylase; (3) 4-coumarate-CoA ligase; (4) cinnamoyl-CoA reductase; (5) cinnamyl alcohol dehydrogenase; (6) *p*-hydroxycinnamoyl-CoA:quinate/shikimate *p*-hydroxycinnamoyltransferase; (7) *p*-coumarate 3-hydroxylase; (8) *p*-hydroxycinnamoyl-CoA:quinate/shikimate *p*-hydroxycinnamoyltransferase; (9) caffeoyl-CoA *O*-methyltransferase; (10) cinnamoyl-CoA reductase; (11) ferulate 5-hydroxylase; (12) caffeic acid *O*-methyltransferase (adapted from Vanholme et al. 2010)

convert it to the monomeric precursors of lignin. These reactions are catalysed by approximately ten enzymes and three cytochrome P450 proteins, i.e. cinnamic acid 4-hydroxylase (C4H), *p*-coumaroyl shikimate 3′-hydroxylase (C3H) and ferulic acid 5-hydroxylase (F5H). The cytochrome P450 proteins are bound to membranes and are attached with endoplasmic reticulum on its outer surface with the help of their N-terminal membrane anchor (Li et al. 2008). Other enzymes such as phenylalanine ammonia lyase (PAL), 4-coumarate-CoA ligase (4CL), caffeic acid *O*-methyltransferase (COMT), caffeoyl-CoA *O*-methyltransferase (CCoAOMT) and cinnamyl alcohol dehydrogenase (CAD) are reported to be soluble in cytosol as shown by various immunocytochemical and biochemical studies. However, some isoforms of PAL and CAD have been reported to be coupled to the vesicles from endoplasmic reticulum and Golgi apparatus (Takabe et al. 1985; Nakashima et al. 1997).

The biosynthetic enzymes of monolignols are localised in the cytosol which implies that their site of synthesis is the cytoplasm of the cell (Liu et al. 2011). From here they are transferred to the cell wall where these monolignol precursors are polymerised and amalgamated in the cell wall.

1.1.1 Transport of Monolignols

The mechanism of transport of monolignol precursors is still a topic of research among the researchers. It has been reported in certain angiosperms and gymnosperms that 4-*O*-β-D-glucosides are formed from monolignols by glycosylation and high levels of these glucosides have been reported in cambial tissues (Whetten and Sederoff 1995; Steeves et al. 2001). It has been believed that these glucosides are either for storage or for transport of monolignols from the cytoplasm to the cell wall and this process is supposed to be aided by uridine diphosphate glucose, coniferyl alcohol glucosyl transferase and coniferin-β-glucosidase (Dharmawardhana et al. 1995; Steeves et al. 2001; Samuels et al. 2002).

One proposed mechanism of transport of monolignols is via endoplasmic reticulum and Golgi-derived vesicles by exocytosis. Certain non-cellulosic compounds like pectins and hemicelluloses have been known for their synthesis in Golgi bodies and their transport also occurs by exocytosis (Lerouxel et al. 2006). It has been suggested by various immunocytochemical, autoradiographic and ultrastructural studies that monolignols follow the same route of transport. In autoradiographic studies, [^{3}H]-Phe, -tyrosine and -cinnamic acid were used to label the developing xylem and it was observed that the radiolabels were found to bind with the Golgi apparatus, rough endoplasmic reticulum and vesicles in plasma membrane and cytoplasm (Pickett-Heaps 1968; Fujita and Harada 1979; Takabe et al. 1985). It was observed by Nakashima et al. (1997) that PAL and CAD were found not only in cytoplasmic matrix of tracheary elements of *Zinnia elegans* but also in vesicles derived from Golgi apparatus and these observations were made by immunocytochemical studies. However, in recent studies [^{3}H]-Phe radiotracer was incorporated in the developing xylem tissue and phenylpropanoid and protein biosynthesis was also inhibited in lodgepole pine (Kaneda et al. 2008). It was observed that the Phe-radiolabel in endoplasmic reticulum and Golgi bodies got associated with proteins and not monolignols and also these vesicles did not contain phenylpropanoids, thereby indicating that vesicular exocytosis is not the only mechanism of monolignol transport. Passive diffusion is another proposed mechanism of monolignol transport. It has been observed that lignins contain several other phenolic compounds which are also present in lignin polymer along with monolignols and such precursors undergo passive diffusion across the plasma membrane (Vanholme et al. 2008; Liu et al. 2011) Nowadays, studies have revealed the involvement of transporters in lignin monomer transport. It has been observed in many plants that specific monomers are deposited at specific sites in the cell wall, thereby indicating the mechanism of deposition to be very organised and regulated (Liu et al. 2011). This suggested that such selective deposition could occur by some active transport mechanism

The transporters mainly involved in this mechanism are ATP-binding cassette (ABC) transporters. Various genomic and proteomic studies disclose that lignified woods show higher expression of these transporters. Microarray studies conducted on *Arabidopsis* stems helped in identification of ABC transporters during lignification and these genes were expressed in coordination with lignin biosynthetic enzymes (Ehlting et al. 2005; Douglas and Ehlting 2005). Similarly, Nilsson et al. (2010) showed the presence of a set of ABC transporters by conducting proteomic studies on leaves, xylem and cambium tissues of poplar trees. The studies conducted so far need more convincing evidences to confirm the exact mechanism involved in the transport of these precursors.

1.1.2 Polymerisation of Monolignols

Polymerisation of the monolignol precursors is the first step after their transport to the cell wall. The polymerisation of these is achieved by dehydrogenation in which phenol of monolignol is oxidised (Vanholme et al. 2010). The reaction is thought to be carried out by different enzymes like peroxidases, laccases, polyphenol oxidases and coniferyl alcohol oxidase (Boerjan et al. 2003). Out of these enzymes, peroxidases and laccases are mainly involved in radical formation. Peroxidases use hydrogen peroxide while laccases use oxygen as the substrate of above-mentioned reaction (Sarkanen and Ludwig 1971; Vanholme et al. 2010). The radicals thus obtained are stable due to delocalisation of unpaired electron (Boerjan et al. 2003). Two such radicals combine to form a dimer and a covalent bond is set up between the monomers. The β-positions of the radicals are the favoured positions where the bond is formed and this results in the formation of mainly β-β, β-*O*-4 and β-5 dimers (Vanholme et al. 2010). The dimer thus formed again gets dehydrogenated to form a phenolic radical which further couples with a monomer radical. This method of polymerisation is known as endwise coupling. The structure of lignin formed after polymerisation depends partly upon the type of monolignol available and the enzyme specificity which is involved in dehydrogenation (Vanholme et al. 2010). The deposition of the lignin polymer starts in middle lamella and corners of the primary walls of the cells. These are known as the nucleation sites and it has been speculated that Ca^{2+}-pectate-bound peroxidases might have a role in lignin deposition (Carpin et al. 2001; Boerjan et al. 2003).

1.2 Distribution of Lignins

Figure 10.2 shows the distribution of lignins in plant kingdom. Lignins are widely distributed in higher plants such as angiosperms and gymnosperms (Weng and Chapple 2010). Gymnosperms mainly contain guaiacyl and syringyl lignins, whereas angiosperms contain all the three types of lignins, viz. syringyl, hydroxyphenyl and guaiacyl lignins (Zhou et al. 2011; Siqueira et al. 2011). Lignins such as

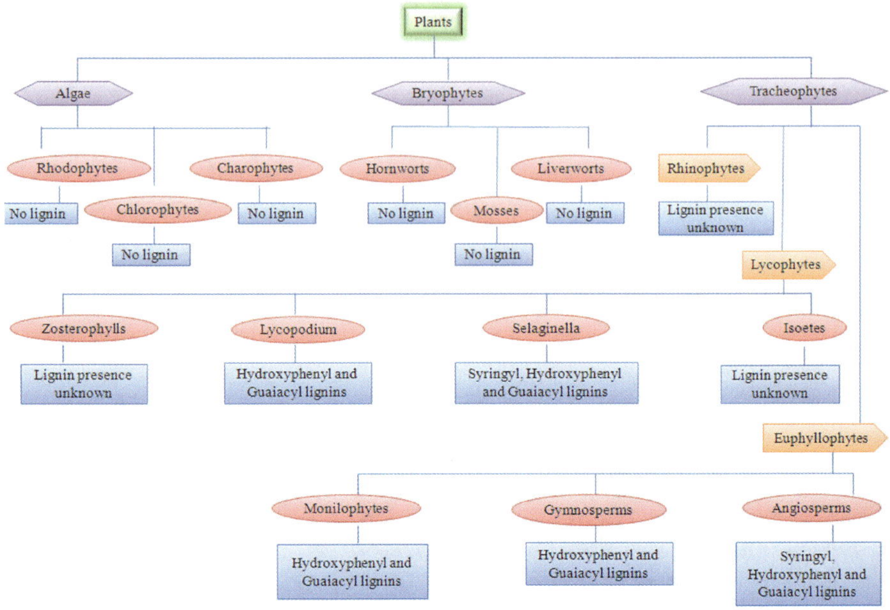

Fig. 10.2 Distribution of lignins in plant kingdom (Weng and Chapple 2010)

syringyl lignins have been shown in lycophyte *Selaginella moellendorffii* by Weng et al. (2008a). Lignin biosynthesis is not found in bryophytes but they show accumulation of soluble phenylpropanoids like lignans and flavonoids as proposed by Basile et al. (1999) and Umezawa (2003). Orthologs of eight main enzymes which are involved in lignin biosynthesis are essential for the synthesis of *p*-coumaryl and coniferyl alcohols. These were found in the genome of moss *Physcomitrella patens*, whereas none was present in *Chlamydomonas reinhardtii* which is green algae (Silber et al. 2008; Weng et al. 2008b; Xu et al. 2009).

Table 10.1 shows the distribution of various types of lignins in different tissues and cells of different plants. Syringyl lignin is predominantly found in fibre and ray parenchyma secondary walls, whereas guaiacyl lignin is present primarily in vessel secondary walls and cell corner middle lamella of hardwood plants (Musha and Goring 1975). In white birch (*Betula papyrifera*) the lignin present in the middle lamella and vessel secondary walls mainly contained guaiacyl lignin, whereas the secondary walls of ray parenchyma and fibre mainly consisted of syringyl lignin, as determined by Fergus and Goring (1970). Contrary to it, Saka and Goring (1988) demonstrated that rays mainly contained guaiacyl lignin. In birch wood, the lignins present in the walls of vessels were rich in guaiacyl units, but the fibre and ray cell walls were found to be rich in syringyl lignin (Hardell et al. 1980). It has been demonstrated in oxford poplar (*Populus maximowiczii* × *Populus berolinensis*) that ferulic acid which is a precursor of guaiacyl lignin is preferentially deposited into vessel elements rather than fibres, while sinapic acid which is a precursor of syringyl lignins preferentially deposited into fibres rather than vessel elements (Terashima

Table 10.1 Distribution of lignins in plants/tissues/cells

S. No.	Type of lignin	Tissue/tissue part/cell type	Plant type	Authors
1	Syringyl lignin	Fibres	Hardwood plants	Musha and Goring (1975)
		Fibres	Poplar (*Populus trichocarpa*)	Zhou et al. (2011)
		Ray parenchyma and fibres	White birch (*Betula papyrifera*)	Fergus and Goring (1970)
		Fibre and ray cell walls	Birch	Hardell et al. (1980)
		Fibres	Poplar (*Populus maximowiczii* × *Populus berolinensis*)	Terashima (1989)
		Bundle sheath fibres	*Erianthus arundinaceus*	Yamamura et al. (2013)
2	Guaiacyl lignin	Vessel cell walls	Birch	Hardell et al. (1980)
		Vessel cell walls	Poplar (*Populus trichocarpa*)	Zhou et al. (2011)
		Middle lamella and vessel's secondary wall	White birch (*Betula papyrifera*)	Fergus and Goring (1970)
		Vessels	Poplar (*Populus maximowiczii* × *Populus berolinensis*)	Terashima (1989)
		Vessels	Aspen (*Populus tremuloides*)	Wolter et al. (1974)
		Vascular bundle sheath	*Erianthus arundinaceus*	Yamamura et al. (2013)
3	*p*-Hydroxyphenyl lignin	Fibre cell walls	Sugarcane	Siqueira et al. (2011)
		Pith	*Phyllostachys pubescens*	Wen et al. (2013)

1989). In aspen (*Populus tremuloides*) callus cultures, pure guaiacyl lignin was found to be associated with vessels (Wolter et al. 1974). Similarly, in poplar wood (*Populus trichocarpa*), Zhou et al. (2011) reported guaiacyl lignin in the vessel cell walls and syringyl lignin in fibre cell walls. The results of experiments performed by Yamamura et al. (2013) suggested the presence of syringyl lignin in the bundle sheath fibres and guaiacyl lignin in the vascular bundle sheath in *Erianthus arundinaceus*. *p*-Hydroxyphenyl lignins were found in the fibre cell walls of sugarcane (Siqueira et al. 2011). More recently, threo- and erythrodiastereomers of *p-hydroxyphenyl*, syringyl and guaiacyltrithioethylpropane compounds were found to exist in the leaf sheath products of *Oryza sativa* by Yamamura et al. (2012). *p*-Hydroxyphenyl units are one of the structures present in lignins from the pith and stem of bamboo as characterised by Wen et al. (2013).

1.3 Physiological Roles of Lignins

Lignin plays many important roles in the biological world and has many fold impacts on all physiological, ecological and economic processes. These are essential components of secondary cell wall thickening which in turn exert a number of effects on plant functions. Plants are in possession of a variety of lignin coordination techniques for monitoring synthesis, accumulation and deposition in response to various intrinsic and extrinsic signals. Depending upon the situation or stress faced, lignin plays protective, sustaining and disruptive roles. Other aspects such as deposition and localisation timing hold control on plant development. It is essential to bear in mind that the lignin depository mechanisms are involved in plant growth, development and defence responses. Lignification is imperative for providing support, water transport and disease resistance in plants. Plants that are unable to synthesise required lignin amounts lose the ability to support plant body and become susceptible to humidity changes as compared to plants with normal lignin content and deposition patterns (Jones et al. 2001; Goujon et al. 2003). Since lignin is known to affect the seed coat specifically seed structure modifications, it controls seed dispersal and dehiscence in many plant species such as *Arabidopsis* (Liljegren et al. 2000a, b). Lignin is involved in the sequestration and fixation of atmospheric carbon dioxide into secondary thickenings of mature plants via carbon cycle. Lignin decomposes at an extremely slow rate as compared to most components of dead vegetation, contributing a major part of humus which replenishes soil fertility and texture. This soil humus is also responsible for an increased photosynthetic productivity of plants by facilitating cation exchange capacity and expansion of the moisture retention capacity in-between flood and drought conditions (Johnston 1964; Fustec et al. 1989). Lignin helps to complex with several composites, from transition metals to insecticides, making lignin a widespread sorbent material for bioremediation requirements. This is true for metal sorption by modified lignin like kraft lignin as seen in *Eucalyptus globulus* (Santos et al. 2011, 2012). They offer a good replacement for heavy metal absorption in a number of cases such as cross-linked lignocatechol gel was prepared by immobilising catechol onto wood lignin followed by cross-linking (Parajuli et al. 2005). Lignocelluloses have been reported in heavy metal removal from contaminated waste waters and agricultural waste products (Lee and Rowell 2004; Harman et al. 2007). Many studies indicate the adhesion and absorption of the heavy metal ions such as Cd, Cu, Ni, Pb and Zn on black liquor (a lignin-based waste product of paper industry). Lignin has a strong pull for these metal ions in the following order of treatment and presence: Pb > Cu > Cd > Zn > Ni (Guo et al. 2008). Kraft lignins have been reported to remove Cu (II) from waste waters (Waltersson 2009). Studies were carried out to understand the adsorption of the heavy metal ions such as copper and lead onto a lignin derivative obtained by treating bagasse soda lignin via Mannich reaction (Liu et al. 2013). Some of the most essential roles played by lignin are concised below.

1.3.1 Mechanical Strength

Cell walls in plants consist of four basic structural units: cellulose, hemicellulose, lignin and pectin. The presence and rearrangement of these elementary units in cellular structure give rise to a remarkably wide range of mechanical properties and efficiency in plants (Gibson 2012). Lignin is most commonly noted for providing support through strengthening of secondary growth in trees making wood a source of timber and fuel (Wardrop 1969; Ros Barcelo 2005). Lignin deposition happens as a result of cell growth accomplished during various growth phases inclusive of secondary thickening. It is the harbinger of intra- and intermolecular adhesion in-between the complex sugar components of the secondary cell wall thickenings. The complex polymer fills in the gaps of cell wall, in-between cellulose, hemicelluloses and pectins specifically in xylem tracheids, vessels and sclereids also in phloem fibres and periderm (Boerjan et al. 2003; Gibson et al. 2010). Lignin develops covalent bonding with hemicellulose and, in turn, forges cross-linkages with different plant polysaccharides, providing mechanistic support to the cell wall by extending the plant as a whole (Chabannes et al. 2001). It is important to understand that mechanical strength was one of the most primal changes that made the transition of plants from aquatic to land habit an evolutionary landmark (Xu et al. 2009; Weng and Chapple 2010; Cesarino et al. 2012).

1.3.2 Transport/Conduction and Waterproofing

Lignin permits water and mineral transport throughout the vascular bundles especially xylem under severe negative pressure without causing any tissue collapse (Jones et al. 2001). The polysaccharide components of plant cell walls are highly hydrophilic and thus permeable to water, whereas lignin is hydrophobic. The cross-linking of polysaccharides by lignin is an obstruction for water absorption to the cell wall making long-distance transport of water and solutes by waterproofing the vascular tissue (Cesarino et al. 2012). Recent reports supported by light and confocal microscopy studies indicate that low lignin-low xylem wood is a poor transporter of water and minerals, henceforth confirming their role for the same (Kitin et al. 2010; Gibson et al. 2010). The induction of lignification helps in waterproofing plant cells as they are impermeable in nature henceforth decreasing the probability for dehydration (Reina et al. 2001).

1.3.3 Pathogen Attack and Wounding

The ability of lignin to resist degradation due to chemically complexity of structure made it act as an effective defensive barrier against herbivores and pathogens (Bonawitz and Chapple 2010; Cesarino et al. 2012). Due to their highly complicated nature, lignin polymers are tricky to degrade and decompose. Plants respond to infection to pathogens by increasing lignifications. In a hypersensitive reaction caused by *Puccinia graminis* in wheat, lignification acts as a defence mechanism (Moerschbacher et al. 1990). This occurs to block the invasion of parasites and serves as a barrier for microorganisms. In *Pinus nigra* lignin deposition is increased by the infection of *Sphaeropsis* (Bonello and Blodgett 2003). RT-PCR and microarray analysis was employed to identify ABA downregulation of main genes responsible for lignin synthesis and deposition belonging to the phenylpropanoid pathway. ABA also downregulates a number of other genes that are involved in activating defence responses and cell signalling (Mohr and Cahill 2004). Lignin polymers curtail the digestive quality of vegetation, targeting herbivores mainly by reducing the nutrition index and desirability of it as fodder (Moore and Jung 2001). This definitely helps in controlling overgrazing and waste land production.

1.3.4 Industrial Uses of Lignins

Though lignin has varied roles in supporting plant functions as explained previously, it is a source of interference in many important industrial processes. However, they have also been utilised for synthesis of many useful by-products. The first reports of commercial applications of lignin can be traced back as early as 1927 by the US-based Marathon Corporation, Rothschild, Wisconsin. The major class of commercial products with promising utility were leather tanning agents recovered as a result of various lignin processes. Highly lignified wood is strong and durable and therefore a great raw material for many applications in industry. Since lignin has a higher calorific value than cellulose, it acts as an excellent fuel wood. More recently, through biorefining, lignins are being used to replace fossil fuels (Waltersson 2009). The kraft process (sulphate pulping of paper) which employs lignin removal tends to provide lignin as a waste product. However, it is used for combustion usually for its high fuel value, acting as an energy source to run mills and its associated processes. High-yielding pulp rich in lignin is used to make newsprint (Harkin 1967; Sjöström 1993). Lignosulphonates (a class of modified lignin) and conversion of lignocelluloses into kraft lignin yield several by-products such as dispersants in the cement industry, water treatment plants, textile industry, as additives in oil, chemical and agriculture industries and as raw materials for production of many commercial chemicals including ethanol, sugars, humic acid, vanillin and DMSO (Vishtal and Kraslawski 2011; Maki et al. 2012). Lignin extracted from willow is used to synthesise expanded polyurethane foam. It is also used for the production of Arboform (a plastic-like substance which is ecofriendly and biodegradable) as reported by the German company, Tecnaro, in 1998. Carbon fibres have been synthesised from lignin instead of fossil oil as primary precursor for the same. As a by-product of pulping process, enormous quantities of lignin exist. Depolymerisation of lignin in to compounds of lower molecular weight is its greatest protruding application discovered till now. Gasification of lignin is also done to produce a mixture of H_2 and CO (Syngas). This Syngas can sequentially be blended into several chemicals like methanol. On the other hand, lignin can be used to get oils through pyrolysis progression but pyrolysis oil made from lignin pyrolysis is extremely oxygenated and unstable in the course of storage and, therefore, needs additional progression and processing (Pandey and Kim 2011). In its place, scenarios of commercialisation of lignin are superior in manufacture of chemicals such as acids, alcohols, phenols and aldehydes. Lignin is possibly the merely practical renewable source for the manufacture of aromatic compounds. In the progress of cost-effective biorefinery procedures for biofuels, an important character is played by valorisation of lignin (Tong et al. 2010).

2 Lignins and Abiotic Stress

Various abiotic stresses such as water, radiation temperature, mechanical injuries and heavy metals cause alteration in normal growth and physiology of plant system. These stresses also have a marked effect on the lignin content, composition and structural rearrangements. This in turn emphasises lignin as a bioindicator as well as successful line of defence against abiotic anomalies. Various abiotic stress responses with specific reference to lignin are discussed as follows.

2.1 *Water Stress (Drought/Flooding)*

Change in the leaf anatomy and ultrastructure is associated with water stress including reduction in size of leaves, thickening of cell wall of leaves, increase in number of large vessels, increase in rate of cutinisation of cell walls and reduction in number of stomata (Lisar et al. 2012).

Gymnosperms and angiosperms synthesise lignin-like polymers in response to stress. These polymers differ from actual lignins, but they induce the formation of lignin structure (Polle et al. 1997). Change in lignin content and other secondary metabolites in trees is expected under variable stresses including nutrient stress, water stress and metal stress (Polle et al. 1997). Increase in protein hydrolysis, decrease in protein synthesis and an increase in lignification of cell wall are associated with several abiotic stresses including water deficit, low temperature and pathogen (Nilsen and Orcutt 1996). Increase in the activities of lignifying enzymes such as phenylalanine ammonia lyases (PAL) and peroxidase in response to water stresses and may lead to an increase in deposition of stress lignins or lignin-like polymers (Moerschbacher et al. 1988; Bruce and West 1989). According to the authors, abiotic and biotic stresses increase the cell wall lignification which is related with reduction in nutrient availability, translocation in plant and plant growth.

Stomatal closure, reduction in transpiration rates, reduced water potential in tissues, decreased photosynthetic rate, inhibition of growth, accumulation of proline, mannitol, sorbitol, ABA (abscisic acid), lignin, formation of free radicals and synthesis of new proteins are certain physiological changes associated with water stress (Lichtenthaler et al. 1981).

Water deficit occurs in environment when uptake of water through the roots of plants is insufficient to meet the water requirement for unhindered growth, transpiration and photosynthesis in shoot (Fan et al. 2006). One of the major environmental factors decreasing plant yield worldwide is drought (Vincent et al. 2005). Water deficit reduces water potential and turgor in plant which lead to difficulty in performing normal physiological function (Lisar et al. 2012). Reduction in anionic peroxidase activity and amplification in the amount of biosynthetic precursors of lignin in xylem tissue may indicate that water deficit leads to reduction in synthesis of lignin in maize plant (Alvarez et al. 2008). Deposition of lignin might be increased in a particular region of root than other, i.e. specific regions of maize roots may possibly react distinctly to water deficit (Fan et al. 2006; Yang et al. 2006; Yoshimura et al. 2008). According to authors, the basal regions of maize which are under the effect of water scarcity show a larger decline in the growth than the other regions and it is coupled with augmented lignin deposition with decrease in cell wall extension (Fan et al. 2006).

Method of proteomic analysis and qualitative real-time PCR were used to identify disparity in expression of aspirant gene under water deficit stress in *Zea mays* (Hu et al. 2009). The expression of two important enzymes of lignin synthesis (cinnamyl alcohol dehydrogenase and caffeate *O*-methyltranferases) was estimated using PCR technique in three drought-forbearing and one drought-susceptible inbred lines under drought stress. The expression of these enzymes were found to increase in there drought-forbearing species in comparison to drought-susceptible lines. A considerable distinction in leaf lignin content was observed amongst the most prominent response to drought stress (Hu et al. 2009).

Activities of lignification-related enzymes (polyphenol oxidase, cinnamyl alcohol dehydrogenase) of the two genotypes of *Trifolium repens* L. exposed to 12 days and second 6 days drought stress, respectively, showed a marked increase (Li et al. 2013).

Gene expression related to cell development and root extension in rice plants was observed to increase with 16 h water stress exposure (Yang et al. 2006). Study revealed that there was an upregulation of lignin biosynthetic genes during in-between and last stages of water stress. Related results were also observed in *Citrullus lanatus* sp. which also showed high resistance to water deficiency (Yoshimura et al. 2008). The studies showed that when the leaves of *Trifolium repens* were exposed to 28 days of water deficit, it led to increased biosynthesis of lignin, accompanied by a hefty increase in activity of PAL and ascorbate peroxidase in early stage (0–14 day). Alternatively, certain enzymatic activities such as guaiacol peroxidase, syringaldazine peroxidase and coniferyl alcohol peroxidase were found to increase during final stage (14–28 day) (Bok-Rye et al. 2007).

It was suggested that in primary roots of *Zea mays*, cell production, size of root elongation zone and whole root elongation capacity are reduced under water deficit conditions (Fraser et al. 1990; Pritchard 1994). The mechanism involved in amending cell wall expansion ability or rate of lignification is a complex multigenic interaction including phenylpropanoids, polysaccharides, reactive oxygen species (ROS), peroxidases, calcium ions concentration, cell wall proteins, wall enzymes and proton pumping in walls. Remarkable wall expansion and segmental development capacity in the tip area up to 0–3 mm of roots under drought condition were observed (Fan et al. 2006; Fan and Neumann 2004). Formation of lignin is stimulated by oxidative polymerisation of monolignols in the presence of hydrogen peroxide; in a plant under 28-day water stress, it was observed that lignin content was increased by 16 % under drought stress and some isoenzymes were highly expressed such as glutathione peroxidase (GPOX) and soluble peroxidase (SPOX) (Boudet 2000). Guaiacol as well as coniferyl alcohol POX activity were linked with lignin biosynthesis in *Picea glauca* (Polle et al. 1994). Authors suggested that increase in activities of enzymes (GPOX, Cyclooxygenase and syringaldazine peroxidase) involved in monolignol polymerisation was associated with lignin content. Drought stress increases lignin content in higher plants which leads to limited production of forage and its utilisation, loss of plant dry matter and dry matter uptake (Guenni et al. 2002).

Change in the quality and content of lignin in *E. globulus* Labill and in hybrids *E. uroglobulus* (*E. globulus* × *E. urograndis*) and *E. urograndis* (*E. urophylla* × *E. grandis*) under drought stress were determined by an experiment conducted by Moura et al. (2011). To meet the aim of the experiment, the plants were divided into three categories, i.e. control, drought and drought healthier. Watering of the drought plant was not done and they were collected when droop symptoms were seen. The control plants were watered daily, whereas plants belonging to drought recovered category were watered when droop symptoms were observed and the plants were collected following recuperation from the drought-induced symptoms. Total lignin content was analysed in basal and tip regions of stem and these were also quantified by GC-MS to determine lignin composition by Rolando et al. (1992). It was reported that in *E. urograndis* under drought stress there was increase in the amount of lignin in basal region, whereas in *E. globulus* there was increase in amount of lignin in apical region and no change was observed in the basal region. *E. uroglobulus* also showed an increase in lignin content in apical region and no significant change in basal region was observed.

Drought can lead to reduced growth of root and shoot and may also lead to decrease in crop yield worldwide (Boyer 1982). Lignin content in plant tissue changes continuously depending on several external factors and physiological state of plant (Polle et al. 1997).

Reduction in growth and grain yield worldwide is the major problem associated with *Glycine max* crops under flood stress (Komastu et al. 2010). Authors reported that soybean under flooding stress showed upregulation of proteins located in cell wall. In a more recent approaches include proteomic analysis of soluble proteins in soybean roots and development mechanisms for monitoring root growth elongation during water stress. An amplification in isoflavone biosynthesis in the root tip maintained root elongation, whereas the enhancement in lignin biosynthesis in above region, i.e. 4–8 mm, may be accountable for inhibition of root elongation. This was due to region-specific regulation of phenylpropanoid metabolism and upregulation of caffeoyl-CoA *O*-methyltransferase, which are involved in the synthesis of lignin, which is further associated with increase in accumulation of lignin and lead to inhibition or reduction in growth rate of specific region (Yamaguchi et al. 2010). Recently, Arasan et al. (2013) characterised and analysed the expression of dirigent gene family (DIR gene), which is correlated to variable abiotic and biotic stresses in *Brassica*. In this study, 29 genes were collected, two from *Brassica rapa* cv. and 27 from *B. rapa* database (Br DIR genes). When the sample plants were exposed to water stress, highest expression of gene was observed at 24 h, at which the acid-soluble lignin content was also recorded to be maximum in the sampled plants. Therefore, it was indicated that there is a correlation between Br DIR-like genes and lignin content. Kasraie et al. (2012) studied the effects of time spraying amino acid on the yield and a number of physiological traits in *Zea mays* L. var. TWC647 under water deficit stress. There was a reported increase in proline accumulation under drought stress and this was correlated to the drastic increase in lignin content. Isoflavones, lignin, phenol and sugar content were reported to be altered by heavy metal application and under water deficit stress in soybean seeds. The seeds were subjected to water stress and these showed an increase in phenol, lignin and isoflavone content. Higher levels of isoflavones, lignin and phenol in seeds are enviable as these compounds are involved in disease resistance and water stress tolerance (Bellaloui 2012).

2.2 Radiation Stress

For normal growth of plants, optimum levels of light, temperature and nutrition are required. If any one of these is depleted or present in excess, plants undergo stress condition. Out of these, light is very important as it is required for photosynthesis, and any change in its intensity leads to stress and production of ROS. Increase in peroxidise activity and build-up of anthocyanin occur to protect the plants against ROS. Also there was decrease in the components of antenna system (Kimura et al. 2003). Lignin synthesis is also reported to be affected by light. It has been observed that the seedlings of *Ebenus cretica* L. shows 2.5 times elevation in lignin content when grown in light as compared to seedlings when grown under dark (Syros et al. 2005). In 3-day-old soybean seedlings, varying conditions of light were used to study the control of lignin biosynthesis. The light treatments were grouped into three categories: In first group, the seedlings were placed in dark. In second group, the seedlings were exposed to light (340 μmol/m^2/s), and in the third group the seedlings were exposed to light (340 μmol/m^2/s, 16 h per days) along with diamine oxidase inhibitors. High level of hydrogen peroxide and lignin is seen in plants grown in light. Also the level of peroxidase and diamine oxidase activity is increased (Andersson-Gunneras et al. 2006). When cultured calli of *Pinus radiata* get differentiated into tracheids, it has been observed that the action of the enzymes cinnamyl alcohol dehydrogenase and phenylalanine ammonia lyase was enhanced, and as a result, the amount of lignin got increased when calli were transferred to a photoperiod of 16 h from dark (Moller et al. 2006). Continuous activity of light (16.7 W/m^2

in mung bean hypocotyls showed increase in the quantity of cationic and anionic peroxidase and laccases which result in increased amount of lignin content (Chen et al. 2002). Under varying intensities of light (60, 160 and 300 μmol/m^2/s), *Phalaenopsis* orchids grown for 1 month at 25 ± 2 °C and there was increase in lignin content in roots and leaves. This was due to increasing light intensity and increased content of lignin which was associated with the induction of phenylalanine ammonia lyase, cinnamyl alcohol dehydrogenase and peroxidase activities (Akgul et al. 2007).

Apart from the role of biosynthesis of lignin in providing rigidity of cell wall, it also provides defence against radiation stress. It was seen that when 7,000 genes of flowering plant *Arabidopsis thaliana* were exposed to high light intensity, there was an increase in expression of 110 genes. Of these genes most of them had role in the biosynthetic pathway of lignin (Kimura et al. 2003). Changes in gene transcript accumulation have been observed in genes that encode lignin biosynthetic enzymes during diurnal cycle and this accumulation was initiated by light if there was a prior period of darkness (Rogers et al. 2005).

The effect of various wavelengths of light such as blue, UV, green, red, yellow and white light on lignin peroxidise in *Phanerochaete chrysosporium* BKM-F-1767 was studied. The results showed that only green light increased the lignin peroxidise production, UV and blue light reduces the activity of lignin peroxidise whereas yellow, red and white light has mixed effect on it (Ramirez et al. 2010).

Atmospheric ozone layer is reduced due to anthropogenic activities which resulted in an elevation of UV radiations reaching the earth's surface. Secondary metabolite production such as lignin, tannins and flavonoids was affected due to UV-B radiation and these secondary metabolites have been reported to play an important role in the evolution of plants from aquatic to terrestrial life and due to this these plants had increased exposure to UV-B radiations (Rozema et al. 1997). Morphogenesis of trichomes was affected by UV-B radiation in *Cucumis sativus* L. cotyledons and induction of lignin accumulation also took place in these structures (Yamasaki et al. 2007). When two different cell cultures of *Camellia sinensis* L. were studied, they varied in their ability for the biosynthesis of phenolic compounds by Chalcone synthase 1 (ChS1) and Chalcone synthase 2 (ChS2). The ChS2 had more capacity as compared to ChS1 for producing phenolic compounds. It had been seen that ChS2 showed more increase of lignin as compared to ChS1 in the intercellular space and cell walls of parenchyma of tracheid elements. When exposed to UV-B radiation, there is formation of a layer on the callus surface which is similar to lignin. UV-B radiation shows restriction to then ChS2 lineage (Zagoskina et al. 2003). This suggests that lignin is involved in the protection of cell in response to radiation stress. There was increase in the thickness of epidermal cell wall of *Chenopodium quinoa* when its cotyledons were exposed to UV-B radiation. This occurred due to increased activity of peroxidase, which in turn results in the accumulation of lignin in cell wall (Hilal et al. 2004). Another report defined the accumulation of bioactive phenolic compounds and lignins in carrot tissue by the combination of injuries and abiotic stresses such as UV-B light and hormones (Heredia and Cisneros-Zevallos 2009).

2.3 *Temperature Stress*

Temperature beyond a threshold level for a period of time sufficient to cause irreversible damage (to plant growth and development) can be defined as temperature stress. A rise in 10–15 °C temperature above optimum conditions is considered as heat stress and temperature below 20 and 0 °C is called as chilling and freezing stress (Chinnusamy et al. 2007). Duration, rate of increase in temperature and intensity of temperature in degrees are the various factors determining the temperature stress (Peet and Willits 1998). Different global circulation models predict that greenhouse gases will gradually increase world's average ambient temperature and melting of glaciers which is a serious threat on plants worldwide due temperature stress (Hall 2001).

2.3.1 Chilling/Freezing

The maximum freezing tolerance of plants is in response to low, non-freezing temperatures (below approximately 10 °C) known by a phenomenon called as 'cold acclimation' (Orvar et al. 2000). An early step in cold acclimatisation is the influx of Ca^{2+} in the cell (Knight et al. 1996; Monroy and Dhindsa 1995). Cell membrane damage occurs mainly when temperatures decrease below 0 °C, as ice is formed in the intercellular spaces of plant tissues because the extracellular ice which causes dehydration to the plant cells (Steponkus 1984; Steponkus et al. 1993). Certain ROS also cause injuries (Suzuki and Mittler 2006) and improvement of cold tolerance in plants can be done by examining changes in lipid composition (Thomashow 1999).

Genes for lignin biosynthesis show highest transcription in the evening or 4–8 h before dawn that corresponds to low-temperature conditions (Rogers and Campbell 2004). High amount of cell wall formation of hypocotyls was observed in *Arabidopsis* during night (Dowson-Day and Millar 1999). Biosynthesis of cell wall along with lignification is initiated during evening and continues till dawn, but cell elongation is stopped during this period of time. Production of phenolic compounds, such as flavonoids and phenylpropanoids compounds, has been given in thermal stress. Plant response to thermal stress is enhanced in the presence of a key enzyme known as phenylalanine ammonia lyase (PAL), which has a major role against cell acclimation against stress in plants. Phenylalanine ammonia lyase (PAL), the principal enzyme of the phenylpropanoid pathway (Kacperska 1993), is considered to be catalysing the transformation, by deamination of L-phenylalanine into *trans-cinnamic* acid which is the major metabolite in the biosynthesis of lignins (Rivero et al. 2001).

Alteration in the plant lignin content was shown in ex vitro Poplar (*Populus tremula* × *P. tremuloides* L. cv. Muhs1) which showed an increase in lignin content when grown at 10 °C, but same was not observed for in vitro seedlings (Hausman et al. 2000). Annual average temperature increase can be positively correlated with the lignin amount in latewood *Picea abies* (Gindl et al. 2000). Variation in the level of lignin precursors of phenylpropanoid pathway and in the activities of related enzymes was observed in plants on exposure to chilling stress. On the contrary, *Brassica napus* plants did not show much changes in the phenols bound to cell wall and lignin content when exposed to varied chilling and freezing temperatures (2 °C for 3 weeks followed by chilling temperature of −5 °C for 18 h, then again at 2 °C for 6 h) (Solecka et al. 1999). Also, higher levels of normal and esterified forms of *p*-coumaric acid, ferulic acid and sinapic acid in the cells of the leaf mesophyll resulted in increase in the activity of phenylalanine ammonia lyase (PAL) enzyme. The amplification in the soluble forms of esterified phenolic acid was shown by Solecka et al. (1999). These phenols are suggested to be transported to vacuoles where they act as substrates of peroxide and protect the plant from ROS (Takahama 1988; Dixon and Paiva 1995; Whetten and Sederoff 1995).

Increase in the levels of esterified forms of ferulic, syringic and *p-hydroxybenzoic* acids was also observed in *Glycine max* root during adaptation to cold stress (10 °C) (Janas et al. 2000). Activity of peroxidase enzyme amplified in wheat (Taşgin et al. 2006; Janda et al. 2007) and *Saccharum* spp. hybrid (Jain et al. 2007) on exposure to cold conditions. Plants have evolved certain mechanism to resist low-temperature injury by scavenging reactive and toxic enzymes and compounds of antioxidant systems (Atici and Nalbantoglu 2003). As a result the peroxidase (POX) and catalase (CAT) converts H_2O_2 to H_2O. Amount of soluble phenols and lignin content were almost similar in leaves and roots of wheat plants at 2 °C. Decrease of lignin and increase of soluble phenols was observed in leaves while, the opposite was observed in roots. This increased soluble phenol content in leaves acted as endogenous antioxidant which helped the plant in tolerance to stress. On other hand, no change in lignin content was observed (Olenichenko and Zagoskina 2005).

Higher activity of enzymes related to cold acclimation as well as upregulation in gene expression was noticed in plants grown at low temperatures. Gene expression coding for cinnamate 3-hydroxylase (C3H) and a cytochrome P450-dependent monooxygenase at the time of cold acclimation of *Rhododendron* report that an increase is involved in the phenylpropanoids and lignin biosynthesis (El Kayal et al. 2006). Alteration in lignin composition and cell wall rigidity was observed and it was proposed that these changes are due to enhanced expression of C3H. Increase in lignin gene expression including CAD (cinnamyl alcohol dehydrogenase) was seen under cold stress in barley leaves (Janska et al. 2011). Lignin biosynthesis was suggested decline and instead of that monolignols were synthesised in the leaves of barley as the peroxidase genes were downregulated.

2.3.2 High Temperature

A long-term exposure can only cause cell injury at moderate temperatures, but at very high temperature, heat stress leads to cell death even within minutes which is due to disintegration of cellular organisation (Schöffl et al. 1999). Aggregation and denaturation of proteins and increased fluidity of membrane lipids are major injuries which are a direct result of high temperatures.

Higher lignin concentration and wider rings were found in latewood trees during a period of high temperature. A direct relationship was made to reconstruct temperatures of the pre-instrumental era by the dendroclimatologists (Eckstein and Aniol 1981; D'Arrigo et al. 1992; Briffa et al. 1998). Low latitudes had been negatively correlated with the gross lignin content of softwoods (Kim et al. 1989). High temperature also determines the wood quality as it is one of major factor influencing amount of chemicals to be spent in paper production (Fengel and Wegener 1984). Overexpression of superoxide dismutase (SOD) in plants affects various physiological processes in such as oxidation of toxic reductants, elimination of H_2O_2, biosynthesis and degradation of lignin in cell walls and catabolism of auxins. The generation of ROS, including singlet oxygen (O_2), superoxide radical (O^{2-}), hydrogen peroxide (H_2O_2) and hydroxyl radical (OH^-), is evoked by heat stress which leads to oxidative stress (Liu and Huang 2000; Mittler 2002; Potters et al. 2007).

Secondary metabolites like phenolics, lignin, flavonoids and anthocyanin play a variety of roles in plants. Soluble phenolics and lignins are accumulated under high-temperature stresses along with improved concentration of phenylalanine ammonia lyase (PAL), but decrease in the peroxidase polyphenyl lyase activity was recorded (Rivero et al. 2001; Taiz and Zeiger 2006). These compounds act as strong antioxidants in plant tissues during stress (Dixon and Paiva 1995; Sgherri et al. 2004). At 35 °C highest PAL activity was considered as response to heat stress in cell acclimation of plants. The intermediates of primary carbon metabolism were used to form secondary metabolites through shikimate, acetate-malonate and acetate-mevalonate pathways. Wasteful products in early research are now considered to serve many roles in plant tolerance to biotic and abiotic stresses. Compounds like terpenoids (e.g. saponins, carotenoids and steroids), phenolics (e.g. flavonoids, tannins, quinines, salicylates and lignins) and alkaloids had also been listed under this category (Hadacek 2002).

Isoenzymes like peroxidase having narrow range of isoelectric points, and each group has a different function in cell. Lignin formation is most likely carried out by acidic (anodic) peroxidases isoenzymes has been studied in (*Ricinus communis* L.) and might be associated with lignification and recovery of cell membrane damage due to heat stress (Gulen and Eris 2004). At 35 °C highest PAL activity was observed in tomato and watermelon, which is considered as response to heat stress in cell acclimation of plants. Under heat stress polysaccharides of cell wall get varied in coffee leaves. Increased oxidative stress causes changes in lignin composition of cell walls in plant tissues (Moura et al. 2010). Determination of enzyme activity and concentration regulation in wheat (*Triticum aestivum* L.) grains was done for diamine oxidase (DAO), peroxidase (POX) and polyamine oxidase (PAO) in comparison with lignin and polyamine concentration. Heat-tolerant (PBW 343) and heat-susceptible (WH 542) varieties were chosen. The activities of enzymes were compared and found that DAO, POX and PAO were higher and independently regulated in heat-tolerant (PBW 343) in comparison with heat-susceptible (WH 542) wheat varieties. Higher enzyme activity had a positive effect in heat-tolerant (PBW 343) variety on lignification in the cells as compared to WH 542 (Chowdhury et al. 2012). Other examples also point out to the stark relation between lignin and temperature stress. The increase in lignin content was also seen in heat-treated pericarp, whereas H_2O_2 concentration declined in comparison to control conditions. This leads to enhancement of resistance in fruits in response to heat stress. Later, it was analysed that lignin and ROS (reactive oxygen species) had important contribution in resistance of fruits against pathogens and physiological disorders due to heat treatment (Yun et al. 2013). Antioxidant and photoprotective responses to elevated CO_2 and heat stress were studied in Holm Oak. It was found that plants grown under current CO_2 concentrations ((350 llÆl)1) (RA) and resprouts from plants grown under elevated CO_2 (RE) showed elevated levels in the cellulose and lignin content, but rarely any difference was seen in for nitrogen, hemicelluloses or soluble sugar content in RE (Pintó-Marijuan et al. 2013).

2.4 Mechanical Injuries

Histological and anatomical studies have shown that changes in the wood are specific for particular type of injury. These changes affect stress resistance and adaptation. Physical barriers such as lignification, formation of cuticle, spines, trichomes and production of tannins and alkaloids serve as the defence mechanisms of various plant species. They help the plant to protect them against various mechanical injuries that were caused by various factors (Delessert et al. 2004). Electron microscopic studies have shown that xylem fibre cell wall thickness is induced in three ways in the injured stem of *Populus* sp.: thickening of cell wall, production of sclereid-like sub-layers and synthesis of additional S2 layers. Under UV-microspectrophotometer, it has been found that the content of lignin is higher in the xylem-modified fibres and asymmetrical distribution of lignin in the middle lamella and secondary wall (Frankenstein et al. 2006). Injuries to the leaves of *Arabidopsis thaliana* caused enhanced expression of genes which regulate biosynthesis of lignin in order to avoid water loss and pathogen infection. The genes that are associated with lignin biosynthesis codes for enzymes 4CL, cinnamyl alcohol dehydrogenase (CAD) and cinnamoyl coenzyme A reductase (CCR). It has been indicated that lignin synthesis takes place in those cells which surround the injured sites (Delessert et al. 2004). In *Eucalyptus gunnii* removal of stem apex causes enhancement in enzyme activities such as sinapyl alcohol dehydrogenase (SA) and cinnamyl alcohol dehydrogenase enzymes which are associated with lignin biosynthesis (Hawkins and Boudet 2003).

During the lignification of tissues, the growth is inhibited because the initial enzyme phenylalanine ammonia lyase that is involved in monolignol biosynthesis pathway directly influences the accumulation of lignin (Depege et al. 1997; Campos et al. 2004). Lignin characteristics differ among plant organs, tissues and cell walls (Grabber et al. 2004). It consists of polyphenolic polymer that is derived from oxidative polymerisation of monolignols including coniferyl, *p*-coumaryl and sinapyl alcohols via a side pathway of phenylalanine metabolism leading to the biosynthesis of lignin (Whetten and Sederoff 1995). Lignin content was increased in rubbed tomato internode in response to mechanical stress. Lignin and suberin are the components of support systems and serve as natural barrier in plants (Vance et al. 1980; Aquino and Mercado 2004). Due to injury, specific reactions take place in wood components such as cambium, bark or xylem (Frankenstein et al. 2006).

Cinnamyl alcohol dehydrogenase induction, SPOX activities and lignification induce mechanical disturbances in the leaves of a 7-day-old seedling of bean by short daily periods of non-injurious wind (Cipollini 1997). It has been reported that accumulation of ferulic acid and lignin occurs in phloem of *Chamaecyparis obtusa* during bark injury after 7 days. Injury also induced lignification in necrotic phloem and resulted in parenchymal zone formation (Kusumoto 2005). After 14 days, lingo-suberised layer was formed between the parenchymatous area and necrotic tissue. After 28 and 56 days of injury, lignin concentration was found to be higher in cell walls of necrotic area. Expression of CaF5H1, that code for F5H, is measured in the detached leaves of *Camptotheca acuminate* plants (Kim et al. 2006). F5H (ferulate 5-hydroxylase) is a cytochrome P450-dependent monooxygenase which promote the syringyl lignin biosynthesis and catalyses the hydroxylation of coniferyl alcohol, ferulic acid and coniferaldehyde. These observations suggest the role of lignins in protection of plants from mechanical stress.

2.5 Heavy Metal Stress

Heavy metal stress and exposure is known to exert twin impacts. Lignin composition, deposition and content undergo changes or lignin starts acting as a metal absorbing matrix. The latter property is of extreme commercial use as described previously and varies according to type of metal ion or limits of exposure. Lignin deposition increased due to excess aluminium (Al) in some plant species and this leads to reduced growth, which is related with lignin accumulation in cell wall. This has been previously recorded in *Triticum aestivum* and *Oryza sativa* (Sasaki et al. 1996; Budikova 1999; Hossain et al. 2002, 2005). The affect of Al was also observed in the root inhibition of *Phragmites australis* as reported by Ederli et al. (2004). Members of the Myrtaceae family demonstrated an increase in lignin content in response to high concentrations of Al (Tahara et al. 2005). As observed in *Camellia sinensis* plants subjected to high Al concentrations, there was a significant reduction in lignin accumulation, indicating an enhanced vegetative growth of the plants under the influence of Al (Lima et al. 2009; Ghanati et al. 2005). In addition to this, it was also seen that Al caused the activation of genes controlling the enzymes of the lignin biosynthesis pathway. Similarly cadmium (Cd) also induced the biosynthesis of lignin in the roots of common reed (Ederli et al. 2004) and the branches and roots of tea (Zagoskina et al. 2007). Cd concentrations ranging from 0.2 mM onwards resulted in increased lignin content in soybean plants (Bhuiyan et al. 2007). This was supplemented by an increase in laccase activity as well. Since, laccases are known to be responsible for lignin biosynthesis during the initial stages of Cd treatment, their activity fluctuations were also of significance. Copper (Cu) is reported to increase the lignin biosynthesis. Thus, Cu deficiency resulted in less lignin (Robson et al. 1981), and surplus Cu exposure resulted in enhanced lignin deposition. Chilli pepper hypocotyls exposed to excess Cu showed high shikimate dehydrogenase levels, enhanced POD activity and a high level accumulation of phenols and lignin (Diaz et al. 2001). Radish roots grown under high Cu concentrations indicated escalated lignin content in comparison with the control plants (Chen et al. 2002). As mentioned previously, laccases that are known to be responsible for the extracellular monolignol polymerisation in soybean roots during early stages of Cu treatments play a very essential role in lignin accumulation (Barnes et al. 1997). When ginseng root cultures were exposed to high Cu concentrations, this led to accumulation of lignin and other phenolic compounds involved in antistress response. Another important aspect is that Cu is structurally vital for laccases (Claus 2004). It mediates an increase in the activity of lignin biosynthesis enzymes, such as PAL and CAD. Another heavy metal resulting in lignin fluctuations at high concentrations is zinc (Zn) as seen in both *Arabidopsis thaliana* and *Thlaspi caerulescens*. An increased expression of genes related to the lignin biosynthetic pathway was reported by Van de Mortel et al. (2006) in response to Zn exposure. Interestingly, both plants are Zn hyper-accumulators; hence, gene expression was even greater in their case. Overall, lignin content and deposition patterns are generally known to increase in response to most heavy metal exposures at high concentrations. Correlation between lignin content with heavy metals and elements such as arsenic, cadmium, lead, titanium, potassium, calcium, magnesium, sodium and silicon concentrations in the reed canary grass was established with plasma optical emission spectrometer. Significantly, when lignin content increased, a decrease in Ca, Cd concentration was recorded, however the amount of Si increased in plants (Poiša et al. 2011). *Lens culinaris* and *Phaseolus mungo* seedlings treated with lead showed reduction in leaf size with enhanced lignin activity, deposition rates and biosynthesis (Haider and Azmat 2012).

3 Conclusion and Future Prospects

Conclusively, lignin is a unique polymer which is one of the most abundant renewable raw materials having a variety of roles throughout the plant kingdom. Apart from providing mechanical strength, lignin content alteration has been cited during stressed conditions, thereby suggesting that various abiotic stresses have the ability to express the genes responsible for their synthesis. Studies can be conducted to decipher the interlinked mechanisms responsible for such fluctuations in lignin content. Hence, in the present times, the potential of lignins must be explored in context to changing environment and antistress response triggered in the host plant. Commercially, maximum amount of lignin is discarded from the pulp and paper industry, but its role in the form of biodiesel and ecofriendly plastics has already been explored. Newer processes and more refined techniques need to be developed to produce more useful by-products from extracted wastes comprising lignin fraction. Another area of interest is environmental remediation via lignin matrices. The use of lignocellulosic materials in bioremediation of pollutants is another upcoming area of research.

References

Akgul M, Copur Y, Temiz S (2007) A comparison of kraft and kraftsodium borohydrate brutia pine pulps. Build Environ 42:2586–2590

Alvarez S, Marsh EL, Schroeder SG, Schachtman DP (2008) Metabolomic and proteomic changes in the xylem sap of maize under drought. Plant Cell Environ 31:325–340

Andersson-Gunneras S, Mellerowicz EJ, Love J, Segerman B, Ohmiya Y, Coutinho PM, Nilsson P, Henrissat B, Moritz T, Sundberg B (2006) Biosynthesis of cellulose-enriched tension wood in *Populus tremula*: global analysis of transcripts and metabolites identifies biochemical and developmental regulators in secondary wall biosynthesis. Plant J 46:349

Aquino EN, Mercado E (2004) Effects of polyphenol oxidase and peroxidase activity, phenolics and lignin content on the browning of cut jicama. Postharvest Biol Technol 33:275–283

Arasan SKT, Park JI, Ahmed NU, Jung HJ, Hur Y, Kang KK, Lim YP, Nou IS (2013) Characterization and expression analysis of dirigent family genes related to stresses in *Brassica*. Plant Physiol Biochem 67C:144–153. doi:10.1016/j.plaphy.2013.02.030

Atici O, Nalbantoglu B (2003) Antifreeze proteins in higher plants. Phytochemistry 64:1187–1196

Barnes JD, Bettarini I, Polle A, Slee N, Raines C, Miglietta F, Raschi A, Fordham M (1997) The impact of elevated CO_2 on growth and photosynthesis in *Agrostis canina* L. ssp. monteluccii adapted to contrasting atmospheric CO_2 concentrations. Oecologia 110:169–178

Basile A, Giordano S, Lopez-Saez JA, Cobianchi RC (1999) Antibacterial activity of pure flavonoids isolated from mosses. Phytochemistry 52:1479–1482

Bellaloui N (2012) Soybean seed phenol, lignin, and isoflavones and sugars composition are altered by Foliar Boron application in soybean under water stress. Food Nutr Sci 3:579–590

Bhuiyan N, Liu W, Liu G, Selvaraj G, Wei Y, King J (2007) Transcriptional regulation of genes involved in the pathways of biosynthesis and supply of methyl units in response to powdery mildew attack and abiotic stresses in wheat. Plant Mol Biol 64:305–318

Boerjan W, Ralph J, Baucher M (2003) Lignin biosynthesis. Annu Rev Plant Biol 54:519–546

Bok-Rye L, Kil-Yong K, Woo-Jin J, Jean-Christophe A, Alain O, Tae-Hwan K (2007) Peroxidases and lignification in relation to the intensity of water-deficit stress in white clover (*Trifolium repens* L.). J Exp Bot 58:1271

Bonawitz ND, Chapple C (2010) The genetics of lignin biosynthesis: connecting genotype to phenotype. Annu Rev Genet 44:337–363

Bonello P, Blodgett JT (2003) *Pinus nigra—Sphaeropsis sapinea* as a model pathosystem to investigate local and systemic effects of fungal infection of pines. Physiol Mol Plant Pathol 63:249–261

Boudet AM (2000) Lignins and lignification: selected issues. Plant Physiol Biochem 38:81–96

Boyer JS (1982) Plant productivity and environment. Science 218:443–444

Briffa KR, Jones PD, Schweingruber FH, Osborn TJ (1998) Influence of volcanic eruptions on Northern Hemisphere summer temperature over the past 600 years. Nature 393:450–455

Bruce RJ, West CA (1989) Elicitation of lignin biosynthesis and isoperoxidase activity by pectic fragments in suspension cultures of castor beans. Plant Physiol 91:889–897

Budikova S (1999) Structural changes and aluminium distribution in maize root tissues. Biol Plan 42:259–266

Campos R, Nonogaki H, Suslow T, Saltveit ME (2004) Isolation and characterization of a wound inducible phenylalanine ammonia-lyase gene (LsPAL1) from Romaine lettuce leaves. Physiol Plant 121:429–438

Carpin S, Crevecoeur M, de Meyer M, Simon P, Greppin H, Penel C (2001) Identification of a Ca2C-pectate binding site on an apoplastic peroxidase. Plant Cell 13:511–520

Cesarino I, Araujo P, Domingues AP Jr, Adilson P, Mazzafera P (2012) An overview of lignin metabolism and its effect on biomass recalcitrance. Braz J Bot 35:303–311

Chabannes M, Ruel K, Yoshinaga A, Chabbert B, Jauneau A, Joseleau JP, Boudet AM (2001) In situ analysis of lignins in transgenic tobacco reveals a differential impact of individual transformations on the spatial patterns of lignin deposition at the cellular and subcellular levels. Plant J 28:271–282

Chen EL, Chen YA, Chen LM, Liu ZH (2002) Effect of copper on peroxidase activity and lignin content in *Raphanus sativus*. Plant Physiol Biochem 40:439–444

Chinnusamy V, Zhu J, Zhu J-K (2007) Cold stress regulation of gene expression in plants. Trends Plant Sci 12:444–451

Chowdhury EM, Choi BS, Park SU, Lim HS, Bae H (2012) Transcriptional analysis of hydroxy-cinnamoyl transferase (HCT) in various tissues of *Hibiscus cannabinus* in response to abiotic stress conditions. Plant Omics 5:305–313

Cipollini DF (1997) Wind-induced mechanical stimulation increases pest resistance in common bean. Oecologia 111:84–90

Claus H (2004) Laccases: structure, reactions, distribution. Micron 35:93–96

D'Arrigo RD, Jacoby GC, Free RM (1992) Tree-ring width and maximum latewood density at the North American tree line: parameters of climatic change. Can J For Res 22:1290–1296

Delessert C, Wilson I, Van Der Straeten D, Dennis E, Dolferus R (2004) Spatial and temporal analysis of the local response to wounding. Plant Mol Biol 55:165–181

Depege N, Thonat C, Coutand C, Julien JL, Boyer N (1997) Morphological responses and molecular modification in tomato plants after mechanical stimulation. Plant Cell Physiol 38:1127–1134

Dharmawardhana DP, Ellis BE, Carlson JE (1995) A β-glucosidase from lodge-pole pine xylem specific for the lignin precursor coniferin. Plant Physiol 107:331–339

Diaz J, Bernal A, Pomar F, Merino F (2001) Induction of shikimate dehydrogenase and peroxidase in pepper (*Capsicum annuum* L.) seedlings in response to copper stress and its relation to lignification. Plant Sci 161:179–188

Dixon RA, Paiva NL (1995) Stress-induced phenylpropanoid metabolism. Plant Cell 7: 1085–1097

Douglas CJ, Ehlting J (2005) *Arabidopsis thaliana* full genome longmer microarrays: a powerful gene discovery tool for agriculture and forestry. Transgenic Res 14:551–561

Dowson-Day MJ, Millar AJ (1999) Circadian dysfunction causes aberrant hypocotyl elongation patterns in Arabidopsis. Plant J 17:63–71

Eckstein D, Aniol RW (1981) Dendroclimatological reconstruction of the summer temperatures for an alpine region. Mitt Forstl Bundesversuchsanst Wien 142:391–398

Ederli L, Reale L, Ferranti F, Pasqualini S (2004) Responses induced by high concentration of cadmium in *Phragmites australis* roots. Physiol Plant 121:66–74

Ehlting J, Mattheus N, Aeschliman DS, Li E, Hamberger B, Cullis IF, Zhuang J, Kaneda M, Mansfield SD, Samuels L, Ritland K, Ellis BE, Bohlmann J, Douglas CJ (2005) Global transcript profiling of primary stems from *Arabidopsis thaliana* identifies candidate genes for missing links in lignin biosynthesis and transcriptional regulators of fiber differentiation. Plant J 42:618–640

El Kayal W, Keller G, Debayles C, Kumar R, Weier D, Teulieres C, Marque C (2006) Regulation of tocopherol biosynthesis through transcriptional control of tocopherol cyclase during cold hardening in *Eucalyptus gunnii*. Physiol Plant 126:212–223

Fan L, Neumann PM (2004) The spatially variable inhibition by water deficit of maize root growth correlates with altered profiles of proton flux and cell wall pH. Plant Physiol 135:2291–2300

Fan L, Linker R, Gepstein S, Tanimoto E, Yamamoto R, Neumann PM (2006) Progressive inhibition by water deficit of cell wall extensibility and growth along the elongation zone of maize

roots is related to increased lignin metabolism and progressive stelar accumulation of wall phenolics. Plant Physiol 140:603–612
Fengel D, Wegener G (1984) Wood chemistry, ultrastructure, reactions. De Gruyter, Berlin
Fergus BJ, Goring DA (1970) The distribution of lignin in birch wood as determined by ultraviolet microscopy. Holzforschung 24:118–124
Frankenstein C, Schmitt U, Koch G (2006) Topochemical studies on modified lignin distribution in the xylem of poplar (*Populus* spp.) after wounding. Ann Bot 97:195–204
Fraser TE, Silk WK, Rost TL (1990) Effects of low water potential on cortical cell length in growing regions of maize roots. Plant Physiol 93:648–651
Fujita M, Harada H (1979) Autoradiographic investigations of cell wall development. II. Tritiated phenylalanine and ferulic acid assimilation in relation to lignification. Mokuzai Gakkaishi 25:89–94
Fustec E, Chauvet E, Gilbert G (1989) Lignin degradation and humus formation in alluvial soils and sediments. Appl Environ Microbiol 55:922–926
Ghanati F, Morita A, Yokota H (2005) Deposition of suberin in roots of soybean 893 induced by excess boron. Plant Sci 168:397–405
Gibson LJ (2012) The hierarchical structure and mechanics of plant materials. J R Soc Interface 9(76):2749–2766
Gibson LJ, Ashby MF, Harley BA (2010) Cellular materials in nature and medicine. Cambridge University Press, Cambridge
Gindl W, Grabner M, Wimmer R (2000) The influence of temperature on latewood lignin content in treeline Norway spruce compared with maximum density and ring width. Trees Struct Funct 14:409–414
Goujon T, Ferret V, Mila I, Pollet B, Ruel K, Burlat V, Joseleau JP, Barriere Y, Lapierre C, Jouanin L (2003) Down-regulation of the AtCCR1 gene in *Arabidopsis thaliana*: effects on phenotype, lignins and cell wall degradability. Planta 217:218–228
Grabber JH, Ralph J, Lapierre C, Barriere Y (2004) Genetic and molecular basis of grass cell-wall degradability. I. Lignin-cell wall matrix interactions. C R Biol 327:455–465
Guenni O, Douglas M, Baruch Z (2002) Responses to drought of five Brachiaria species. Biomass production, leaf growth, root distribution, water use and forage quality. Plant Soil 243:229–241
Gulen H, Eris A (2004) Effect of heat stress on peroxidase activity and total protein content in strawberry plants. Plant Sci 166:739–744
Guo XY, Zhang SZ, Shan XQ (2008) Adsorption of metal ions on lignin. J Hazard Mater 151: 134–142
Hadacek F (2002) Secondary metabolites as plant traits: current assessment and future perspectives. Crit Rev Plant Sci 21:273–322
Haider S, Azmat R (2012) Failure of survival strategies in adaption of heavy metal environment in *Lens culinaris* and *Phaseolus mungo*. Pak J Bot 44:1959–1964
Hall AE (2001) Crop responses to environment. CRC Press, Boca Raton, FL, pp 24–26
Hardell HL, Leary OJ, Stoll M, Westermark U (1980) Variation in lignin structure in defined morphological parts of birch. In: Lin SY, Dence C (eds) Chemistry. Springer, Berlin, Svensk papperstidning, pp 83:71–74
Harkin JM (1967) Lignin production and detection in wood. Forest Products Laboratory, Forest Service U.S. Department of Agriculture University of Wisconsin
Harman G, Patrick R, Spittler T (2007) Removal of heavy metals from polluted waters using lignocellulosic agricultural waste products. Ind Biotechnol 3:366–374
Hausman JF, Evers D, Thiellement H, Jouve L (2000) Compared responses of poplar cuttings and in vitro raised shoots to short-term chilling treatments. Plant Cell Rep 19:954–960
Hawkins S, Boudet A (2003) 'Defence lignin' and hydroxycinnamyl alcohol dehydrogenase activities in wounded *Eucalyptus gunnii*. Eur J Plant Pathol 33:91–104
Heredia J, Cisneros-Zevallos L (2009) The effect of exogenous ethylene and methyl jasmonate on pal activity, phenolic profiles and antioxidant capacity of carrots (*Daucus carota*) under different wounding intensities. Postharvest Biol Technol 51:242–249

Higuchi T (1985) Biosynthesis of lignin. In: Higuchi T (ed) Biosynthesis and biodegradation of wood components. Academic, New York, pp 141–160

Hilal M, Parrado MF, Rosa M, Gallardo M, Orce L, Marta Massa E, Gonzalez JA, Prado FE (2004) Epidermal lignin deposition in quinoa cotyledons in response to UV-B radiation. Photochem Photobiol 79:205–210

Hossain MT, Mori R, Wakabayashi KSK (2002) Growth promotion and an increase in cell wall extensibility by silicon in rice and some other Poaceae seedlings. J Plant Res 115:23–27

Hossain MA, Zakir AK, Hossain M, Kihara T, Koyama H, Hara T (2005) Aluminum-induced lipid peroxidation and lignin deposition are associated with an increase in H_2O_2 generation in wheat seedlings. J Soil Sci Plant Nutr 51(2):223–230

Hu Y, Li WC, Xu YQ, Li GJ, Liao Y, Fu FL (2009) Differential expression of candidate genes for lignin biosynthesis under drought stress in maize leaves. J Appl Genet 50(3):213–223

Jain R, Shrivastava A, Solomon S, Yadav R (2007) Low temperature stress-induced biochemical changes affect stubble bud sprouting in sugarcane (*Saccharum spp*. hybrid). Plant Growth Regul 53:17–23

Janas KM, Cvikrova M, Palagiewicz A, Eder J (2000) Alterations in phenylpropanoid content in soybean roots during low temperature acclimation. Plant Physiol Biochem 38:587–593

Janda T, Szalai G, Lesko K, Yordanova R, Apostol S, Popova LP (2007) Factors contributing to enhanced freezing tolerance in wheat during frost hardening in the light. Phytochemistry 68:1674–1682

Janska A, Aprile A, Zamecnik J, Cattivelli L, Ovesna J (2011) Transcriptional responses of winter barley to cold indicate nucleosome remodelling as a specific feature of crown tissues. Funct Integr Genomics 11:307–325

Johnston HH (1964) The relationship of brown humus to lignin. Plant Soil 21:191–200

Jones L, Ennos AR, Turner SR (2001) Cloning and characterization of irregular xylem4 (irx4): a severely lignin-deficient mutant of *Arabidopsis*. Plant J 26:205–216

Kacperska A (1993) Water potential alteration—a prerequisite or a triggering stimulus for the development of freezing tolerance in overwintering herbaceous plants? In: Li PH, Christerson L (eds) Advances in plant cold hardiness. CRC Press, Boca Raton, pp 73–91

Kaneda M, Rensing KH, Wong JC, Banno B, Mansfield SD, Samuels AL (2008) Tracking monolignols during wood development in lodgepole pine. Plant Physiol 147:1750–1760

Kasraie P, Nasri M, Khalatbari M (2012) The effects of time spraying amino acid on water deficit stress on yield, yield component and some physiological characteristics of grain corn (TWC647). Ann Biol Res 3:4282–4286

Kim WJ, Campbell AG, Koch P (1989) Chemical variation in lodgepole pine with latitude, elevation, and diameter class. For Prod J 39:7–12

Kim YJ, Kim DG, Lee SH, Lee I (2006) Wound-induced expression of the ferulate 5-hydroxylase gene in *Camptotheca acuminata*. Biochim Biophys Acta 1760:182–190

Kimura M, Yamamoto YY, Seki M, Sakurai T, Sato M, Abe T, Yoshida S, Manabe K, Shinozaki K, Matsui M (2003) Identification of *Arabidopsis* genes regulated by high light-stress using cDNA microarray. Photochem Photobiol 77:226–233

Kitin P, Voelker SL, Meinzer FC, Beeckman H, Strauss SH, Lachenbruch B (2010) Tyloses and phenolic deposits in xylem vessels impede water transport in low-lignin transgenic poplars: a study by cryo-fluorescence microscopy. Plant Physiol 154:887–898

Knight H, Trewavas AJ, Knight MR (1996) Cold calcium signaling in *Arabidopsis* involves two cellular pools and a change in calcium signature after acclimation. Plant Cell 8:489–503

Komastu S, Kobayashi Y, Nishizawa K, Naryo Y, Furukawa K (2010) Comparative proteomics analysis of differentially expressed protein in soy bean cell wall during flooding stress. Amino Acids 39:1435–1449

Kusumoto D (2005) Concentrations of lignin and wall-bound ferulic acid after wounding in the phloem of *Chamaecyparis obtusa*. Trees Struct Funct 19:451–456

Lee BG, Rowell RM (2004) Removal of heavy metals ions from aqueous solutions using lignocellulosic fibers. J Nat Fibr 1(1):97–108

Lerouxel O, Cavalier DM, Liepman AH, Keegstra K (2006) Biosynthesis of plant cell wall polysaccharides—a complex process. Curr Opin Plant Biol 9:621–630

Lewis NG, Yamamoto E (1990) Lignin—occurrence, biogenesis and biodegradation. Annu Rev Plant Physiol Plant Mol Biol 41:455–496

Li X, Weng JK, Chapple C (2008) Improvement of biomass through lignin modification. Plant J 54:569–581

Li Z, Peng Y, Ma X (2013) Different response on drought tolerance and post-drought recovery between the small-leafed and the large-leafed white clover (*Trifolium repens* L.) associated with antioxidative enzyme protection and lignin metabolism. Acta Physiol Plant 35(1): 213–222

Lichtenthaler HK, Buschmann C, Doll M, Fietz HJ, Bach T, Kozel U, Meier D, Rahmsdorf U (1981) Photosynthetic activity, chloroplast ultrastructure, and leaf characteristics of high-light and low-light plants and of sun and shade leaves. Photosynth Res 2:115–141

Liljegren S, Ditta G, Eshed Y, Savidge B, Bowman J, Yanofsky M (2000a) Control of fruit dehiscence in Arabidopsis by the SHATTERPROOF MADS-box genes. Nature 404:766–769

Liljegren S, Kempin S, Chen A, Roeder A, Guimil S, Khammungkhune T, Yanofsky M (2000b) A bHLH gene, INDEHISCENT1, is required for fruit dehiscence and mediates the fruitful phenotype. In: 11th international conference on Arabidopsis research, University of Madison Press, Madison, WI

Lima JD, Mazzafera P, Moraes WS, Silva RB (2009) Chá: aspectos relacionados a qualidade e perspectivas. Ciên Rural 39:1270–1278

Lisar SYS, Motafakkerazad R, Hossain MM, Rahman IMM (2012) Water stress in plants: causes,effect and responses. In: Ismail MR (ed) Water stress. InTech Publishers pp 1–15

Liu X, Huang B (2000) Heat stress injury in relation to membrane lipid peroxidation in creeping bent grass. Crop Sci 40:503–510

Liu C-J, Miao Y-C, Zhang K-W (2011) Sequestration and transport of lignin monomeric precursors. Molecules 16:710–727

Liu X, Zhu H, Qin C, Zhou J, Zhao JR, Wang S (2013) Heavy metal ions on lignin derivative. BioResources 8:2257–2269

Maki K, Kouisni L, Paleologou M, Haolt-Hindle P (2012) The FPInnovations lignin demonstration plant: process and lignin products. Bio Economy Technology Centre Thunder Bay, Ontario Chemical Engineering & Technology, 2012 TAAPI PEERS Conference, U.S.A. pp 1863–1871

Mittler R (2002) Oxidative stress, antioxidants and stress tolerance. Trends Plant Sci 7:405–410

Moerschbacher BM, Noll U, Flott B, Reisner HJ (1988) Lignin biosynthetic enzymes in stem rust infected, resistant and susceptible near-isogenic wheat lines. Physiol Mol Plant Pathol 33: 33–46

Moerschbacher BM, Noll U, Gorrichon L, Reisener H-J (1990) Specific inhibition of lignification breaks hypersensitive resistance of wheat to stem rust. Plant Physiol 93:465–470

Mohr P, Cahill D (2004) Suppression by abscisic acid of lignin production and monolignol pathway gene expression in interactions of Arabidopsis with oomycete and bacterial pathogens. In Phytopathology: abstracts of the American Society of Plant Biologists annual meeting American Society of Plant Biologists, St. Paul, MN, p 72

Moller R, Ball R, Henderson A, Modzel G, Find J (2006) Effect of light and activated charcoal on tracheary element differentiation in callus cultures of *Pinus radiata* D. Don. Plant Cell Tiss Org Cult 85:161–171

Monroy AF, Dhindsa RS (1995) Low temperature signal transduction: induction of cold acclimation—specific genes of alfalfa by calcium at 25°C. Plant Cell 7:321–331

Moore KJ, Jung HJG (2001) Lignin and fiber digestion. J Range Manage 54:420–430

Moura JCMS, Bonine CAV, Viana JOF, Dornelas MC, Mazzafera P (2010) Abiotic and biotic stresses and changes in the lignin content and composition in plants. J Integr Plant Biol 52: 360–376

Moura JCMS, Bonine CAV, Viana JOF, Dornelas MC, Mazzafera P (2011) Drought stress and changes in the lignin content and composition in Eucalyptus. BMC Proc 5(suppl 7):P103

Musha Y, Goring DAI (1975) Distribution of syringyl and guaiacyl moieties in hardwoods as indicated by ultraviolet microscopy. Wood Sci Technol 9:45–58

Nakashima J, Mizuno T, Takabe K, Fujita M, Saiki H (1997) Direct visualization of lignifying secondary wall thickenings in *Zinnia elegans* cells in culture. Plant Cell Physiol 38:818–827

Nicholson RL, Hammerschmidt R (1992) Phenolic compounds and their role in disease resistance. Annu Rev Phytopathol 30:369–389

Nilsen ET, Orcutt DM (1996) The physiology of plants under stress: abiotic factors. Wiley, New York, p 689

Nilsson R, Bernfur K, Gustavsson N, Bygdell J, Wingsle G, Larsson C (2010) Proteomics of plasma membranes from poplar trees reveals tissue distribution of transporters, receptors, and proteins in cell wall formation. Mol Cell Proteomics 9:368–387

Olenichenko N, Zagoskina N (2005) Response of winter wheat to cold: production of phenolic compounds and L-phenylalanine ammonia lyase activity. Appl Biochem Microbiol 41: 600–603

Orvar BL, Sangwan V, Omann F, Dhindsa R (2000) Early steps in cold sensing by plant cells: the role of actin cytoskeleton and membrane fluidity. Plant J 23:785–794

Pandey MP, Kim CS (2011) Lignin depolymerization and conversion: a review of thermochemical methods. Wiley, Weinheim

Parajuli D, Inoue K, Ohto K, Oshima T, Murota A, Funaoka M, Makino K (2005) Adsorption of heavy metals on crosslinked lignocatechol: a modified lignin gel. React Funct Polym 62: 129–139

Peet MM, Willits DH (1998) The effect of night temperature on greenhouse grown tomato yields in warm climate. Agric For Meteorol 92:191–202

Pickett-Heaps JD (1968) Xylem wall deposition: radioautographic investigations using lignin precursors. Protoplasma 65:181–205

Pintó-Marijuan MP, Joffre R, Casals I, De Agazio M, Zacchini M, García-Plazaola JI, Esteban R, Aranda X, Guárdia M, Fleck M (2013) Antioxidant and photoprotective responses to elevated CO_2 and heat stress during holm oak regeneration by resprouting, evaluated with NIRS (near-infrared reflectance spectroscopy). Plant Biol 15:5–17

Poiša L, Adamovičs A, Platače R, Teirumnieka E (2011) Evaluation of the factors that affect the lignin content in the reed canarygrass (*Phalaris arundinacea* L.) in Latvia. BioTechnology 1:224–231

Polle A, Otter T, Seifert F (1994) Apoplastic peroxidases and lignification in needles of Norway spruce (*Picea abies* L.). Plant Physiol 106:53–60

Polle A, Otter T, Sandermann H Jr (1997) Biochemistry and physiology of lignin synthesis. In: Rennenberg H, Eschrich W, Ziegler H (eds) Trees—contributions to modern tree physiology. Backhuys, Leiden, pp 455–475

Potters G, Pasternak TP, Guisez Y, Palme KJ, Jansen MAK (2007) Stress-induced morphogenic responses: growing out of trouble? Trends Plant Sci 12:98–105

Pritchard J (1994) The control of cell expansion in roots. New Phytol 127:3–26

Ralph J, Brunow G, Boerjan W (2007) Lignins. In: Rose F, Osborne K (eds) Encyclopedia of life sciences. Wiley, Chichester. doi:10.1002/9780470015902.a0020

Ramirez DA, Munoz SV, Atehortua L, Michel FC Jr (2010) Effects of different wavelengths of light on lignin peroxidase production by the white-rot fungi *Phanerochaete chrysosporium* grown in submerged cultures. Bioresour Technol 101:9213–9220

Raven JA (1984) Physiological correlates of the morphology of early vascular plants. Bot J Linn Soc 88:105–126

Reina JJ, Dominguez E, Heredia A (2001) Water sorption-desorption in conifer cuticles: the role of lignin. Physiol Plant 112:372–378

Rivero RM, Ruiz JM, Garcia PC, Lopez-Lefebre LR, Sanchez E, Romero L (2001) Resistance to cold and heat stress: accumulation of phenolic compounds in tomato and watermelon plants. Plant Sci 160:315–321

Robson AD, Hartley RD, Jarvis SC (1981) Effect of copper deficiency on phenolic and other constituents of wheat cell walls. New Phytol 89:361–371

Rogers LA, Campbell MM (2004) The genetic control of lignin deposition during plant growth and development. New Phytol 164:17–30
Rogers LA, Dubos C, Cullis IF, Surman C, Poole M, Willment J, Mansfield SD, Campbell MM (2005) Light, the circadian clock, and sugar perception in the control of lignin biosynthesis. J Exp Bot 56:1651–1663
Rolando C, Monties B, Lapierre C (1992) Thioacidolysis. In: Dence CW, Lin SY (eds) Methods in lignin chemistry. Springer, Heidelberg, pp 334–349
Ros Barcelo A (2005) Xylem parenchyma cells deliver the H_2O_2 necessary for lignification in differentiating xylem vessels. Planta 220(5):747–756
Rozema J, van de Staaij J, Bjorn LO, Caldwell M (1997) UV-B as an environmental factor in plant life: stress and regulation. Trends Ecol Evol 12:22–28
Saka S, Goring D (1988) The distribution of lignin in white birch wood as determined by bromination with TEM-EDXA. Holzforschung 42:149–153
Samuels AL, Rensing K, Douglas CJ, Mansfield S, Dharmawardhana P, Ellis B (2002) Cellular machinery of wood production: differentiation of secondary xylem in *Pinus contorta* var. latifolia. Planta 216:72–82
Santos RB, Capanema EA, Balakshin MY, Chang HM, Jameel H (2011) Effect of hardwoods characteristics on kraft pulping process: emphasis on lignin structure. BioResources 6(4): 3623–3637
Santos RB, Hart P, Jameel H, Chang HM (2012) Kinetics of hardwood carbohydrate degradation during kraft pulp cooking. Ind Eng Chem Res 51:12192–12198
Sarkanen KV, Ludwig CH (1971) Lignin: occurrence, formation, structure, and reactions. Wiley, New York
Sasaki M, Yamamoto Y, Matsumoto H (1996) Lignin deposition induced by aluminum in wheat (*Triticum aestivum*) roots. Physiol Plant 96:193–198
Schöffl F, Prandl R, Reindl A (1999) Molecular responses to heat stress. In: Shinozaki K, Yamaguchi-Shinozaki K (eds) Molecular responses to cold, drought, heat and salt stress in higher plants. R.G. Landes, Austin, TX, pp 81–98
Sgherri C, Stevanovic B, Navari-Izzo F (2004) Role of phenolic acid during dehydration and rehydration of *Ramonda serbica*. Physiol Plant 122:478–485
Silber MV, Meimberg H, Ebel J (2008) Identification of a 4-coumarate: CoA ligase gene family in the moss, Physcomitrella patens. Phytochemistry 69:2449–2456
Siqueira G, Milagres AMF, Carvalho W, Koch G, Ferraz A (2011) Topochemical distribution of lignin and hydroxycinnamic acids in sugarcane cell walls and its correlation with the enzymatic hydrolysis of polysaccharides. Biotechnol Biofuels 4:1–9
Sjöström E (1993) Wood chemistry: fundamentals and applications. Academic, New York. ISBN 0-12-647480-X
Solecka D, Boudet A-M, Kacperska A (1999) Phenylpropanoid and anthocyanin changes in low-temperature treated winter oilseed rape leaves. Plant Physiol Biochem 37:491–496
Steeves C, Forster H, Pommer U, Savidge R (2001) Coniferyl alcohol metabolism in conifers. I. Glucosidic turnover of cinnamyl aldehydes by UDPG: coniferyl alcohol glucosyltransferase from pine cambium. Phytochemistry 57:1085–1093
Steponkus P (1984) Role of the plasma membrane in freezing injury and cold acclimation. Annu Rev Plant Physiol 35:543–584
Steponkus PL, Uemura M, Webb MS (1993) Membrane destabilization during freeze-induced dehydration. Curr Top Plant Physiol 10:37–47
Stewart WN, Rothwell GW (1993) Paleobotany and the evolution of plants. Cambridge University Press, New York, NY
Suzuki N, Mittler R (2006) Reactive oxygen species and temperature stresses: a delicate balance between signaling and destruction. Physiol Plant 126:45–51
Syros TD, Yupsanis TA, Economou AS (2005) Expression of peroxidases during seedling growth in Ebenus cretica L. as affected by light and temperature treatments. Plant Growth Regul 46:143–151

Tahara K, Norisada M, Hogetsu T, Kojima K (2005) Aluminum tolerance and aluminum-induced deposition of callose and lignin in the root tips of Melaleuca and Eucalyptus species. J For Res 10:325–333

Taiz L, Zeiger E (2006) Plant physiology. Sinauer Associates, Sunderland, MA

Takabe K, Fujita M, Harada H, Saiki H (1985) Autoradiographic investigations of lignification in the cell walls of cryptomeria (*Cryptomeria japonica* D. Don). Mokuzai Gakkaishi 31:613–619

Takahama U (1988) Hydrogen peroxide-dependent oxidation of flavonoids and hydroxycinnamic acid derivatives in epidermal and guard cells of *Tradescantia virginiana* L. Plant Cell Physiol 29:475–481

Taşgin E, Atici O, Nalbantoğlu B, Popovac LP (2006) Effects of salicylic acid and cold treatments on protein levels and on the activities of antioxidant enzymes in the apoplast of winter wheat leaves. Phytochemistry 67:710–715

Terashima. N (1989) An improved radiotracer method for studying formation and structure of lignin. In: Lewis NG, Paice MG (eds) Plant cell wall polymers: biogenesis and biodegradation. ACS Symposium Series No. 399, pp 148–159 American Chemical Society, Washington DC

Thomashow MF (1999) Plant cold acclimation: freezing tolerance genes and regulatory mechanisms. Annu Rev Plant Biol 50:571–599

Tong XL, Ma Y, Li YD (2010) Biomass into chemicals: conversion of sugars to furan derivatives by catalytic processes. Appl Catal A-Gen 385:1–13

Umezawa T (2003) Diversity in lignan biosynthesis. Phytochem Rev 2:371–390

Van de Mortel JE, Villanueva LA, Schat H, Kwekkeboom J, Coughlan S, Moerland PD, van Themaat EVL, Koornneef M, Aarts MGM (2006) Large expression differences in genes for iron and zinc homeostasis, stress response, and lignin biosynthesis distinguish roots of *Arabidopsis thaliana* and the related metal hyperaccumulator *Thlaspi caerulescens*. Plant Physiol 142:1127–1147

Vance CP, Kirk TK, Sherwood RT (1980) Lignification as a mechanism of disease resistance. Annu Rev Phytopathol 18:259–288

Vanholme R, Morreel K, Ralph J, Boerjan W (2008) Lignin engineering. Curr Opin Plant Biol 11:278–285

Vanholme R, Demedts B, Morreel K, Ralph J, Boerjan W (2010) Lignin biosynthesis and structure. Plant Physiol 153:895–905

Vincent D, Lapierre C, Pollet B, Cornic G, Negroni L, Zivy M (2005) Water deficits affect caffeate O-methyltransferase, lignification and related enzymes in maize leaves. A proteomic investigation. Plant Physiol 137:949–960

Vishtal A, Kraslawski K (2011) Challenges of lignins. BioResources 6(3):3547–3568

Waltersson J (2009) The metal binding properties of kraft lignin. Degree Project, ECTS 30.0

Wardrop AB (1969) The structure of the cell wall in lignified collenchyma of *Eryngium* sp. Aus J Bot 17(2):229–240

Wen JL, Sun SL, Xue BL, Sun RC (2013) Quantitative structural characterization of the lignins from the stem and pith of bamboo (*Phyllostachys pubescens*). Holzforschung. doi:10.1515/hf-2012-0162

Weng J-K, Chapple C (2010) The origin and evolution of lignin biosynthesis. New Phytol 187:273–285

Weng J-K, Banks JA, Chapple C (2008a) Parallels in lignin biosynthesis. Commun Integr Biol 1:20–22

Weng J-K, Li X, Stout J, Chapple C (2008b) Independent origins of syringyl lignin in vascular plants. Proc Natl Acad Sci U S A 105:7887–7892

Whetten R, Sederoff R (1995) Lignin biosynthesis. Plant Cell 7:1001–1013

Wolter KE, Harkin JM, Kirk TK (1974) Guaiacyl lignin associated with vessels in aspen callus cultures. Plant Physiol 31:140–143

Xu Z, Zhang D, Hu J, Zhou X, Ye X, Reichel KL, Stewart NR, Syrenne RD, Yang X, Gao P (2009) Comparative genome analysis of lignin biosynthesis gene families across the plant kingdom. BMC Bioinformatics 10(suppl 11):S3

Yamaguchi M, Valliyodam B, Zhang J, Lenoble ME, Yu O, Roger E, Nguyen HT, Sharp RE (2010) Regulation of growth response to water stress in the soya bean primary roots. Proteomic analysis reveals regions specific regulation of phenylpropanoid metabolism and control of free ion in elongation zone. Springer 33:223–243

Yamamura M, Hattori T, Suzuki S, Shibata D, Umezawa T (2012) Microscale thioacidolysis method for the rapid analysis of β-O-4 substructures in lignin. Plant Biol 29:419–423

Yamamura M, Noda S, Hattori T, Shino A, Kikuchi J, Takabe K, Tagane S, Gau M, Uwatoko N, Mii M, Suzuki S, Shibata D, Umezawa T (2013) Characterization of lignocellulose of *Erianthus arundinaceus* in relation to enzymatic saccharification efficiency. Plant Biotechnol 30:25–35

Yamasaki S, Noguchi N, Mimaki K (2007) Continuous UV-B irradiation induces morphological changes and the accumulation of polyphenolic compounds on the surface of cucumber cotyledons. J Radiat Res 48:443–454

Yang L, Wang CC, Guo WD, Li XB, Lu M, Yu CL (2006) Differential expression of cell wall related genes in the elongation zone of rice roots under water deficit. Rus J Plant Physiol 53:390–395

Yoshimura K, Masuda A, Kuwano M, Yokota A, Akashi K (2008) Programmed proteome response for drought avoidance/tolerance in the root of a C-3 xerophyte (wild watermelon) under water deficits. Plant Cell Physiol 49:226–241

Yun Z, Gao H, Liu P, Liu S, Luo T, Jin S, Xu Q, Xu J, Cheng Y, Deng X (2013) Comparative proteomic and metabolomic profiling of citrus fruit with enhancement of disease resistance by postharvest heat treatment. BMC Plant Biol 13:44. doi:10.1186/1471-2229-13-44

Zagoskina NV, Dubravina GA, Alyavina AK, Goncharuk EA (2003) Effect of ultraviolet (UV-B) radiation on the formation and localization of phenolic compounds in tea plant callus cultures. Rus J Plant Physiol 50:270–275

Zagoskina N, Goncharuk E, Alyavina A (2007) Effect of cadmium on the phenolic compounds formation in the callus cultures derived from various organs of the tea plant. Rus J Plant Physiol 54:237–243

Zeyen RJ, Carver TLW, Lyngkjaer MF (2002) Epidermal cell papillae. In: Belanger RR, Bushnell WR (eds) The powdery mildew: a comprehensive treatise. APS Press, St Paul, MN, pp 107–125

Zhou C, Li Q, Chiang VL, Lucia LA, Griffis DP (2011) Chemical and spatial differentiation of syringyl and guaiacyl lignins in poplar wood via time of flight secondary ion mass spectrometry. Anal Chem 83:7020–7026

Chapter 11
Humic Substances and Plant Defense Metabolism

Ricardo L.L. Berbara and Andrés C. García

Contents

R.L.L. Berbara (✉)
Agronomy Institute, Department of Soil Sciences, Federal Rural University of Rio de Janeiro (UFRRJ), Rio de Janeiro 2389000, Brazil
e-mail: rberbara@hotmail.com

A.C. García
Agronomy Institute, Department of Soil Sciences, Federal Rural University of Rio de Janeiro (UFRRJ), Rio de Janeiro 2389000, Brazil

Department of Chemistry, Agricultural University of Havana, San Jose de las Lajas, Cuba

P. Ahmad and M.R. Wani (eds.), *Physiological Mechanisms and Adaptation Strategies in Plants Under Changing Environment: Volume 1*, DOI 10.1007/978-1-4614-8591-9_11,

1 Introduction

Soil organic matter is one of the most frequently discussed soil science topics in the scientific literature. Humic substances (HS) are the humified organic matter fraction in soils and play a key role in various soil and plant functions. HS influence the adaptation of plants to conditions of environmental stress by increasing the nutrient availability. They also have a direct effect on plant metabolic processes related to growth and development.

The actions of HS, both in soil and plants, are directly related to their structural characteristics. Using simple dilution procedures, three HS fractions have been identified: (a) humic acids (HA), which are soluble in basic medium and precipitate in an acidic medium (pH ~ 2.5), (b) fulvic acids (FA), which are soluble in both basic and acidic media; and (c) humins, which are not extractable from soil or sediment either in a basic or an acidic medium (Schnitzer 1978).

The effects of HS on higher plants and their development can occur either indirectly (e.g., through increased soil fertility or reduced soil compaction) or directly (by increasing total biomass production) (Nardi et al. 2002, 2007). These authors reported that humic fractions that were smaller in size exhibited greater structural flexibility and had a greater effect on the Krebs cycle in maize. Muscolo et al. (2007) suggested that the positive effects of HS in plants were more closely related to the structural characteristics of the HS than to their molecular mass. Canellas et al. (2010) argued that the molecular size of HS was not the primary factor in the stimulation of root growth in maize; rather, the hydrophobicity index was reported to be the predominant factor (Dobbss et al. 2010).

Although there may be discrepancy about the mechanisms by which HS acts on plants, still most authors are in agreement about the positive effects of HS on plants and the link between HS and increased metabolic efficiency. Some recent studies have reported that HA stimulated a phenylpropanoid metabolic system in plants and induced phenyl (tyrosine) ammonia lyase (PAL/TAL) activity (Schiavon et al. 2010). There are also reports of HA having direct effects on the generation of radical oxygen species (ROS) and the activity of catalyst enzymes in maize (Cordeiro et al. 2011). Likewise, the protective effects of HS or HS-based biostimulants have been reported for plants growing under conditions of hydric and saline stress (Vasconcelos et al. 2009; Aydin et al. 2012; García et al. 2012a).

Overall, the studies show that HS cause changes in the physiological mechanisms and adaptive processes of plants under various environmental conditions. Understanding the relationships between HS and plant physiological processes allows us to explore new techniques for managing humified organic matter in ecosystems. In agricultural systems, for example, the application of HS could be an ecologically sustainable method for improving the production of crops and other food plants. In this chapter, the scientific basis for the use of HS in agriculture is discussed.

2 Characteristics and Effects of HS

HS are recognized as the most widely distributed components of organic matter on the planet, and they are present both in terrestrial and aquatic environments. They are formed through the chemical and biological degradation of plant and animal remains and by microbial activity (Schnitzer 1978). According to Schnitzer (1978), the formation of HS can be summarized as four hypotheses: (a) the plant transformation hypothesis, (b) the chemical polymerization hypothesis, (c) the cell autolysis hypothesis, and (d) the microbial synthesis hypothesis. It is not easy to determine which hypothesis is most valid, and it is possible that all four occur simultaneously; with soil conditions determining which process is dominant. According to Ghabbour and Davies (2001), the first HS hypothesis suggests that HS include an extraordinary amount of complex, amorphous, heterogeneous, and chemically reactive molecules that are produced during biomass decomposition due to the chemical reactions that occur randomly in a large pool of organic molecules. In 2012, Nebbioso and Piccolo (2012) proposed that HS are formed through a process in which heterogeneous molecules associate according to shape, size, chemical affinity, hydrophobicity, and structure. The formation of HS is limited by the strength of the interactions that stabilize the associations between molecules within its supramolecular structure.

Regardless of the specific biosynthesis mechanisms involved in the formation of HS, researchers agree that HS are formed in terrestrial environments by the decomposition of plant and animal material deposited in the soil. This decomposition process helps in explaining the chemical and structural heterogeneity of humified materials. Figure 11.1 shows the structural diversity of a humified solid fraction isolated from a vermicompost. Ligninic fragments, fatty acids, and nitrogenous compounds derived from plant and animal materials can be observed.

2.1 *Structural Characteristics of HS, HS Fractions, and Their Effects*

Although it is possible to isolate three basic HS fractions in soil (HA, FA, and HU), HA and FA have been the most frequently studied fractions in scientific publications for examining the functions of HS. The structural characteristics of these fractions vary according to their source material and the time of transformation or formation time of organic matter for that material. Canellas et al. (2012) compared humic fractions of vermicomposts with their synthetic derivatives isolated from various Brazilian soil types. They used ^{13}C-MNR analysis to demonstrate how structure varies according to the source material of the fraction. The authors reported that a higher amount of polar and alkylic structures was present in the vermicompost fractions compared with their derivatives and that the content of these structures was similar to that of the fractions isolated from soils. The humic fractions

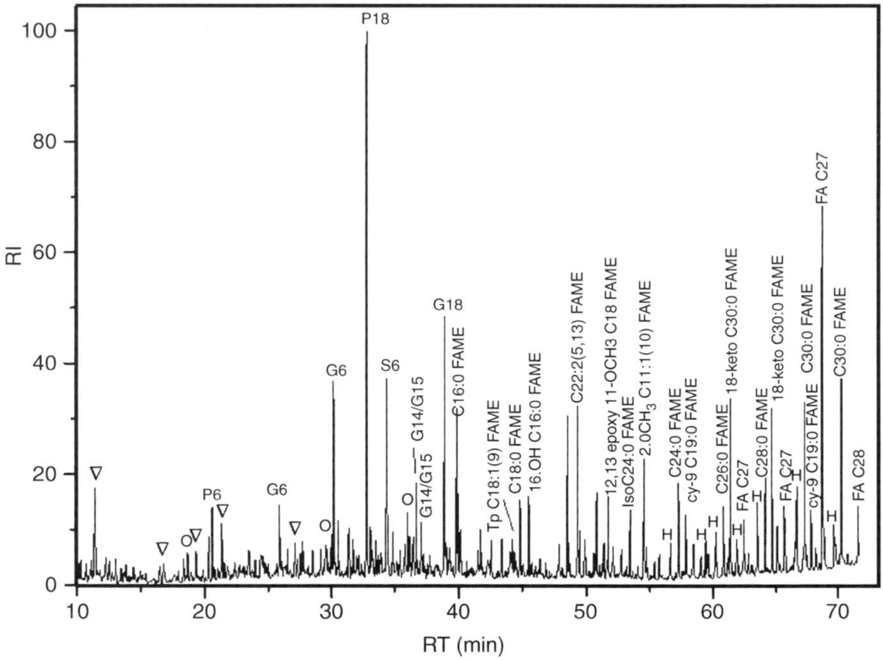

Fig. 11.1 TIC of a solid residual fraction from a manure vermicompost (off-line Py-TMAH-CG-MS). Carbohydrate derivatives (S), nitrogen compounds (O), lignin derivatives, fatty acid methyl esters (FAME), fatty alcohols (FA), hydrocarbons (H), terpene compounds (Tp)

isolated from oxisols were more hydrophobic and aromatic than those from other soils. Moreover, the humic fractions stimulated the emission of lateral roots and the activity of H^+-ATPases in corn plants (*Zea mays* L.), regardless of the original source of the humic material. *O*-alkyl, methoxy/*N*-alkyl, and hydrophobicity index (HB/HI) structures explained 88 % of the effect on enzyme activity.

The structural characteristics of HA from vermicomposts at different stages of maturation and the effect of HA on plants were also examined. During vermicomposting, HA had a lower amount of carbohydrate structures and the preservation of alkylic and Aryl structures were reduced. There were no changes in the molecular weight of the HA molecules; however, increases in hydrophobic structures occurred with increased maturation time of the vermicompost. HA from 60-day vermicompost had effects on the emergence of lateral roots and proton pump induction in the plants, in spite of the structural features of HA (Aguiar et al. 2013).

Studies that characterize humic fractions from different source materials are abundant in the literature. The high variability in the structural characteristics of HS has made molecular structure a benchmark for assessing the quality of organic matter and for monitoring changes in HS that occur during the soil management or composting. Table 11.1 shows some examples of humic fractions isolated from composted sources and soil. The characterization of HS using physicochemical techniques helps explain the evolution of organic matter over time and the effects of exogenous actions, such as fertilizer application and other management activities.

Table 11.1 Studies of humic fractions showing the original source material, structural characterization based on different physicochemical techniques, and the main results of the structural analyses

HS	Source	Characterization technique	Main observations	References
HA, FA	Soil	Elemental composition	↑ Aromatic condensation with ↑ soil depth	Gondar et al. (2005)
		UV–vis	↓ *O*-alkyl-C with ↑ soil depth	
		FTIR		
		^{13}C-RMN		
HA, FA	Soil	^{1}H-RMN ^{13}C-RMN	HA are + aromatic, contain less amounts of C-carbohydrate and H aliphatic compared to FA	Tao et al. (1999)
HA, FA	Soil	Elemental composition UV–vis FTIR Fluorescence ^{1}H-RMN	FA with ↓ C levels, ↑ O levels, ↓ E_4/E_6, relation, ↓ degree of aromaticity compared to HA	Dobbss et al. (2009)
HS	Soil	DRIFT	Changes in functional groups with the use of fertilizers	Ferrari et al. (2011)
		^{1}H-RMN	↑ Aromatic amino acids, lignins, and fatty acids/esters in soil fertilized with farm manure	
FA	CPT	FTIR ^{13}C-RMN	↓ Aliphatic alcohol structures, ↑ aromatic structures during decomposition	Baddi et al. (2004)
HA	CPT	Elemental composition	↓ H, H/C, C/N; ↑ N and S	Amir et al. (2010)
		FTIR	↑ Esterified peptide structures	
		^{13}C-RMN	↑ Oxidized aromatic structures and nitrogenated alkyl	
HA	VCT	Elemental composition	↓ C, H, C/N, C/O, structure similar to protein, and carbohydrate structures	Li et al. (2011)
		FTIR Electronic microscopy	↑ O, N, C/H, aromaticity, and polycondensation	

2.2 *Structure-Function Relations in the Interaction of HS with Metallic Elements*

Another very important property of HS related to structural characteristics is their ability to interact with metal ions. Interactions between HS and ions influence soil characteristics and fertility, thereby impacting plants. HS can form complex compounds with metal ions which have varying stabilities and chemical characteristics. Interactions between HS and metals affect soil properties for the reason that some nutrient availability processes and chemical forms of metallic elements are governed by HS. Therefore, important plant development processes such as precipitation-dissolution, ion exchange, mobility, transport and accumulation, and the chemical

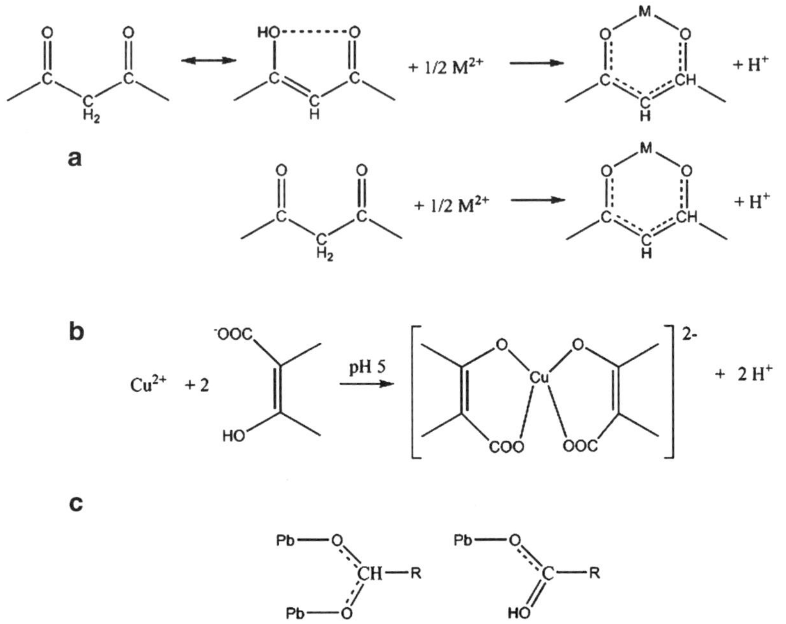

Fig. 11.2 Proposed chemical reactions and compounds for the interaction of HS functional groups with the metal ions Pb^{2+} and Cu^{2+}. (**a**) Piccolo and Stevenson (1982), (**b**) Schnitzer (1978), (**c**) Jerzykiewicz (2004)

and biochemical activity of metals are largely determined by HS (Senesi et al. 1986). One important structural feature of HS that allows for their interaction with metal ions is their high number of oxygenated functional groups (CO_2H_2, OH phenols, C=O). The presence of these groups allows HS to establish more stable complex bonds than those present in the soil, making metallic elements more available (Schnitzer 1978). With heavy metals, for example, HS form complex compounds in the following order of stability: $Pb^{2+} > Cu^{2+} > Ni^{2+} > Co^{2+} > Zn^{2+} > Cd^{2+} > Fe^{2+} > Mn^{2+} > Mg^{2+}$ (Irving and Williams 1953).

Several studies have reported the interaction of different humic fractions with metallic elements. HA have been demonstrated to form complex compounds with Al^{3+} ions in soils (Gerke 1994). FA have a high capacity to form bonds with Cu^{2+} and Ca^{2+} (Iglesias et al. 2003), and HA and HU are able to form complex compounds with Cu^{2+} in the soil (Plaza et al. 2005; Alvarez-Puebla et al. 2004). Lead is one of the most widely studied elements with regard to the formation of complex compounds with HS (Filella and Town 2001). Some of these studies have led to the proposal of new mechanisms to explain the interaction of HS with metal ions, as shown in Fig. 11.2.

3 Interactions Between Plant Root Systems and HS

The interactions that occur between HS and plant root systems are of great importance for understanding the modes of action of HS. Two fundamental issues are important for the studies of the HS–root relationship. The first issue is about understanding what happens in plant's environment where HS are present, and the second issue is related to the fact that most experiments examining the effects of HS in plants use root application methods. HS fractions interact directly with root structures. Studies of HA and FA marked with ^{14}C isotopes have shown that these HS fractions are associated in greater quantities with the cell wall within the first few hours of HS–root interaction (3 h) and subsequently (18 h) become part of the soluble component of the cells. Of the different HS fractions, HA are associated in higher numbers with the cell wall at the roots, while more FA are incorporated into the cell (Vaughan and Ord 1981). Observing HS–root interactions allow us to understand how these substances are assimilated and how they affect plant processes at the leaf level. Experiments with wheat plants have shown that of the total ^{14}C associated with HS that is assimilated by plants through roots; only 5 % is transported to leaf tissues (Vaugham and Linehan 1976).

Additionally, with regard to HS–root interactions, there are reports in the literature of physical interactions (specifically with HA) that change the functionality of cell membranes. HA supramolecular colloidal clusters in solution can migrate to the surface of the roots and cause the clogging of pores and root transpiration sites. These HA–root interactions and the formation of layers of agglomerates may be governed by electrostatic and van der Waal interactions. This phenomenon causes reductions in hydraulic conductivity, leaf organ growth, transpiration, and resistance to hydric stress and the mechanism of action is known as colloidal stress (Asli and Neumann 2010).

Evidence of the agglomeration phenomenon of HA in roots was recently captured by light microscopy, as shown in Fig. 11.3. HA–root interactions have also been demonstrated using ^{13}C-NMR spectroscopy, which showed that HA that formed agglomerates on root surfaces had lower structural complexity than exogenously applied HA. At the same time, it was determined that these HA interaction and agglomeration events are detected by ROS generation mechanism in the leaf

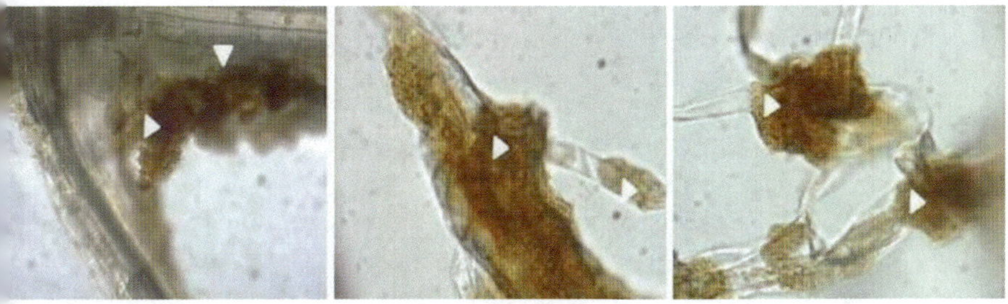

Fig. 11.3 Agglomeration effects of HA isolated from cattle manure vermicompost on the roots of rice plants grown in nutrient solution (×100 magnification). Layers of HA agglomerates on the root surface and emerging root hairs can be observed

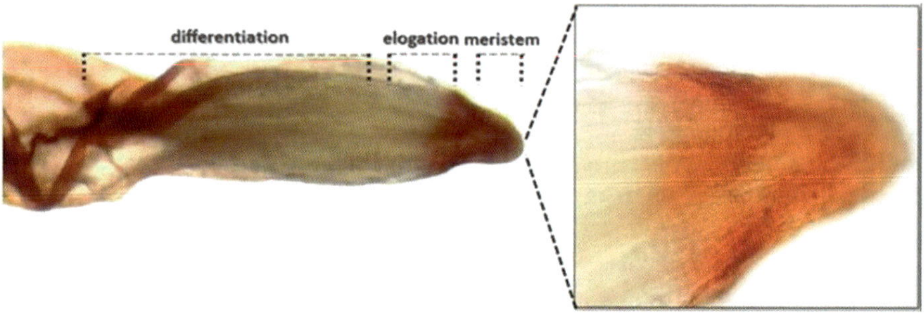

Fig. 11.4 Location of $O_2^{\bullet -}$ ions determined by histochemical techniques following the application of HA to a rice plant root growing in a nutrient solution

and root tissues, triggering the activity of antioxidative metabolic enzymes. For the first time, it was observed that this type of interaction could be related to the modes of action of HS in plants and that growth and development could be controlled through the regulation of REDOX homeostasis and other metabolic processes that are stimulated by HS (García et al. 2012b).

3.1 ROS and the Process of Root Growth and Development

In recent years, the involvement of ROS in metabolic processes associated with plant growth and development has been reported (Foreman et al. 2003; Marino et al. 2012). In contrast to what was previously thought, ROS can regulate root growth processes through independent pathways to phytohormones such as auxins. Studies conducted on *Arabidopsis* roots identified the transcription factor UPBEAT1 (UPB1), which regulates the balance between cell proliferation and differentiation. It was found that UBP1 directly regulates a series of peroxidases (POX) that modulate the quantity of ROS in regions of cell proliferation and elongation as soon as differentiation processes begin. UPB1 disruption results in a change in ROS balance and a delay in the root differentiation process (Tsukagoshi et al. 2010). ROS production in plant roots has also been observed when HA is applied to the roots, as demonstrated in Fig. 11.4.

Additionally, it is now understood that signaling enzymes such as NADHP oxidases and phospholipases D are of great importance in the formation of root hairs because ROS produced by NADHP oxidases activate Ca^{2+} channels in the apical plasma membrane, stimulating Ca^{2+} influxes which are linked to lateral root growth (Šamaj et al. 2004). Knowledge about the involvement of ROS in membrane permeability and polarization processes opens new avenues for understanding the mechanisms of plant adaptation in high-stress environments. Plant signaling mechanisms in which ROS target the activation of Ca^{2+} channels in cell membranes have been reported (Kurusu et al. 2013). This is one of the most important ROS-mediated steps in the regulation of plant stress, hormonal signaling, polar growth, and development

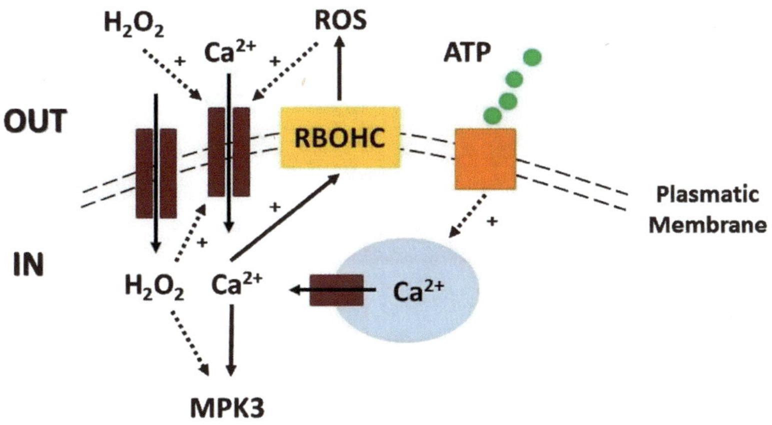

Fig. 11.5 Diagram showing a possible pathway for MPK3 induced by the action of extracellular ATP and ROS. Modified from Demidchik et al. (2009)

(Mori and Schroeder 2004). In ROS signaling processes, low OH concentrations induce Ca^{2+} pumping, while high OH concentrations induce Ca^{2+} incorporation through passive mechanisms (Zepeda-Jazo et al. 2011). With regard to apical meristem development, the maintenance and establishment of OH concentrations depend on the mechanisms that maintain ROS homeostasis. Oxidizing environments induce the reduction of cell proliferation, while less oxidizing (reducing) environments induce mitosis and cell differentiation (De Tullio et al. 2010) (Fig. 11.5).

4 Antioxidative Response to HS in Plants

4.1 Functions, Characteristics, and Pathways of ROS Action in Plants

ROS ($^1O_2$, $O_2^{\bullet-}$, OH, and H_2O_2) were initially recognized in plants as toxic chemical species produced through aerobic metabolism (Mittler et al. 2011). However, in addition to being the by-products of antistress metabolism, they also have roles in signal transduction (Miller et al. 2010). Current studies show that ROS signaling functions are involved in most existing plant metabolic processes. ROS plays an important role in cellular transduction mechanisms that control metabolic processes such as growth regulation and development, biotic and abiotic stress responses, and cell death (Suzuki et al. 2012). Under conditions of homeostasis and stress, ROS are found at different concentrations in plant tissues. During normal growth processes, ROS content in plant cells is low (approximately 240 μM s^{-1} of $O_2^{\bullet-}$ and 0.5 μM of H_2O_2). In contrast, in plants under stress, $O_2^{\bullet-}$ content increases to 240–270 μM s^{-1} and H_2O_2 content increases to 5–15 μM (Polle 2001). The $O_2^{\bullet-}$ anion is produced in the thylakoids and is the product of aerobic respiration, with approximately 1–2 % of O_2 consumed by plants being transformed into $O_2^{\bullet-}$. $O_2^{\bullet-}$ is one of the first ROS formed in plants, with a mean half-life of approximately 4.2 µs. H_2O_2 is formed by the reduction and dismutation of $O_2^{\bullet-}$. H_2O_2 has increasingly been recognized as the most important ROS signaling messenger due to its ability to cross cell membranes and its mean half-life (~1 ms) (Gill and Tuteja 2010; Mittler 2002).

Using ROS for signaling is advantageous for plants because plant cells have a high capacity for producing ROS and controlling internal ROS levels. A complex enzyme system, which is present in most cellular compartments, is used for this purpose. Recent studies have shown that *Rbohs*, a group of membrane proteins encoded by a family of ten genes in *Arabidopsis* (*AtRboh*), play a key role in ROS action and production mechanisms (Suzuki et al. 2012; Mittler 2002; Mittler et al. 2004). These findings are of great importance because *Rbohs* have numerous functions in plants and are known to regulate signaling mechanisms in response to abiotic stress (Kwak et al. 2003; Miller et al. 2009). Moreover, the role of *Rbohs* in *Rboh-ROS* regulation mechanisms, cell elongation processes, and root hair growth has been confirmed by Takeda et al. (2008).

4.2 ROS Generation in Response to HS

As previously discussed, the production of ROS in response to certain stimuli is clearly an efficient mechanism for regulating many metabolic pathways in plants. While there is currently a limited number of studies on plant ROS generation in response to HS, interest in HS–ROS relationship has persisted for several decades. Soil HS fractions (HA, FA, and water-soluble fractions) have been shown to stimulate the production of $O_2^{\bullet-}$ anions in vitro. A study of the xanthine/xanthine-oxidase system showed that the level of $O_2^{\bullet-}$ production depended on the humic fraction. FA and water-soluble fractions were the least effective in stimulating xanthine/xanthine-oxidase and the production of $O_2^{\bullet-}$. HA were the most effective fraction in stimulating the production of $O_2^{\bullet-}$ (Vaughan and Ord 1982). The knowledge that HS stimulate $O_2^{\bullet-}$ production through regulation of the xanthine/xanthine-oxidase system is important, given the key role that this system plays in defense mechanisms against biotic stressors (Berner and Van der Westhuizen 2010) and abiotic stress from heavy metals (Corpas et al. 2008).

The production of ROS following the application of HA was also observed in maize (*Zea mays* L.). HA isolated from soil were applied to the roots of maize plants grown with high and low N-NO_3^- concentrations. Using fluorescence techniques, the presence of ROS was detected in the roots of plants grown under three different growth conditions (low N-NO_3^-, high N-NO_3^-, and no N-NO_3^-). Under these conditions, the production of ROS following the application of HA did not impede the stimulation of lateral root growth and increased root biomass, suggesting that ROS production resulting from HA can act as an intermediary agent in the action processes of HS in plants (Cordeiro et al. 2011). The implications of these results led us to reconsider our understanding of the action mechanisms of HS in plants. During the

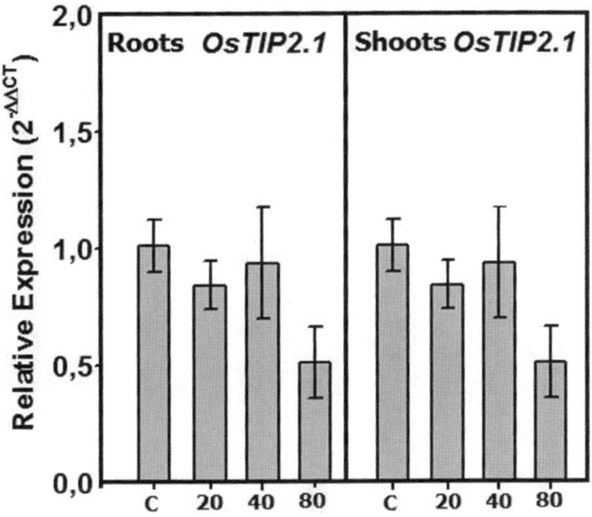

Fig. 11.6 *OsTIP2.1* gene expression in the leaves and roots of rice plants after 8 h of root application of HA under hydroponic growing conditions

growth of maize coleoptiles, ROS are released in the cell walls (specifically, OH is formed from $O_2^{\bullet-}$ on the cell wall), suggesting that ROS may increase the extensibility of the wall and replace auxins as growth inducers (Schopfer et al. 2002).

ROS production following the application of HA was also observed in rice plants (*Oryza sativa* L.) treated with HA from manure vermicompost. Root applications of HA at concentrations of 20, 40, and 80 mg L^{-1} regulated ROS ($O_2^{\bullet-}$ and H_2O_2) production in the leaves and roots at 8 and 24 h of contact. The levels of $O_2^{\bullet-}$ and H_2O_2 in both the leaves and roots were dependent on the concentration of HA. Although root application of HS induced ROS production in these studies, lipid peroxidation was low or nonexistent in some of the treatments, suggesting that the membrane peroxidation phenomenon was not the only result of HA-induced ROS. Another important result was found in the same study. Tonoplast aquaporin genes (*OsTIPs*) of the leaves and roots were also regulated by the application of HA, suggesting that ROS and *OsTIPs* are involved in the action mechanisms of HS inside the plant cells (García et al. 2012b). Therefore, it is possible that the production of ROS in plants due to the application of HS involves the regulation of TIPs in the cell and that there is a relationship with nitrogen metabolism. Figure 11.6 shows that for *OsTIP2.1* isoform, more diluted or concentrated solutions of HS exerted a repressing effect on the expression of these genes in both the leaves and roots.

Although the role of TIPs in HS action in plants is not yet completely understood, it is known that the *TIP2.1* isoforms are related to N-NH_4^+ transport. For example, some studies report that *AtTIP2.1* isoforms are induced in plants under long periods of nitrogen deficiency and short periods of N-NH_4^+ supplementation (Lopez et al. 2003). The role of TIPs in ammonium transport has also been noted (Loque et al. 2005). The effects of HS on the regulation of aquaporins, especially tonoplast, are of great importance because aquaporins play various roles in metabolite transport at the cellular level and are regulated in response to conditions of abiotic stress, such as hydric stress and high salinity (Li et al. 2008).

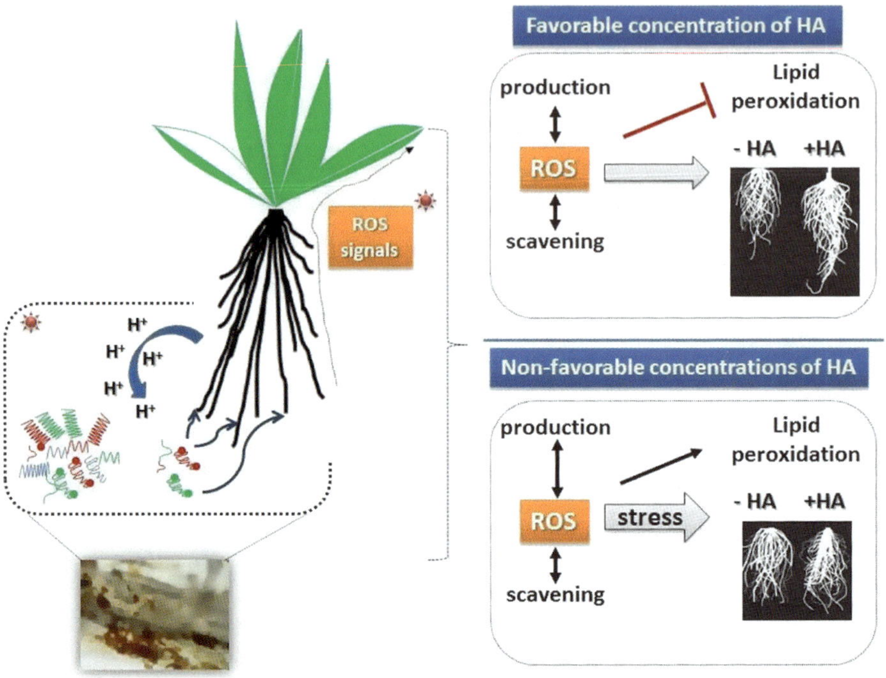

Fig. 11.7 Schematic representation of HS action in ROS production metabolism at different HS concentrations according to the results of past studies. HS–root interaction is shown on the *left*; production and signaling in the plant and the impacts on root growth are shown on the *right*

The results of studies of HS action in plants have shown that ROS production, particularly the production of H_2O_2, is dependent on the concentration of HS. The specific nature of HS–root interactions seems to be the key for triggering the physiological events in plants. It has been observed that in rice plants treated with moderate HS concentrations, ROS production does not cause lipid peroxidation, thereby favoring the processes of growth and lateral root formation. However, when plants are treated with elevated concentrations of HS, a high rate of ROS production can lead to lipid peroxidation and negatively affect the growth and root development, as shown in Fig. 11.7.

4.3 Relationship of ROS with Other Metabolic Pathways That Respond to HS

Understanding HS-induced ROS production in plants requires knowledge of the most current concepts related to oxidative mechanisms in plants. Given the evidence that ROS production occurs after the application of HS to plant roots, the mechanisms underlying the physiological responses to HS must also be explained. At the biochemical level, one of the most widely studied modes of action of HS in

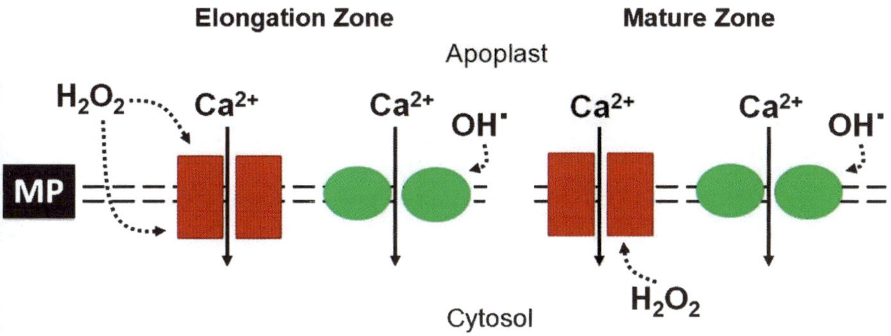

Fig. 11.8 Effects of ROS on Ca^{2+} channels in plasma membrane in the region of elongation and mature epidermis roots of *Arabidopsis* plants. Modified from Demidchik et al. (2007)

plants is similar to that of auxins. Currently, a large number of researchers have recognized that HS enter the plants through root system, where they mimic auxins and are recognized by hormonal receptors in cells. Other studies have reported the fragments of HS with structures similar to auxins that enter plants and exert auxin-like effects (Nardi et al. 2002). Root growth and an increased number of secondary roots are the most visible morphological effects of these auxin-like effects of HS. Therefore, what is the role of ROS in the biochemical-physiological mechanisms of HS action in plants?

Although auxins play the most crucial role in the regulation of growth and root development, redox regulation also plays a relevant role. The formation and preservation of the root apical meristem require ROS homeostasis in plant tissues. Redox regulation controls biochemical processes along the root and regulates the activity of auxins that influence root growth (De Tullio et al. 2010). Transcriptome studies of *Arabidopsis* plants have shown that auxin signaling and homeostasis are modified by ROS. In the apoplast, ROS may regulate the transcripts of auxin receptors and auxin/indole-3-acetic acid (Aux/IAA) transcriptional repressors through mechanisms that remain unknown (Blomster et al. 2011).

Other biochemical effects that are reported to have occurred as the result of HS action in plants are related to stimuli at the membrane level, with several studies reporting the stimulation of H^+-ATPase activity (Canellas et al. 2010; Mora et al. 2010). However, ROS have been shown to function as signaling molecules, acting through mechanisms that involve membrane hyperpolarization, activation of Ca^{2+} channels, and intracellular signaling to increase the growth of secondary roots. $O_2^{\bullet-}$ anions produced outside cells can be transformed to H_2O_2 and OH radicals. H_2O_2, for example, is capable of crossing the membrane and accumulating intracellularly. Through mechanisms that are not yet understood, H_2O_2 can stimulate Ca^{2+} channels intra- or extracellularly. The stimulus required to open the Ca^{2+} channels can increase cytosolic Ca^{2+}, stimulating NADPH oxidases and inducing MPK3 (Demidchik et al. 2009). This type of response also occurs in different regions of the root. In the region of elongation and mature root epidermis, high concentrations of H_2O_2 in the apoplast and the cytosol, respectively, can activate Ca^{2+} channels through membrane hyperpolarization (Demidchik et al. 2007), as demonstrated in Fig. 11.8.

5 Other HS-Induced Responses in Plant Metabolism

The greatest impacts of HS action in plants are due to the ability of HS to stimulate various metabolic pathways. Photosynthesis is the fundamental metabolic process underlying the production of all O_2 and organic matter on the planet; therefore, studies on the influence of HS in the regulation of photosynthesis are of great importance. To study the effects of HS on photosynthesis, three humic fractions (HA, FA, and HU) were isolated from soil samples and tested in *Pachira macrocarpa* plants. The three HS fractions stimulated the activity of chlorophyllases (a) and (b). FA showed greater stimulation of chlorophyllase (a) activity, while HA showed greater stimulation of chlorophyllase (b) activity (Yang et al. 2004). The effects of HS on photosynthetic processes were also shown in a study of the application of HA to lettuce (*Lactuca sativa* L.) plants. HA isolated from forest soils were applied to the roots in nutrient solution at concentrations of 100 and 1,000 mg L^{-1} and were found to stimulate the photosynthetic activity and augment the chlorophyll content and conductance of mesophyll cells (Haghighi et al. 2012).

Given the effects of HS on photosynthesis, we expected that carbon (C) metabolism would also be affected as a consequence of changes in photosynthetic activity and pigment content. Four HA fractions from soils were shown to stimulate the activity of enzymes related to C metabolism in maize plants (*Zea mays* L.). The four HU fractions stimulated the activity of enzymes belonging to the metabolism of glycolytic pathway (glucokinase (GK), phosphoglucoisomerase (PGI), PPi-dependent phosphofructokinase (PFK), pyruvate kinase (PK)) and Krebs cycle [citrate synthase (CS), malate dehydrogenase (MDH), and isocitrate and $NADP^+$-isocitrate dehydrogenase ($NADP^+$-IDH)] (Nardi et al. 2007). However, in some studies, HS have been shown to modify the activity of enzymes in C and N metabolism, depending on the source material of HS. In studies evaluating the effect of humic acids from forest soils and pasture on the activity of enzymes from C and N metabolism in *Pinus nigra* callus, forest-derived HA were found to inhibit PGI activity and inhibit phosphoenolpyruvate carboxykinase (PEPC), glutamate dehydrogenase (GDH), MDH, and glutamine synthetase (GS) activity. On the contrary, HA from pasture soils stimulated these enzymes (Muscolo et al. 2007).

Furthermore, any actions on primary metabolism in plants can have repercussions for secondary metabolism. The primary metabolites are precursors that induce the activity of secondary metabolism enzymes. The metabolic pathway of phenylpropanoid compost synthesis is one of the most specialized linkages between metabolic functions in plants. Phenylalanine ammonia lyase (PAL) and tyrosine ammonia lyase (TAL), the two enzymes that initiate phenylpropanoid biosynthesis, use phenylalanine and tyrosine, respectively, which are substrates derived from primary metabolism (Ferrer et al. 2008). HA from vermicompost have also been shown to exert effects on the synthetic pathway of phenylpropanoids through the regulation of PAL/TAL enzymes. In maize plants, humic acid concentrations of 0.5, 1.0, and 2 mg (C) L^{-1} stimulated the expression of genes encoded for PAL and the enzymatic activity of PAL and TAL. After 48 h of treatment, the levels of phenols and flavonoids in the plants increased in response to the stimulus exerted on the metabolic pathway (Schiavon et al. 2010).

6 Studies of the Effect of HS Using Large-Scale Gene Expression Techniques

Some studies have used large-scale genetic analysis techniques (LGAT) to show the complex effects which HS may exert on the function and adaptive mechanisms of plants. The LGATs that have been used to study the effects of HS in plants include cDNA-AFLP and microarray analysis. Trevisan et al. (2011) applied the cDNA-AFLP technique in the study of *Arabidopsis* plants treated with HS. A combination of 160 primers was used and a total of 133 genes were found to be involved in the effects exerted on the plants by HS. The cDNA-AFLP technique demonstrated that, of the numerous genes involved in the HS–plant interaction, many were related to metabolic and developmental processes, as well as RNA or transcription processes. The authors showed that of all upregulated transcripts following the application of HS, 34 % belonged to metabolic processes and 9 % to stress stimuli processes. The authors noted that using of cDNA-AFLP technique allowed them to confirm that HS affect plants through the complex action mechanisms which involve both auxin-like action and other signaling mechanisms independent of auxins.

In addition, the effects of the application of HA to *Brassica napus* plants have also been studied using microarray techniques. Some results about the function of the group of differentially expressed genes were similar to those reported in the study that used cDNA-AFLP technique. The results of the microarray analysis study showed that four fundamental metabolic pathways (fatty acids, phytohormones, senescence, and ion development and transport) were represented by a low number of differentially expressed genes. However, other metabolic pathways were more specifically affected by the action of humic acids both in the leaves and roots. Of the total number of differentially expressed genes, 10.6 % belonged to general cellular metabolism processes, 6.6 % to nitrogen and sulfur metabolism processes, 6.1 % to carbon and photosynthesis metabolism, and 6.1 % to stress response (Jannin et al. 2012).

LGATs are robust and highly accurate techniques. The studies described above used two different LGAT techniques, HS from different sources, and different plant species. Nevertheless, both studies found that HS effects were exerted not only through known traditional pathways, such as auxin type pathways, but also through other auxin-independent pathways. Genes that are involved in the functioning of other important pathways in plants, such as nitrogen uptake and photosynthesis, are expressed in the response of plants to HS application. However, both techniques revealed that a large number of genes or transcripts that respond to HS are related to stress response. These results corroborate the findings of recent reports demonstrating the generation of ROS in plants treated with HS (García et al. 2012b; Cordeiro et al. 2011).

7 Evidence of Protective Effects of HS in Plants Growing Under High-Stress Conditions

The diverse effects of HS in plants suggest that two distinct metabolic events control their modes of action. The regulation of REDOX homeostasis, which is related to cell signaling and hormonal control, is one of the most apparent results of the application of HS to plants. However, the mechanisms explaining the link between ROS and auxins in regulating antistress responses are still not well understood (Tognetti et al. 2012). Compounds such as nitric oxide (NO) have been shown to play an intermediary role in the action of HS in plants. Zandonadi et al. (2010) reported that humic acids from vermicompost stimulated NO biosynthesis in maize plants. The authors showed that humic acids stimulated NO production at lateral root emergence sites. The stimulation of NO biosynthesis by HS suggests that this mode of action is involved in homeostatic regulation. NO has antioxidant properties and acts as a signaling molecule in the synthesis of enzymes related to ROS catalysis. NO has been shown to play an important role in plant resistance to abiotic stress (e.g., hydric stress, high salinity, or high concentrations of heavy metals) (Siddiqui et al. 2011).

NO has also been shown to be an important intermediary in the action pathways of abscisic acid (ABA) in stomata regulation (Huang et al. 2013). Similarly, two HS fractions of different molecular weights have also been shown to regulate stomata opening. Besides, HS have been shown to exert effects on the opening of stomata in *Pisum sativum* L. plants through an auxin-like action and phospholipase A_2 stimulation (Russell et al. 2006).

7.1 Evidence of Antistress Protective Effects of HS

In addition of examining the role of HS in regulating the primary and secondary metabolism in plants, some studies have discussed the possibility of using these substances to mitigate the effects of abiotic stresses such as saline soils, hydric stress, and concentrations of heavy metals. In one study, bean plants (*Phaseolus vulgaris* L.) were grown in soils artificially salinized with salts from various sources, with the doses of 0.05 and 0.1 % w:w. Without the application of HA, high doses of salt caused the death of plants, while plants grown under the same saline conditions in the presence of HA survived. The application of HA improved the plant's growth and development and the assimilation of mineral elements such as phosphorus nitrogen (Aydin et al. 2012). Leaf applications of HS in tomato plants (*Lycopersicon esculentum* L.) were also tested under natural levels of soil salinity. Plants that received leaf applications of HS at two different physiological stages (10 and 15 days after transplanting) showed improved conditions and internal fruit qualities, compared to plants grown in saline soils without HS application. The internal pH of the fruit, BRIX degrees, malic acid, vitamin C, and total soluble salts were all higher than in the plants not treated with HS (Pérez et al. 2011).

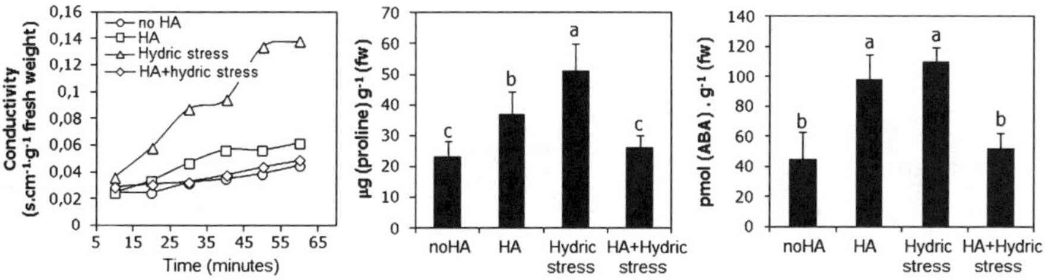

Fig. 11.9 Behavior patterns in rice plants treated with HA under normal growing conditions and conditions of hydric stress (due to loss of water). Level of ABA, a key hormone in stress metabolism signaling. Proline, which is an antioxidant amino acid that responds to stress events and root membrane permeability, was measured by using the conductivity of plant tissues in deionized water

These results indicate that HS stimulate the enzymatic activity of peroxidases (POX), thereby decreasing the amount of H_2O_2 in the leaf and root tissues. These studies demonstrated the inhibition of stress-induced lipid peroxidation in plants treated with humic acids and a preservation of cell membrane permeability was observed (García et al. 2012a). The same HA were also applied to the leaves of rice plants grown in soil under drought conditions. An increase in POX enzymatic activity was observed in response to the application of HA. Plant stress tolerance increased, as shown by improvement in growth and development of the plants, even under drought conditions (Hernández et al. 2012).

The plants to which HA was applied under conditions of hydric stress, permeability of the cell membrane was similar to that of plants grown in normal conditions. Proline content in mature root tissues increased when HA was applied; however, in plants grown under conditions of hydric stress and treated with HA, proline levels were similar to the levels found in plants under normal growth conditions. Moreover, ABA production in mature root tissues was not stimulated by the application of HA, and under conditions of stress, ABA content was similar to that found in plants grown in normal conditions. These results clearly suggest that HA has protective effects in plants experiencing hydric stress. While further studies are needed to determine the mechanisms behind these effects, the protective effects appear to be exerted through ABA-independent pathways, as shown in Fig. 11.9.

8 Potential Use of HS in Agricultural Systems

Based on the results discussed so far on the impact of application of HS to plants under stress, application of HS is becoming a routine in agriculture and research on the effects of HS should be targeted at developing new technologies for sustainable agriculture. Leaf application of HS-based extracts from compost is a method that has been tested in recent years. Using vermicomposts as raw material to obtain HS in an aqueous medium is an ecological and low-cost method which is

Fig. 11.10 The preparation and leaf application of a HS-based liquid extract from vermicompost in small-scale urban maize farming

environmentally safe and readily available for agricultural applications. Most of these HS extracts are dark in color and have the additional advantage of containing minerals, natural phytohormones, and microorganisms that contribute to plant growth. Small- and medium-scale leaf applications are viable and easily implemented methods that do not require the use of large-scale technology and human resources, as shown in Fig. 11.10.

8.1 Use of Liquid Extracts of Vermicompost (Liquid Humus)

Salinity is a major problem that affects soil and consequently influences the crop yields. High soil salinity produces a condition of physiological stress in plants known as saline stress, which affects the plant's biological productivity and reduces agricultural yields. The application of liquid humus to the leaves may be a viable alternative method for mitigating the negative effects of saline stress on food plants.

A liquid humus (Liplant®) obtained from bovine manure vermicompost was applied to the leaves of tomato plants (*Lycopersicon esculentum* Mill.) grown in salinized soils during the optimal and suboptimal planting seasons. Applied doses of 1:50 (v:v) of liquid humus during both seasons resulted an increased crop yields, compared to plants that did not receive humus applications. The application of liquid humus did not change the internal attributes of the fruit, demonstrating that the liquid humus was innocuous when applied under these conditions (Pérez et al. 2009, 2011). Additionally, the application of the liquid humus Liplant to tomatoes grown under normal conditions in red ferralitic soil stimulated several physiological parameters in the plants. Leaf applications of humus at a rate of 1 L ha^{-1} increased the number of roots per plant, root length, leaf number, and biomass (Terry et al. 2012). Other studies have shown that liquid humus increases the uptake of K, P, and N in tomato plants and enhances the photosynthetic pigment content, resulting in an increased agricultural yields and higher fruit quality (Tejada et al. 2008).

Liquid humus was applied to bean plants cultivated in oxisols using tillage system which resulted in increased leaf surface area, liquid assimilation rate, biomass, and total agricultural yield. The highest yields were obtained when the liquid humus was applied at a rate of 1:60 (v:v) (Del Valle et al. 2012). It was also used in the production of watermelon seedlings (*Citrullus lanatus* cv. "Crimson Sweet"). Liquid humus when applied to the leaves in 22.5 mL m^{-2} doses resulted in the development of seedlings with greater leaf surface area and a more developed root system (Silvia-Matos et al. 2012). In strawberry plants (*Fragaria* × *ananassa* Duch.) this product improved several physiological and agricultural parameters. The application of liquid humus was associated with increased leaf surface area (10.1–18.9 %) and dry mass (13.9–27.2 %). In addition, the application of the humus was associated with a 5.7–12.1 % reduction in fruit albinism, 8.5–11.2 % decrease in fruit malformation, and an increase in total yield to the tune of 26.5 % (Singh et al. 2010).

Several studies have examined the application of liquid humus for agricultural use. Most of the observed effects have been positive in increasing yields and improving fruit quality. The results of these studies show that the application of liquid humus is a sustainable option with numerous advantages for the production of food plants under both normal and adverse environmental conditions.

9 Conclusions and Future Prospects

The structure–function–properties relationship should be the basis of future studies of HS and their effects on the environment. Due to the heterogeneous structural characteristics and numerous sources of origin of HS, every study, whether dealing with living organisms or natural systems, must include the structural characterization of HS as its premise. Regarding the mode of action of HS in plants, we observed that in spite of experimental design, HS have effects on several different metabolic processes in plants, as demonstrated through the use of large-scale gene sequencing techniques. The main actions of HS can potentially be observed in each specific metabolic pathway. However, we still lack a complete explanation of how these actions create more effective plant responses that protect plants against possible stress.

Thus far, the identification of a relationship between the stimuli produced by HS in plant defense mechanisms and primary metabolism suggests that future studies of HS should focus on new elements in the modes of action of HS. Current ideas about mechanisms that explain the action of HS in plants need to be expanded beyond auxin-like effects and other known effects. To understand the heterogeneity of HS actions, future research must consider ROS as a chemical species of great importance in metabolic signaling processes in plants. It has been shown that HS can exert noticeable effects on the ROS production metabolism, thereby regulating the REDOX homeostasis in plants.

Finally, the results discussed in this study and the proposed framework for future studies should aid in the development of new ideas about the application of HS that will guide future studies on the use of humified organic matter in agricultural ecosystems, particularly under the conditions of high stress.

Acknowledgments To the TWAS/CNPq for the grant to Andres Calderin Garcia and FAPERJ/Prioridade Rio and CNPq/UNIVERSAL to Ricardo L. Berbara.

References

Aguiar NO, Olivares FL, Novotny EH, Dobbss LB, Balmori DM, Santos-Júnior LG, Chagas JG, Façanha AR, Canellas LP (2013) Bioactivity of humic acids isolated from vermicomposts at different maturation stages. Plant Soil 362:161–174

Alvarez-Puebla RA, Valenzuela-Calahorro C, Garrido JJ (2004) Cu (II) retention on a humic substance. J Colloid Interface Sci 270:47–55

Amir S, Jouraiphy A, Meddich A, Gharous M, Winterto P, Hafidi M (2010) Structural study of humic acids during composting of activated sludge-green waste: elemental analysis, FTIR and ^{13}C NMR. J Hazard Mater 177:524–529

Asli S, Neumann PM (2010) Rhizosphere humic acid interacts with root cell walls to reduce hydraulic conductivity and plant development. Plant Soil 336:313–322

Aydin A, Kant C, Turan M (2012) Humic acid application alleviates salinity stress of bean (*Phaseolus vulgaris* L.) plants decreasing membrane leakage. Afr J Agric Res 7:1073–1086

Baddi GA, Hafidi M, Cegarra J, Alburquerque JA, González J, Gilard V, Revel JC (2004) Characterization of fulvic acids by elemental and spectroscopic (FTIR and ^{13}C-NMR) analyses during composting of olive mill wastes plus straw. Bioresour Technol 93:285–290

Berner JM, Van der Westhuizen AJ (2010) Inhibition of xanthine oxidase activity results in the inhibition of Russian wheat aphid-induced defense enzymes. J Chem Ecol 36:1375–1380

Blomster T, Salojärvi J, Sipari N, Brosché M, Ahlfors R, Keinänen M, Overmyer K, Kangasjärvi J (2011) Apoplastic reactive oxygen species transiently decrease auxin signaling and cause stress-induced morphogenic response in Arabidopsis. Plant Physiol 157:1866–1883

Canellas LP, Piccolo A, Dobbss LB, Spaccini R, Olivares FL, Zandonadi DB, Façanha AR (2010) Chemical composition and bioactivity properties of size-fractions separated from a vermicompost humic acid. Chemosphere 78:457–466

Canellas LP, Dobbss LB, Oliveira AL, Chagas JG, Aguiar NO, Rumjanek VM, Novotny EH, Olivares FL, Spaccini R, Piccolo A (2012) Chemical properties of humic matter as related to induction of plant lateral roots. Eur J Soil Sci 63:315–324

Cordeiro FC, Santa-Catarina C, Silveira V, de Souza SR (2011) Humic acid effect on catalase activity and the generation of reactive oxygen species in corn (*Zea Mays* L). Biosci Biotechnol Biochem 75:70–74

Corpas FJ, Palma JM, Sandalio LM, Valderrama R, Barroso JB, del Río LA (2008) Peroxisomal xanthine oxidoreductase: characterization of the enzyme from pea (Pisum sativum L.) leaves. J Plant Physiol 165:1319–1330

De Tullio MC, Jiang K, Feldman LJ (2010) Redox regulation of root apical meristem organization: connecting root development to its environment. Plant Physiol Biochem 48:328–336

Del Valle GH, Hernández GO, Izquierdo FG, Fortes NA (2012) Influence of the no till and the liquid extract of vermicompost in indicators and indexes growth in common bean (*Phaseolus vulgaris* L.) cv. cc-25-9. Rev Cie Téc Agr 21:86–90

Demidchik V, Shabala SN, Davies JM (2007) Spatial variation in H_2O_2 response of Arabidopsis thaliana root epidermal Ca^{2+} flux and plasma membrane Ca^{2+} channels. Plant J 49:377–386

Demidchik V, Shang Z, Shin R, Thompson E, Rubio L, Laohavisit A, Mortimer JC, Chivasa S, Slabas AR, Glover BJ, Schachtman DP, Shabala SN, Davies JM (2009) Plant extracellular ATP signalling by plasma membrane NADPH oxidase and Ca^{2+} channels. Plant J 58:903–913

Dobbss LB, Rumjaneck VM, Baldotto AM, Velloso ACX, Canellas LP (2009) Caracterização química e espectroscópica de ácidos húmicos e fúlvicos isolados da camada superficial de latossolos brasileiros. R Bras Ci Solo 33:51–63

Dobbss L, Canellas LP, Olivares FL, Aguiar NO, Peres LEP, Spaccini R, Piccolo A (2010) Bioactivity of chemically transformed humic matter from vermicompost on plant root growth. J Agric Food Chem 127:1–10

Ferrari E, Francioso O, Nardi S, Saladini M, dal Ferro N, Morari F (2011) DRIFT and HR MAS NMR characterization of humic substances from a soil treated with different organic and mineral fertilizers. J Mol Struct 998:216–224

Ferrer JL, Austin MB, Stewart C Jr, Noel JP (2008) Structure and function of enzymes involved in the biosynthesis of phenylpropanoids. Plant Physiol Biochem 46:356–370

Filella M, Town RM (2001) Heterogeneity and lability of Pb(II) complexation by humic substances: practical interpretation tools. Fresenius J Anal Chem 370:413–418

Foreman J, Demidchik V, Bothwell JHF, Mylona P, Henk M, Torresk MA, Linstead P, Costa S, Brownlee C, Jonesk JDG, Davies JM, Dolan L (2003) Reactive oxygen species produced by NADPH oxidase regulate plant cell growth. Nature 422:442–446

García AC, Berbara RLL, Farias LP, Izquierdo FG, Hernández OL, Campos RH, Castro RN (2012a) Humic acids of vermicompost as an ecological pathway to increase resistance of rice seedlings to water stress. Afr J Biotechnol 11:3125–3134

García AC, Santos LA, Izquierdo FG, Sperandio MVL, Castro RN, Berbara RLL (2012b) Vermicompost humic acids as an ecological pathway to protect rice plant against oxidative stress. Ecol Eng 47:203–208

Gerke J (1994) Aluminum complexation by humic substances and aluminum species in the soil solution. Geoderma 63:165–175

Ghabbour EA, Davies G (2001) Humic substances. Structures, models and functions. In: The Royal Society of Chemistry, Thomas Graham House, Science Park, Milton Road, Cambridge CB4 OW, UK Registered Charity No. 207890

Gill SS, Tuteja N (2010) Reactive oxygen species and antioxidant machinery in abiotic stress tolerance in crop plants. Plant Physiol Biochem 48:909–930

Gondar D, Lopez R, Fiol S, Antelo JM, Arce F (2005) Characterization and acid–base properties of fulvic and humic acids isolated from two horizons of an ombrotrophic peat bog. Geoderma 126:367–374

Haghighi M, Kafi M, Fang P (2012) Photosynthetic activity and N metabolism of lettuce as affected by humic acid. J Veg Sci 18:182–189

Hernández R, García A, Portuondo L, Muñiz S, Berbara R, Izquierdo F (2012) Protección antioxidativa de los ácidos húmicos extraídos de vermicompost en arroz (Oryza sativa L.) var. IACuba30. Rev Protección Veg 27:102–110

Huang AX, Sheb XP, Zhang YY, Zhao JL (2013) Cytosolic acidification precedes nitric oxide removal during inhibition of ABA induced stomatal closure by fusicoccin. Russ J Plant Physiol 60:60–68

Iglesias A, Lopez R, Fiol S, Antelo JM, Arce F (2003) Analysis of copper and calcium–fulvic acid complexation and competition effects. Water Res 37:3749–3755

Irving H, Williams RJP (1953) The stability of transition-metal complexes. J Chem Soc 3:3192–3210

Jannin L, Arkoun M, Ourry A, Laîné P, Goux D, Garnica M, Fuentes M, San Francisco S, Baigorri R, Cruz F, Houdusse F, Garcia-Mina JM, Yvin JC, Etienne P (2012) Microarray analysis of humic acid effects on Brassica napus growth: involvement of N, C and S metabolisms. Plant Soil 359:297–319

Jerzykiewicz M (2004) Formation of new radicals in humic acids upon interaction Pb (II) ions. Geoderma 122:305–309

Kurusu T, Kuchitsu K, Nakano M, Nakayama Y, Iida H (2013) Plant mechanosensing and Ca^{2+} transport. Trends Plant Sci 18(4):227–233

Kwak JM, Mori IC, Pei ZM, Leonhardt N, Torres MA, Dangl JL, Bloom RE, Bodde S, Jones JD, Schroeder JI (2003) NADPH oxidase AtrbohD and AtrbohF genes function in ROS-dependent ABA signaling in Arabidopsis. EMBO J 22:2623–2633

Li G-W, Peng Y-H, Yu X, Zhang M-H, Cai W-M, Sun W-N, Su W-A (2008) Transport functions and expression analysis of vacuolar membrane aquaporins in response to various stresses in rice. J Plant Physiol 165:1879–1888

Li X, Xing M, Yang J, Huang Z (2011) Compositional and functional features of humic acid-like fractions from vermicomposting of sewage sludge and cow dung. J Hazard Mater 185:740–748

Lopez F, Bousser A, Sissoëff I, Gaspar M, Lachaise B (2003) Diurnal regulation of water transport and aquaporin gene expression in maize roots: contribution of PIP2 proteins. Plant Cell Physiol 44:1384–1395

Loque D, Ludewig U, Yuan L, von Wiren N (2005) Tonoplast intrinsic proteins AtTIP2;1 and AtTIP2;3 facilitate NH_3 transport into the vacuole. Plant Physiol 137:671–680
Marino D, Dunand C, Puppo A, Pauly N (2012) A burst of plant NADPH oxidases. Trends Plant Sci 17:9–15
Miller G, Schlauch K, Tam R, Cortes D, Torres MA, Shulaev V, Dangl JL, Mittler R (2009) The plant NADPH oxidase RBOHD mediates rapid systemic signaling in response to diverse stimuli. Sci Signal 84:45
Miller G, Suzuki N, Ciftci-Yilmaz S, Mittler R (2010) Reactive oxygen species homeostasis and signaling during drought and salinity stresses. Plant Cell Environ 33:453–467
Mittler R (2002) Oxidative stress, antioxidants and stress tolerance. Trends Plant Sci 7:405–410
Mittler R, Vanderauwera S, Gollery M, Van BF (2004) Reactive oxygen gene network of plants. Trends Plant Sci 9:490–498
Mittler R, Vanderauwera S, Suzuki N, Miller G, Tognetti VB, Vandepoele K, Gollery M, Shulaev V, Breusegem FV (2011) ROS signaling: the new wave? Trends Plant Sci 16:300–309
Mora V, Bacaicoa E, Zamarreño AM, Aguirre E, Garnica M, Fuentes M, García-Mina JM (2010) Action of humic acid on promotion of cucumber shoot growth involves nitrate-related changes associated with the root-to-shoot distribution of cytokinins, polyamines and mineral nutrients. J Plant Physiol 167:633–642
Mori IC, Schroeder JI (2004) Reactive oxygen species activation of plant Ca^{2+} channels. A signaling mechanism in polar growth, hormone transduction, stress signaling, and hypothetically mechanotransduction. Plant Physiol 135:702–708
Muscolo A, Sidari M, Attiná E, Francioso O, Tugnoli V, Nardi S (2007) Biological activity of humic substances is related to their chemical structure. Soil Sci Soc Am J 71:75–85
Nardi S, Pizzeghello D, Muscolo A, Vianello A (2002) Physiological effects of humic substances on higher plants. Soil Biol Biochem 34:1527–1536
Nardi S, Muscolo A, Vaccaroa S, Baiano S, Spaccini R, Piccolo A (2007) Relationship between molecular characteristics of soil humic fractions and glycolytic pathway and krebs cycle in maize seedlings. Soil Biol Biochem 39:3138–3146
Nebbioso A, Piccolo A (2012) Advances in humeomics: enhanced structural identification of humic molecules after size fractionation of a soil humic acid. Anal Chim Acta 720:77–79
Pérez JJR, Izquierdo FG, Escobar IMR, Mayoral JAL (2009) Liquid casting effect on tomato yields in saline soils of Cuban east region. Centro Agrícola 36:57–61
Pérez JJR, Izquierdo FG, Escobar IMR, Ruisánchez Y, Mayoral JAL, Amador BM, Espinoza FH, Fabré TB, Amador CA, Silvera CMO, Morales YA, Milanés JYR (2011) Efectos del humus líquido sobre algunos parámetros de calidad interna en frutos de tomate cultivados en condiciones de estrés salino [Effects of liquid humus on some parameters of internal quality of tomato fruits grown under salt stress conditions]. Centro Agrícola 38:57–61
Piccolo A, Stevenson FJ (1982) Infrared spectra of Cu^{2+}, Pb^{2+} and Ca^{2+} complexes of soil humic substances. Geoderma 27:195–208
Plaza C, D'Orazio V, Senesi N (2005) Copper (II) complexation of humic acids from the first generation of EUROSOILS by total luminescence spectroscopy. Geoderma 125:177–186
Polle A (2001) Dissecting the superoxide dismutase–ascorbate peroxidase–glutathione pathway in chloroplasts by metabolic modeling. Computer simulations as a step towards flux analysis. Plant Physiol 126:445–462
Russell L, Stokes AR, Macdonald H, Muscolo A, Nardi S (2006) Stomatal responses to humic substances and auxin are sensitive to inhibitors of phospholipase A2. Plant Soil 283:175–185
Šamaj J, Baluška F, Menzel D (2004) New signalling molecules regulating root hair tip growth. Trends Plant Sci 9:217–220
Schiavon M, Pizzeghello D, Muscolo A, Vaccaro S, Francioso O, Nardi S (2010) High molecular size humic substances enhance phenylpropanoid metabolism in maize (*Zea mays* L.). J Chem Ecol 36:662–669
Schnitzer M (1978) Humic substances: chemistry and reactions. In: Schnitzer M, Khan SU (eds) Soil organic matter. Elsevier, Amsterdam

Schopfer P, Liszkay A, Bechtold M, Frahry G, Wagner A (2002) Evidence that hydroxyl radicals mediate auxin-induced extension growth. Planta 214:821–828

Senesi N, Sposito G, Martin JP (1986) Copper (ii) and iron (iii) complexation by soil humic acids: an IR and ESR study. Sci Total Environ 55:351–362

Siddiqui MH, Al-Whaibi MH, Basalah MO (2011) Role of nitric oxide in tolerance of plants to abiotic stress. Protoplasma 248:447–455

Silvia-Matos RRS, Cavalcante IHL, Júnior GBS, Albano FG (2012) Foliar spray of humic substances on seedling production of Watermelon cv. Crimson Sweet. J Agron 11:60–64

Singh R, Gupta RK, Patil RT, Sharma RR, Asrey R, Kumar A, Jangra KK (2010) Sequential foliar application of vermicompost leachates improves marketable fruit yield and quality of strawberry (*Fragaria x ananassa* Duch.). Sci Hortic 124:34–39

Suzuki N, Koussevitzky S, Mittler R, Miller G (2012) ROS and redox signalling in the response of plants to abiotic stress. Plant Cell Environ 35:259–270

Takeda S, Gapper C, Kaya H, Bell E, Kuchitsu K, Dolan L (2008) Local positive feedback regulation determines cell shape in root hair cells. Science 319:1241–1244

Tao Z-Y, Zhang J, Zhai J-J (1999) Characterization and differentiation of humic acids and fulvic acids in soils from various regions of China by nuclear magnetic resonance spectroscopy. Anal Chim Acta 395:199–203

Tejada M, Gonzalez JL, Hernandez MT, Garcia C (2008) Agricultural use of leachates obtained from two different vermicomposting processes. Bioresour Technol 99:6228–6232

Terry E, Díaz de Armas MM, Padrón JR, Tejeda T, Zea ME, Camacho-Ferre F (2012) Effects of different bioactive products used as growth stimulators in lettuce crops (*Lactuca sativa* L.). J Food Agric Environ 10:386–389

Tognetti VB, Mühlenbock P, van Breusegem F (2012) Stress homeostasis the redox and auxin perspective. Plant Cell Environ 35:321–333

Trevisan S, Botton A, Vaccaro S, Vezzaro A, Quaggiotti S, Nardi S (2011) Humic substances affect Arabidopsis physiology by altering the expression of genes involved in primary metabolism, growth and development. Environ Exp Bot 74:45–55

Tsukagoshi H, Busch W, Benfey PN (2010) Transcriptional regulation of ROS controls transition from proliferation to differentiation in the root. Cell 143:606–616

Vasconcelos FAC, Zhang X, Ervin EH, Kiehl CJ (2009) Enzymatic antioxidant responses biostimulants maize and soybean subjected to drought. Sci Agric 66:395–402

Vaugham D, Linehan DJ (1976) The growth of wheat plants in humic acid solutions under axenic conditions. Plant Soil 44:445–449

Vaughan D, Ord BG (1981) Uptake and incorporation of ^{14}C-labelled soil organic matter by roots of Pisum sativum L. J Exp Bot 32:679–687

Vaughan D, Ord BG (1982) An in vitro effect of soil organic matter fractions and synthetic humic acids on the generation of superoxide radicals. Plant Soil 66:113–116

Yang C, Wang M, Lu Y, Chang I, Chou C (2004) Humic substances affect the activity of chlorophyllase. J Chem Ecol 8:1561–1573

Zandonadi DB, Santos MP, Dobbss LB, Olivares FL, Canellas LP, Binzel ML, Okorokova-Façanha AL, Façanha AR (2010) Nitric oxide mediates humic acids-induced root development and plasma membrane H^+-ATPase activation. Planta 231:1025–1036

Zepeda-Jazo I, Velarde-Buendía AM, Enríquez-Figueroa R, Bose J, Shabala S, Muñiz-Murguía J, Pottosin II (2011) Polyamines interact with hydroxyl radicals in activating Ca^{2+} and K^+ transport across the root epidermal plasma membranes. Plant Physiol 157:2167–2180

Chapter 12
Mitochondrial Respiration: Involvement of the Alternative Respiratory Pathway and Residual Respiration in Abiotic Stress Responses

Silvana Chocobar-Ponce, Carolina Prado, Rosana González, Mariana Rosa, and Fernando E. Prado

Contents

1 Introduction

To stay alive and maintain their cellular environment, all cells need energy that is obtained through a process called cellular respiration. Energy of cells comes from bonds of ATP molecule that is obtained from energy-rich molecules, which are broken down into smaller components, releasing their accumulated energy (Nelson and Cox 2008). Many organic molecules can be utilised as energy-source molecules, but glucose and related hexoses represent most used fuel molecules for most eukaryotic organisms of kingdoms Animalia, Plantae, Fungi and Protista as well as many of the Monera kingdom (prokaryotes). Heterotrophic organisms, i.e. animals,

S. Chocobar-Ponce • C. Prado • R. González • M. Rosa • F.E. Prado (✉)
Laboratorio de Fisiología Vegetal, Facultad de Ciencias Naturales e IML, Universidad Nacional de Tucumán, Miguel Lillo 205, CP 4000, San Miguel de Tucumán, Argentina
e-mail: prad@arnet.com.ar; fepra@csnat.unt.edu.ar

P. Ahmad and M.R. Wani (eds.), *Physiological Mechanisms and Adaptation Strategies in Plants Under Changing Environment: Volume 1*, DOI 10.1007/978-1-4614-8591-9_12,

fungi, most bacteria and many protists, obtain their *preformed* fuel molecules, already made by other organisms. Meanwhile plants and some protists and bacteria are autotrophs, manufacturing their own fuel molecules (Ballard and Whitlock 2004). Both aerobic and anaerobic respirations provide energy for essential metabolic processes in all living cells. The first one, named respiration simply, is present in most animals, plants and some prokaryotic organisms whereas the latter, also named fermentation, mostly occurs in bacteria, yeast and some prokaryotes. In spite of universality of aerobic respiration, anaerobic respiration also occurs in cells under hypoxic or anoxic conditions such as germinating seeds and waterlogged roots. Metabolic pathways of both respiration types are similar in virtually all eukaryotic organisms and in most prokaryotic ones (Plaxton and Podestá 2006).

Aerobic respiration comprises three metabolic pathways: (1) glycolysis, (2) tricarboxylic acid cycle (TCA) and (3) oxidative phosphorylation coupled to electron transport chain (ETC) (Hatefi 1985). Glycolysis occurs in the cytosol whereas latter two take place inside mitochondrion (Nunes-Nesi et al. 2008). Feedback integration between mitochondrial functions and cellular metabolic processes allows energy to get a dynamic equilibrium during cell homeostasis (Collins et al. 2002; Benard et al. 2006). In particular, phosphotransfer systems, which are able to transfer phosphoryl groups between different cellular compartments in absence of major changes in cytosolic levels of adenine nucleotides or changes in ATP/ADP ratio (Dzeja and Terzic 1998), have become inefficient mechanisms to coupling mitochondrial ATP production with cellular ATP consumption and/or ATP sensing (Abas et al. 2000). Therefore, mitochondria can exert a regulatory control of cellular processes which are essential to living cells (Dzeja et al. 2001). Alteration of mitochondrial functionality leads to a disruption of energy cell homeostasis increasing cellular degenerative states (Lesnefsky et al. 2001). Thus, preservation of mitochondrial functionality allows cell to withstand metabolic alterations (Wallace 1999).

Enzymes and proteins that participate in mitochondrial respiratory pathway are well known, but their control, regulation and interactions among them are less well understood (Havelund et al. 2013). Apparently functional plasticity of mitochondria may be enough to respond to different mismatches occurring into cell metabolism (Cortassa et al. 2009). Moreover, mitochondria have a great adaptability to stress and also have ability to perceive and generate different stress signals. Then, they can trigger protective mechanisms and/or act as stress end effectors to protect their functionality, ATP production and even cell survival (Collins et al. 2002; Halestrap and Brosnan 2008). Modulation of ATP production, ion transport, cytochrome *c* release, redox state and free radical generation have been already analysed in relation to cell protection in mammalian mitochondria (Gustafsson and Gottlieb 2008; Menze et al. 2009). Thus, it may be expected that cytoprotection capacity of mitochondria could be useful to enhance stress tolerance of cells. However, modulatory approaches of mitochondrial respiration seem to exert cell protective effects after a short-term perturbation of mitochondria homeostasis. After that, cytoprotective signal events begin and entry of mitochondria into stress-resistant state is priming (Dzeja et al. 2001). Paradoxically, respective contributions to enhance mitochondrial stress tolerance and cell health still remain fertile ground for further debate and discovery

(Johnson et al. 2012). Against background described earlier, this review uncovers how unique functional features of plant mitochondria participate in plant cell responses to abiotic stress.

2 Structural Organisation of Eukaryotic Mitochondria

Mitochondria have evolved from an α-proteobacterial symbiont which endosymbiotically was incorporated into a protocell over 1.5 billion years ago. Despite their common ancestral origin, mitochondria morphology varies among eukaryotic cells (Embley and Martin 2006). Mitochondria of most animals, algae and fungi generally appear as ellipsoid-shaped organelles and greatly vary in size, which ranges between 1 and 2 μm length. They also greatly vary among organisms and tissues depending on cell type and physiological state (Logan 2006). Certainly in higher plant cells, mitochondria appear as spherical-shaped organelles, but this may not always be the model and so in some plants they appear to be cup- or filament-shaped organelles (Logan et al. 2003). Like animals, plant mitochondria are not uniform or static entities within the cell; they can vary in number, composition and function between different cells, organs and species over development and in response to different stresses (Carrie et al. 2012). Frequent fusions and fissions as well as high motilities also occur in plant mitochondria (Arimura et al. 2004). Furthermore, observations of mitochondria of certain animals made by transmission electron microscopy (TEM) have revealed that they have a long filamentous morphology while in certain algae (*Chlorella*, *Chlamydomonas*), mitochondria appear as single-branched organelles: *mitochondrial reticulum*. It is generally accepted that mitochondrial reticule structure is infrequent in higher plant cells because profiles of branched mitochondria are rarely observed in ultrathin sections by TEM observations (Newcomb 1990).

Mitochondria morphology can be quite variable among species and cell types, either within the same species or from same cell type in different metabolic states. Typical eukaryotic mitochondria have two membranes: a smooth outer membrane that surrounds an inner membrane forming numerous invaginations, which delimits a protein-rich core named *matrix*. Both membranes differ in composition and function. Outer membrane contains less protein and more lipid than inner one, also being permeable to all molecules of 10,000 Da, or less due to presence of large channel-forming proteins on its surface (Mannella 1992). Early studies on mitochondria structure performed by TEM allowed development of two arrangement models of the inner membrane: *Palade's model* (Palade 1952) and *Sjostrand's model* (Sjostrand 1953). *Palade's model*, also known as *baffle model*, considers that mitochondrial cristae are formed by baffles that protrude from inner membrane into the matrix. Contrarily, *Sjostrand's model* also called *septa model* proposes that matrix is divided by inner membrane in many distinct forming septa compartments. Analyses of high-resolution scanning electron microscopy (HRSEM) and electron microscopic 3D tomography of ultrathin slices of mitochondria contributed to establish that both models are not entirely correct (Mannella 2006). In fact, obtained results demonstrate that tubular cristae rather than lamellar ones are predominant. Also they are structurally distinct from the rest of inner membrane. In this new model of mitochondrial membrane architecture, intracristae spaces are connected to parts of the inner membrane very close to outer membrane (inner boundary membrane) through narrow tubules instead of large opening connections occurring in the *baffle model* suggested by Palade (1952). This structural motif originally named *pediculi cristae* and latter *crista junction* was first suggested by Daems and Wisse (1966). Most cristae have more than one *crista junction* which can be arranged either on same side of the inner boundary membrane or on opposite side when cristae cross the matrix completely. Electron microscopic tomography analysis also revealed that *crista junctions* have a determined size and morphology and are independent of mitochondria source and/or fixation procedure, being suggested that *crista junctions* constitute structural components in all eukaryotic mitochondria (Perkins and Frey 2000). Further, internal architecture, i.e. *crista junctions* and intercristae spaces, changes widely depending upon external conditions (Mannella 2006). According to Hackenbrock's topological model, mitochondrial inner membrane can exist in two conformational states: *orthodox* and *condensed* (Hackenbrock 1966). *Orthodox* state (expanded matrix) has inner membrane predominantly tubular with scarce cristae interconnections and one or two *crista junctions*. *Condensed* state (contracted matrix) has dilated open intracristal spaces that appear as irregularly shaped compartments which are connected to each other and to external intermembrane space by numerous intercristal connections and *crista junctions* (Logan 2006). By computer simulation studies, it was recognised that in *condensed* state inner membrane shape influences diffusion of ions and metabolites. Simulation also indicates that narrow *crista junctions* reduce diffusion between intracristal and intermembrane compartments. This fact leads to a reduced ADP concentration, and so a reduced ATP production occurs. Contrarily, in the *orthodox* state, reduced bulk of ADP can be concentrated into smaller intercristal spaces, helping to minimise negative effects of decreasing ADP diffusion on ATP synthesis. In fact, computer-simulated results suggest that changes of inner membrane shape can be a useful mechanism to regulate ATP production (Mannella 2006). Whether this mechanism is operating in vivo still remains unclear. Mitochondria with a much reduced inner membrane system and electron light matrix have been observed in cotyledons and embryos of some dry seeds. Such mitochondria undergo structural and functional changes during imbibition (water uptake) and germination (radicle protrusion) processes (Howell et al. 2006). In this way, Logan et al. (2001) demonstrated that upon imbibition in dry quiescent maize embryos, the biogenesis of mitochondria is enhanced. Interestingly, prior germination (within 24 h of imbibition), these mitochondria have an *orthodox* state, while after root protrusion (after 48 h of imbibition), they appear in *condensed* state. Therefore, it could be speculated that intrinsic capacity to switch from *orthodox* to *condensed* state that occurs during mitochondria biogenesis in germinating seeds is indicative of a metabolic-linked *orthodox/condensed* function. Thus, *orthodox/condensed* oscillation may reflect functional changes into mitochondrial biochemistry pathways, for example, switching to be reliant on electron provision from external NADH dehydrogenases to newly assembled TCA cycle (Logan et al. 2001).

From new structural insights, it is clear that there are as many as six discrete compartments in plant mitochondria: outer membrane, inner boundary membrane, cristal membrane, intracristal space, intermembrane space and matrix (Frey and Mannella 2000). In most eukaryotic organisms, structural basis of mitochondrial energy conversion and ATP synthesis is a cooperative work among five integral membrane protein complexes. Classical respiratory chain comprises four major protein complexes (I–IV) which interact among them through a small lipid molecule, the ubiquinone (UQ), also called coenzyme Q, and a small protein, cytochrome *c*, involved in the transfer of electrons to reduce O_2 to form H_2O, and another associated with ATP generation (complex V), called ATP synthase. Complex I, also called NADH dehydrogenase complex, catalyses electron transfer from NADH to UQ. Complex II, also called succinate dehydrogenase, is responsible for the transfer of electrons from succinate to UQ. Complex III, generally called bc_1 complex or ubiquinone–cytochrome *c* oxidoreductase, catalyses electron transfer from ubiquinol (UQH_2) to cytochrome *c* (Cyt *c*). Complex IV, also called cytochrome *c* oxidase or simply COX, catalyses electron transfer from cytochrome *c* to O_2. Altogether, they are referred to as oxidative phosphorylation system (OXPHOS) (Sunderhaus et al. 2010). Complexes I, III and IV coupled electron transport to proton pumping from the matrix to intermembrane space. Generated H^+ gradient can be used by complex V to catalyse the formation of ATP from ADP (Boekema and Braun 2007). All of these complexes coexist in the inner mitochondrial membrane. Complexes I, II and IV are thought to be embedded in mitochondrial membrane in such a way that they are in contact with both intermembrane space and matrix, while complex II is associated with matrix side only. Binding site for NADH of complex I faces the matrix (Alberts et al. 2002). In the last decade, structural features of individual respiratory complexes from various organisms have been determined (Boekema and Braun 2007).

To explain arrangement of respiratory chain complexes, two different models were proposed: *liquid state* and *solid state*. In *liquid state* model, respiratory complexes are randomly arranged in lipid bilayer inner membrane and can freely diffuse laterally (Hackenbrock et al. 1986). In earliest *solid state* model, proposed over 50 years ago (Chance and Williams 1955), a higher organisation is assumed and so electron transfer from one complex to next occurs through preset pathways. Kinetic evidences derived from studies performed on bovine heart mitochondria suggest that both models coexist (Bianchi et al. 2004). However, isolation of respiratory supercomplexes represents an additional support of *solid state* model (Schäfer et al. 2006). Different respiratory supercomplexes, also called *respirasomes*, have been isolated from bacteria, yeasts, mammals and plants (Marques et al. 2007; Ramírez-Aguilar et al. 2011). Despite importance of *respirasomes* into respiratory electron chain, their functional significance is still largely unknown. However, a crucial function proposed for *respirasomes* seems to be stabilisation of individual respiratory complexes (Schägger et al. 2004).

Plant respiratory chain is by and large similar to that of mammals and microbes but possesses some distinctive features. In an overview of plant respiratory chain, it can see classical complexes I, II, III and IV, as well as ATP synthase (complex V). There are also different non-energy-conserving bypass systems such as alternative oxidase (AOX), non-H^+-pumping NADH/NADPH dehydrogenases (NDs) and uncoupling protein homologues (UCPs). The latter constitute a subfamily within mitochondrial anion carrier family that can function either directly dissipating H^+ gradient across inner membrane or indirectly by catalysing a free fatty acid (FFA)-induced H^+ recycling (Hourton-Cabassa et al. 2004). Different carriers for transport of substrates, cofactors and products are also localised (Millar et al. 2011). Like other eukaryotic mitochondria, distinct *respirasomes* (supercomplexes) could be described in plant mitochondria. Regarding organisation of mitochondrial respiratory chain, complexes I and IV as well as AOX and NDs do not form part of plant supercomplexes (Eubel et al. 2003; Boekema and Braun 2007). By using blue native gel electrophoresis technique (BN-PAGE), three supercomplexes, i.e. dimeric ATP synthase, III_2I_1 supercomplex and dimeric III_2I_1 supercomplex, were identified in the respiratory chain of non-green tissues of different species (*Arabidopsis thaliana*, *Phaseolus vulgaris*, *Solanum tuberosum* and *Hordeum vulgare*). In other studies carried out with mitochondria from *Spinacia oleracea* leaves and thermogenic appendix of *Arum maculatum*, a special supercomplex family different to *A. thaliana* supercomplex model was found. These special supercomplexes have the following compositions: I_1III_2, III_2IV_1 and $I_1III_2IV_1$, being the two last less abundant (Sunderhaus et al. 2010). Presence of supercomplexes into plant respiratory chain can affect alternative respiration because electrons might be channelled within supercomplexes, reducing access of alternative enzymes to their electron-donating substrates. For example, supercomplex I_1III_2 channelises directly electrons from complex I to complex III giving a restricted access of AOX to UQH_2 and then probably acts as modulator of AOX activity (Sunderhaus et al. 2010).

Based on that plant respiratory supercomplexes are relatively labile in comparison with their mammalian and fungal counterparts (Eubel et al. 2003), it has been speculated that *respirasome* architecture in higher plant mitochondria might correspond to a more dynamic structure that allows reorganisation of complexes I, III, and IV to efficiently execute other tasks not related to proposed substrate channelling between them, as a functional plant *respirasome* advantage (Bianchi et al. 2004). This assumption seems plausible for two major reasons: (1) plant respiratory complexes also have other plant-specific non-respiratory functions. For example, complex III is a bifunctional protein complex involved in processing targeting signals for a vast number of nuclear-encoded protein precursors that are matrix targeted (Braun and Schmitz 1995) and (2) NDs and AOX might interact with supercomplexes III_2+IV_4, I_1+III_2 and/or even with free complexes I_1 and III_2 to transfer electrons as a flexible mechanism to accommodate two interconnected ATP-generating processes in mitochondria and chloroplasts, as well as other functions of respiratory complexes (Raghavendra and Padmasree 2003).

It will be important in a near future to carry out new studies to get a comprehensive elucidation of supercomplex structure of respiratory chain in higher plants.

For this purpose it will be obligatory to analyse situations in mitochondria from different plant tissues and from many species, specifically in which alternative respiratory pathway can be strongly induced by applying specific stresses or by genetic manipulations (Eubel et al. 2003).

3 The Unique Genome of Plant Mitochondria

Mitochondria are conspicuous cell organelles responsible of many essential functions. They are not uniform or static entities within the cell and can vary in number, composition and function between different cells, organs or species, as well as in response to different environmental stresses (Garesse and Vallejo 2001; Jacoby et al. 2012). Mitochondria functionality largely is dependent of nuclear-encoded proteins. Conjunction of mitochondrial and nuclear genes gives own mitochondrial genetic system for replication of mitochondrial genome (mtDNA), transcription, processing of mRNA and protein translation (Boore 1999). Requirements of mitochondrial activity of different tissues vary from a primary energy metabolism-oriented organelle to a biosynthetic-oriented organelle (Johnson et al. 2007). In fact, mitochondria are tuned by the nucleus to play specific functions into different cells (Ikemura 1985). Plant mitochondria have many functional similarities with fungal and animal mitochondria, but also there are significant variations among them (Fauron et al. 2004).

Most of sequenced mtDNAs of animals including the human one are very small (14–18 kbp). With few exceptions they have 37 genes: 2 corresponding to rRNAs (16S and 12S), 13 for mitochondrial proteins and 22 corresponding to tRNAs. Mitochondrial ribosomes of animals and yeast seem to have 16S and 12S rRNAs as unique components, whereas in bacterial ribosomes three rRNAs (23S, 16S and 5S) are present (Fauron et al. 2004). Noteworthy, despite generalised rRNA pattern of 16S and 12S, the 5S ribosomal rRNA has also been found in mitochondria from some animals (Magalhães et al. 1998). Fungal mtDNAs are somewhat larger (70–100 kbp), while all known mtDNAs of higher plants are bigger and range in size from 208 kbp in white mustard (*Brassica hirta*) (Kubo and Newton 2008) to over 2,000 kbp in certain members of the Cucurbitaceae family, reaching a maximum size of 2,900 kbp in muskmelon (*Cucumis melo*) (Alverson et al. 2011). Even within a single family of plants (e.g. Cucurbitaceae), mtDNA is very variable up to over eightfold when comparing 330 kbp of watermelon (*Citrullus vulgaris*) with 2,900 kbp of muskmelon (Alverson et al. 2010). Like bacteria, also plant mtDNAs encode a 5S ribosomal rRNA. In plant mitochondria, tRNAs are encoded by nuclear and mitochondrial genomes, whereas in animals all tRNAs are encoded by mitochondrial genome only (Adams and Palmer 2003).

Mitochondrial DNAs of flowering plants (angiosperms) exhibit multiple other distinctive or unique features that distinguish them from other eukaryotes (Kubo and Newton 2008). Angiosperm mtDNAs contain abundant intergenic repeat sequences and many introns that mainly correspond to group II introns. Some of these introns are *trans*-spliced (Bonen 2008). Complex structure of angiosperm mtDNA often arises by frequent and active homologous recombination at large repeat regions, creating a multipartite, highly redundant organisation of subgenomic molecules (Chang et al. 2013). Another recombination type that is characteristic of mtDNA of flowering plants involves sporadic, low-frequency illegitimate events at smaller repeats, resulting in substoichiometric DNA molecules that may replicate autonomously and eventually lead to cytoplasmic male sterility (Abdelnoor et al. 2003). Recombination activity and maintenance of subgenomic molecules are supposed to be under nuclear control (Shedge et al. 2007). When comparing mono- and dicotyledonous mtDNAs, one can see that only genic exons are conserved. Comparisons between rice and maize, or between *Arabidopsis* and rapeseed (*Brassica napus*), reveal that most of intergenic spaces are not conserved. On a short evolutive time scale, mtDNAs frequently take up exogenous DNA but also lose portions of their own DNA, leading to many DNA rearrangements. Incorporation of noncoding sequences via intracellular transfer of plastid (chloroplast) and nuclear DNA into mitochondrial genome and mitochondrial DNA into nuclear genome (Wang et al. 2012a), as well as incorporation of unidentified open reading frames (ORFs) by horizontal gene transfer (HGT) (Talianova and Janousek 2011), is also occurring frequently. Interestingly, angiosperm mtDNAs exhibit differential gene losses, indicating that the process of gene transfer to the nucleus continues to the present (Adams and Palmer 2003; Wang et al. 2012a).

In higher plants whole mtDNA appears commonly assembled as a circular chromosome that in earlier years was termed *master circle* or *master chromosome* (Lonsdale et al. 1988). In more recent years, other studies have revealed that in vivo an extended mtDNA organisation, different of *master circle*, also occurs. In this mtDNA model, circular, linear and complex heterogeneous DNA populations are present. Moreover, a very low number of copies of the master chromosome occur in each mitochondrion; even master chromosome can be present only in certain cells. It is likely for recombinant events between large and small repeat sequences to be responsible of great diversity of cycles observed in flowering plant mtDNAs (Backert et al. 1997). Furthermore, mtDNA is usually organised in membrane-associated nucleoids, which are located in the matrix (Dai et al. 2005). Despite that mitochondria of higher plants contain considerably bigger mtDNAs, they do not encode a proportionately higher number of genes (Bullerwell and Gray 2004). For example, mtDNA of *A. thaliana* (~367 kbp) is 22 times as large as human mtDNA (16 kbp) and encodes 33 proteins, whereas human mtDNA encoded 13 proteins (2.7-fold less) (Unseld et al. 1997). In an overall context, genes located on mtDNA of flowering plants differ slightly between species but generally encode products that are directly or indirectly involved in both oxidative phosphorylation and ATP production (Sugiyama et al. 2005). They may be organised in gene clusters or dispersed over complete mtDNA, giving rise to both mono- and polycistronic transcripts. Transcripts of many mitochondrial mRNAs undergo systematic conversions of cytidine to uridine (C-to-U editing) and more rarely uridine to cytidine (U-to-C editing), restoring conserved codons (Castandet and Araya 2011). RNA editing are sequence alterations into RNA molecule so that transcript sequences differ from their DNA template. It describes targeted sequence alterations of RNA molecule by nucleotide insertion/deletion or conversion of mechanisms giving a great RNA and protein diversity, which results in specific amino acid substitutions, deletions and changes in gene expression (Tang et al. 2012). Some mitochondrial tRNAs are transcribed from nuclear genes and imported into mitochondria (Adams and Palmer 2003). To date, mtDNAs of a few higher plant species (nearly 30) have been completely sequenced with free access (Mower et al. 2012, GenBank's Entrez Organelle Genome Database: http://www.ncbi.nlm.nih.gov/sites/entrez?db5genome).

4 Functional Organisation of Plant Mitochondria

Structural basis of OXPHOS in plant mitochondria comprises classical five respiratory complexes (I–V) and so-called alternative components, i.e. AOX and NDs (Sunderhaus et al. 2010). Besides universally distributed H^+-pumping NADH dehydrogenase (complex I), it is evident that, in addition to it, other dehydrogenases in the inner mitochondrial membrane of plant cells are also able to oxidise mitochondrial and cytosolic NADH and NADPH without pumping protons. There are at least four additional non-H^+-pumping type II NADH/NADPH dehydrogenases (NDs). Two NDs are located on external face of the inner membrane (facing intermembrane space), one being able to oxidise NADH and NADPH the other. Similarly, other two NDs are located on internal face facing the matrix (Elhafez et al. 2006). Like fungi, protists, some primitive animals and certain bacteria, plant mitochondria can oxidise external NADH and NADPH via NDs (Matus-Ortega et al. 2011). However, ability to oxidise cytosolic NADH/NADPH is depending upon existence of voltage-dependent anion channels (VDACs) in the outer mitochondrial membrane (Vander Heiden et al. 2000). Relative rates of NADH and NADPH oxidation vary in mitochondria of different tissues and then must be assumed that there are, at least, two different NDs (Zottini et al. 1993). Supporting this assumption, higher inhibition of NADPH oxidation by diphenyleneiodonium (DPI), an inhibitor of one-electron transfer flavoenzymes (O'Donnell et al. 1994), compared with NADH oxidation commonly is observed (Roberts et al. 1995). Furthermore, oxidation of mitochondrial NADH can also occur without inducing an NADPH oxidation (Menz and Day 1996). These findings allow one to assume that both NDs have different reaction mechanisms. Thus, NADH-dependent ND activity probably transfers electrons directly to UQ, differing of NADPH-dependent ND activity (Rasmusson et al. 2008). Activity of plant NDs is rotenone insensitive (Peckmann et al. 2012), requiring micromolar concentrations of Ca^{2+} for their normal activities both external NADH- and NADPH-dependent dehydrogenases as well as internal NADPH-dependent isoform. In contrast, internal NADH-dependent ND does not require Ca^{2+} ions for its activity. In rotenone-insensitive NADH-dependent NDs isolated from bacteria and yeast, either Ca^{2+}-dependent activity or appreciable NADPH oxidation is observed (Geisler et al. 2007). The H^+-pumping NADH dehydrogenase (complex I) is a multimeric enzyme complex with a molecular mass ranging between 550 and 900 kDa, carrying prosthetic groups: flavin mononucleotide (FMN), iron–sulphur centres (FeS) and ubiquinone (UQ) (Friedrich et al. 1995). Contrarily, non-H^+-pumping NADH/NADPH dehydrogenases (NDs) are enzymes constituted by a single polypeptide chain, molecular mass of 50–60 kDa and adenine dinucleotide (FAD) as prosthetic group. Presence of FAD converts NDs in potential generators of reactive oxygen species (ROS) (Møller 2001). In mammalian mitochondria, NADH-dependent ND activity is not found. However, it has been reported that mitochondria from bovine heart are able to oxidise external NADH through a rotenone-sensitive dehydrogenase. Rotenone is a natural pesticide that specifically inhibits H^+-pumping NADH dehydrogenase (complex I) activity (Runswick et al. 1991). Whether this activity is due to a non-H^+-pumping NADH dehydrogenase (ND) is still uncertain. However, mammalian cells generally lack system to transport cytosolic NADH inside mitochondria, being able to oxidise cytosolic NADH either directly by a Ca^{2+}-dependent glycerol-3-phosphate (G3P) dehydrogenase (Hansford 1994) or indirectly through both malate–oxaloacetate (MAL–OOA) and G3P shuttles (Møller 2001). Activities of these shuttles are depending upon high substrate (malate, G3P and OAA) concentrations to corresponding cytosolic enzymes, so mammalian dehydrogenases are unable to oxidise external NADH in absence of such substrates (Rasmusson et al. 2008). In yeast occur both G3P shuttle and external mitochondrial NADH-dependent dehydrogenase, but they are not Ca^{2+} dependent (Rigoulet et al. 2004). Presence of mitochondrial FAD-dependent and cytosolic NAD^+-dependent G3P dehydrogenases in *A. thaliana* seedlings leads to assume that G3P shuttle can be operating in plants and thus contributing to regulate cytosolic NADH/NAD^+ homeostasis (Shen et al. 2006).

Plant external NDs are well-known enzymes but their physiological role still remains unclear. Nevertheless, it has been reported that external NDs can be induced either under certain physiological conditions or by different environmental stresses (Rasmusson et al. 2008). Currently it is accepted that variations in redox state of cytosolic NADH/NADPH pool induced by changes in Ca^{2+} concentration are regulated by NDs activities. Therefore, ND activities could also influence metabolic fluxes among primary carbon metabolism, nitrogen assimilation, biosynthetic pathways, growth, intracellular redox shuttles and NADPH-dependent pathogenic responses (Liu et al. 2008; Smith et al. 2011). NDs also provide additional electron entry points to mitochondrial respiratory chain such as electron transfer flavoprotein:quinone oxidoreductase (ETFQ–OR), L-galactono-1,4-lactone (GalL) dehydrogenase and G3P dehydrogenase (Pineau et al. 2008). Increased electron entry points lead to a substantial increase of both branching and complexity of respiratory electron chain (Rasmusson and Møller 2011). In mitochondria of plants, many fungi and protozoans also occur in an additional terminal oxidase called *alternative oxidase* (AOX) (Vanlerberghe and McIntosh 1997). This enzyme is a quinol oxidase located on the inner mitochondrial membrane that acts as alternative component into mitochondrial respiratory chain. AOX catalyses both oxidation of UQH_2 and reduction of O_2 to H_2O and thus constitutes a branch in respiratory chain and a terminal oxidase for electrons' exit (Millenaar and Lambers 2003). It is not linked to H^+-gradient building complexes and dissipates free energy released during electron flow as heat (Moore and Siedow 1991). Interestingly, in a recent work carried out in mitochondria from potato tubers, it has been found that product of O_2 reduction by AOX is hydrogen peroxide (H_2O_2) and not H_2O, as was previously surmised (Bhate and Ramasarma 2009). However, in a singular letter signed by some of most important AOX investigators, it is affirmed that product of AOX activity is H_2O (Møller et al. 2010). AOX activity is not inhibited by cyanide (CN), azide, sulphide and nitric oxide (NO), but these compounds strongly inhibit COX activity. Although presence of nitric oxide synthase (NOS) in mitochondria is still doubtful (Gas et al. 2009), it has been demonstrated that respiratory activity contributes to mitochondrial NO generation (Gupta et al. 2010). Moreover, NO production results in aconitase activity inhibition and AOX activity induction (Gupta et al. 2012a). Further enhanced production of mitochondrial NO has been found in absence of AOX, pointing a role of AOX in regulating NO generation (Cvetkovska and Vanlerberghe 2012). AOX activity is also induced by salicylic acid (SA) (Matos et al. 2009), light, probably, by a phytochrome-mediated effect (Ribas-Carbó et al. 2008), jasmonic acid (Fung et al. 2004), ethylene, H_2O_2 and Ca^{2+} (Wang et al. 2012b). By contrast, salicylhydroxamic acid (SHAM) and *n*-propyl gallate (*n*PG) are strong inhibitors of AOX activity (Vanlerberghe 2013). Further, AOX activity is also decreased by linoleic acid (Sluse et al. 1998) (Fig. 12.1). Based on that AOX activity bypasses two H^+-pumping sites on respiratory electron chain, it can reduce mitochondrial ATP production by as much as 60 % (Rasmusson et al. 2008). AOX-induced loss of respiratory energy emphasises importance of alternative respiratory pathway (AP) in ATP yield of total respiration. Indeed it is thought that in leaves, AOX supports photosynthetic activity because it dissipates only reducing equivalents derived from chloroplasts, whereas cellular energy status is not affected (Yoshida et al. 2006). In non-photosynthetic tissues, AOX activity seems to participate in both cellular redox regulation and other related metabolic pathways (van Dongen et al. 2011).

Since its early proposed function in thermogenic tissues of *Arum* plants (Meeuse 1975), functions of AOX have increased continuously without reaching the last one. Thereby, it has been also suggested that AOX can act as ROS scavenger to protect cells against oxidative stress (Justine et al. 2001). In that way, AOX may be necessary to modulate initiation of programmed cell death (PCD) pathway responsive to mitochondrial respiratory status (Vanlerberghe et al. 2002). AOX also dampens mitochondrial generation of ROS, presumably by preventing over-reduction of ETC components, i.e. UQ (Sugie et al. 2006). It is noteworthy that prior to 2000, it is thought that AOX had a limited distribution in plants and some fungi and protists, but in 2003 and 2004, a surprising paradigm shift has occurred when AOXs from prokaryotic and animal species were reported for the first time (Stenmark and Nordlund 2003; McDonald and Vanlerberghe 2004). Today it is assumed that AOX has a widespread distribution, being found in all life kingdoms, excepting the Monera kingdom (Archaea, domain) (McDonald et al. 2009). Since AOX is found practically in all species, there remain many open questions related to its respiratory function. Presence of several NDs on dehydrogenase side of respiratory chain linked to AOX on oxidative side determines that plant mitochondrial respiratory chain has a unique highly branched organisation of electron flux, with occurring NDs at front

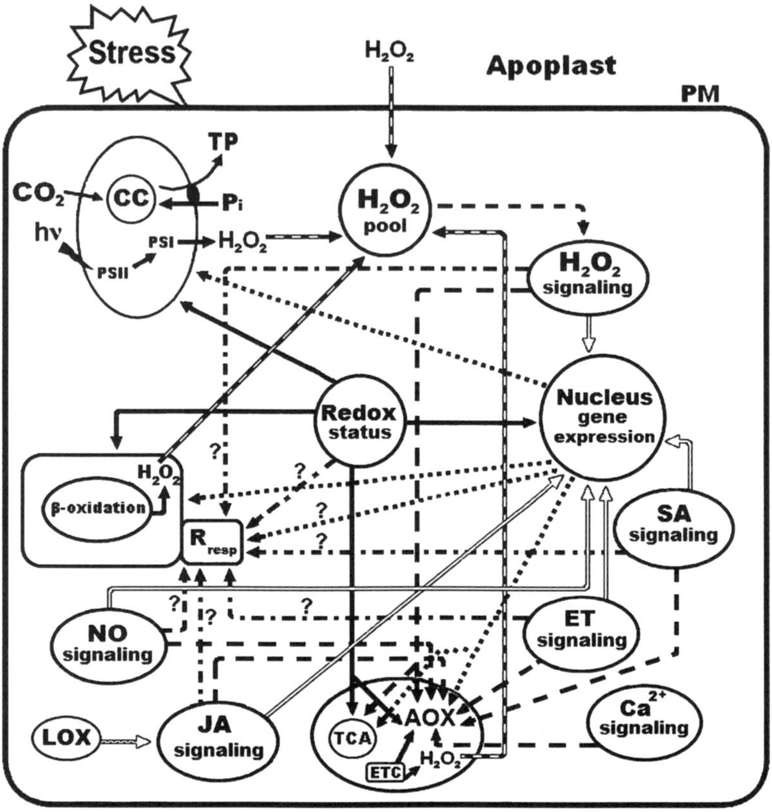

Fig. 12.1 Hypothetical model showing participation of different signal molecules including Ca^{2+} ions and H_2O_2 to regulate AOX activity and probably also R_{resp} in stressed plants. Model also shows a simplified pattern of signalling pathways toward nuclear transduction for triggering gene expression, as well as redox regulatory pathways. *AOX* alternative oxidase, *CC* Calvin cycle, *ET* ethylene, *ETC* electron transport chain, *H_2O_2* hydrogen peroxide, *JA* jasmonic acid, *LOX* lipoxygenase, *NO* nitric oxide, *Pi* inorganic phosphate, *PM* plasma membrane, *PSI* photosystem I, *PSII* photosystem II, *R_{resp}* residual respiration, *SA* salicylic acid, *TCA* tricarboxylic acid cycle, *TP* triose phosphate. *Dashed line*, AOX regulation; *dotted* and *dashed line*, probably R_{resp} regulation; *dotted line*, nuclear regulation; *dashed double line*, sites of H_2O_2 generation; *solid line*, redox regulation; *double solid line*, gene expression pathways

end and AOX at ending (Rasmusson et al. 2004). Thus, during oxidation of NADH from the Krebs cycle, one, two or three sites of energy conservation may be bypassed: (1) when CP and NDs are involved, (2) when CP and AOX are involved or (3) when AOX and NDs are involved. These dissipative pathways allow to sustain functionality of ETC even if membrane potential ($\Delta\psi$) reaches high values (Fernie et al. 2004), which uncouples generation of transmembrane proton electrochemical gradient ($\Delta\mu H^+$), i.e. the proton-motive force for ATP synthesis. Finally, this situation leads to a reduction of ATP content (Jarmuszkiewicz et al. 2000), but NDs and AOX activities are not directly controlled by cellular adenylate status. Together,

AOX and NDs activities allow mitochondria to modulate respiratory function to face challenge resulting from high diurnal and nocturnal variability in energy demand due to functional interconnectivity between mitochondria and chloroplasts (Lee et al. 2010). Thus, extent to which these pathways are acting will affect more or less cellular energy yield (Sweetlove et al. 2007).

Besides AOX and ND activities, plant mitochondria also contain a UCP-like protein, homologue to animal uncoupling protein and so-called plant-uncoupling mitochondrial protein (PUMP) (Vercesi et al. 2006). PUMP also localises in inner membrane and uncouples ATP production because it transports H^+ from intermembrane space to matrix, which is accompanied by an increased H^+ leakage (Garlid et al. 2000). Similar to fatty acid (FA)-dependent anionic uniporter, mitochondrial PUMP reenters H^+ into the matrix by a fatty acid-cycling process, which bypasses ATP synthase and dissipates proton-motive force (Almeida et al. 2002). Thereby, energy contained in $\Delta\mu H^+$ is dissipated as heat, and thus energetic metabolism becomes less efficient (Vercesi et al. 2006). PUMP activity is dependent on pH value, redox status of UQ and mitochondrial concentration of Mg^{2+} (Navet et al. 2005), and further, it increases in presence of FFAs and decreases with ADP, GDP, ATP and GTP (van Dongen et al. 2011). To date five different PUMPs have been identified in plant mitochondria with different tissue distribution and roles (Hourton-Cabassa et al. 2004). Although two energy-dissipating systems (AOX and PUMP) produce same final effect, i.e. decreased ATP synthesis, both act at different levels in overall energy transduction pathway. AOX acts as a safety valve to diminish sudden increases of reduction state of UQ pool when reducing power and/or phosphate potential (ATP/ADP ratio) of cell become overloaded rapidly, while PUMP can correct respiratory energy imbalance through a slower modulation of phosphate potential via direct consumption of H^+ gradient when activated by FFAs (Rasmusson et al. 2009). To avoid a wasteful use of substrates when AOX and PUMP operate simultaneously, it is probably that activities of both systems be regulated differently and also probably respond differently to endogenous and exogenous stimuli (Millar et al. 2011). In support of these assumptions, recently it has been communicated that PUMP and AOX are functionally connected by FFAs (e.g. linoleic acid). Fatty acids increase PUMP activity and decrease AOX one (Borecký and Vercesi 2005). In that context, Almeida et al. (1999) raise a crucial question: *do they work in concert, simultaneously, or in a temporal sequence according to particular physiological state of cell?*

5 AOX: A Structure and Function Puzzle?

From identification of its activity, AOX has been studied extensively. Today, AOX is most studied enzyme of mitochondrial electron chain with more than 600 papers published until 2012 (Prado FE, personal revision). However, despite well-known physiological and metabolic features, total functions of AOX are still quite incomplete, leaving considerable uncertainty about cellular events that may have contributed to its intriguing puzzle of functions. It is accepted that in all plant AOXs have a closely similar composition of amino acids, just over 280, deduced from gene sequences (Siedow and Umbach 2000). By hydropathy plot comparison of several AOX amino acid sequences established that AOX has three conserved features, i.e. two α-helical hydrophobic regions of about 20 amino acids that correspond to four-helix membrane-spanning domains (MSD), an interhydrophobic region of about 40 amino acids containing an helical amphipathic region that is located in the intermembrane space and links both helical MSD, and two hydrophilic regions with about 100 amino acids each (Andersson and Nordlund 1999; Albury et al. 2009). SDS-PAGE of AOX without treatment with a thiol-reducing agent, i.e. dithiothreitol (DTT), shows a protein band between 60 and 75 kDa that corresponds to oligomeric enzyme, also called *oxidised form* (–S–S–). While in presence of DTT, AOX occurs as homodimer with identical monomeric subunits of about 32 kDa, *reduced form* (–SH HS–) (Umbach and Siedow 1993). These authors also demonstrated using DTT and a cross-linking reagent (succinimidyl-4-(*N*-maleimidomethyl) cyclohexane-1-carboxylate, SMCC) that enzyme monomers can be linked by disulfide bonds or covalently. Further, *oxidised form* is less active than *reduced form* (Umbach et al. 1994). Studies carried out in fungi (Yukioka et al. 1998) and protozoan (Jarmuszkiewicz et al. 1997) reveal that active AOX occurs as monomeric enzyme, being insensitive to modulation by organic acids. Oligomeric difference between plants and non-plant organisms is not catalytically significant (Siedow and Umbach 2000).

From earliest immunoblotting studies in the 1980s with both poly- and monoclonal antibodies against AOX isolated from *Sauromatum guttatum* (Elthon and McIntosh 1987; Elthon et al. 1989), up to more recent studies in the 2000s, using powerful analytical tools such as electron paramagnetic resonance spectroscopy (EPR) (Berthold et al. 2002) and Fourier transform infrared spectroscopy (FTIR) (Maréchal et al. 2009), it has come a long way to get elucidate structure of AOX molecule. Structurally AOX corresponds to a protein with a nonheme diiron carboxylate centre, having iron ions (Fe^{2+} and Fe^{3+} mixed valent) bound to glutamate and histidine residues into a four-helix MSD (Albury et al. 2010). Based on EPR analysis and gene sequence data, two structural models of AOX have been proposed. The first one utilises few amino acid sequences and classifies AOX as a member of diiron family proteins, which also includes ribonucleotide reductase (RNR) R2 subunit and hydroxylase component of methane monooxygenase (Siedow et al. 1995). Moreover, the Siedow's model considers AOX as an integral mitochondrial inner membrane protein with two helical MSD connected by an interhydrophobic region at intermembrane space. Subsequently, Andersson and Nordlund (1999) from additional AOX sequences pointed out that this model is not in agreement with structural features of other well-characterised diiron proteins and then proposed another model of AOX structure. Although this second model also considers AOX as a diiron protein, it differs in precise ligation sphere of diiron centre. For instance, one of C-terminal Glu-*X*-*X*-His motifs identified by Siedow's study containing a Glu-270 (Siedow et al. 1995) appeared not to be fully conserved in newly identified sequences and consequently seemed unlikely to play a role to bind iron ions. Instead, the Andersson and Nordlund model uses a third Glu-*X*-*X*-His motif to coordinate iron. This third motif contains a Glu-217, which is located in the intermembrane space according to the Siedow's model. Since such a choice implies that transmembrane helices can no longer be retained, Andersson and Nordlund (1999) proposed that AOX is an interfacial rather than a transmembrane protein. However, modelling of AOX active site indicates that an interhelical region is required to form diiron centre (Albury et al. 2002). Meanwhile, current model considers AOX as a monotopic protein integrated into single lipid layer of the inner mitochondrial membrane (Albury et al. 2010). AOX catalytic site that reduces O_2 to H_2O and/or H_2O_2 also contains a binding site for O_2-reducing substrate, ubiquinol (UQH_2) (Moore et al. 1995). In addition, studies of site-directed mutagenesis support extensively this AOX model (Albury et al. 2002; Umbach et al. 2002; Crichton et al. 2010). On the other hand, it has been suggested that several amino acid residues situated at the matrix end of two membrane-spanning helices are key components to enzyme activity. Among these residues, Tyr-253 and Tyr-275 situated in close proximity to residues that have been identified through inhibitor (SHAM) resistance screens are most important. Tyr-253 becomes important in quinone binding and Tyr-275 in close spatial proximity to diiron centre (Berthold et al. 2000). Mutation studies have revealed that Tyr-275, unlike Tyr-253, is essential for AOX activity and it is potentially involved in electron transport (Albury et al. 2002). Thus, numerous studies conducted to date, have allowed obtain new insights to a better and more complete understanding on structure, catalytic activity and regulation of active site of mitochondrial AOXs from plants fungi and protists (Albury et al. 2009).

Plants are continuously exposed to changes in environmental conditions (stress) when they inhabit in nature (Searle and Turnbull 2011). Stress responses of plants include a wide variety of biochemical, physiological and morphological changes ranging from cell metabolism to whole plant growth. Stress-induced changes are dependent upon stress factor and plant age (Bako 2006), including regulation of gene expression (Baena-González 2010), restructuration of inner and outer shapes of target cells (Hilal et al. 1998), reduction of growth rate (Fiorani et al. 2005), decrease of photosynthetic activity (Chaves et al. 2009), alteration of carbon allocation (Pizarro and Bisigato 2010), synthesis of primary (soluble sugars, sugar alcohols, polyols, fructans, proline, glycine betaine, proteins) (Coueé et al. 2006) and secondary (flavonoids, anthocyanins, phenolics, lignin, tannins) (Edreva et al. 2008) metabolites and induction of adventitious organs, such as roots, shoots or hairs (Santos Macedo et al. 2012). In all plants, stress adaptations are energy-consuming processes, spatial and temporally associated to metabolic changes, which lead to an imbalance between carbon allocation and respiratory electron transport (Baena-González 2010). Thus, modulation of AOX activity can play a key role during reprogramming cellular metabolism to stress tolerance (Sieger et al. 2005; Clifton et al. 2006). In this way, it is known that AOX is involved in cellular homeostasis redox and flexibility of carbon balance of plants under stress conditions (Rasmusson et al. 2009). Also it has been reported that activation of AOX occurs during early stress period and can exert either positive or negative feedback effects on different signalling pathways to overcome environmental adversity (Arnholdt-Schmitt et al.

2006). Although AOX activation seems to be more frequent response of mitochondrial respiratory chain to environmental stresses (Clifton et al. 2006), there is also evidence that stress-induced loss of AOX activity has no dramatic effects (Giraud et al. 2008). For example, in *A. thaliana* a decreased AOX expression gives a 20 % growth reduction under low-temperature stress (Fiorani et al. 2005), while in tobacco cells reduced AOX expression prevents stress-induced cell death (Robson and Vanlerberghe 2002). By contrast, overexpression of AOX in a normal line of tobacco cells triggers ozone (O_3) sensitivity (Pasqualini et al. 2007). However, in a tobacco cell line with cytoplasmic male sterility, increased AOX enhances ozone tolerance (Dutilleul et al. 2003). Further, because under field conditions plants are more likely to be exposed simultaneously to various stresses instead of just one as occurs under controlled conditions, it can be expected that role of AOX becomes more important (Mittler 2006; Giraud et al. 2008). In fact, AOX mainly would serve to maintain cell energy charge and a constant growth under different environmental stresses (Moore et al. 2002). A common stress response is the production of ROS that leads to oxidative damage (Fujita et al. 2006; Gechev et al. 2006). Mitochondria and chloroplasts are main places of cellular ROS generation that increases due to stress-induced inhibition or over-reduction of ETC. Oxidative damage can affect different molecules such as lipids, proteins and nucleic acids, which can result in inhibition of important mitochondrial enzymes of respiratory chain such as NADH dehydrogenase complex (complex I) and succinate dehydrogenase (complex II) as well as ATP synthase (complex V) (Zhang et al. 1990). ROS-catalysed lipid peroxidation generates aldehydic molecules such as 4-hydroxy-2 nonenal (HNE) that is mostly cytotoxic product derived from lipid peroxidation (Uchida 2003). HNE inhibits activity of pyruvate dehydrogenase (PDC) and 2-oxoglutarate dehydrogenase (OGDC) through modification of lipoic acid residues (Millar and Leaver 2000). Moreover, uncontrolled ROS production can lead to cell death (Yao et al. 2002). An efficient mitochondrial mechanism to suppress ROS is carried out by antioxidant enzyme Mn superoxide dismutase (MnSOD). This enzyme catalyses dismutation of superoxide radical ($O_2^{\bullet-}$) into H_2O_2 and O_2, being toxic H_2O_2 neutralised by other two antioxidant enzymes: (1) ascorbate peroxidase (APX) and (2) glutathione peroxidase (GPX) (Chew et al. 2003; Navrot et al. 2006). Involvement of AOX in controlling ROS production under stress conditions is well demonstrated (Blokhina and Fagerstedt 2010; Gupta et al. 2012b). According to current knowledge, AOX possesses special functional features that convert it in an important enzyme involved in responses to oxidative stress. Furthermore, it was hypothesised that under stress conditions, AOX does not only act as a stress-responsive enzyme, because it also acts as stress-regulator enzyme and thus is involved in defining cellular stress responses (Arnholdt-Schmitt et al. 2006). Main evidences supporting AOX role in defining stress responses are as follows: (1) lack of AOX activity gives altered stress defence, (2) AOX expression is triggered by different signals, which indicates that it is a common stress response, and (3) AOX activity can dampen induction of PCD during mitochondrial dysfunction occurring in stressed plants (Robson and Vanlerberghe 2002; Van Aken et al. 2009a). Depending upon stress acting, signalling pathway can be retrograde when stress signal perturbs mitochondrial respiratory chain functions, which triggers subsequent signals to the nucleus, or anterograde when stress signal induces AOX expression without affecting mitochondrial function (Li et al. 2013). However, in *A. thaliana* plants recently it has been demonstrated that six *cis*-acting regulatory elements (CAREs) located in the region of promoter co-regulate both AOX and ND genes (Ho et al. 2008). This fact leads to assume that activities of mitochondrial enzymes are inseparable parts of an integrated functional system rather than a functional system based on individual activities. In agreement with this assumption, increases of AOX and ND activities lead to increased function of alternative pathway (AP) (Van Aken et al. 2009b). However, precise roles that play enzymes of AP in stressed plants are very difficult to establish because they also participate in key metabolic processes under non-stressed conditions (Rasmusson and Møller 2011).

Because AOX is a regulatory target of stress responses, many stress signals regulate its expression and synthesis (Millenaar and Lambers 2003; Vidal et al. 2007). AOX synthesis is regulated transcriptionally in response to stress and during plant development by several cellular conditions including elevated ROS level, organic acids and sugars (Wagner 1995; Millenaar et al. 2002). Post-transcriptional regulation of AOX occurs through oxidation/reduction of intermonomeric disulfide bonds (Umbach and Siedow 1993). A second AOX activation mechanism involves α-keto acids such as glyoxylate, hydroxypyruvate and notably pyruvate (Pastore et al. 2001). Activation by α-keto acids becomes significant when intermonomeric disulfide bonds are reduced. Moreover α-keto acids have less activating effect on oxidised AOX (Rhoads et al. 1998). Regulation of AOX by pyruvate has been questioned due to observation that normal concentration of pyruvate in vivo seems to be higher than pyruvate concentration required to give a maximal AOX stimulation in vitro (Millenaar et al. 1998). Furthermore, AOX activity also increases by increasing concentration of substrate (UQH_2) inside inner membrane (Moore and Siedow 1991). AOX activity is also regulated by pH. It has been demonstrated that pH effect depends upon plant species (Millenaar and Lambers 2003). In almost all analysed AOXs, optimum pH ranges between 6.4 and 7.2 (Elthon et al. 1986; Hoefnagel et al. 1997), reaching a relatively low value (6.25) in *Vigna unguiculata* (Lima-Júnior et al. 2000). Pyruvate activation is also pH dependent (Jarmuszkiewicz et al. 2002). In this way, in some plants optimum pH varies significantly in presence and absence of pyruvate (Lima-Júnior et al. 2000), whereas in others optimum pH is not affected by presence or absence of pyruvate (Hoefnagel et al. 1997). Indeed due to presence of H^+-pumping enzymes, pH of mitochondria will be basic with a value nearly 8. Thus, it is reasonable to assume that as soon as activity of H^+-pumping enzymes decreases by inhibition or low energy demand, a decrease in pH will occur, thereby activating AOX (Millenaar and Lambers 2003). Effect of pH on AOX activity under stress conditions has been tentatively attributed to a possible stress-induced pH signal that could induce a reversible protonisation/deprotonisation process of enzyme (Jarmuszkiewicz et al. 2002). In fact, identification of enigmatic pH signal could help understand conundrum on role of AOX in environmental changing conditions. In recent years, however, a great deal of progress has been made in understanding post-translational regulation of AOX due to discovery of a specific mitochondrial thioredoxin isoform (PtTrxh2), which has the capacity to reduce disulfide bonds between monomers of enzyme homodimer to give a free thiol more active AOX, also enabling pyruvate activation (Gelhaye et al. 2004; Martí et al. 2011; Yoshida et al. 2013). Extent that these biochemical traits can affect integration of all regulatory processes of AOX activity on total AP function under stress conditions is not yet fully known. Therefore, there are still many unresolved questions about functionality of mitochondrial AP in stressed plants (Millenaar and Lambers 2003). According to a generalised opinion, components of AP must act coordinately to avoid over-reduction of respiratory chain, helping to balance redox status of mitochondrial matrix and cytosol in response to stress signals (Vanlerberghe et al. 2009).

Regarding AOX-like proteins present in several animal phyla, both DNA sequence and gene expression analyses have conclusively established that AOX is present in at least four different animals. However, AOX-like animal genes were characterised from eight phyla, i.e. Chordata, Hemichordata, Echinodermata, Nematoda, Mollusca, Annelida, Cnidaria and Porifera (McDonald and Vanlerberghe 2004, 2006). Presence of AOX activity in animals is revealed as very significant because numerous animal organisms are exposed to potent inhibitors of COX activity such as CN, NO and sulphide (Greishaber and Völkel 1998; Cooper 2003). Moreover, these compounds can be absorbed and accumulated by animals due to their relative abundance in a particular environment or simply from food chain because they also can be normal components of different metabolic processes (Poulton 1990). Because AOX function decreases ATP yield, occurrence of AOX activity in animals could be disadvantageous under circumstances of high energy demand such as strong muscular activity (e.g. fast running), when comparing with immobile organisms, i.e. plants. However, ongoing researches suggest that expression of AOX is beneficial to animals, including those that are fast moving (e.g. flies) (Sanz et al. 2010). An AOX cDNA isolated from the chordate *Ciona intestinalis* Linnaeus (a sea squirt) was expressed in cultured human kidney cells and localised in mitochondria. This allotopically expressed AOX conferred a CN-resistant respiration to cells, turning it to an *n*PG-sensitive form, which indicates that expressed AOX is catalytically active (Hakkaart et al. 2006). Similarly, expression of this cDNA using a short hairpin RNA (shRNA) sequence in COX-deficient human cells either from COX_{15} mutation or COX_{10} silencing gene demonstrated that AOX expression was well tolerated by two cell types to compensate growth defect and enhanced oxidant sensitivity that exhibit COX-deficient human cells (Dassa et al. 2009). Innocuousness of this AOX and data obtained with common fruit fly (*Drosophila melanogaster*) (Fernández-Ayala et al. 2009) could lead to assume that expression of AOX in human cells could really be beneficial for a broad spectrum of disorders that affect mitochondrial respiratory chain. So, AOX would have the potential to circumvent or correct functional defects of one or several components of respiratory chain if they are induced by stress conditions or even having a genetic origin (Rustin and Jacobs 2009). On the other hand, presence of AOX activity in some pathogenic parasites (e.g. *Trypanosoma brucei*, *T. vivax*, *Cryptosporidium parvum*, *Blastocystis hominis*) has become interesting to human health. This has led that attention is focused on AOX as a potential drug target or to development of more effective chemotherapeutic drugs against pathogenic diseases but that in turn are less toxic to man and also animals (Roberts et al. 2004; Suzuki et al. 2004). Being occurrence of AOX activity in the Animalia kingdom a relatively recent event, there is no set of structural and functional data as big as that found to other eukaryotic organisms. Thus AOX-protein evolutive pathway, as well as phylogenetic relationships among AOX-containing organisms, is still a largely incomplete and unclear puzzle.

6 Role of AOX in Integrating Cell Metabolism Under Stress Conditions

AP that does not produce ATP is an energetically unfavourable pathway; however, it is widely distributed in practically all kingdoms of life suggesting that it plays a significant role in living organisms. Although there are still some doubts about real significance of AOX into respiratory metabolism under normal conditions, apparently this does not occur under stress conditions where AOX seems to play an important regulatory role to maintain electron flow through respiratory chain, and thus ensuring the maintenance of mitochondrial respiratory functionality as well as cellular homeostasis. In this sense, increased levels of AOX have been found in plants exposed to both abiotic and biotic stresses such as low temperature (Fiorani et al. 2005; Armstrong et al. 2008), drought (Bartoli et al. 2005; Ribas-Carbó et al. 2005; Wang and Vanlerberghe 2013), heavy metals (Castro-Guerrero et al. 2008; Prado et al. 2010; Panda et al. 2013), salinity (Smith et al. 2009; Martí et al. 2011) and pathogen attack (Hanqing et al. 2010; Liao et al. 2012; Cvetkovska and Vanlerberghe 2013). Moreover in soybean seedling roots, analyses of AOX expression indicate that AP through terminal AOX activity has an active participation as modulator of cellular redox status during root aging (Millar et al. 1998). One of most important functions of AOX under stress conditions is to limit endogenous ROS formation (Maxwell et al. 1999). Under stress conditions besides stress-induced increase of AP, a determined basal level of cytochrome pathway (CP) is always required. To explain this feature, it has been suggested that during stress an increased exchange of respiratory and TCA cycle substrates is occurring (Van Aken et al. 2009a). Also it was hypothesised that AOX participates actively to maintain mitochondrial O_2 homeostasis of stressed cells (McDonald and Vanlerberghe 2006). This can take place due to large difference in O_2 affinity between COX and AOX activities. K_m values for O_2 range from approximately 0.1 μM in COX to 1–20 μM in AOX (10- to 200-fold higher) (Gupta et al. 2009). Based on these K_m values, it is assumed that when a higher O_2 concentration occurs, AOX activity reduces its excess, without limiting O_2 supply to CP. Concomitantly to this O_2 decrease, mitochondrial ROS production is decreased (Skutnik and Rychter 2009). Indeed, potential of AOX to regulate both ROS generation and energy balance serves to maintain a constant growth under environmental stress conditions (Moore et al. 2002). In field-growing plants, leaves are frequently exposed to excessive sun radiation which may affect AP (Rosa et al. 2009a). Role of AP to protect photosynthetic electron transport chain (PETC) against harmful effects of excessive sunlight has been demonstrated in several plant species (Bartoli et al. 2005; Zhang et al. 2012). Agreeing with this fact in a yellow-variegated mutant of *A. thaliana* subjected to photooxidative stress, an increased expression of an AOX isoform (AOX1) has been demonstrated (Yoshida et al. 2008). This fact has attracted much attention in relation to interactions between AOX and photosynthesis because leaves are most important ROS generator under light excess (Zhang et al. 2012). In this context it has been reported that AOX inhibition in leaves of stressed plants causes over-reduction of PSII complex (Bartoli et al. 2005). Further biochemical studies demonstrated that AOX pathway optimised photosynthesis by dissipating excess of reducing equivalents from chloroplasts through (1) operation of specific metabolite shuttles related to cellular redox state (MAL–OAA shuttle) and ROS (AS–GSH cycle) and (2) modulating enzyme activities of the Calvin cycle and antioxidant metabolism (Raghavendra and Padmasree 2003).

Growth adaptation to changing environment is genetically determined; thus any genetic variation also will affect AP activity (Arnholdt-Schmitt et al. 2006). Therefore, adaptive cell programmes to different stresses depend upon differential AOX gene expression that also provides growth markers to stressed plants (Clifton et al. 2006). Different studies have shown that environment affects AP respiration linked to variable plant growth behaviour (Sieger et al. 2005), being differential activity of mitochondrial AOX crucial to optimise metabolic efficiency for a well-adapted regulation of growth and development (Yoshida et al. 2011). Further, a direct link between growing tissues and AOX activity was also reported (Hilal et al. 1998). On the other hand, AOX activity also modulates strength of mitochondrial stress signalling pathway (Vanlerberghe et al. 2009; Vanlerberghe 2013).

According to Arnholdt-Schmitt et al. (2006), *metabolic optimisation* refers to modulation of programmes existing in the cell in relation to their initiation and/or realisation, but it must be not assumed as *programme induction*, which corresponds to an earlier event. Because AOX activity is also involved in production of signalling molecules, i.e. H_2O_2, and NO (Tischner et al. 2004; Bhate and Ramasarma 2009), it remains to be elucidated whether AOX directly per se contributes to *programme induction* (Arnholdt-Schmitt et al. 2006). A multiplicity of AOX isoforms have been isolated from different plant species encoded by multigene families with five genes in *A. thaliana*, four genes in *Saccharum officinarum* (sugarcane) and *Lycopersicon esculentum* (tomato) and three genes in *Zea mays* (maize), *Oryza sativa* (rice), *Glycine max* (soybean) and *Vitis vinifera* (vitis) (Thirkettle-Watts et al. 2003; Ho et al. 2007; Costa et al. 2009). AOX genes can be expressed constitutively in non-stressed plants or inductively in stressed plants (Mhadhbi et al. 2013). In stressed plants, AOX would play two main functions: (1) maintenance of mitochondrial respiratory capacity and (2) preservation of mitochondria and chloroplasts functionality (Millenaar and Lambers 2003). Gene analyses show that AOX isoforms are expressed in different organs at different growth stages, which supports broad metabolic controls exerted by AOX during the plant growth, but not associated to oxidative stress (Atkin and Macherel 2009; van Dongen et al. 2011). Agreeing with these findings, previous microarray studies carried out with an AOX antisense line of *A. thaliana* showed that carbon metabolism outside of both mitochondrion and chloroplast compartments is depending of AOX activity (Umbach et al. 2005). Overall genetic studies of AOX indicate that differential expression of AOX genes is correlated with specific role that each gene has to fulfil for normal development or under stress condition, being also linked to metabolic pathway signals to integrate a *regulatory apparatus*, ultimately responsible to assess specificity of each expressed AOX isoform. Although available data allow support of all proposed physiological functions of AOX, today a simple model to unify all results is not available. Thus, it can be hypothesised that AOX may exert a ROS-based control of mitochondrial stress signalling pathway that allows to define signal strength that ultimately determines cell survival or death (Vanlerberghe et al. 2009). So, we can assume that AOX ultimately can either regulate or suppress ROS generation in non-stressed and stressed plants and thereby protect them against oxidative damage.

Researches in recent years have hinted importance of AP into metabolism of photosynthetic fixed carbon in both photosynthesising and non-photosynthesising tissues that are greater than once imagined (van Dongen et al. 2011; Prado et al. 2013). Worth mentioning, this process is not only suggested function of AP, through AOX activity, but is arguable that it is one most important. Inhibition of CP in stressed plants leads to increased $NADH/NAD^+$ ratio, whereas TCA cycle slows down (Araújo et al. 2012). This fact gives significant lower amount of available TCA intermediates (carbon skeletons) for metabolic processes (Fernie et al. 2004; Atkin and Macherel 2009). In these conditions AP via modulation of redox state of respiratory chain and ROS generation (Gupta et al. 2009) may maintain the carbon flux through TCA cycle to provide carbon skeletons for cellular biosynthetic processes involved in plant stress responses (Borecký and Vercesi 2005; Gandin et al. 2009).

7 Metabolic Interconnectivity Between AOX and Residual Respiration in Carbohydrate Metabolism Under Abiotic Stress

Despite well-known CP and AP respiratory pathways (Millar et al. 2011), in many plants during measurement of tissue respiration, addition of cyanide and SHAM (inhibitors of COX and AOX respirations) does not show a full inhibition of O_2 consumption (Atkin et al. 2002; Zottini et al. 2002; Prado et al. 2010). This enigmatic respiratory component has been called residual respiration (R_{resp}) (Ribas-Carbó et al. 1997). Although R_{resp} is known for many years (Johnson-Flanagan and Owens 1985), its origin still remains unclear. In a more recent work on this enigmatic respiration, it was found that an AOX-lacking leaf might induce a residual O_2-consuming process, which would not normally be seen in an AOX-containing leaf (Guy and Vanlerberghe 2005). However, R_{resp} is not observed in isolated mitochondria and is believed to be extramitochondrial in origin, possibly due to peroxisome/glyoxysome fatty acid oxidation (Møller et al. 1988). In agreement with this assumption, it has been demonstrated that in glyoxysomes of soybean cotyledons occurs a β-oxidation of fatty acids in which $FADH_2$ is oxidised directly by O_2 to produce H_2O_2 that is broken down by catalase (CAT) (Atkin et al. 2002). Because this process is not affected by respiratory inhibitors, it can be assumed that inhibitor-insensitive O_2-dependent peroxidation of fatty acids by lipoxygenase may contribute really to R_{resp} (Atkin et al. 2002). However, whether R_{resp} occurs in absence of SHAM and cyanide is unknown.

In almost all abiotic stresses, a clear trend toward increase of mitochondrial respiration has been found (Prado et al. 2011; Jacoby et al. 2012). This trend indicates that high respiration rates in stressed plants can be either beneficial or detrimental to growth rate, depending on plant tolerance strategy (Atkin and Macherel 2009). A high respiration rate signifies that more ATP is produced, providing vital energy for growth and defence processes (Zsigmond et al. 2008). By contrast, cost of high respiration rate is that carbon is expended on respiration instead of being allocated to biosynthetic pathways, leading to major alterations into carbohydrate metabolism (Flexas et al. 2006; Araújo et al. 2013). Thus, this fact can become limiting of growth capacity and tolerance efficiency (Krasensky and Jonak 2012). In many plants subjected to different environmental stresses like drought, cold, salinity and heavy metals, both carbohydrates and alternative respiration pathway play critical roles during growth and development (Ribas-Carbó et al. 2005). Stimulation of AP leads to an increased metabolic demand which allows consuming excess of carbon to correct imbalance between carbohydrate supply and demand, thus controlling anabolism and growth (Millenaar and Lambers 2003). Relationships between carbohydrate metabolism and AP have been extensively studied but have not been completely clarified. Whereas Millenaar et al. (2002) showed that relative contribution of AP respiration increased with decreasing sugar concentrations, Padmasree et al. (2002) demonstrated that AP is directly depending on level of soluble sugars. In contrary, González-Meler et al. (2001) reported that AP function is non-sugar dependent. It has been also reported that under stress condition, sucrose synthesis might act as an effective sink for excess of ATP through AP function (Solomos and Laties 1975). Moreover all sugar signalling pathways interact with stress pathways to modulate carbohydrate metabolism (Ho et al. 2001; Gupta and Kaur 2005). Thus, these findings tend to indicate that carbohydrate metabolism and AP are coupled and integrated processes, which are necessary to support both catabolic and anabolic pathways occurring in plants exposed to changing environmental conditions (Sieger et al. 2005).

In this context, it was demonstrated that abiotic stresses such as heavy metals and mineral deficiency reduce both glycolysis rate and pyruvate content, being accompanied by an increased contribution of the oxidative pentose phosphate pathway (OPPP) and R_{resp} (Rakhmankulova et al. 2003; Prado et al. 2013). In these conditions, plants either can allocate the fixed carbon to synthesis of protective metabolites, i.e. proline, sugar alcohols, polyols, soluble sugars and glycine betaine, within the primary metabolism (Rathinasabapathi 2000), or channelise it toward secondary metabolism to enhance synthesis of protective secondary metabolites (Seigler 1998). However, how stress factors AP and R_{resp} are involved in both tolerance of plants and synthesis of secondary metabolites remains still largely unknown. It has been suggested that dissipation of redox-equivalent excess generated during photosynthesis occurs through CP and AP pathways, while sucrose synthesis is backed up by CP alone. Indeed, mitochondrial respiration through both CP and AP pathway can optimise photosynthetic carbon assimilation by modulating level of cellular metabolites related to intracellular redox status (triose phosphate, phosphoglyceric acid, malate and oxaloacetate) (Osorio et al. 2013) and sucrose biosynthesis status (glucose-6-phosphate) (Padmasree et al. 2002). In summary, it could be assumed that in stressed plants, when CP decreases, AP, through AOX activity linked to R_{resp}, must have a key role in the regulation of intracellular concentration of soluble carbohydrates, as well as in redirectioning of carbon metabolism to enhance cellular defences against occurring stress. In agreement with this assumption in a recent study carried out in our laboratory, we showed that values of alternative respiration pathway to total respiration (AP/T_{resp}) ratio and R_{resp} activity in Cr-stressed leaves of *Salvinia minima* plants denote clearly an increasing relative contribution of AP and R_{resp} to stress defence with increasing Cr(VI) concentrations in growth solution (Prado et al. 2013). Although exact role of R_{resp} still remains unclarified, there are some evidences linking it to drought tolerance (Shugaeva et al. 2007), as well as in channelisation of carbon skeletons toward secondary metabolism under mineral deficiency (Rakhmankulova et al. 2003). Regarding the latter, both AOX and R_{resp} activities have been related to synthesis of highly polymerised phenylpropanoid-derived compounds, i.e. suberin and lignin (Johnson-Flanagan and Owens 1985; Santos Macedo et al. 2012). Of interest, in a well-conduced study, a member of our laboratory staff (M. Hilal) and co-workers demonstrated by the first time, by using an anti-AOX monoclonal antibody, a similar location for both lignin deposition and synthesised AOX protein in cross-sections from differentiation zone of soybean seedling roots exposed to saline stress (Hilal et al. 1997). Phenylpropanoid-derived compounds are frequently named secondary metabolites and often are referred to as compounds that have no fundamental role in maintenance of life processes in plants. However, secondary metabolites are important to plant life due to facilitating interactions between plants and environment giving them adaptation and defence (Ramakrishna and Ravishankar 2011). Among secondary metabolites, soluble and polymerised phenolics are most broadly distributed in higher plants, and they are synthesised in response to many environmental stresses, such as UV-B radiation, wounding, salinity, drought, nutrient deficiency, high light, low temperature, heavy metals, pathogen attack and herbicide treatment (Dixon and Paiva 1995; Prado et al. 2011). Lignin is most important polymerised phenolic in vascular plants, being a heteropolymer occurring in secondary cell walls (Boerjan et al. 2003). Main components of lignin are monolignols (*p*-hydroxycinnamyl alcohol monomers), i.e. coniferyl, sinapyl and *p*-coumaryl alcohols. These monolignols once incorporated into lignin-growing polymer originate three different lignin types: (1) guaiacyl lignin (G, main component coniferyl alcohol), (2) syringyl lignin (S, main component sinapyl alcohol), and (3) graminaceous lignin (H, main component *p*-coumaryl alcohol). In softwoods (gymnosperms and ferns), G type predominates, whereas G and S types are present in hardwoods (angiosperms). H type occurs in grass or annual plants (Vanholme et al. 2010). Suberin is another polyphenolic compound that increases under stress conditions

(Pollard et al. 2008). Normally, it is a cell wall constituent found throughout plant, mainly in cell walls of external and internal tissues such as Casparian strip in root endodermis and cork cells of periderm. Structurally, suberin is a complex lipophilic polymer, containing long-chain acids, hydroxy acids and long-chain aliphatic alcohols. It is composed by a hydrophobic fatty acid-derived domain (aliphatic suberin) and a *p*-hydroxycinnamic domain (aromatic suberin). These domains are covalently linked by ester bonds. Moreover, aromatic domain is covalently linked to structural cell wall polysaccharides (Ranathunge et al. 2011). In almost all abiotic stresses, formation of lignin and suberin occurs through well-characterised peroxidase (POX)-mediated and H_2O_2-dependent processes (Ros Barceló 1997; Bernards et al. 2004; Ralph et al. 2004). Although acting POXs can catalyse oxidation of *p*-coumaryl, coniferyl and sinapyl monolignols (Boerjan et al. 2003), suberin-related enzymes show a marked substrate preference for *p*-hydroxycinnamates (e.g. ferulic acid and derivatives) (Almagro et al. 2009). Both processes catalysing POXs correspond to apoplastic H_2O_2-consuming guaiacol peroxidase type (G-POX), in both anionic and cationic forms (Fecht-Christoffers et al. 2006). Regarding origin of H_2O_2 consumed during oxidative polymerisation reactions, available data are not totally clear being assumed that it can be generated by different mechanisms (Karlsson et al. 2005). Although many studies indicate that this H_2O_2 is provided by a membrane-bound NADPH oxidase (NOX) that catalyses formation of $O_2^{\bullet-}$ which spontaneously produces H_2O_2 (Ros Barceló et al. 2002), some others indicate that this conversion may be mediated by a tissue-specific cupro-zinc superoxide dismutase (CuZn-SOD) located in cell walls that catalyse enzymatic dismutation of $O_2^{\bullet-}$ in O_2 and H_2O_2 (Karlsson et al. 2005) or, alternatively, by another dismutase enzyme, i.e. MnSOD (Corpas et al. 2006). Moreover, other enzyme-based H_2O_2-generating systems include cell wall cationic POXs (Almagro et al. 2009), poly(di) amine oxidases (Angelini and Federico 1989) and oxalate oxidase (Hu et al. 2003). Hydrogen peroxide can also be produced non-enzymatically during auto-oxidation of apoplastic phenolics (Takahama et al. 2001).

Despite well-studied mechanisms on origin of H_2O_2 involved in POX-mediated synthesis of suberin and lignin, its generation seems to be related to R_{resp} activity (Johnson-Flanagan and Owens 1985) and also, probably, with AOX activity (Santos Macedo et al. 2012). In agreement with these findings, we observed significant increases of both R_{resp} and AOX activities, in consonance with increased contents of soluble and polymerised phenolics in floating and submerged leaves of *S. minima* exposed to increasing Cr(VI) concentrations (Prado 2012; Prado et al. 2013). We also observed in floating leaves of Cr-stressed plants positive correlations between soluble phenolics content, G-POX activity, generation of H_2O_2 and AOX and R_{resp} activities (Prado 2012; Prado et al. 2013). Regarding our findings Yamasaki et al. (1997) had previously proposed that flavonols inside vacuole participate in a POX–flavonol quenching system to reduce intravacuolar imported H_2O_2. Since a high amount of H_2O_2 is needed to cell wall lignification, POX–flavonol H_2O_2 scavenger system can also become important during normal plant growth. Therefore it can be assumed that phenolics also play an important role as cellular non-enzymatic antioxidants (Skórzyńska-Polit et al. 2004). Noteworthy, AOX-induced H_2O_2 production till now is matter of a controversial debate (Bhate and Ramasarma 2009; Møller et al. 2010).

It is well known that in plants, carbohydrate disturbances, mainly sucrose, produced by abiotic stresses lead to alterations of carbon allocation to energy generation and provision of carbon skeletons (Devi et al. 2007; Stobrawa and Lorenc-Plucińska 2007; Rosa et al. 2009a). Since sucrose is the link between assimilated photosynthetic carbon in chloroplasts and metabolic carbon consumption distant in space and time from them, understanding how environmental stresses can affect both the AP and sucrose metabolism at cell level becomes very important to elucidate mechanisms of stress tolerance. Sucrose metabolism is controlled mainly by three enzymes: (1) sucrose synthase (SS), (2) invertase (INV) and (3) sucrose phosphate synthase (SPS) (Winter and Huber 2000). The first two are responsible of sucrose hydrolysis and latter of its synthesis. Sucrose hydrolysis catalysed by INV has a higher ATP cost than SS-catalysed hydrolysis. INV-derived products require two molecules of ATP to produce two hexose phosphates, while SS-derived products require one ATP to give one hexose phosphate, while another reaction product, i.e. uridine diphosphate glucose (UDPG), serves to conserve energy of glycosidic binding (Geigenberger 2003). Resulting UDPG either can serve as substrate for both cellulose and callose synthesis or participate in carbohydrate moiety synthesis of glycolipids and glycoproteins and even in sucrose resynthesis (Kleczkowski et al. 2004). Since increased AP leads to a decreased ATP generation (Yoshida et al. 2006), it can be assumed that SS activity (soluble and membrane bound) must play main role into sucrose metabolism of stressed plants. Similarly, it has been communicated that decreased glucose with an increase fructose reflects a low INV/SS ratio that is considered an adaptive mechanism to save cellular energy (Shi et al. 2008). In agreement with these findings, obtained results in Cr-stressed leaves of *S. minima* plants showed increase and decrease in both SS and INV activities, as well as a decreased glucose concentration with significant increase of fructose content (Prado et al. 2013). On the other hand, it has been proposed that SS activity alone can maintain a constant sucrose concentration in *Erythronium americanum* bulbs growing under elevated CO_2 and O_3 concentrations, independent of variable amount of carbohydrates that is translocated to bulbs (Gandin et al. 2009). Soluble SS activity has also been correlated with starch synthesis and overall sink strength (Geigenberger 2011). In this sense, it was hypothesised that SS through regulation of glycolytic intermediate pool, and AP through modulation of AOX activity by pyruvate, can adjust accumulation of starch in rhythm with sink growth capacity (Gandin et al. 2009). However, increased SS-dependent starch synthesis is not a common response of plants to environmental stress (Rosa et al. 2009b), and thus it can be assumed that in both synthesis and mobilisation of starch, different metabolic pathways are operating (Devi et al. 2007; Prado 2012). Interestingly, a mitochondrial SS type (M SS) recently discovered seems to have no sucrolytic activity (Subbaiah et al. 2006), and thus its role becomes a metabolic conundrum until now. Overlapping SS activity, plants possess three well-known INV types which according to their location are named apoplastic (CW INV), cytosolic (C INV) and vacuolar (V INV), respectively (Roitsch and González 2004). To three INV types, recently, a fourth type, i.e. A/N INV, located in both mitochondria (M INV) and chloroplasts (P INV) was added (Szarka et al. 2008; Vargas et al. 2008). Available data indicate that all INVs participate in important metabolic pathways, and also in carbohydrate partitioning regulation due to both substrate and reaction products are simultaneously metabolites and signal molecules in stressed- and non-stressed plants (Sturm and Tang 1999; Essmann et al. 2008; Ruan et al. 2010). In this way, it has been demonstrated that CW INV, C INV and V INV, individually or in combination with phytohormones such as cytokinin, auxin, gibberellin, abscisic acid and ethylene, are involved in many regulatory processes related to growth, carbon allocation, development and secondary metabolism, among most important (Proels and Roitsch 2009; Ruan et al. 2010). Circumstantial support of such functions is evident in apparent capacity of INVs to amplify effects of sugars on gene expression. Moreover, repression of photosynthetic genes is associated with elevated INV activity (Koch 2004). Furthermore, depending upon INV-acting type, INV activity could also alter its regulatory effect on sugar-sensing system related to SS activity. Indeed if plasma membrane can signal sugar transfer to sugar-sensing systems, then hydrolysis of sucrose catalysed by CW INV has the potential to double sugar source input (twofold more hexose molecules than that SS-catalysed hydrolysis). On the other hand, in tissues where V INV is an important contributor to sucrose cleavage, there can be a significant temporal delay in input to metabolic sources of signals, and also there can be differences in compartmentalisation of glucose versus fructose (Koch et al. 1996). Sucrose being a stimulator of SS expression and activity, and considering that INV activity is modulated by hexose pool (Keurentjes et al. 2008), leads one to think that two enzymes can affect differentially the sugar-sensing system (Fig. 12.2).

Sucrose-synthesising enzyme, SPS, plays an important role in carbohydrate metabolism because it controls both sucrose synthesis and sucrose ↔ starch interconversion in response to changes in sucrose demand imposed by stress conditions (Geigenberger et al. 1999). In contrast to SS and INV, SPS activity seems to be, in general, more affected by environmental stresses than the first ones (Prado et al. 2011). Indeed, it decreased under heavy metal stress (Mishra and Dubey 2008) whereas increased under low temperature (Hurry et al. 1994), salinity (Rosa et al. 2009b), hypoxia (Albrecht and Mustroph 2003) and drought (Fresneau et al. 2007). Moreover under drought (Fresneau et al. 2007) and heavy metal, i.e. Cr(VI) (Prado et al. 2013) stresses, SPS activity seems to be affected differentially either with increases or decreases, according to plant species (drought) or plant tissue (Cr). Lack of positive correlations between SPS, SS and INV activities in both photosynthetic and non-photosynthetic organs of Cr-stressed *S. minima* plants leads us to assume that a complex set of metabolic and gene expression interactions regarding sucrose biosynthesis and carbon allocation takes place in stressed plants (Prado et al. 2013). Together, previous evidences and our data would suggest that both AOX and R_{resp} may serve to channel excess of carbohydrates toward phenylpropanoid pathway in stressed plants (Fig. 12.3).

High concentrations of soluble sugars, different from most common sucrose, glucose and fructose, have also been found in stressed plants. Monosaccharides (mannose, galactose and glucose), disaccharides (trehalose, maltose and melibiose) and oligosaccharides (raffinose, stachyose, verbascose), as well as fructose-derived oligo- and polysaccharides such as kestose, neokestose, inulin and levan, may also

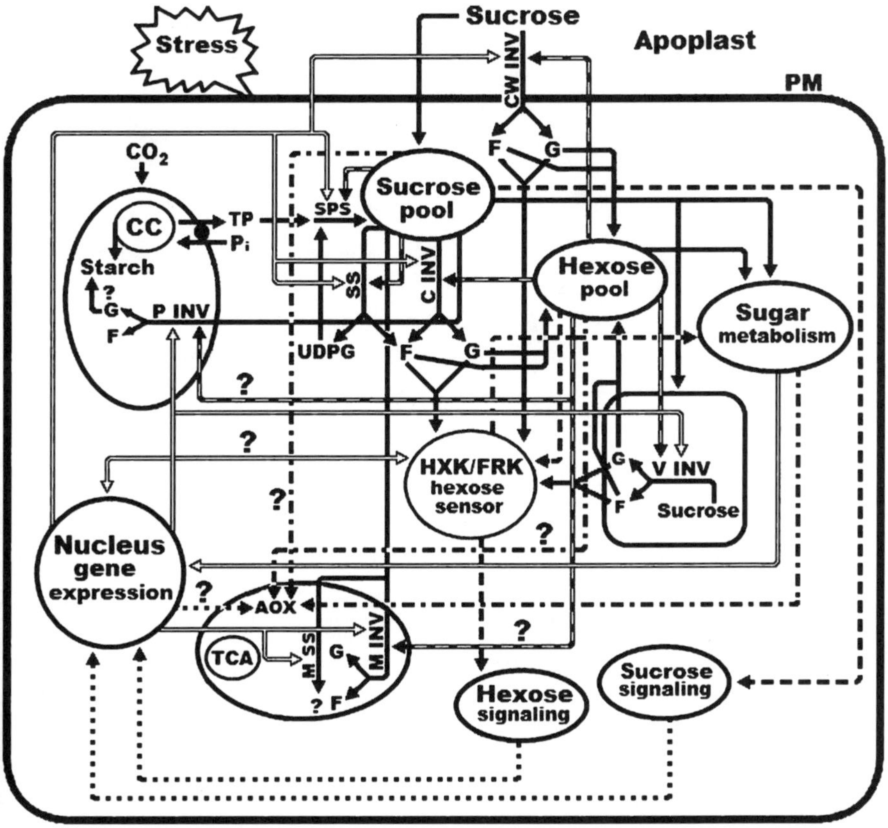

Fig. 12.2 Simplified model of sucrose and hexose sensing and signal pathways in stressed plants. Model also includes regulating relationships among sucrose-related enzymes, i.e. INV, SS and SPS with both sucrose and hexose pools, as well as possible crosstalk between AOX with sugars, sugar sensing and sugar metabolic regulatory pathways. Although both hexokinase (HXK) and fructokinase (FRK) sensors modulate sugar signalling through different and independent pathways (Granot et al. 2013), in this hypothetical model, they were considered together to facilitate interpretation. *G* glucose, *F* fructose, *FRK* fructokinase, *HXK* hexokinase, *C INV* cytosolic invertase, *CW INV* cell wall invertase, *M INV* mitochondrial invertase, *M SS* mitochondrial sucrose synthase, *P INV* plastidial invertase, *SS* sucrose synthase, *SPS* sucrose phosphate synthase, *UDPG* uridine diphosphate glucose. *Solid line*, sucrose and hexoses metabolic and sensing interrelationships; *dashed line*, sugars signalling pathways; *dashed double line*, sucrose- and hexose-dependent enzymatic regulatory pathways; *double solid line*, gene expression pathways; *dotted* and *dashed line*, known and probable sugar regulatory pathways of AOX activity; *dotted line*, probable sugar-sensing pathways involved in gene expression related to AOX

play important roles in plant defence (Gilbert et al. 1997; Rizhsky et al. 2004; Livingston et al. 2009). Positive effects of exogenous trehalose to reduce photo-oxidative damage occur through interactions of this sugar with both sucrose metabolism that is significantly modified (Bae et al. 2005) and sugar-dependent

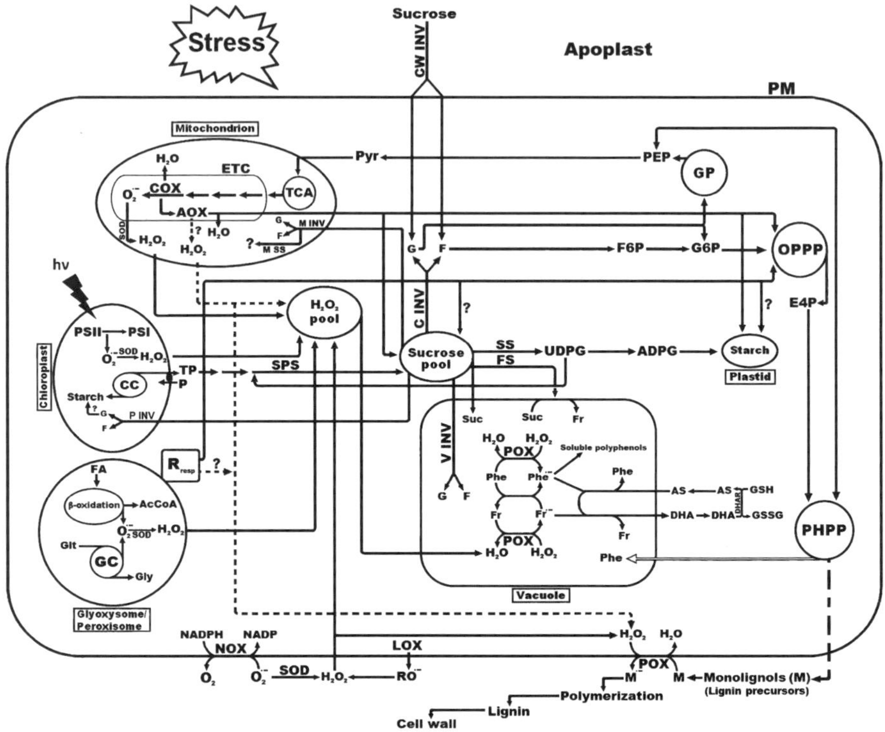

Fig. 12.3 Schematic model depicting possible involvement of AOX and R_{resp} into carbohydrate metabolism and secondary metabolites synthesis in plants exposed to abiotic stresses. Model also indicated main enzymes related to sucrose metabolism as well as typical sites of H_2O_2 production and involvement of vacuolar phenolics and fructans as ROS scavenger molecules. Model is based on available data and our own results. *AcCoA* acetyl coenzyme A, *ADPG* adenosine diphosphate glucose, *AS* ascorbic acid, *COX* cytochrome *c* oxidase, *DHA* dehydroascorbic acid, *DHAR* dehydroascorbic acid reductase, *E4P* erythrose-4-phosphate, *F6P* fructose-6-phosphate, *FA* fatty acid, *Fr* fructan, $Fr^{\bullet-}$ fructan free radical, *FS* fructan synthase, *G6P* glucose-6-phosphate, *GC* glyoxylic acid cycle, *GP* glycolytic pathway, *Glt* glycolate, *Gly* glycine, *GSH* reduced glutathione, *GSSG* oxidised glutathione, *LOX* lipoxygenase, $M^{\bullet-}$ monolignol free radical, *NOX* NADPH oxidase, *OPPP* oxidative pentose phosphate pathway, $O_2^{\bullet-}$ superoxide free radical, *PEP* phosphoenolpyruvic acid, *Phe* phenolic, $Phe^{\bullet-}$ phenoxyl free radical, *PHPP* phenylpropanoid pathway, *POX* peroxidase, *Pyr* pyruvic acid, $RO^{\bullet-}$ carbonyl free radical, *SOD* superoxide dismutase. *Solid line*, metabolic pathways; *dashed line*, probable involvement of AOX and R_{resp} in lignin synthesis

signalling pathways (Avonce et al. 2004). Similarly, mannose and galactose directly participate in the synthesis of antioxidant ascorbic acid (Szarka et al. 2013). These findings coupled to new insights into stress signalling pathways and activities of M INV and P INV, as well as enigmatic M SS, have increased complexity of metabolic and signalling network between AP through AOX activity and R_{resp}, occurring in plants exposed to different environmental stresses (Fig. 12.2).

8 Conclusion and Future Prospects

Organisation of cooperating enzymes into multienzyme complexes (metabolons) such as Calvin cycle, TCA cycle and phenylpropanoid biosynthetic pathway is a pervasive feature of metabolism (Jørgensen et al. 2005). A number of functional and regulatory roles are facilitated by organisation of metabolism into macromolecular complexes. These complexes can facilitate direct flux between branch pathways that compete for common metabolites. In particular, this mechanism participates in regulatory network of fluxes, *metabolic channels*, between two plant metabolisms (Winkel 2004). Therefore, it is still a difficult question to know whether organisation and regulation of such *metabolic channels* is a temporary or persistent event, so as to understand their functionality and importance into stressed cells. Therefore, despite made research effort, there remain many open questions regarding relationships between AP and R_{resp}. To answer these questions, an enormous biochemical and physiological work will be necessary made to define key regulatory steps into metabolic pathways and also specify roles of both constitutive and inductive enzymes in these pathways to provide a fuller understanding of metabolic interactions occurring under changing environmental conditions. It is noted that to get these realisations fully, it will also be necessary to obtain a complete elucidation of expression and regulation of genes corresponding to proteins involved in these metabolic processes. Also analyses of mitochondrial retrograde signals to control extra-mitochondrial gene expression can also serve to obtain new insights on regulatory aspects of mitochondrial respiratory relationships. Assessing of cellular and organellar metabolic fluxes in relation to AOX and R_{resp} activities will represent an important research frontier for next years.

> To reach light into intricate and complex tunnel of plant metabolism and decipher the stress conundrum, will be need to travel a long way yet.
>
> .... to victory always! (Ernesto Guevara)

References

Abas L, Bogoyevitch MA, Guppy M (2000) Mitochondrial ATP production is necessary for activation of the extracellular-signal regulated kinases during ischaemia/reperfusion in rat myocyte-derived H9c2 cells. Biochem J 349:119–126

Abdelnoor RV, Yule R, Elo A, Christensen AC, Meyer-Gauen G, Mackenzie SA (2003) Substoichiometric shifting in the plant mitochondrial genome is influenced by a gene homologous to MutS. Proc Natl Acad Sci U S A 100:5968–5973

Adams KL, Palmer JD (2003) Evolution of mitochondrial gene content: gene loss and transfer to the nucleus. Mol Phylogenet Evol 29:380–395

Alberts B, Johnson A, Lewis J, Raff M, Roberts K, Walter P (2002) Molecular biology of the cell, 4th edn. Garland Science, New York

Albrecht G, Mustroph A (2003) Localization of sucrose synthase in wheat roots: increased in situ activity of sucrose synthase correlates with cell wall thickening by cellulose deposition under hypoxia. Planta 217:252–260

Albury MS, Affourtit C, Crichton PG, Moore AL (2002) Structure of the plant alternative oxidase. Site-directed mutagenesis provides new information on the active site and membrane topology. J Biol Chem 277:1190–1194

Albury MS, Elliott C, Moore AL (2009) Towards a structural elucidation of the alternative oxidase in plants. Physiol Plant 137:316–327

Albury MS, Elliott C, Moore AL (2010) Ubiquinol-binding site in the alternative oxidase: mutagenesis reveals features important for substrate binding and inhibition. Biochim Biophys Acta 1797:1933–1939

Almagro L, Gómez Ros LV, Belchi-Navarro S, Bru R, Ros Barceló A, Pedreño MA (2009) Class III peroxidases in plant defence reactions. J Exp Bot 60:377–390

Almeida AM, Jarmuszkiewicz W, Khomsi H, Arruda P, Vercesi AE, Sluse FE (1999) Cyanide-resistant, ATP-synthesis-sustained, and uncoupling-protein-sustained respiration during post-harvest ripening of tomato fruit. Plant Physiol 119:1323–1329

Almeida AM, Navet R, Jarmuszkiewicz W, Vercesi AE, Sluse-Goffart CM, Sluse FE (2002) The energy-conserving and energy-dissipating processes in mitochondria isolated from wild type and nonripening tomato fruits during development on the plant. J Bioenerg Biomembr 34:487–498

Alverson AJ, Wei X, Rice DW, Stern DB, Barry K, Palmer JD (2010) Insights into the evolution of mitochondrial genome size from complete sequences of *Citrullus lanatus* and *Cucurbita pepo* (*Cucurbitaceae*). Mol Biol Evol 27:1436–1448

Alverson AJ, Rice DW, Dickinson S, Barry K, Palmer JD (2011) Origins and recombination of the bacterial-sized multichromosomal mitochondrial genome of cucumber. Plant Cell 23:2499–2513

Andersson ME, Nordlund P (1999) A revised model of the active site of alternative oxidase. FEBS Lett 449:17–22

Angelini R, Federico R (1989) Histochemical evidence of polyamine oxidation and hydrogen peroxide generation in the cell wall. J Plant Physiol 135:212–217

Araújo WL, Nunes-Nesi A, Nikoloski Z, Sweetlove LJ, Fernie AR (2012) Metabolic control and regulation of the tricarboxylic acid cycle in photosynthetic and heterotrophic plant tissues. Plant Cell Environ 35:1–21

Araújo WL, Nunes-Nesi A, Fernie AR (2013) On the role of plant mitochondrial metabolism and its impact on photosynthesis in both optimal and sub-optimal growth conditions. Photosynth Res. doi:10.1007/s11120-013-9807-4

Arimura S, Yamamoto J, Aida GP, Nakazono M, Tsutsumi N (2004) Frequent fusion and fission of plant mitochondria with unequal nucleoid distribution. Proc Natl Acad Sci U S A 101:7805–7808

Armstrong AF, Badger MR, Day DA, Barthet MM, Smith PMC, Millar AH, Whelan J, Atkin OK (2008) Dynamic changes in the mitochondrial electron transport chain underpinning cold acclimation of leaf respiration. Plant Cell Environ 31:1156–1169

Arnholdt-Schmitt B, Costa JH, Fernandes de Melo D (2006) AOX—a functional marker for efficient cell reprogramming under stress? Trends Plant Sci 11:282–287

Atkin OK, Macherel D (2009) The crucial role of plant mitochondria in orchestrating drought tolerance. Ann Bot 103:581–597

Atkin OK, Zhang Q, Wiskich JT (2002) Effect of temperature on rates of alternative and cytochrome pathway respiration and their relationship with the redox poise of the quinone pool. Plant Physiol 128:212–222

Avonce N, Leyman B, Mascorro-Gallardo JO, Van Dijck P, Thevelein JM, Iturriaga G (2004) The Arabidopsis trehalose-6-P synthase AtTPS1 gene is a regulator of glucose, abscisic acid, and stress signaling. Plant Physiol 136:3649–3659

Backert S, Nielsen BL, Börner T (1997) The mystery of the rings: structure and replication of mitochondrial genomes from higher plants. Trends Plant Sci 2:477–483

Bae H, Herman E, Bailey B, Bae HJ, Sicher R (2005) Exogenous trehalose alters *Arabidopsis* transcripts involved in cell wall modification, abiotic stress, nitrogen metabolism, and plant defense. Physiol Plant 125:114–126

Baena-González EB (2010) Energy signaling in the regulation of gene expression during stress. Mol Plant 3:300–313

Bako SP (2006) Effects of plant age, ascorbate and kinetin applications on integrity of the photosynthetic pigment complex in maize (*Zea mays* L.) plants grown under heat stress. Asian J Plant Sci 5:357–362

Ballard JWO, Whitlock MC (2004) The incomplete natural history of mitochondria. Mol Ecol 13:729–744

Bartoli CG, Gomez F, Gergoff G, Guiamét JJ, Puntarulo S (2005) Up-regulation of the mitochondrial alternative oxidase pathway enhances photosynthetic electron transport under drought conditions. J Exp Bot 56:1269–1276

Benard G, Faustin B, Passerieux E, Galinier A, Rocher C, Bellance N, Delage JP, Casteilla L, Letellier T, Rossignol R (2006) Physiological diversity of mitochondrial oxidative phosphorylation. Am J Physiol Cell Physiol 291:C1172–C1182

Bernards MA, Summerhurst DK, Razem FA (2004) Oxidases, peroxidases and hydrogen peroxide: the suberin connection. Phytochem Rev 3:113–126

Berthold DA, Andersson ME, Nordlund P (2000) New insight into the structure and function of the alternative oxidase. Biochim Biophys Acta 1460:241–254

Berthold DA, Voevodskaya N, Stenmark P, Gräslund A, Nordlund P (2002) EPR studies of the mitochondrial alternative oxidase—evidence for a diiron carboxylate center. J Biol Chem 277:43608–43614

Bhate RH, Ramasarma T (2009) Evidence for H_2O_2 as the product of reduction of oxygen by alternative oxidase in mitochondria from potato tubers. Arch Biochem Biophys 486:165–169

Bianchi C, Genova ML, Parenti Castelli G, Lenaz G (2004) The mitochondrial respiratory chain is partially organized in a supercomplex assembly: kinetic evidence using flux control analysis. J Biol Chem 279:36562–36569

Blokhina O, Fagerstedt KV (2010) Reactive oxygen species and nitric oxide in plant mitochondria: origin and redundant regulatory systems. Physiol Plant 138:447–462

Boekema EJ, Braun HP (2007) Supramolecular structure of the mitochondrial oxidative phosphorylation system. J Biol Chem 282:1–4

Boerjan W, Ralph J, Baucher M (2003) Lignin biosynthesis. Annu Rev Plant Biol 54:519–546

Bonen L (2008) *Cis*- and *trans*-splicing of group II introns in plant mitochondria. Mitochondrion 8:26–34

Boore JL (1999) Animal mitochondrial genomes. Nucleic Acids Res 27:1767–1780

Borecký J, Vercesi AE (2005) Plant uncoupling mitochondrial protein and alternative oxidase: energy metabolism and stress. Biosci Rep 35:271–286

Braun HP, Schmitz UK (1995) Are the 'core' proteins of the mitochondrial bc_1 complex evolutionary relics of a processing protease? Trends Biochem Sci 20:171–175

Bullerwell CE, Gray MW (2004) Evolution of the mitochondrial genome: protist connections to animals, fungi and plants. Curr Opin Microbiol 7:528–534

Carrie C, Murcha MW, Giraud E, Ng S, Zhang MF, Narsai R, Whelan J (2012) How do plants make mitochondria? Planta. doi:10.1007/s00425-012-1762-3

Castandet B, Araya A (2011) RNA editing in plant organelles. Why make it easy? Biochemistry (Mosc) 76:924–931

Castro-Guerrero NA, Rodríguez-Zavala JS, Marín-Hernández A, Rodríguez-Enríquez S, Moreno-Sánchez R (2008) Enhanced alternative oxidase and antioxidant enzymes under Cd^{2+} stress in *Euglena*. J Bioenerg Biomembr 40:227–235

Chance B, Williams GR (1955) A method for the localization of sites for oxidative phosphorylation. Nature 176:250–254

Chang S, Wang Y, Lu J, Gai J, Li J, Chu P, Guan R, Zhao T (2013) The mitochondrial genome of soybean reveals complex genome structures and gene evolution at intercellular and phylogenetic levels. PLoS One 8:e56502. doi:10.1371/journal.pone.0056502

Chaves M, Flexas J, Pinheiro C (2009) Photosynthesis under drought and salt stress: regulation mechanisms from whole plant to cell. Ann Bot 103:551–560

Chew O, Whelan J, Millar AH (2003) Molecular definition of the ascorbate-glutathione cycle in Arabidopsis mitochondria reveals dual targeting of antioxidant defenses in plants. J Biol Chem 278:46869–46877

Clifton R, Millar AH, Whelan J (2006) Alternative oxidases in Arabidopsis: a comparative analysis of differential expression in the gene family provides new insights into function of non-phosphorylating bypasses. Biochim Biophys Acta 1757:730–741

Collins TJ, Berridge MJ, Lipp P, Bootman MD (2002) Mitochondria are morphologically and functionally heterogeneous within cells. EMBO J 21:1616–1627

Cooper CE (2003) Competitive, reversible, physiological? Inhibition of mitochondrial cytochrome oxidase by nitric oxide. IUBMB Life 55:591–597

Corpas FJ, Fernández-Ocaña A, Carreras A, Valderrama R, Luque F, Esteban FJ, Rodríguez-Serrano M, Chaki M, Pedrajas JR, Sandalio ML, del Río LA, Barroso JB (2006) The expression of different superoxide dismutase forms is cell-type dependent in olive (*Olea europaea* L.) leaves. Plant Cell Physiol 47:984–994

Cortassa S, O'Rourke B, Winslow RL, Aon MA (2009) Control and regulation of integrated mitochondrial function in metabolic and transport networks. Int J Mol Sci 10:1500–1513

Costa JH, Fernandes de Melo D, Gouveia Z, Cardoso HG, Peixe A, Arnholdt-Schmitt B (2009) The alternative oxidase family of *Vitis vinifera* reveals an attractive model to study the importance of genomic design. Physiol Plant 137:553–565

Coueé I, Sulmon C, Gouesbet G, El Amrani A (2006) Involvement of soluble sugars in reactive oxygen species balance and responses to oxidative stress in plants. J Exp Bot 57:449–459

Crichton PG, Albury MS, Affourtit C, Moore AL (2010) Mutagenesis of the plant alternative oxidase reveals features important for oxygen binding and catalysis. Biochim Biophys Acta 1797:732–737

Cvetkovska M, Vanlerberghe GC (2012) Alternative oxidase modulates leaf mitochondrial concentrations of superoxide and nitric oxide. New Phytol 195:32–39

Cvetkovska M, Vanlerberghe GC (2013) Alternative oxidase impacts the plant response to biotic stress by influencing the mitochondrial generation of reactive oxygen species. Plant Cell Environ 36:721–732

Daems WT, Wisse E (1966) Shape and attachment of the cristae mitochondriales in mouse hepatic cell mitochondria. J Ultrastruct Res 16:123–140

Dai H, Lo YS, Litvinchuk A, Wang YT, Jane WN, Hsiao LJ, Chiang KS (2005) Structural and functional characterizations of mung bean mitochondrial nucleoids. Nucleic Acids Res 33:4725–4739

Dassa EP, Dufour E, Gonçalves S, Jacobs HT, Rustin P (2009) The alternative oxidase, a tool for compensating cytochrome *c* oxidase deficiency in human cells. Physiol Plant 137:427–434

Devi R, Munjral N, Gupta AK, Kaur N (2007) Cadmium induced changes in carbohydrate status and enzymes of carbohydrate metabolism, glycolysis and pentose phosphate pathway in pea. Environ Exp Bot 61:167–174

Dixon RA, Paiva NL (1995) Stress-induced phenylpropanoid metabolism. Plant Cell 7:1085–1097

Dutilleul C, Garmier M, Noctor G, Mathieu C, Chetrit P, Foyer CH, De Paepe R (2003) Leaf mitochondria modulate whole cell redox homeostasis, set antioxidant capacity, and determine stress resistance through altered signaling and diurnal regulation. Plant Cell 15:1212–1226

Dzeja PP, Terzic A (1998) Phosphotransfer reactions in the regulation of ATP-sensitive K^+ channels. FASEB J 12:523–529

Dzeja PP, Holmuhamedov EL, Ozcan C, Pucar D, Jahangir A, Terzic A (2001) Mitochondria gateway for cytoprotection. Circ Res 89:744–746

Edreva A, Velikova V, Tsonev T, Dagnon S, Gürel A, Aktaş L, Gesheva E (2008) Stress-protective role of secondary metabolites: diversity of functions and mechanisms. Gen Appl Plant Physiol 34:67–78

Elhafez D, Murcha MW, Clifton R, Soole KL, Day DA, Whelan J (2006) Characterization of mitochondrial alternative NAD(P)H dehydrogenases in *Arabidopsis*: intraorganelle location and expression. Plant Cell Physiol 47:43–54

Elthon TE, McIntosh L (1987) Identification of the alternative terminal oxidase of higher plant mitochondria. Proc Natl Acad Sci U S A 84:8399–8403

Elthon TE, Stewart CR, Mc Coy CA, Bonner WD Jr (1986) Alternative respiratory path capacity in plant mitochondria effect of growth temperature, the electrochemical gradient, and assay pH. Plant Physiol 80:378–383

Elthon TE, Nickels RL, McIntosh L (1989) Monoclonal antibodies to the alternative oxidase of higher plant mitochondria. Plant Physiol 89:1311–1317

Embley TM, Martin W (2006) Eukaryotic evolution, changes and challenges. Nature 440:623–630

Essmann J, Bones P, Weis E, Scharte J (2008) Leaf carbohydrate metabolism during defense. Intracellular sucrose-cleaving enzymes do not compensate repression of cell wall invertase. Plant Signal Behav 3:885–887

Eubel H, Jänsch L, Braun HP (2003) New insights into the respiratory chain of plant mitochondria. Supercomplexes and a unique composition of complex II. Plant Physiol 133:274–286

Fauron C, Allen J, Clifton S, Newton K (2004) Plant mitochondrial genomes. In: Daniell H, Chase C (eds) Molecular biology and biotechnology of plant organelles. Chloroplasts and mitochondria. Springer, Dordrecht, pp 151–177

Fecht-Christoffers MM, Führs H, Braun HP, Horst WJ (2006) The role of hydrogen peroxide-producing and hydrogen peroxide-consuming peroxidases in the leaf apoplast of cowpea in manganese tolerance. Plant Physiol 140:1451–1463

Fernández-Ayala DJM, Sanz A, Vartiainen S, Kemppainen K, Babusiak M, Mustalahti E, Costa R, Tuomela T, Zeviani M, Chung J, O'Dell KMC, Rustin P, Jacobs HT (2009) Expression of the *Ciona intestinalis* alternative oxidase (AOX) in *Drosophila* complements defects in mitochondrial oxidative phosphorylation. Cell Metab 9:449–460

Fernie AR, Carrari F, Sweetlove LJ (2004) Respiratory metabolism: glycolysis, the TCA cycle and mitochondrial electron transport. Curr Opin Plant Biol 7:254–261

Fiorani F, Umbach AL, Siedow JN (2005) The alternative oxidase of plant mitochondria is involved in the acclimation of shoot growth at low temperature. A study of Arabidopsis AOX1 a transgenic plants. Plant Physiol 139:1795–1805

Flexas J, Bota J, Galmés J, Medrano H, Ribas-Carbó M (2006) Keeping a positive carbon balance under adverse conditions: responses of photosynthesis and respiration to water stress. Physiol Plant 127:343–352

Fresneau C, Ghashghaie J, Cornic G (2007) Drought effect on nitrate reductase and sucrose-phosphate synthase activities in wheat (*Triticum durum* L.): role of leaf internal CO_2. J Exp Bot 58:2983–2992

Frey TG, Mannella CA (2000) The internal structure of mitochondria. Trends Biochem Sci 25:319–324

Friedrich T, Steinmüller K, Weiss H (1995) The proton-pumping respiratory complex I of bacteria and mitochondria and its homologue in chloroplasts. FEBS Lett 479:1–5

Fujita M, Fujita Y, Noutoshi Y, Takahashi F, Narusaka Y, Yamaguchi-Shinozaki K, Shinozaki K (2006) Crosstalk between abiotic and biotic stress responses: a current view from the points of convergence in the stress signaling networks. Curr Opin Plant Biol 9:436–442

Fung RWM, Wang CY, Smith DL, Gross KC, Tian MS (2004) MeSA and MeJA increase steady-state transcript levels of alternative oxidase and resistance against chilling injury in sweet peppers (*Capsicum annuum* L.). Plant Sci 166:711–719

Gandin A, Lapointe L, Dizengremel P (2009) The alternative respiratory pathway allows sink to cope with changes in carbon availability in the sink-limited plant *Erythronium americanum*. J Exp Bot 60:4235–4248

Garesse R, Vallejo CG (2001) Animal mitochondrial biogenesis and function: a regulatory cross-talk between two genomes. Gene 263:1–16

Garlid KD, Jabůrek M, Ježek P, Vařecha M (2000) How do uncoupling proteins uncouple? Biochim Biophys Acta 1459:383–389

Gas E, Flores-Pérez U, Sauret-Güeto S, Rodríguez-Concepción M (2009) Hunting for plant nitric oxide synthase provides new evidence of a central role for plastids in nitric oxide metabolism. Plant Cell 21:18–23

Gechev TS, Van Breusegem F, Stone JM, Denev I, Laloi C (2006) Reactive oxygen species as signals that modulate plant stress responses and programmed cell death. Bioessays 28:1091–1101
Geigenberger P (2003) Response of plant metabolism to too little oxygen. Curr Opin Plant Biol 6:247–256
Geigenberger P (2011) Regulation of starch biosynthesis in response to a fluctuating environment. Plant Physiol 155:1566–1577
Geigenberger P, Reimholz R, Deiting U, Sonnewald U, Stitt M (1999) Decreased expression of sucrose phosphate synthase strongly inhibits the water stress-induced synthesis of sucrose in growing potato tubers. Plant J 19:119–129
Geisler DA, Broselid C, Hederstedt L, Rasmusson AG (2007) Ca^{2+}-binding and Ca^{2+}-independent respiratory NADH and NADPH dehydrogenases of *Arabidopsis thaliana*. J Biol Chem 282:28455–28464
Gelhaye E, Rouhier N, Gérard J, Jolivet Y, Gualberto J, Navrot N, Ohlsson PI, Gunnar W, Hirasawa M, Knaff DB, Wang H, Dizengremel P, Meyer Y, Jacquot JP (2004) A specific form of thioredoxin *h* occurs in plant mitochondria and regulates the alternative oxidase. Proc Natl Acad Sci U S A 101:14545–14550
Gilbert GA, Wilson C, Madore MA (1997) Root-zone salinity alters raffinose oligosaccharide metabolism and transport in Coleus. Plant Physiol 115:1267–1276
Giraud E, Ho LHM, Clifton R, Carroll A, Estavillo G, Tan YF, Howell KA, Ivanova A, Pogson BJ, Millar AH, Whelan J (2008) The absence of ALTERNATIVE OXIDASE1a in Arabidopsis results in acute sensitivity to combined light and drought stress. Plant Physiol 147:595–610
González-Meler MA, Giles L, Thomas RB, Siedow JN (2001) Metabolic regulation of leaf respiration and alternative pathway activity in response to phosphate supply. Plant Cell Environ 24:205–215
Granot D, David-Schwartz R, Kelly G (2013) Hexose kinases and their role in sugar-sensing and plant development. Front Plant Sci 4:44. doi:10.3389/fpls.2013.00044
Greishaber MK, Völkel S (1998) Animal adaptations for tolerance and exploitation of poisonous sulfide. Annu Rev Physiol 60:33–53
Gupta AK, Kaur N (2005) Sugar signalling and gene expression in relation to carbohydrate metabolism under abiotic stresses in plants. J Biosci 30:761–776
Gupta KJ, Zabalza A, van Dongen JT (2009) Regulation of respiration when the oxygen availability changes. Physiol Plant 137:383–391
Gupta KJ, Fernie AR, Kaiser WM, van Dongen JT (2010) On the origins of nitric oxide. Trends Plant Sci 16:160–168
Gupta KJ, Igamberdiev AU, Mur LA (2012a) NO and ROS homeostasis in mitochondria: a central role for alternative oxidase. New Phytol 195:1–3
Gupta KJ, Shah JK, Brotman Y, Jahnke K, Willmitzer L, Kaiser WM, Bauwe H, Igamberdiev AU (2012b) Inhibition of aconitase by nitric oxide leads to induction of the alternative oxidase and to a shift of metabolism towards biosynthesis of amino acids. J Exp Bot 63:1773–1784
Gustafsson AB, Gottlieb RA (2008) Heart mitochondria: gates of life and death. Cardiovasc Res 77:334–343
Guy RD, Vanlerberghe GC (2005) Partitioning of respiratory electrons in the dark in leaves of transgenic tobacco with modified levels of alternative oxidase. Physiol Plant 125:171–180
Hackenbrock CR (1966) Ultrastructural bases for metabolically linked mechanical activity in mitochondria. I. Reversible ultrastructural changes with change in metabolic steady state in isolated liver mitochondria. J Cell Biol 30:269–297
Hackenbrock CR, Chazotte B, Gupte SS (1986) The random collision model and a critical assessment of diffusion and collision in mitochondrial electron transport. J Bioenerg Biomembr 18:331–368
Hakkaart GA, Dassa EP, Jacobs HT, Rustin P (2006) Allotopic expression of a mitochondrial alternative oxidase confers cyanide resistance to human cell respiration. EMBO Rep 7:341–345
Halestrap AP, Brosnan JT (2008) From metabolic cycles to compartmentation: another first for Krebs. Biochemist 30:24–28

Hanqing F, Kun S, Mingquan L, Hongyu L, Xin L, Yan L, Yifeng W (2010) The expression, function and regulation of mitochondrial alternative oxidase under biotic stresses. Mol Plant Pathol 11:429–440

Hansford RG (1994) Role of calcium in respiratory control. Med Sci Sports Exerc 26:44–51

Hatefi Y (1985) The mitochondrial electron transport and oxidative phosphorylation system. Annu Rev Biochem 54:1015–1069

Havelund JF, Thelen JJ, Møller IM (2013) Biochemistry, proteomics, and phosphoproteomics of plant mitochondria from non-photosynthetic cells. Front Plant Sci 4:51. doi:10.3389/fpls.2013.00051

Hilal M, Castagnaro A, Moreno H, Massa EM (1997) Specific localization of the respiratory alternative oxidase in meristematic and xylematic tissues from developing soybean roots and hypocotyls. Plant Physiol 115:1499–1503

Hilal M, Zenoff AM, Ponessa G, Moreno H, Massa EM (1998) Saline stress alters the temporal patterns of xylem differentiation and alternative oxidase expression in developing soybean roots. Plant Physiol 117:695–701

Ho SL, Chao YC, Tong WF, Yu SM (2001) Sugar coordinately and differentially regulates growth- and stress-related gene expression via a complex signal transduction network and multiple control mechanisms. Plant Physiol 125:877–890

Ho LHM, Giraud E, Lister R, Thirkettle-Watts D, Low J, Clifton R, Howell KA, Carrie C, Donald T, Whelan J (2007) Characterization of the regulatory and expression context of an alternative oxidase gene provides insights into cyanide-insensitive respiration during growth and development. Plant Physiol 143:1519–1533

Ho LHM, Giraud E, Uggalla V, Lister R, Glen A, Thirkettle-Watts D, Van Aken O, Whelan J (2008) Identification of regulatory pathways controlling gene expression of stress responsive mitochondrial components in Arabidopsis. Plant Physiol 147:1858–1873

Hoefnagel MHN, Rich PR, Zhang Q, Wiskich JT (1997) Substrates kinetics of the plant mitochondrial alternative oxidase and the effects of pyruvate. Plant Physiol 115:1145–1153

Hourton-Cabassa C, Matos AR, Zachowski A, Moreau F (2004) The plant uncoupling protein homologues: a new family of energy-dissipating proteins in plant mitochondria. Plant Physiol Biochem 42:283–290

Howell KA, Millar AH, Whelan J (2006) Ordered assembly of mitochondria during rice germination begins with promitochondrial structures rich in components of the protein import apparatus. Plant Mol Biol 60:201–223

Hu X, Bidney DL, Yalpani N, Duvick JP, Crasta O, Folkerts O, Lu G (2003) Overexpression of a gene encoding hydrogen peroxide-generating oxalate oxidase evokes defense responses in sunflower. Plant Physiol 133:170–181

Hurry VM, Malmberg G, Gardestrom P, Öquist G (1994) Effects of a short-term shift to low temperature and of long-term cold hardening on photosynthesis and ribulose-1,5-bisphosphate carboxylase/oxygenase and sucrose phosphate synthase activity in leaves of winter rye (*Secale cereale* L.). Plant Physiol 106:983–990

Ikemura T (1985) Codon usage and tRNA content in unicellular and multicellular organisms. Mol Biol Evol 2:13–34

Jacoby RP, Li L, Huang S, Lee CP, Millar AH, Taylor NL (2012) Mitochondrial composition, function and stress response in plants. J Integr Plant Biol 54:887–906

Jarmuszkiewicz W, Wagner AM, Wagner MJ, Hryniewiecka L (1997) Immunological identification of the alternative oxidase of *Acanthamoeba castellanii* mitochondria. FEBS Lett 411:110–114

Jarmuszkiewicz W, Almeida AM, Vercesi AE, Sluse FE, Sluse-Goffart CM (2000) Proton re-uptake partitioning between uncoupling protein and ATP synthase during benzohydroxamic acid-resistant state 3 respiration in tomato fruit mitochondria. J Biol Chem 275:13315–13320

Jarmuszkiewicz W, Hryniewiecka L, Sluse FE (2002) The effect of pH on the alternative oxidase activity in isolated *Acanthamoeba castellanii* mitochondria. J Bioenerg Biomembr 34:221–226

Johnson DT, Harris RA, French S, Blair PV, You J, Bemis KG, Wang M, Balaban RS (2007) Tissue heterogeneity of the mammalian mitochondrial proteome. Am J Physiol Cell Physiol 292:689–697

Johnson BN, Charan RA, LaVoie MJ (2012) Recognizing the cooperative and independent mitochondrial functions of Parkin and PINK1. Cell Cycle 11:2775–2776

Johnson-Flanagan AM, Owens JN (1985) Peroxidase activity in relation to suberization and respiration in white spruce (*Picea glauca* [Moench] Voss) seedling roots. Plant Physiol 79:103–107

Jørgensen K, Rasmussen AV, Morant M, Nielsen AH, Bjarnholt N, Zagrobelny M, Bak S, Møller BL (2005) Metabolon formation and metabolic channeling in the biosynthesis of plant natural products. Curr Opin Plant Biol 8:280–291

Justine Y, Yip H, Vanlerberghe GC (2001) Mitochondrial alternative oxidase acts to dampen the generation of active oxygen species during a period of rapid respiration induced to support a high rate of nutrient uptake. Physiol Plant 112:327–333

Karlsson M, Melzer M, Prokhorenko I, Johansson T, Wingsle G (2005) Hydrogen peroxide and expression of hipI-superoxide dismutase are associated with the development of secondary cell walls in *Zinnia elegans*. J Exp Bot 56:2085–2093

Keurentjes JJB, Sulpice R, Gibon Y, Steinhauser MC, Fu J, Koornneef M, Stitt M, Vreugdenhil D (2008) Integrative analyses of genetic variation in enzyme activities of primary carbohydrate metabolism reveal distinct modes of regulation in *Arabidopsis thaliana*. Genome Biol 9:R129

Kleczkowski LA, Geisler M, Ciereszko I, Johansson H (2004) UDP-glucose pyrophosphorylase. An old protein with new tricks. Plant Physiol 134:912–918

Koch KE (2004) Sucrose metabolism: regulatory mechanisms and pivotal roles in sugar sensing and plant development. Curr Opin Plant Biol 7:235–246

Koch KE, Wu V, Xu J (1996) Sugar and metabolic regulation of genes for sucrose metabolism: potential influence of maize sucrose synthase and soluble invertase responses on carbon partitioning and sugar sensing. J Exp Bot 47:1179–1185

Krasensky J, Jonak C (2012) Drought, salt, and temperature stress-induced metabolic rearrangements and regulatory networks. J Exp Bot 63:1593–1608

Kubo T, Newton KJ (2008) Angiosperm mitochondrial genomes and mutations. Mitochondrion 8:5–14

Lee CP, Eubel H, Millar AH (2010) Diurnal changes in mitochondrial function reveal daily optimization of light and dark respiratory metabolism in *Arabidopsis*. Mol Cell Proteomics 9:2125–2139

Lesnefsky EJ, Moghaddas S, Tandler B, Kerner J, Hoppel CL (2001) Mitochondrial dysfunction in cardiac disease: ischemia-reperfusion, aging, and heart failure. J Mol Cell Cardiol 33:1065–1089

Li CR, Liang DD, Li J, Duan YB, Li H, Yang YC, Qin RY, Li L, Wei PC, Yang JB (2013) Unravelling mitochondrial retrograde regulation in the abiotic stress induction of rice *ALTERNATIVE OXIDASE 1* genes. Plant Cell Environ 36:775–788

Liao YW, Shi K, Fu LJ, Zhang S, Li X, Dong DK, Jiang YP, Zhou YH, Xia XJ, Liang WS, Yu JQ (2012) The reduction of reactive oxygen species formation by mitochondrial alternative respiration in tomato basal defense against TMV infection. Planta 235:225–238

Lima-Júnior A, Fernandes de Melo D, Costa JH, Orellano EG, Jolivet Y, Jarmuszkiewicz W, Sluse F, Dizengremel P, Silva Lima M (2000) Effect of pH on CN-resistant respiratory activity and regulation on *Vigna unguiculata* mitochondria. Plant Physiol Biochem 38:765–771

Liu YJ, Norberg FE, Szilagyi A, De Paepe R, Akerlund HE, Rasmusson AG (2008) The mitochondrial external NADPH dehydrogenase modulates the leaf NADPH/$NADP^+$ ratio in transgenic *Nicotiana sylvestris*. Plant Cell Physiol 49:251–263

Livingston DP III, Hincha DK, Heyer AG (2009) Fructan and its relationship to abiotic stress tolerance in plants. Cell Mol Life Sci 66:2007–2023

Logan DC (2006) The mitochondrial compartment. J Exp Bot 57:1225–1243

Logan DC, Millar AH, Sweetlove LJ, Hill SA, Leaver CJ (2001) Mitochondrial biogenesis during germination in maize embryos. Plant Physiol 125:662–672

Logan DC, Scott I, Tobin AK (2003) The genetic control of plant mitochondrial morphology and dynamics. Plant J 36:500–509

Lonsdale DM, Brears T, Hodge TP, Melville SE, Rottmann WH (1988) The plant mitochondrial genome: homologous recombination as a mechanism for generating heterogeneity. Philos Trans R Soc Lond B 319:149–163

Magalhães PJ, Andreu AL, Schon SA (1998) Evidence for the presence of 5S rRNA in mammalian mitochondria. Mol Biol Cell 9:2375–2382

Mannella CA (1992) The 'ins' and 'outs' of mitochondrial membrane channels. Trends Biochem Sci 17:315–320

Mannella CA (2006) The relevance of mitochondrial membrane topology to mitochondrial function. Biochim Biophys Acta 1762:140–147

Maréchal A, Kido Y, Kita K, Moore AL, Rich PR (2009) Identification of three redox states of recombinant *Trypanosoma brucei* alternative oxidase by FTIR spectroscopy and electrochemistry. J Biol Chem 284:31827–31833

Marques I, Dencher NA, Videira A, Krause F (2007) Supramolecular organization of the respiratory chain in *Neurospora crassa* mitochondria. Eukaryot Cell 6:2391–2405

Martí MC, Florez-Sarasa I, Camejo D, Ribas-Carbó M, Lázaro JJ, Sevilla F, Jiménez A (2011) Response of mitochondrial thioredoxin PsTrxo1, antioxidant enzymes, and respiration to salinity in pea (*Pisum sativum* L.) leaves. J Exp Bot 62:3863–3874

Matos AR, Mendes AT, Scotti-Campos P, Arrabaça JD (2009) Study of the effects of salicylic acid on soybean mitochondrial lipids and respiratory properties using the alternative oxidase as a stress-reporter protein. Physiol Plant 137:485–497

Matus-Ortega MG, Salmerón-Santiago KG, Flores-Herrera O, Guerra-Sánchez G, Martínez F, Rendón JL, Pardo JP (2011) The alternative NADH dehydrogenase is present in mitochondria of some animal taxa. Comp Biochem Physiol D 6:256–263

Maxwell DP, Wang Y, McIntosh L (1999) The alternative oxidase lowers mitochondrial reactive oxygen production in plant cells. Proc Natl Acad Sci U S A 96:8271–8276

McDonald AE, Vanlerberghe GC (2004) Branched mitochondrial electron transport in the Animalia: presence of alternative oxidase in several animal phyla. IUBMB Life 56:333–341

McDonald AE, Vanlerberghe GC (2006) Origins, evolutionary history, and taxonomic distribution of alternative oxidase and plastoquinol terminal oxidase. Comp Biochem Physiol D 1:357–364

McDonald AE, Vanlerberghe GC, Staples JF (2009) Alternative oxidase in animals: unique characteristics and taxonomic distribution. J Exp Biol 212:2627–2634

Meeuse BJD (1975) Thermogenic respiration in aroids. Annu Rev Plant Physiol 26:117–126

Menz RI, Day DA (1996) Identification and characterization of an inducible NAD(P)H dehydrogenase from red beetroot mitochondria. Plant Physiol 112:607–613

Menze MA, Boswell L, Toner M, Hand SC (2009) Occurrence of mitochondria-targeted late embryogenesis abundant (LEA) gene in animals increases organelle resistance to water stress. J Biol Chem 284:10714–10719

Mhadhbi H, Fotopoulos V, Mylona PV, Jebara M, Aouani ME, Polidoros AN (2013) Alternative oxidase *Aox1* gene expression in roots of *Medicago truncatula* is a genotype-specific component of salt stress tolerance. J Plant Physiol 170:111–114

Millar AH, Leaver CJ (2000) The cytotoxic lipid peroxidation product, 4-hydroxy-2-nonenal, specifically inhibits decarboxylating dehydrogenases in the matrix of plant mitochondria. FEBS Lett 481:117–121

Millar AH, Atkin OK, Menz RI, Henry B, Farquhar G, Day DA (1998) Analysis of respiratory chain regulation in roots of soybean seedlings. Plant Physiol 117:1083–1093

Millar AH, Whelan J, Soole KL, Day DA (2011) Organization and regulation of mitochondrial respiration in plants. Annu Rev Plant Biol 62:79–104

Millenaar FF, Lambers H (2003) The alternative oxidase: *in vivo* regulation and function. Plant Biol 5:2–15

Millenaar FF, Benschop JJ, Wagner AM, Lambers H (1998) The role of the alternative oxidase in stabilizing the in vivo reduction state of the ubiquinone pool and the activation state of the alternative oxidase. Plant Physiol 118:599–607

Millenaar FF, Gonzalez-Meler MA, Siedow JN, Wagner AM, Lambers H (2002) Role of sugars and organic acids in regulating the concentration and activity of the alternative oxidase in *Poa annua* roots. J Exp Bot 53:1081–1088

Mishra P, Dubey RS (2008) Effect of aluminium on metabolism of starch and sugars in growing rice seedlings. Acta Physiol Plant 30:265–275

Mittler R (2006) Abiotic stress, the field environment and stress combination. Trends Plant Sci 11:15–19

Møller IM (2001) Plant mitochondria and oxidative stress: electron transport, NADPH turnover, and metabolism of reactive oxygen species. Annu Rev Plant Physiol Plant Mol Biol 52:561–591

Møller IM, Berczi A, van der Plas LHW, Lambers H (1988) Measurement of the activity and capacity of the alternative pathway in intact tissues: identification and problems and possible solutions. Physiol Plant 73:642–649

Møller IM, Rasmusson AG, Siedow JN, Vanlerberghe GC (2010) The product of the alternative oxidase is still H_2O. Arch Biochem Biophys 495:93–94

Moore AL, Siedow JN (1991) The regulation and nature of the cyanide-resistant alternative oxidase of plant mitochondria. Biochim Biophys Acta 1059:121–140

Moore AL, Umbach AL, Siedow JN (1995) Structure-function relationships of the alternative oxidase of plant mitochondria: a model of the active site. J Bioenerg Biomembr 27:367–377

Moore AL, Albury MS, Crichton PG, Affourtit C (2002) Function of the alternative oxidase: is it still a scavenger? Trends Plant Sci 7:478–481

Mower JP, Sloan DB, Alverson AJ (2012) Plant mitochondrial genome diversity: the genomics revolution. In: Wendel JH (ed) Plant genome diversity volume 1: plant genomes, their residents, and their evolutionary dynamics. Springer, New York, pp 123–144

Navet R, Douette P, Puttine-Marique F, Sluse-Goffart CM, Jarmuszkiewicz W, Sluse FE (2005) Regulation of uncoupling protein activity in phosphorylating potato tuber mitochondria. FEBS Lett 579:4437–4442

Navrot N, Collin V, Gualberto J, Gelhaye E, Hirasawa M, Rey P, Knaff DB, Issakidis E, Jacquot JP, Rouhier N (2006) Plant glutathione peroxidases are functional peroxiredoxins distributed in several subcellular compartments and regulated during biotic and abiotic stresses. Plant Physiol 142:1364–1379

Nelson DL, Cox MM (2008) Lehninger principles of biochemistry, 5th edn. W.H. Freeman & Co, New York

Newcomb W (1990) Mitochondrial structure. In: Dennis DT, Turpin DH (eds) Plant physiology, biochemistry and molecular biology. Longman Scientific & Technical, England, pp 103–105

Nunes-Nesi A, Sulpice R, Gibon Y, Fernie AR (2008) The enigmatic contribution of mitochondrial function in photosynthesis. J Exp Bot 59:1675–1684

O'Donnell VB, Smith GCM, Jones OTG (1994) Involvement of phenyl radicals in iodonium compound inhibition of flavoenzymes. Mol Pharmacol 46:778–785

Osorio S, Vallarino JG, Szecowka M, Ufaz S, Tzin V, Angelovici R, Galili G, Fernie AR (2013) Alteration of the interconversion of pyruvate and malate in the plastid or cytosol of ripening tomato fruit invokes diverse consequences on sugar but similar effects on cellular organic acid, metabolism, and transitory starch accumulation. Plant Physiol 161:628–643

Padmasree K, Padmavathi L, Raghavendra AS (2002) Essentiality of mitochondrial oxidative metabolism for photosynthesis: optimization of carbon assimilation and protection against photoinhibition. Crit Rev Biochem Mol Biol 37:71–119

Palade G (1952) The fine structure of mitochondria. Anat Rec 114:427–451

Panda SK, Sahoo L, Katsuhara M, Matsumoto H (2013) Overexpression of alternative oxidase gene confers aluminum tolerance by altering the respiratory capacity and the response to oxidative stress in tobacco cells. Mol Biotechnol. doi:10.1007/s12033-012-9595-7

Pasqualini S, Paolocci F, Borgogni A, Morettini R, Ederli L (2007) The overexpression of an alternative oxidase gene triggers ozone sensitivity in tobacco plants. Plant Cell Environ 30:1545–1556

Pastore D, Trono D, Laus MN, Di Fonzo N, Passarella S (2001) Alternative oxidase in durum wheat mitochondria. Activation by pyruvate, hydroxypyruvate and glyoxylate and physiological role. Plant Cell Physiol 42:1373–1382

Peckmann K, von Willert DJ, Martin CE, Herppich WB (2012) Mitochondrial espiration in ME-CAM, PEPCK-CAM, and C_3 succulents: comparative operation of the cytochrome, alternative, and rotenone-resistant pathways. J Exp Bot 63:2909–2919

Perkins GA, Frey TG (2000) Recent structural insight into mitochondria gained by microscopy. Micron 31:97–111

Pineau B, Layoune O, Danon A, De Paepe R (2008) L-galactono-1,4-lactone dehydrogenase is required for the accumulation of plant respiratory complex I. J Biol Chem 283: 32500–32505

Pizarro LC, Bisigato AJ (2010) Allocation of biomass and photoassimilates in juvenile plants of six Patagonian species in response to five water supply regimes. Ann Bot 106:297–307

Plaxton WC, Podestá FE (2006) The functional organization and control of plant respiration. Crit Rev Plant Sci 25:159–198

Pollard M, Beisson F, Li Y, Ohlrogge JB (2008) Building lipid barriers: biosynthesis of cutin and suberin. Trends Plant Sci 13:236–246

Poulton JE (1990) Cyanogenesis in plants. Plant Physiol 94:401–405

Prado C (2012) Respuestas bioquímicas y anatomofisiólogicas de *Salvinia minima* al cromo. Su potencial uso como agente fitorremediador de ambientes acuáticos. PhD thesis, Universidad Nacional de Tucumán, Tucumán

Prado C, Rodríguez-Montelongo L, González JA, Pagano EA, Hilal M, Prado FE (2010) Uptake of chromium by *Salvinia minima*: effect on plant growth, leaf respiration and carbohydrate metabolism. J Hazard Mater 177:546–553

Prado C, Podazza G, Pagano E, Prado FE, Rosa M (2011) Heavy metals—functional and metabolic interactions between carbohydrates and secondary metabolites in plants. A review. In: Brar SK (ed) Hazardous materials: types, risks and control. Nova, New York, pp 1–52

Prado C, Rosa M, Pagano E, Prado FE (2013) Metabolic interconnectivity among alternative respiration, residual respiration, carbohydrates and phenolics in leaves of *Salvinia minima* exposed to Cr(VI). Environ Exp Bot 87:32–38

Proels RK, Roitsch T (2009) Extracellular invertase LIN6 of tomato: a pivotal enzyme for integration of metabolic, hormonal, and stress signals is regulated by a diurnal rhythm. J Exp Bot 60:1555–1567

Raghavendra AS, Padmasree K (2003) Beneficial interactions of mitochondrial metabolism with photosynthetic carbon assimilation. Trends Plant Sci 8:546–553

Rakhmankulova ZF, Fedyaev VV, Podashevka OA, Usmanov IY (2003) Alternative respiration pathways and secondary metabolism in plants with different adaptive strategies under mineral deficiency. Russ J Plant Physiol 50:206–212

Ralph J, Bunzel M, Marita MJ, Hatfield RD, Lu F, Kim H, Schatz PF, Grabber JH, Steinhart H (2004) Peroxidase-dependent cross-linking reactions of *p*-hydroxycinnamates in plant cell walls. Phytochem Rev 3:79–96

Ramakrishna A, Ravishankar GA (2011) Influence of abiotic stress signals on secondary metabolites in plants. Plant Signal Behav 6:1720–1731

Ramírez-Aguilar SJ, Keuthe M, Rocha M, Fedyaev VV, Kramp K, Gupta KJ, Rasmusson AG, Schulze WX, van Dongen JT (2011) The composition of plant mitochondrial supercomplexes changes with oxygen availability. J Biol Chem 286:43045–43053

Ranathunge K, Schreiber L, Franke R (2011) Suberin research in the genomics era—new interest for an old polymer. Plant Sci 180:399–413

Rasmusson AG, Møller IM (2011) Mitochondrial electron transport and plant stress. In: Kempken F (ed) Plant mitochondria: advances in plant biology volume, 1st edn. Springer, New York, pp 357–381

Rasmusson AG, Soole KL, Elthon TE (2004) Alternative NAD(P)H dehydrogenases of plant mitochondria. Annu Rev Plant Biol 55:23–39

Rasmusson AG, Geisler DA, Møller IM (2008) The multiplicity of dehydrogenases in the electron transport chain of plant mitochondria. Mitochondrion 8:47–60

Rasmusson AG, Fernie AR, van Dongen JT (2009) Alternative oxidase: a defence against metabolic fluctuations? Physiol Plant 137:371–382

Rathinasabapathi B (2000) Metabolic engineering for stress tolerance: installing osmoprotectant synthesis pathways. Ann Bot 86:709–716

Rhoads DM, Umbach AL, Sweet CR, Lennon AM, Rauch GS, Siedow JN (1998) Regulation of the cyanide-resistant alternative oxidase of plant mitochondria. Identification of the cysteine

residue involved in α-keto acid stimulation and intersubunit disulfide bond formation. J Biol Chem 273:30750–30756
Ribas-Carbó M, Lennon AM, Robinson SA, Giles L, Berry JA, Siedow JN (1997) The regulation of electron partitioning between the cytochrome and alternative pathways in soybean cotyledon and root mitochondria. Plant Physiol 113:903–911
Ribas-Carbó M, Taylor NL, Giles L, Busquets S, Finnegan PM, Day DA, Lambers H, Medrano H, Berry JA, Flexas J (2005) Effects of water stress on respiration in soybean leaves. Plant Physiol 139:466–473
Ribas-Carbó M, Giles L, Flexas J, Briggs W, Berry JA (2008) Phytochrome-driven changes in respiratory electron transport partitioning in soybean (*Glycine max*. L.) cotyledons. Plant Biol 10:281–287
Rigoulet M, Aguilaniu H, Avéret N, Bunoust O, Camougrand N, Grandier-Vazeille X, Larsson C, Pahlman IL, Manon S, Gustafsson L (2004) Organization and regulation of the cytosolic NADH metabolism in the yeast *Saccharomyces cerevisiae*. Mol Cell Biochem 256–257:73–81
Rizhsky L, Liang H, Shuman J, Shulaev V, Davletova S, Mittler R (2004) When defense pathways collide. The response of Arabidopsis to a combination of drought and heat stress. Plant Physiol 134:1683–1696
Roberts TH, Fredlund KM, Møller IM (1995) Direct evidence for the presence of two external NAD(P)H dehydrogenases coupled to the electron transport chain in plant mitochondria. FEBS Lett 373:307–309
Roberts CW, Roberts F, Henriquez FL, Akiyoshi D, Samuel BU, Richards TA, Mihous W, Kyle D, McIntosh L, Hill GC, Chaudhuri M, Tzipori S, McLeod R (2004) Evidence for mitochondrial-derived alternative oxidase in the apicomplexan parasite *Cryptosporidium parvum*: a potential anti-microbial agent target. Int J Parasitol 34:297–308
Robson CA, Vanlerberghe GC (2002) Transgenic plant cells lacking mitochondrial alternative oxidase have increased susceptibility to mitochondria-dependent and -independent pathways of programmed cell death. Plant Physiol 129:1908–1920
Roitsch T, González MC (2004) Function and regulation of plant invertases: sweet sensations. Trends Plant Sci 9:606–613
Ros Barceló A (1997) Lignification in plant cell walls. Int Rev Cytol 176:87–132
Ros Barceló A, Pomar F, Lópes-Serrano M, Martinez P, Pendreño MA (2002) Developmental regulation of the H_2O_2-producing system and basic peroxidase isoenzyme in the *Zinnia elegans* lignifying xylem. Plant Physiol Biochem 40:325–332
Rosa M, Hilal M, González JA, Prado FE (2009a) Low-temperature effect on enzyme activities involved in sucrose–starch partitioning in salt-stressed and salt-acclimated cotyledons of quinoa (*Chenopodium quinoa* Willd.) seedlings. Plant Physiol Biochem 47:300–307
Rosa M, Prado C, Podazza G, Interdonato R, González JA, Hilal M, Prado FE (2009b) Soluble sugars: metabolism, sensing, and abiotic stress. A complex network in the life of plants. Plant Signal Behav 4:388–393
Ruan YL, Jin Y, Yang YJ, Li GJ, Boyer JS (2010) Sugar input, metabolism, and signaling mediated by invertase: roles in development, yield potential, and response to drought and heat. Mol Plant 3:942–955
Runswick MJ, Fearnley IM, Skehel JM, Walker JE (1991) Presence of an acyl carrier protein in NADH-ubiquinone oxidoreductase from bovine heart mitochondria. FEBS Lett 286:121–124
Rustin P, Jacobs HT (2009) Respiratory chain alternative enzymes as tools to better understand and counteract respiratory chain deficiencies in human cells and animals. Physiol Plant 137:362–370
Santos Macedo E, Sircar D, Cardoso HG, Peixe A, Arnholdt-Schmitt B (2012) Involvement of alternative oxidase (AOX) in adventitious rooting of *Olea europaea* L. microshoots is linked to adaptive phenylpropanoid and lignin metabolism. Plant Cell Rep 31:1581–1590
Sanz A, Fernández-Ayala DJM, Stefanatos RK, Jacobs HT (2010) Mitochondrial ROS production correlates with, but does not directly regulate lifespan in drosophila. Aging 2:220–223
Schäfer E, Seelert H, Reifschneider NH, Krause F, Dencher NA, Vonck J (2006) Architecture of active mammalian respiratory chain supercomplexes. J Biol Chem 281:15370–15375

Schägger H, de Coo R, Bauer MF, Hofmann S, Godinot C, Brandt U (2004) Significance of respirasomes for the assembly/stability of human respiratory chain complex I. J Biol Chem 279:36349–36353

Searle SY, Turnbull MH (2011) Seasonal variation of leaf respiration and the alternative pathway in field-grown *Populus*×*canadensis*. Physiol Plant 141:332–342

Seigler DS (1998) Plant secondary metabolism. Kluwer Academic, Boston

Shedge V, Arrieta-Montiel M, Christensen AC, Mackenzie SA (2007) Plant mitochondrial recombination surveillance requires unusual RecA and MutS homologs. Plant Cell 19:1251–1264

Shen WY, Wei YD, Dauk M, Tan YF, Taylor DC, Selvaraj G, Zou JT (2006) Involvement of a glycerol-3-phosphate dehydrogenase in modulating the NADH/NAD^+ ratio provides evidence of a mitochondrial glycerol-3-phosphate shuttle in *Arabidopsis*. Plant Cell 18:422–441

Shi K, Fu LJ, Dong DK, Zhou YH, Yu JQ (2008) Decreased energy synthesis is partially compensated by a switch to sucrose synthase pathway of sucrose degradation in restricted root of tomato plants. Plant Physiol Biochem 46:1040–1044

Shugaeva NA, Vyskrebentseva EI, Orekhova SO, Shugaev AG (2007) Effect of water deficit on respiration of conducting bundles in leaf petioles of sugar beet. Russ J Plant Physiol 54:329–335

Siedow JN, Umbach AL (2000) The mitochondrial cyanide-resistant oxidase: structural conservation amid regulatory diversity. Biochim Biophys Acta 1459:432–439

Siedow JN, Umbach AL, Moore AL (1995) The active site of the cyanide resistant oxidase from plant mitochondria contains a binuclear iron center. FEBS Lett 362:10–14

Sieger SM, Kristensen BK, Robson CA, Amirsadeghi W, Eng SE, Abdel-Mesih A, Møller IM, Vanlerberghe GC (2005) The role of alternative oxidase in modulating carbon use efficiency and growth during macronutrient stress in tobacco cells. J Exp Bot 56:1499–1515

Sjostrand FS (1953) Electron microscopy of mitochondria and cytoplasmic double membranes. Nature 171:30–31

Skórzyńska-Polit E, Drążkiewicz M, Wianowska D, Maksymiec W, Dawidowicz AL, Tukiendorf A (2004) The influence of heavy metal stress on the level of some flavonols in the primary leaves of *Phaseolus coccineus*. Acta Physiol Plant 26:247–254

Skutnik M, Rychter AM (2009) Differential response of antioxidant systems in leaves and roots of barley subjected to anoxia and post-anoxia. J Plant Physiol 166:926–937

Sluse FE, Almeida AM, Jarmuszkiewicz W, Vercesi AE (1998) Free fatty acids regulate the uncoupling protein and alternative oxidase activities in plant mitochondria. FEBS Lett 433:237–240

Smith CA, Melino VJ, Sweetman C, Soole KL (2009) Manipulation of alternative oxidase can influence salt tolerance in *Arabidopsis thaliana*. Physiol Plant 137:459–472

Smith CA, Barthet M, Melino V, Smith P, Day D, Soole K (2011) Alterations in the mitochondrial alternative NAD(P)H dehydrogenase NDB4 lead to changes in mitochondrial electron transport chain composition, plant growth and response to oxidative stress. Plant Cell Physiol 52:1222–1237

Solomos T, Laties GG (1975) The mechanism of ethylene and cyanide action in triggering the rise in respiration in potato tubers. Plant Physiol 55:73–78

Stenmark P, Nordlund P (2003) A prokaryotic alternative oxidase present in the bacterium *Novosphingobium aromaticivorans*. FEBS Lett 552:189–192

Stobrawa K, Lorenc-Plucińska G (2007) Changes in carbohydrate metabolism in fine roots of the native European black poplar (*Populus nigra* L.) in a heavy-metal-polluted environment. Sci Total Environ 373:157–165

Sturm A, Tang GQ (1999) The sucrose-cleaving enzymes of plants are crucial for development, growth and carbon partitioning. Trends Plant Sci 4:401–407

Subbaiah CC, Palaniappan A, Duncan K, Rhoads DM, Huber SC, Sachs MM (2006) Mitochondrial localization and putative signaling function of sucrose synthase in maize. J Biol Chem 281:15625–15635

Sugie A, Naydenov N, Mizuno N, Nakamura C, Takumi S (2006) Overexpression of wheat alternative oxidase gene Waox1a alters respiration capacity and response to reactive oxygen species under low temperature in transgenic Arabidopsis. Genes Genet Syst 81:349–354

Sugiyama Y, Watase Y, Nagase M, Makita N, Yagura S, Hirai A, Sugiura M (2005) The complete nucleotide sequence and multipartite organization of the tobacco mitochondrial genome: comparative analysis of mitochondrial genomes in higher plants. Mol Genet Genomics 272:603–615

Sunderhaus S, Klodmann J, Lenz C, Braun HP (2010) Supramolecular structure of the OXPHOS system in highly thermogenic tissue of *Arum maculatum*. Plant Physiol Biochem 48:265–272

Suzuki T, Nihei C, Yabu Y, Hashimoto T, Suzuki M, Yoshida A, Nagai K, Hosokawa T, Minagawa N, Suzuki S, Kita K, Ohta N (2004) Molecular cloning and characterization of *Trypanosoma vivax* alternative oxidase (AOX) gene, a target of the trypanocide ascofuranone. Parasitol Int 53:235–245

Sweetlove LJ, Fait A, Nunes-Nesi A, Williams T, Fernie AR (2007) The mitochondrion: an integration point of cellular metabolism and signalling. Crit Rev Plant Sci 26:17–43

Szarka A, Horemans N, Passarella S, Tarcsay Á, Örsi F, Salgó A, Bánhegyi G (2008) Demonstration of an intramitochondrial invertase activity and the corresponding sugar transporters of the inner mitochondrial membrane in Jerusalem artichoke (*Helianthus tuberosus* L.) tubers. Planta 228:765–775

Szarka A, Bánhegyi G, Asard H (2013) The inter-relationship of ascorbate transport, metabolism and mitochondrial, plastidic respiration. Antioxid Redox Signal. doi:10.1089/ars.2012.5059

Takahama U, Oniki T, Hirota S (2001) Phenolic compounds of brown scales of onion bulbs produce hydrogen peroxide by autooxidation. J Plant Res 114:395–402

Talianova M, Janousek B (2011) What can we learn from tobacco and other solanaceae about horizontal DNA transfer? Am J Bot 98:1231–1242

Tang W, Fei Y, Page M (2012) Biological significance of RNA editing in cells. Mol Biotechnol 52:91–100

Thirkettle-Watts D, McCabe TC, Clifton R, Moore C, Finnegan PM, Day DA, Whelan J (2003) Analysis of the alternative oxidase promoters from soybean. Plant Physiol 133:1158–1169

Tischner R, Planchet E, Kaiser WM (2004) Mitochondrial electron transport as a source for nitric oxide in the unicellular green alga *Chlorella sorokiniana*. FEBS Lett 576:151–155

Uchida K (2003) 4-Hydroxy-2-nonenal: a product and mediator of oxidative stress. Prog Lipid Res 42:318–343

Umbach AL, Siedow JN (1993) Covalent and noncovalent dimers of the cyanide-resistant alternative oxidase protein in higher plant mitochondria and their relationship to enzyme activity. Plant Physiol 103:845–854

Umbach AL, Wiskich JT, Siedow JN (1994) Regulation of alternative oxidase kinetics by pyruvate and intermolecular disulfide bond redox status in soybean seedling mitochondria. FEBS Lett 348:181–184

Umbach AL, González-Meler MA, Sweet CR, Siedow JN (2002) Activation of the plant mitochondrial alternative oxidase: insights from site-directed mutagenesis. Biochim Biophys Acta 1554:118–128

Umbach AL, Fiorani F, Siedow JN (2005) Characterization of transformed Arabidopsis with altered alternative oxidase levels and analysis of effects on reactive oxygen species in tissue. Plant Physiol 139:1806–1820

Unseld M, Marienfeld JR, Brandt P, Brennicke A (1997) The mitochondrial genome of *Arabidopsis thaliana* contains 57 genes in 366,924 nucleotides. Nat Genet 15:57–61

Van Aken O, Giraud E, Clifton R, Whelan J (2009a) Alternative oxidase: a target and regulator of stress responses. Physiol Plant 137:354–361

Van Aken O, Zhang B, Carrie C, Uggalla V, Paynter E, Giraud E, Whelan J (2009b) Defining the mitochondrial stress response in *Arabidopsis thaliana*. Mol Plant 2:1310–1324

van Dongen JT, Gupta KJ, Ramírez-Aguilar SJ, Araújo WL, Nunes-Nesi A, Fernie AR (2011) Regulation of respiration in plants: a role for alternative metabolic pathways. J Plant Physiol 168:1434–1443

Vander Heiden MG, Chandel NS, Li XX, Schumacker PT, Colombini M, Thompson CB (2000) Outer membrane permeability can regulate coupled respiration and cell survival. Proc Natl Acad Sci U S A 97:4666–4671

Vanholme R, Demedts B, Morreel K, Ralph J, Boerjan W (2010) Lignin biosynthesis and structure. Plant Physiol 153:895–905

Vanlerberghe GC (2013) Alternative oxidase: a mitochondrial respiratory pathway to maintain metabolic and signaling homeostasis during abiotic and biotic stress in plants. Int J Mol Sci 14:6805–6847

Vanlerberghe GC, McIntosh L (1997) Alternative oxidase: from gene to function. Annu Rev Plant Physiol Plant Mol Biol 48:703–734

Vanlerberghe GC, Robson CA, Yip JYH (2002) Induction of mitochondrial alternative oxidase in response to a cell signal pathway down-regulating the cytochrome pathway prevents programmed cell death. Plant Physiol 129:1829–1842

Vanlerberghe GC, Cvetkovska M, Wang J (2009) Is the maintenance of homeostatic mitochondrial signaling during stress a physiological role for alternative oxidase? Physiol Plant 137:392–406

Vargas WA, Pontis HG, Salerno GL (2008) New insights on sucrose metabolism: evidence for an active A/N-Inv in chloroplasts uncovers a novel component of the intracellular carbon trafficking. Planta 227:795–807

Vercesi AE, Borecký J, de Godoy MI, Arruda P, Cuccovia IM, Chaimovich H (2006) Plant uncoupling mitochondrial proteins. Annu Rev Plant Biol 57:383–404

Vidal G, Ribas-Carbó M, Garmier M, Dubertret G, Rasmusson AG, Mathieu C, Foyer CH, De Paepe R (2007) Lack of respiratory chain complex I impairs alternative oxidase engagement and modulates redox signalling during elicitor-induced cell death in tobacco. Plant Cell 19:640–655

Wagner AM (1995) A role for active oxygen species as second messenger in the induction of alternative oxidase gene expression in *Petunia hybrida* cells. FEBS Lett 368:339–342

Wallace DC (1999) Mitochondrial diseases in man and mouse. Science 283:1482–1488

Wang J, Vanlerberghe GC (2013) A lack of mitochondrial alternative oxidase compromises capacity to recover from severe drought stress. Physiol Plant. doi:10.1111/ppl.12059

Wang D, Rousseau-Gueutin M, Timmis JN (2012a) Plastid sequences contribute to some plant mitochondrial genes. Mol Biol Evol 29:1707–1711

Wang H, Huang J, Liang X, Bi Y (2012b) Involvement of hydrogen peroxide, calcium, and ethylene in the induction of the alternative pathway in chilling-stressed Arabidopsis callus. Planta 235:53–67

Winkel BS (2004) Metabolic channeling in plants. Annu Rev Plant Biol 55:85–107

Winter H, Huber SC (2000) Regulation of sucrose metabolism in higher plants: localization and regulation of activity of key enzymes. Crit Rev Plant Sci 19:31–67

Yamasaki H, Sakihama Y, Ikehara N (1997) Flavonoid-peroxidase reaction as a detoxification mechanism of plant cells against H_2O_2. Plant Physiol 115:1405–1412

Yao N, Tada Y, Sakamoto M, Nakayashiki H, Park P, Tosa Y, Mayama S (2002) Mitochondrial oxidative burst involved in apoptotic response in oats. Plant J 30:567–579

Yoshida K, Terashima I, Noguchi K (2006) Distinct roles of the cytochrome pathway and alternative oxidase in leaf photosynthesis. Plant Cell Physiol 47:22–31

Yoshida K, Watanabe C, Kato Y, Sakamoto W, Noguchi K (2008) Influence of chloroplastic photooxidative stress on mitochondrial alternative oxidase capacity and respiratory properties: a case study with Arabidopsis yellow variegated 2. Plant Cell Physiol 49:592–603

Yoshida K, Watanabe CK, Terashima I, Noguchi K (2011) Physiological impact of mitochondrial alternative oxidase on photosynthesis and growth in *Arabidopsis thaliana*. Plant Cell Environ 34:1890–1899

Yoshida K, Noguchi K, Motohashi K, Hisabori T (2013) Systematic exploration of thioredoxin target proteins in plant mitochondria. Plant Cell Physiol 54:875–892. doi:10.1093/pcp/pct037

Yukioka H, Inagaki S, Tanaka R, Katoh K, Miki N, Mizutani A, Masuko M (1998) Transcriptional activation of the alternative oxidase gene of the fungus *Magnaporthe grisea* by a respiratory-inhibiting fungicide and hydrogen peroxide. Biochim Biophys Acta 1442:161–169

Zhang Y, Marcillat O, Giulivi C, Ernster L, Davies KJ (1990) The oxidative inactivation of mitochondrial electron transport chain components and ATPase. J Biol Chem 265:16330–16336

Zhang LT, Zhang ZS, Gao HY, Meng XL, Yang C, Liu JG, Meng QW (2012) The mitochondrial alternative oxidase pathway protects the photosynthetic apparatus against photodamage in Rumex K-1 leaves. BMC Plant Biol 12:40. doi:10.1186/1471-2229-12-40

Zottini M, Mandolino G, Zannoni D (1993) Oxidation of external NAD(P)H by mitochondria from taproots and tissue cultures of sugar beet (*Beta vulgaris*). Plant Physiol 102:579–585

Zottini M, Formentin E, Scattolin M, Carimi F, Lo Schiavo F, Terzi M (2002) Nitric oxide affects plant mitochondrial functionality in vivo. FEBS Lett 515:75–78

Zsigmond L, Rigó G, Szarka A, Székely G, Ötvös K, Darula Z, Medzihradszky KF, Koncz C, Koncz Z, Szabado L (2008) Arabidopsis PPR40 connects abiotic stress responses to mitochondrial electron transport. Plant Physiol 146:1721–1737

Index

P. Ahmad and M.R. Wani (eds.), *Physiological Mechanisms and Adaptation Strategies in Plants Under Changing Environment: Volume 1*, DOI 10.1007/978-1-4614-8591-9,

D

N

O

P

Printed by Printforce, the Netherlands